Francesco Scaravilli (Ed.)

The Neuropathology of HIV Infection

With 165 Figures

Springer-Verlag
London Berlin Heidelberg New York
Paris Tokyo Hong Kong
Barcelona Budapest

Francesco Scaravilli, MD, PhD, FRCPath
Reader in Neuropathology, Institute of Neurology, The National
Hospital, Queen Square, London WC1N 3BG, UK

Front Cover illustration: Multinucleated giant cell (MGC) in the white matter in a case of HIV encephalitis.
It is surrounded by cells which include macrophages and some of its nuclei are linked by 'bridges'.

ISBN-13: 978-1-4471-1959-3   e-ISBN-13: 978-1-4471-1957-9
DOI: 10.1007/978-1-4471-1957-9

British Library Cataloguing in Publication Data
The Neuropathology of HIV Infection
    I. Scaravilli, Francesco
    616.97
    ISBN 3-540-19739-7

Library of Congress Cataloging-in-Publication Data
The neuropathology of HIV infection / [edited by] Francesco
    Scaravilli.
        p. cm.
    Includes index.
    ISBN 3-540-19739-7.—ISBN 0-387-19739-7
    1. AIDS (Disease)—Complications.   2. Nervous system—Infections.
    I. Scaravilli, Francesco, 1939–
    [DNLM: 1. Acquired Immunodeficiency Syndrome—complications.
    2. Acquired Immunodeficiency Syndrome—pathology.   3. HIV
    Infections—complications.   4. HIV Infections—pathology.
    5. Nervous System—pathology.   6. Nervous System Diseases—
    pathology.   WD 308 A288545]
    RC607.A26A3632   1992
    616.97'9207—dc20
    DNLM/DLC
    for Library of Congress   92-2222

© Springer-Verlag London Limited 1993
Softcover reprint of the hardcover 1st edtion 1993

Typeset by Best-set Typesetter Ltd
28/3830-543210 Printed on acid-free paper

Foreword

The discovery in the mid 1980s of the high frequency of involvement of the nervous system in AIDS initiated a new endeavour in neuropathology. It was quickly realised that many parts of the central and peripheral nervous systems, and muscle, could be involved and in a variety of ways – by inflammation, tumour, degeneration, demyelination, vascular occlusion and haemorrhage. A number of such processes could be seen in the same individual, but the emphasis varied; and important differences were found in the frequency of particular manifestations between paediatric and adult cases.

The major question posed by this richness of pathological expression is 'By what mechanisms do these diverse consequences follow from the primary infection?' Answers at one level came quickly. Immunosuppression produced by the virus permitted opportunistic infections to develop with resulting destruction of neural tissue by the usual mechanisms. But at another level the incompleteness of our understanding of how these mechanisms operate in detail was highlighted. Evidence for direct viral involvement of the parenchyma was also found, but the mechanism of other processes proved more difficult to elucidate.

In the past five years much has been learnt about HIV involvement of the nervous system. Moreover, as the neuropathological changes underlying the clinical syndromes have been explored, it has become apparent that an understanding of their pathogenesis is likely to illuminate other neurological disorders sharing some of their pathological features.

This book will thus interest a wide readership. It contains concise, up-to-date accounts of our present knowledge of the epidemiology and neurology of HIV infection and the neuropathological consequences of it in all their diversity. Dr Scaravilli is to be congratulated on assembling a group of co-authors who have already made significant contributions to this important field.

1993

W.I. McDonald
Professor of Clinical Neurology

Preface

When I was asked to organize a book on the neuropathology of AIDS, two questions came to mind before deciding to take on this task. Firstly, was there a need for such a book? Others have been written on this subject, and indeed the most recent was published less than two years ago at the time of writing. Secondly, should the task of writing be in the hands of a single author, or a small team, or should it be a truly multiauthor work?

In any rapidly expanding field, and the neuropathology of AIDS certainly comes into this category, there is a continuing need to bring together recent findings in the several disciplines involved. If the aim can also be to stimulate discussion, to elicit criticism and to originate new ideas, then the publication of another book would seem to be justified. The last few years have seen an increasing interest in neuropathological aspects of AIDS – for example the changes in the cortical grey matter which may prove to have some relevance to the interpretation of the AIDS – dementia complex.

Considering the second question, it was felt that, even in a relatively limited topic such as the neuropathology of AIDS, it would be an advantage to have the input of experts in specific fields, allowing a truly multidisciplinary approach to the subject. In the selection of the co-authors I have sought the advice of friends and colleagues, but the final choice was entierely mine. Once assigned their topic the authors were free to develop it as they wished, and I have tried to keep my influence as discrete as possible. Inevitably there is some overlap, but this often highlights differing opinions and uncertainties.

A chapter on the general pathological lesions in AIDS has been included. Neurological illness takes place at or towards the end of a long and distressing period characterized by infections, tumours and malnutrition, which certainly cause changes in many organs. Knowledge of these changes may sometimes give a clue to otherwise unexplained findings in the nervous system.

1993   *Francesco Scaravilli*

Contents

Contributors

H. Budka
Professor of Neuropathology,
Department of Neuropathology,
Institute of Neurology
(Obersteiner Institut),
Vienna, Austria

M.M. Esiri
Reader in Neuropathology,
The Radcliffe Infirmary,
Oxford OX2 6HE, UK

G.N. Fuller
MRC Research Fellow in Neurology,
Charing Cross and Westminster Hospital
Medical School,
Dean Ryle Street,
London SW1P 2AP, UK

F. Gray
Professor of Neuropathology,
Département de Pathologie
(Neuropathologie),
Hôpital Henri Mondor,
94010 Créteil, France

J.N. Harcourt-Webster
Consultant Histopathologist,
Westminster Hospital,
17 Horseferry Road,
London SW1P 2AR, UK

M.J.G. Harrison
Professor of Clinical Neurology,
Department of Neurology,
Middlesex Hospital,
Mortimer Street,
London W1N 8AA, UK

J.M. Jacobs
Reader,
Department of Neuropathology,
Institute of Neurology,
The National Hospital,
Queen Square,
London WC1N 3BG, UK

B. Kendall
Consultant Radiologist,
Institute of Neurology,
The National Hospital,
Queen Square,
London WC1N 3BG, UK

P.G.E. Kennedy
Professor of Clinical Neurology,
Department of Neurology,
Institute of Neurological Sciences,
University of Glasgow,
Southern General Hospital,
Glasgow G51 4TF, UK

J. Mikol
Professor of Neuropathology,
Hôpital de Lariboisière (Paris VII),
Paris, France

M. Mintz
Division of Pediatric Neurology of the
Department of Neurosciences,
UMD-New Jersey Medical School,
185 South Orange Avenue,
Newark, NJ 07103-2757, USA

P.C.K. Petito
Professor of Pathology and Director of
Neuropathology,
The New York Hospital-Cornell Medical
Center,
525 East 68th Street,
New York NY 10021, USA

F. Scaravilli
Professor of Neuropathology,
Institute of Neurology,
The National Hospital,
Queen Square,
London WC1N 3BG, UK

L.R. Sharer
Associate Professor of Pathology
(Neuropathology),
UMD-New Jersey Medical School,
185 South Orange Avenue,
Newark, NJ 07103-2757,
USA

E. Sinclair
Department of Neuropathology,
The National Hospital,
Queen Square,
London WC1N 3BG, UK

J.N. Weber
Professor in Sexually Transmitted Diseases
and Communicable Diseases,
Department of Genito-Urinary Medicine,
St Mary's Hospital,
Praed Street,
London WC2, UK

I.V.D. Weller
Professor of Genito-Urinary Medicine,
Academic Department of Genito-Urinary
Medicine,
University College and Middlesex School of
Medicine,
James Pringle House,
The Middlesex Hospital,
London W1N 8AA, UK

C.A. Wiley
Associate Professor of Neuropathology,
Department of Pathology M-012,
University of California San Diego,
La Jolla, CA 92093, USA

1 HIV Infection: A Cellular Approach

J.N. Weber

The Human Immunodeficiency Viruses

The human immunodeficiency viruses types 1 and 2 (HIV-1 and HIV-2) are RNA viruses which belong to the lentivirus group of the retrovirus family (Weiss, 1985). In common with other members of this group which infect animals (visna/maedi virus, equine infectious anaemia virus, caprine arthritis/encephalitis virus), HIV is a non-transforming virus (that is, not directly oncogenic) which replicates through the generation of a proviral DNA intermediate by the action of the retroviral enzyme RNA-directed DNA polymerase (reverse transcriptase, RT). HIV is a virus capable of causing lysis in infected cells. There is a visible cytopathic effect in vitro, which can be used for assays of HIV infectivity. In vivo, a persistent infection is established following integration of proviral DNA into the host genome. Infection lasts for the life of the cell, and after sufficient cells are infected, infection lasts for the life of the host. All retroviral infections appear to be persistent.

The lentiviruses are genotypically and phenotypically distinct from the other two groups of retroviruses which infect humans: the tumour-causing oncornaviruses (human T-cell leukaemia/lymphoma viruses, HTLV-I and HTLV-II), and the (probably) non-pathogenic 'foamy' spumaviruses (human foamy virus, HFV). These other retroviruses will not be addressed further.

Structure of HIV and Viral Protein Products

The stylized structure of HIV-1 is shown in Figure 1.1. Two single-stranded copies of RNA are non-covalently linked to the core proteins, closely associated with the RNA-directed DNA polymerase (RT), within an outer envelope lipid and protein membrane. The viral envelope is derived from the host cell and contains a virally-encoded transmembrane glycoprotein (gp41) which is non-covalently anchoring the extravirion spike glycoprotein gp120. The proviral genomes of HIV-1 and HIV-2 are 9–10 kb long. In common with all animal retroviruses (including the oncornaviruses), the HIV genome contains three

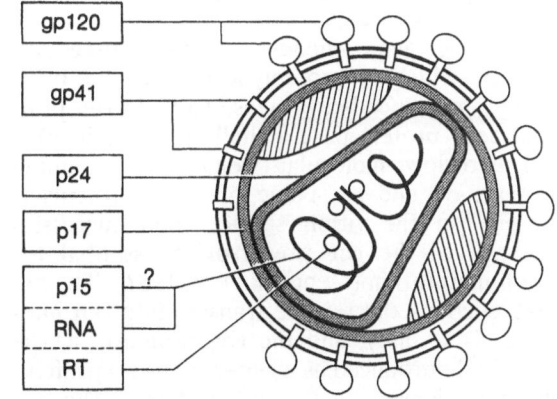

Figure 1.1. The location of the viral product. (After Gelderblom et al, 1987.)

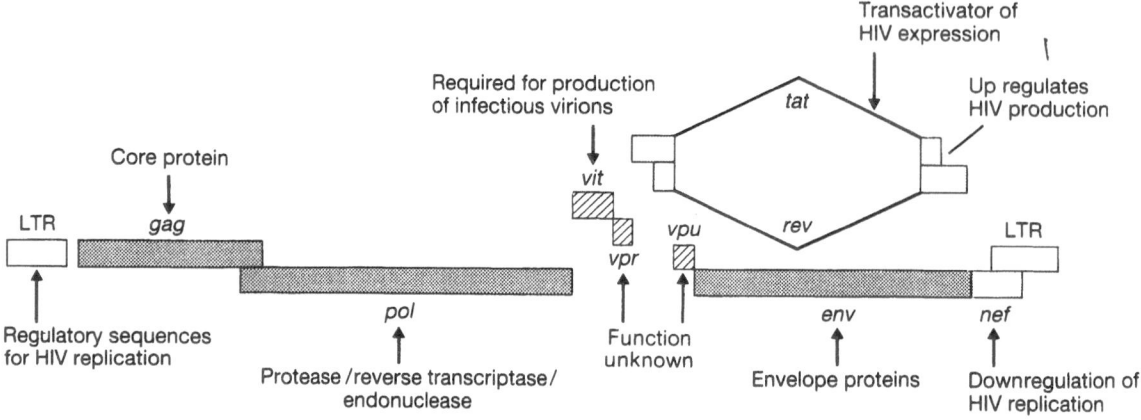

Figure 1.2. Genetic structure of human immunodeficiency virus types 1 (HIV-1). (After Peterlin and Luciw, 1988.)

principal structural genes, *gag* (encoding core proteins), *pol* (encoding for the reverse transcriptase and other enzymes), and *env* (encoding for the spike and transmembrane glycoproteins) (Figure 1.2). The genome is bound at both the 5' and the 3' end by a non-coding long terminal repeat region (LTR), which contains the promotor and regulatory gene product acceptor sequences. Sequencing data from several different HIV-1 strains has shown that there is considerable diversity in the genome of HIV isolates, with up to 26 per cent variance in the *env* gene between geographically-distinct isolates (Wain-Hobson, 1989). The *gag* and *pol* genes are more conserved between isolates than *env*, and this is consistent with the diversity seen in other animal retroviruses (Saag et al, 1988). However, the genetic diversity seen in HIV isolates, where no two viruses have identical sequences, is higher than for any other known retrovirus; in fact the isolates are more diverse than those seen with influenza A viruses.

The *gag* gene encodes a 55 kD precursor protein, which is proteolytically cleaved by a virally (*pol*)-encoded protease to yield the principal core protein p24, a shell-like protein p18, and two low molecular weight RNA-binding proteins p7 and p8. None of these proteins is present on the virion surface, and successful cleavage of the *gag* precursor is essential for virion maturation (Debouck et al, 1987). The *pol* gene encodes the reverse transcriptase, an integrase (DNA-ase), and the *gag* protease. These are initally produced as a *gag–pol* fusion protein, which is cleaved by the *pol* protease. The *env* gene encodes a precursor protein of 160 kD, which is cleaved by a cellular protease to give the heavily-glycosylated outer spike protein gp120,

and the non-covalently linked anchoring transmembrane protein gp41. All of the structural gene products and their precursors are immunogenic in natural human infection (Weber and Jeffries, 1990).

Other non-structural genes have been identified in the HIV genome which encode for proteins and whose function it is to regulate viral replication. These genes were first identified as novel to HIV, but have subsequently been demonstrated in other human and animal retroviruses. The *tat* gene (trans-activator) encodes a 23 kD nuclear protein which accelerates the production of viral proteins (including itself) by as much as 1000 times (Sodroski et al, 1985). The *cis*-acting element required for *tat* activity is the trans-acting response element (TAR), which is defined as the region from positions +19 to +42 in the 5' LTR (Hauber et al, 1987). Interaction of TAR and *tat* is thought to facilitate transcriptional elongation, increase mRNA stability, and enhance translation.

In contrast, the *nef* gene encodes a cytoplasmic 27 kD protein which probably retards viral replication (Luciw et al, 1987). The *nef* protein may be responsible for the ability of the virus to turn off growth and generate clinical, and possibly viral, latency. The mechanism of *nef* action is unknown, and indeed its action remains controversial, but as the *nef* product is exclusively cytoplasmic, its action on the integrated HIV genome is likely to be through a second messenger system (Terwilliger et al, 1986). The *nef* gene has homology with the *ras* proto-oncogene, and some investigators have claimed that *nef* can phosphorylate protein kinase-C, which would provide one explanation for its mechanism of action in the cytoplasm (Guy et al, 1987).

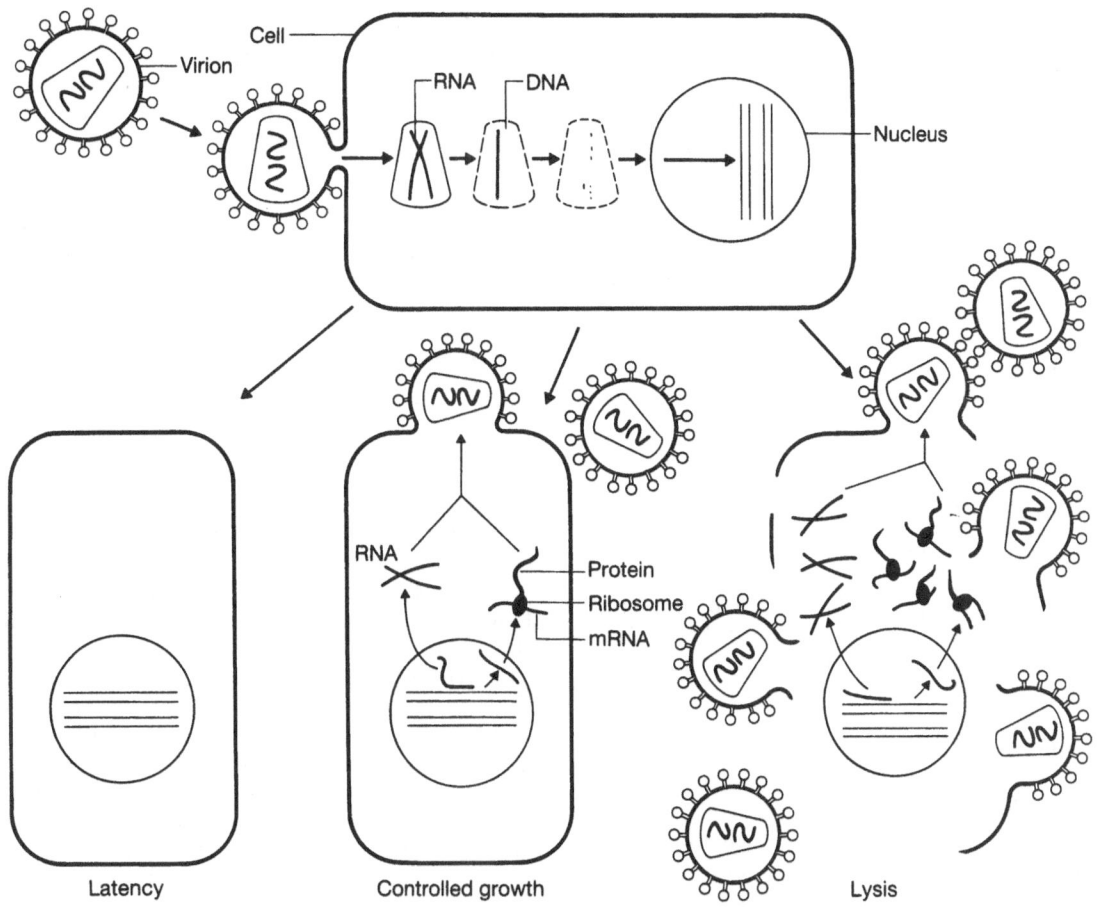

Figure 1.3. The life cycle of HIV. (After Haseltine et al, 1988.)

The *rev* gene (regulator of virion protein expression) encodes a 17 kD viral protein, which acts as a novel molecular switch in the nucleus to determine selectively whether regulatory non-structural proteins or complete viral particles are made (Terwilliger et al, 1987). Other regulatory genes determine how much mature virus is re-leased (*vpu*), and if that virus will be highly in-fectious (*vif* – virion infectivity factor) (Sodroski et al, 1986; Strebel et al, 1988). Regulatory proteins therefore affect the production not only of structural gene proteins, but also of regulatory gene proteins, including themselves. This level of organizational complexity had not previously been recognized or expected in what are the small genomes of the animal retroviruses.

HIV-infected cells arrested in the Go phase have little, if any, viral transcription. This block appears to occur after entry of HIV, in early transcription of integrated DNA (Pang et al, 1990). After cellular activation, which in T-cells is

linked to T-cell growth and to expression of the interleukin-2 receptor, nuclear factors which activate interleukins and their receptors also lead to HIV transcription (Nabel and Baltimore, 1987). Initially, low levels of short viral tran-scripts are synthesized with accumulation of *tat* protein. On reaching a threshold, *tat* greatly increases the expression of TAR-containing RNA transcripts, while *nef* decreases cellular signalling and HIV transcription (Terwilliger et al, 1986). Controlled viral replication depends upon a balance being maintained between these positive and negative regulators, such that when the infected cell dies, enough virions will be released to infect other cells (Figure 1.3). Like all para-sites, viruses are most successful when viral re-plication leads to persistent infection, rather than cell death. In this respect, the regulatory genes of HIV confer a considerable advantage to the virus in the control of the rate of replication by viral genes.

Simian Immunodeficiency Viruses

In parallel to the emergence of the AIDS epidemic, it became clear from 1982–3 that a similar range of diseases was being observed in captive macaques in primate centres in the USA and elsewhere (Daniel et al, 1985). Animals were dying of an acquired immunodeficiency syndrome, which was inevitably termed 'simian AIDS' (SAIDS). In 1984 a novel lentivirus was isolated from an infected macaque, and this virus was ultimately termed 'simian immunodeficiency virus' (SIVmac) (Daniel et al, 1988). Macaques are Old World monkeys from Asia, and are not naturally infected with SIVmac in the wild. Thus, this virus did not originate in these animals, but must have contaminated them following their arrival in the primate centres (Schneider and Hunsmann, 1988). Widespread sero-epidemiology of wild-caught African, Asian and South American primates and monkeys has yielded considerable information which is pertinent to the development of the HIV epidemic, and to the pathogenesis of AIDS.

Wild monkeys infected by SIVs have only been found in sub-Saharan Africa; no monkeys in Asia or New World monkeys are infected naturally with SIVs (Schneider and Hunsmann, 1988). Within Africa, SIV infection is found most commonly in the African green monkey (or vervet), and is called SIVagm. The African green monkey is extremely widespread in Africa, and over 50 per cent of the species are seropositive for SIVagm (Fukasawa et al, 1988). In the wild, infected monkeys do not appear to develop any SIVagm-associated illness. Experimental infection of vervets with SIVagm in captivity also does not lead to disease and similarly, inoculation of SIVagm into macaques does not currently appear to be pathogenic (Gardner and Luciw, 1988). By virtue of the widespread infection of vervets by SIVagm therefore, this species may well be the 'natural' host of this virus, and in this host, SIVagm is non-pathogenic.

Other African monkeys infected by SIVs include baboons and mangabeys. The SIV isolated from the sooty mangabey monkey, SIVsm, is non-pathogenic in the mangabey, but causes SAIDS when inoculated into Asian macaques (Desrosiers, 1988). SIVsm is thus probably the origin of SIVmac (Mulder, 1988). These viruses are closely related to HIV-2, both in terms of genomic organization and cross-hybridization, as well as almost total antigenic cross-reactivity. Thus, HIV-2 can be seen to have a very close counterpart in a wild animal population, and, at least historically, was probably a zoonosis.

This association has led to a search for an African monkey (or other animal) harbouring a virus equivalent for HIV-1. Recently, Peeters et al (1989) have described the isolation of an HIV-1-like virus from wild chimpanzees in Gabon. If confirmed, this could lead to evidence for a zoonotic origin for HIV-1 in Africa.

Use of the SIV Model

SIVmac infects T-lymphocytes through binding the same receptor as HIV-1 and HIV-2. Infection leads to an AIDS-like disease. The virus is now fully characterized, and can therefore be used as a model for AIDS vaccine development. Studies using a formalin-fixed inactivated SIV virus as an immunogen have led to the development of strong immune responses in immunized animals, which have then been protected against challenge with live homologous strains of SIVman. These challenge experiments have now been repeated worldwide, and it does, indeed, seem likely that the SIV model in macaques will be an important step towards the development of an AIDS vaccine. Recently, Kestler et al (1990) have described a molecular clone of SIVmac which is capable of inducing SIV-related disease, including neurological disease, in experimentally-infected macaques. This will open up unique and important opportunities for the detailed study of the contribution of viral genes to pathogenesis of infection, latency and disease.

Strain Variations of HIV

It is now clear that HIV is constantly evolving in vivo, and that the properties of diverse HIV isolates differ both phenotypically and genotypically. Isolates taken from patients with asymptomatic HIV infection tend to grow more slowly and give rise to lower titres of RT activity (slow/low viruses) compared to isolates taken from subjects with more advanced disease (fast/high viruses). The latter isolates more readily establish continuous viral replication in transformed, permanent CD4-positive lymphocytoid and monocytoid cell lines (Evans and Levy, 1989). Even isolates taken longitudinally from the same patient show differences in their replicating ability and cytopathic effect (Evans and Levy,

1989), particularly in their ability to form syncytia. Isolates taken in the later stages of the disease have an expanded host cell tropism in vitro, faster replication rates, and a greater cytopathic effect. A large number of related but genotypically distinguishable variants evolve in parallel in vivo, and co-exist during chronic infection. The immune mechanisms underlying the emergence of rapid/high viruses with progression of clinical disease have not been determined. However, there is no evidence that person-to-person spread leads to more pathogenic viruses being spread. It is most likely that a mixture of slow/low and fast/high viral genomes are transmitted, but that the immune response dampens down the fast/high ones preferentially, as these viruses will replicate most rapidly, and hence be presented more readily to the immune system (Bangham and McMichael, 1990).

Cellular Tropism of HIV

In common with all macromolecules, retroviruses like HIV can only enter cells through binding to a specific cellular receptor. In the case of HIV-1, HIV-2 and the SIVs, early research in 1983–4 showed that these viruses can only productively infect T-lymphocytes in vitro (Klatzmann et al, 1984a). As AIDS was associated with decline of CD4-positive T-lymphocytes in the peripheral blood, known as the T-helper cells, two groups then postulated that the receptor for the entry of HIV might be the CD4 surface molecule itself, which is found on the surface of, and characterizes, T-helper cells (Dalgliesh et al, 1984; Klatzmann et al, 1984b). Only cells which expressed CD4 were infectable by HIV. Initally, this was confirmed by blocking HIV infection with antibodies to CD4. Later, Maddon et al (1986) showed that the transfection of the CD4 cDNA into cells which did not normally express it (HeLa cells), to generate a HeLa(CD4-positive) line, induced susceptibility to infection by HIV. Thus, transfection of CD4 alone conferred susceptibility to HIV infection, showing that this was indeed the receptor for HIV on the cell surface.

The outer envelope spike protein of HIV, gp120, binds to CD4 with high affinity ($Ka > 10^{-9}$ Molar), which is of similar affinity to antibody–antigen binding. Following gp120–CD4 binding, there is a conformational change – as yet poorly characterized, and then the viral envelope binds to the cell surface by a pH independent mechanism. This implies that HIV enters the cell by direct fusion of the viral envelope with the cell membrane, and not by active endocytosis of the receptor. It is still possible that the fusion of HIV to the cell surface post-CD4–gp120 binding is mediated through a second receptor, as not all human cells are infectable by HIV, even after transfection of CD4. All sub-primate non-human cells transfected with CD4 are still resistant to HIV infection (Maddon et al, 1986).

Thus, the tropism of HIV in vivo appears to be defined wholly by the expression of CD4 on the surface of susceptible cells. In vivo, CD4 expression on the cell surface is seen predominantly in T-helper lymphocytes, and less frequently on monocytes, macrophages and B-lymphocytes. For all practical purposes, the tropism of HIV is defined by the distribution of its receptor, CD4, and the decline of CD4-positive cells in HIV infection can be seen as a direct consequence of the action of HIV on susceptible cells.

However, an additional experiment has shown that this model is still too simplistic. If human CD4 is transfected into a mouse T-cell line, HIV will bind to the cell, but will not enter or replicate. The whole HIV genome can be transfected into murine cells where it leads to productive infection, showing that the block to entry of HIV is at the cell surface, and not due, for example, to inefficient transcription. Most human cells which express CD4 naturally or by transfection can be infected by HIV, but cells from other species are not infectable by HIV, even when human CD4 is present on the cell surface. The implication of this observation is that another molecule(s) is required in addition to CD4, to allow efficient entry of HIV into the cell. This molecule is probably widely distributed on human cells, and would certainly have a lower affinity for HIV than does CD4. The nature of this molecule is still unclear, but failure to identify it to date has led to an inability to develop a murine transgenic model for HIV infection.

Fine variation in the tropism of viral isolates of HIV has been noted. Some isolates grow better in CD4-positive monocytes/macrophages than in peripheral blood T-cells, and some vice versa. Chen has now mapped the region of the virus which is responsible for the difference in tropism, and has shown that a small region of the envelope protein gp120 leads to monocytotropism after CD4-binding (O'Brien et al, 1990). It is possible that this region of gp120 is interacting with the second component of entry. As isolates from the CNS have been predominantly monocytotropic, the significance of this fine tropism for CNS disease should be investigated. This could probably be best achieved by use of the molecular clone of SIVmac: sequential sacrifice of animals after infection, and isolation and sequencing of virus from the brain would enable the course of the monocytotropism variants to be studied. Aside from a differential ability to infect monocytes versus T-cells, no other CNS specific variants of HIV have been described, and there is little evidence for a CNS-tropic virus as the cause of disease.

Pathogenesis of HIV Infection

Although the loss of CD4-positive T-lymphocytes is central to the immune defect in AIDS, the precise mechanisms responsible for the cell death remain to be determined. Possible mechanisms of HIV disease in the CNS include:

1. Direct cytopathic effect on CD4-positive cells
2. Direct infection of CD4-negative cells
3. Infection of monocytes/macrophages
4. Contribution of other infectious co-factors
5. Immune response to infected cells

Direct Cytopathic Effect of CD4-Positive Cells

When HIV-infected cells expressing the gp120 protein on their cell surfaces are co-cultivated with CD4 receptor-bearing cells, fusion between infected and uninfected cells takes place, with the formation of giant multinucleated cells known as syncytia. In vitro, syncytial production is cytopathic, and can be used as a marker of viral replication. Not all cells infected by HIV will form syncytia, and the generation of these distinctive giant cells is a feature of CD4 expression, and also probably of the putative second component of entry (see above). Syncytial formation in vitro requires a highly-productive infection, or a large number of virions. The role of syncytial production in vivo is unclear, but it is unlikely to be a major mechanism of pathogenesis in the destruction of the CD4-positive immune system.

High rates of viral replication independent of the formation of multinucleated giant cells can cause cell death (Weber et al, 1987). As many as 500 uninfected cells can combine to form a syncytium, with subsequent death of the original infected cell, as well as that of uninfected 'bystander' T-cells. Accumulation of unintegrated linear and circular viral DNA, as well as defects in post-translational processes, may result in the presence of large amounts of DNA and RNA which can then interfere with the processing of heteronuclear cellular RNA, leading to the arrest of normal cellular processes and the death of the cell (Rosenburg and Fauci, 1990). As noted previously, the formation of multinucleated giant cells precedes cell death in vitro, but not all chronically infected T-cells exhibit cytopathology. This resistance to cell killing has been attributed to a reduced interaction of the viral envelope with CD4 molecules, due perhaps to low levels of surface CD4 expression in some cells, or as a result of post-transcriptional down-regulation of CD4 by the viral glycoprotein.

Interaction of the virus envelope protein gp120 with CD4 can, in itself, be cytotoxic, causing cell fusion and death. Furthermore, since cells which express surface viral proteins but do not release virus show no signs of degeneration, release of viral particles may result in alterations of cell membrane permeability and subsequent cell death.

Direct Infection of CD4-Negative Cells

In relation to the tropism of HIV in vivo, a number of experiments have shown that HIV can infect CD4-negative cells in vitro (Cheng-Mayer et al, 1985). Cell lines derived from human malignant gliomas do not express CD4 by immunofluorescence, neither is there detectable mRNA for CD4 (Cheng-Mayer et al, 1985). However, these cell lines may be infected by HIV in vitro, and infection cannot be blocked by antibodies to CD4, nor by soluble recombinant CD4. It requires rather more virus to produce infection, and very few infectious virions are produced by the cells. The HIV genome is present in the cells, but few viral proteins are expressed on the cell surface and, therefore, the cells do not appear positive by immunocytochemistry. Thus, in vitro, CNS tissue may be directly infectable by HIV, independently of CD4. However, there is little evidence at present that CD4-negative cells are significantly infected by HIV or SIV in vivo.

Infection of Monocytes/Macrophages

As noted above, the virus variants isolated from the CNS in HIV infection appear to be tropic for monocytes/macrophages. Monocytotropic variants of HIV appear to become more common with time in HIV infection, presumably as a result of viral variability in infected subjects. Entry of HIV into macrophages is dependent on the expression of CD4, although the presence of large amounts of high-affinity Fc-receptors on macrophages could lead to increased intake of HIV through binding of antibody-coated virus to these receptors (Cheng-Mayer et al, 1985). This process also seems to be CD4 dependent, and so is unlikely to be a major mechanism of entry in vivo.

However, macrophages are terminally differentiated and do not replicate, and thus there is little productive HIV infection from these cells. Rather, they form reservoirs of infection, in which there is little cytopathic effect and little cell killing. It is known that HIV infection of macrophages can induce increased production of the cytokine tumour necrosis factor (TNF). TNF, in turn, can positively up-regulate transcription of integrated HIV (Rosenburg and Fauci, 1990). Thus, HIV-infected macrophages may lead to a positive feedback cycle of TNF production and increased HIV replication. The involvement of other cytokines in this cycle may be an important mechanism in the indirect action of HIV in the CNS.

Contribution of Other Infectious Co-factors

The quest for co-factors in the evolution of HIV-related disease has become something of a grail for AIDS research. As other infectious agents are so commonly found in AIDS and HIV infection, it is not surprising that many agents have come under scrutiny for their role as a possible co-factor. These have ranged from cytomegalovirus, HTLV-1, other herpes viruses, syphilis, hepatitis B virus, and more recently *Mycoplasma fermentans*. I believe that the Kestler et al experiment cited above (Kestler et al, 1990), in which it was shown that the molecular clone of SIVmac can produce fatal disease in macaques, comes as close as possible to the demonstration that co-factors are not a prerequisite for HIV-related disease.

Immune Response to Infected Cells

Despite a considerable amount of data on the humoral and cellular immune response to HIV

antigens in natural infection, the contribution of these parameters to the causation of disease is unknown. Many antibodies to HIV are produced throughout the course of HIV infection, but none appears to fix complement, or lead to lysis of infected cells. While antibody-dependent cellular cytotoxicity has been recorded in HIV infection, its presence or titre does not correspond to disease progression. In terms of cytotoxic T-cell response to viral antigens, many epitopes have now been defined, but it is not yet possible to define whether the cytotoxic T-lymphocyte reponse is detrimental or beneficial to HIV-infected cells, or to the host in general terms. Once again, it is likely that these questions can only be addressed in animal models.

References

Bangham CR, McMichael AJ. (1990) Virology: nosing ahead in the cold war. Nature 344:16.

Cheng-Mayer C, Rutka JT, Rosenblum ML et al. (1985) Human immunodeficiency virus can productively infect cultured human glial cells. Proc Natl Acad Sci USA 84: 3526–30.

Dalgliesh AG, Beverley PCL, Clapham PR et al. (1984) The CD4 (T4) antigen is an essential component of the receptor for the AIDS retrovirus. Nature 312:767–8.

Daniel MD, Letvin NL, King NW et al. (1985) Isolation of T-cell tropic HTLV-III-like retrovirus from macaques. Science 228:1201–4.

Daniel MD, Li Y, Naidu YM et al. (1988) Simian immunodeficiency virus from African green monkeys. J Virol 62:4123–8.

Debouck C, Gorniak JG, Strickler JE et al. (1987) Human immunodeficiency virus protease expressed in *Escherichia coli* exhibits autoprocessing and specific maturation of the gag precursor. Proc Natl Acad Sci USA 84:8903–6.

Desrosiers RC. (1988) Simian immunodeficiency viruses. Annu Rev Microbiol 42:607–25.

Evans LA, Levy JA. (1989) Characteristics of HIV infection and pathogenesis. Biochim Biophys Acta 989:237–54.

Fukasawa M, Miura T, Hasegawa A et al. (1988) Sequence of simian immunodeficiency virus from African green monkey, a new member of the HIV/SIV group. Nature 333:457–61.

Gardner MB, Luciw PA. (1988) Simian immunodeficiency viruses and their relationship to the human immunodeficiency viruses. AIDS 2 (suppl 1):s3–10.

Gelderblom HR, Hausmann EH, Ozel M et al. (1987) Fine structure of human immunodeficiency virus (HIV) and immunolocalization of structural proteins. Virology 156: 171–6.

Guy B, Kieny MP, Riviere Y et al. (1987) Nature 330:266–9.

Haseltine WA, Wong-Staal F. (1988) The molecular biology of the AIDS virus. Sci Am 259:52–62.

Hauber J, Perkins A, Heimer EP et al. (1987) Transactivation of human immunodeficiency virus gene expression is mediated by nuclear events. Proc Natl Acad Sci USA 84:6364–8.

Kestler H, Kodama T, Ringer D et al. (1990) Induction of AIDS in rhesus macaques by molecularly cloned SIV. Science 248:1109–12.

Klatzmann D, Champagne E, Chamaret S et al. (1984a) T-lymphocyte T4 molecule behaves as the receptor for human retrovirus LAV. Nature 312:767–8.

Klatzmann D, Champagne E, Chamaret S et al. (1984b) T-lymphocyte T4 molecule behaves as the receptor for human retrovirus LAV. Nature 225:59–63.

Luciw PA, Cheng-Mayer C, Levy JA. (1987) Mutational analysis of the human immunodeficiency virus: the orf-B region down-regulates virus replication. Proc Natl Acad Sci USA 84:1434–8.

Maddon PJ, Dalgleish AG, McDougal JS et al. (1986) The T4 gene encodes the AIDS virus receptor and is expressed in the immune system and the brain. Cell 47:333–48.

Mulder C. (1988) Human AIDS virus is not from monkeys. Nature 333:396.

Nabel G, Baltimore D. (1987) An inducible transcription factor activates expression of human immunodeficiency virus in T-cells. Nature 326:711–13.

O'Brien WA, Koyanagi Y, Namazie A et al. (1990) HIV-1 tropism for mononuclear phagocytes can be determined by regions of gp-120 outside the CD4-binding domain. Nature 348:69–73.

Pang S, Koyanagi Y, Miles S et al. (1990) High levels of unintegrated HIV-1 DNA in brain tissue of AIDS dementia patients. Nature 343:85–9.

Peeters M, Honore C, Huet T et al. (1989) Isolation and partial characterization of an HIV related virus occurring naturally in chimpanzees in Gabon. AIDS 3:625–31.

Peterlin BM, Luciw PA. (1988) Molecular biology of HIV. AIDS 2 (suppl 1): s29–40.

Rosenburg Z, Fauci A. (1990) Cytokine induction of HIV expression. Immunol Today 11:176.

Saag MS, Hahn BH, Gibbons J et al. (1988) Extensive variation of human immunodeficiency virus type-1 in vivo. Nature 334:440–4.

Schneider J, Hunsmann G. (1988) Simian lentiviruses: the SIV group. AIDS 2:1–9.

Sodroski J, Rosen C, Wong-Staal F et al. (1985) Trans-acting transcriptional regulation of human T-cell leukemia virus type III long terminal repeat. Science 227:171–3.

Sodroski J, Goh WC, Rosen C et al. (1986) Replicative and cytopathic potential of HTLV-III/LAV with sor gene deletions. Science 231:1549–53.

Strebel K, Klimkait T, Martin MA. (1988) A novel gene of HIV-1, VPU, and its 16-kilodalton product. Science 241: 1221–3.

Terwilliger E, Sodroski JG, Rosen CA et al. (1986) Effects of mutations within the 31 orf open reading frame region of human T-cell lymphotropic virus type III (HTLV-III/LAV) on replication and cytopathogenicity. J Virol 60:754–60.

Terwilliger E, Berghoff R, Sia R et al. (1987) The art gene product of human immunodeficiency virus is required for replication. J Virol 62:655–8.

Wain-Hobson S. (1989) HIV genome variability in vivo. AIDS 3 (suppl 1):s13–18.

Weber JN, Jeffries DJ. (1990) Humoral immune responses to HIV. In: Gottlieb M, Pinching AJ, Jeffries DJ, eds. Current opinion on HIV. A Wiley; London, 1990.

Weber JN, Clapham PR, Weiss RA et al. (1987) Human immunodeficiency virus infection in two cohorts of homosexual men: neutralising sera and association of anti-gag antibody with prognosis. Lancet i:119–22.

Weiss R. (1985) Human T-cell retroviruses. In: Weiss R et al, eds. RNA tumor viruses, 2nd edn, vol 2, Supplements and appendices. Cold Spring Harbor Laboratory: New York, 405–85.

2 Epidemiology and Natural History of HIV Infection: An Overview

I.V.D. Weller

Introduction

In the summer of 1981 reports of 5 cases of *Pneumocystis carinii* pneumonia (PCP) and 26 cases of Kaposi's sarcoma (KS) amongst homosexual men in Los Angeles, California and New York represented the first description of the acquired immune deficiency syndrome (AIDS) (Friedman-Kien et al, 1981, 1982; Gottlieb et al, 1981a,b). It was not realized at this time, of course, that the causative agent was already present in persons in four other continents. The virus, variously named, was first discovered in 1983 (Barré-Sinoussi et al, 1983; Gallo et al, 1983). The first antibody tests were developed in 1984. In 1986 the sub-committee of the International Committee for the Taxonomy of Viruses suggested that the generic name for the virus should be the human immunodeficiency virus (HIV) (Biberfield et al, 1987). By this time the CD4 antigen had been identified as the cellular receptor for HIV and the first trials of the antiviral agent zidovudine were demonstrating efficacy in patients with symptomatic disease (Fischl et al, 1987).

Definition

AIDS is now defined as an illness characterized by one or more of the indicator diseases. In the absence of another cause of immune deficiency and without laboratory evidence of HIV infection (if the patient has not been tested or the results are inconclusive), certain diseases when definitely diagnosed are indicative of AIDS (Table 2.1). However, regardless of the presence of other causes of immune deficiency, if there is laboratory evidence of HIV infection, certain other indicator diseases also constitute a diagnosis of AIDS (Table 2.2). These indicator diseases may in some cases require only a presumptive diagnosis (Centers for Disease Control, 1987).

Table 2.1. Disease diagnostic of AIDS without laboratory evidence of HIV

Candidiasis – oesophageal, pulmonary
Cryptococcosis – extrapulmonary
Cytomegalovirus disease – disseminated
Cryptosporidiosis – diarrhoea persisting >1 month
Herpes simplex virus infection
  Mucocutaneous ulceration lasting >1 month
  Pulmonary, oesophageal infection
Kaposi's sarcoma – patient aged <60
Primary cerebral lymphoma – patient aged <60
Lymphoid interstitial pneumonia – child aged <13
Mycobacterium avium–intracellulare } disseminated
Mycobacterium kansasii
Pneumocystis carinii pneumonia
Progressive multifocal leukoencephalopathy
Cerebral toxoplasmosis

Modes of Transmission of HIV

HIV has been isolated from lymphocytes in peripheral blood, cell-free plasma, semen, cer-

Table 2.2. Diseases diagnostic of AIDS if laboratory evidence of HIV exists

Diseases diagnosed definitively
Recurrent/multiple bacterial infections – child aged under 13
Coccidioidomycosis – disseminated
HIV encephalopathy
Histoplasmosis – disseminated
Isosporiasis – diarrhoea persisting >1 month
Kaposi's sarcoma at any age
Primary cerebral lymphoma at any age
Non-Hodgkin's lymphoma – diffuse, undifferentiated B-cell type, or unknown phenotype
Any disseminated mycobacterial disease caused by an organism other than *M. tuberculosis*
Mycobacterial tuberculosis – extrapulmonary
Salmonella septicaemia – recurrent
HIV wasting syndrome

Diseases diagnosed presumptively
Candidiasis – oesophageal
Cytomegalovirus retinitis with visual loss
Kaposi's sarcoma
Lymphoid interstitial pneumonia – child aged under 13
Mycobacterial disease (acid-fast bacilli; species not identified by culture) – disseminated
Pneumocystis carinii pneumonia
Cerebral toxoplasmosis

vical secretions, cerebrospinal fluid, tears, saliva, urine and breast milk. The major modes of transmission are through anal and vaginal intercourse, sharing of contaminated needles, transfusion of infected blood and blood products, infusion of infected factor VIII and vertical transmission from mother to fetus in utero.

Global Epidemiology

By March 1991 over 334 000 cases from over 150 countries had been reported to the World Health Organization (WHO) (WHO/CDC, 1990) (Table 2.3). WHO estimates that worldwide there are

Table 2.3. World Health Organization global statistics: cumulative AIDS cases reported to WHO as of March 1991

Region	No. of cases	Case rate (1988 reported cases per 100 000 population)[a]
African	83 249	0–36.0
Americas	199 237	0–49.1
Eastern Mediterranean	675	0–1.4
European	47 779	0–5.6
South East Asia	148	0
Western Pacific	3 127	0–3
	334 215	

[a] Varies according to country.

now between 6 and 10 million people infected with HIV. The reporting of AIDS cases in Europe, North America and Oceania is based primarily on the WHO/CDC definition. However, reporting in Africa and some countries in Central and South America is based largely on the WHO clinical definition. AIDS in an adult is defined by the existence of at least two of a group of major signs associated with at least one minor sign, in the absence of known causes of immunosuppression such as cancer, severe malnutrition or other recognized aetiologies:

Major signs
1. Weight loss ≥10 per cent of body weight
2. Chronic diarrhoea >1 month
3. Prolonged fever >1 month (intermittent or constant)

Minor signs
1. Persistent cough for >1 month
2. Generalized pruritic dermatitis
3. Recurrent herpes zoster
4. Oropharyngeal candidiasis
5. Chronic progressive and disseminated herpes simplex infection
6. Generalized lymphadenopathy

The presence of generalized KS or cryptococcal meningitis are sufficient by themselves for the diagnosis of AIDS.

Similarly, paediatric AIDS is suspected in an infant or child presenting with at least two of the following major signs associated with at least two of the following minor signs, again in the absence of known causes of immunosuppression such as cancer, severe malnutrition or other recognized aetiologies:

Major signs
1. Weight loss or abnormally slow growth
2. Chronic diarrhoea >1 month
3. Prolonged fever >1 month

Minor signs
1. Generalized lymphadenopathy
2. Oropharyngeal candidiasis
3. Repeated common infections (otitis, pharyngitis, etc)
4. Persistent cough
5. Generalized dermatitis
6. Confirmed maternal HIV infection

The WHO has identified three distinctive epidemiological patterns of HIV infection globally.

Pattern 1 countries include North America, Western Europe, Australia and New Zealand. In these countries the epidemic began in the late 1970s and early 1980s amongst homosexual men and injecting drug users. However, heterosexual transmission does occur and is increasing. In Europe, increasing cases of AIDS amongst intravenous drug users, particularly from Southern Europe, are being reported at such a rate that by the end of 1991 it has been estimated that there will be more cases of AIDS amongst injecting drug users in Europe than in homosexual and bisexual men.

Pattern 2 countries include most of sub-Saharan Africa and areas of the Caribbean, and in these regions HIV transmission is predominantly heterosexual and perinatal, with an additional contribution from unscreened blood transfusions and inadequate injection procedures. The overall male to female ratio is approximately 1:1 as compared to 9:1 in pattern 1 countries. In South America, which would have fitted pattern 1 in the early 1980s, there has been a considerable increase in heterosexual cases (pattern 1/2).

As far as concerns North Africa, the Middle East, Eastern Europe and most of the countries in Asia and Oceania other than Australia and New Zealand (pattern 3), it would appear that HIV was introduced late in the global pandemic. Therefore, cases reported only represent a small fraction of the world total. However, rapid changes are taking place, particularly in Asia, where there has been a considerable increase in HIV prevalence, particularly amongst intravenous drug users and prostitutes.

Natural History of Infection: Classification

Centers for Disease Control (CDC) Classification System for HIV Infection

Persons infected with HIV, during a mean incubation period of about 8–10 years to severe disease expression, present with a variety of manifestations bridging the gap between asymptomatic infection and the later severe immunodeficiency and life-threatening secondary infectious diseases or cancer. Initially a large number of terms were used to describe and assess patients with early and late disease manifestations of HIV infection in terms of their symptoms, signs and laboratory findings. In 1986 the Centers for Disease Control proposed a classification system to facilitate and clarify communication about the infection and disease (Table 2.4).

Table 2.4. Centers for Disease Control (CDC) classification system for HIV infection

Group I	Acute infection
Group II	Asymptomatic infection
Group III	Persistent generalized lymphadenopathy
Group IV	Other disease
Subgroup A	Constitutional disease (fever >1 month, weight loss >10 per cent base line, diarrhoea >1 month)
Subgroup B	Neurological disease (dementia, myelopathy, peripheral neuropathy; attributed to HIV)
Subgroup C	Secondary infectious diseases
	C1 Those specified in CDC surveillance definition
	C2 Others: oral hairy leucoplakia, multidermatomal herpes zoster, recurrent salmonellosis bacteraemia, nocardiosis, tuberculosis, oral candidiasis.
Subgroup D	Secondary cancers (Kaposi's sarcoma, non-Hodgkin's lymphoma, primary cerebral lymphoma)
Subgroup E	Other conditions, such as lymphoid interstitial pneumonitis

Adapted from CDC (1986b).

The classification in four principal groups is hierarchical in that persons classified in a particular group should not be reclassified in a preceding group if clinical findings resolve, since clinical improvement does not necessarily reflect changes in the severity of the underlying disease. Constitutional disease (subgroup IVA) is defined as one or more of the following: fever persisting for more than one month; involuntary weight loss of more than 10 per cent of baseline or diarrhoea persisting more than one month; and the absence of a concurrent illness or condition other than HIV infection to explain the findings. Subgroup E includes clinical findings or diseases not classifiable in the other groups that may be attributed to HIV infection, and/or may be indicative of a defect in cell-mediated immunity. This includes patients with chronic lymphoid interstitial pneumonitis. Also included are those patients whose signs or symptoms could be attributed either to HIV infection or to another co-existing disease not classified elsewhere and patients with other clinical illnesses, the course or management of which may be complicated or altered by HIV infection. Examples include: patients with constitutional symptoms not meeting the criteria for subgroup IVA; patients with infectious diseases not listed in subgroup IVC; and patients with neoplasms not listed in subgroup IVD. This classification was not meant to imply any change in the definition of AIDS described previously and used for national and international reporting.

Progression to Symptomatic Disease

A series of important cohort studies now give some indication of the natural history of infection with HIV and the increasing risk of progression to symptomatic disease. The San Francisco General Hospital cohort study of 288 seropositive homosexual men reported a 41 per cent progression rate from estimated seroconversion date to any CDC group IV disease at 6 years (Moss et al, 1988). Progression rates to AIDS in cohorts of homosexual and bisexual men reported in the literature have varied widely. In one study the 3 year cumulative incidence rate of AIDS was 34 per cent for men in New York City, 17 per cent for men in Washington DC and 8 per cent for men in Denmark (Moss et al, 1988). It was suggested that these differences occurred because some cohorts had been infected longer than others, and

it is well known that the risk of AIDS for HIV infected persons increases with increasing time since infection.

The results of studies where there is an estimated or known seroconversion date are shown in Table 2.5. Unfortunately, many of the studies do not report standard errors along with progression rates. The estimated cumulative rates at 5 years range from 11–23 per cent with a median of 14 per cent.

However, it is clear that there is still some variability across cross-studies even when date of seroconversion has been controlled for. As a result there has been much discussion in the literature concerning the question as to whether certain 'at risk' groups have a lower or greater incidence of progression than others. This has been suggested for haemophiliac patients where progression rates have ranged from 5–16 per cent at 5 years. However, this may be due to a high incidence of KS in homosexual men which is rarely seen in haemophiliac patients. Progression rates to PCP, therefore, may provide a more equitable comparison of homosexual and haemophiliac cohorts. Results of studies reporting Kaplan Meier progression rates to PCP are

Table 2.5. Cumulative progression rates (%) to AIDS from known (K) or estimated (E) date of seroconversion in studies of homosexual men and haemophiliacs

		Year								
		1	2	3	4	5	6	7	8	9
Studies in homosexual men										
Biggar, 1990	International (K) (n = 476)	0.4	3.1	6.6	11.5	22.8	36.4	–	–	–
Hessol et al, 1988	SF[a] City Clinic (K) (n = 181)	–	–	4.0	9.0	14.0	22.0	34.0	38.0	42.0
Moss et al, 1988	SF General Hospital (E) (n = 288)	–	–	–	5.0	11.0	22.0	–	–	–
Munoz et al, 1988	MACS[b]: all subjects (E) (n = 1523)	0.5	4.0	11.2	15.8	20.5	–	–	–	–
Phair et al, 1988	MACS: seroconvertors (K) (n = 277)	–	–	8.0	–	–	–	–	–	–
Schecter et al, 1988	Vancouver (n = 96)	–	–	–	–	13.0	–	–	–	–
Kelly et al, 1990	London (E) (n = 172)	0.6	1.8	7.2	11.6	14.3	–	–	–	–
Studies in haemophiliacs										
Biggar et al, 1990	International (K) (n = 173)	0	0.6	4.2	10.9	15.8	23.9	–	–	–
Giesecke et al, 1988	Sweden (K) (n = 98)	0	0	0	1.0	5.0	–	–	–	–
Goedert et al, 1986	Maryland, US (K) (n = 304)	0	0.3	3.2	6.2	6.2	11.2	17.2	31.2	41.2
Lee et al, 1988	London (K) (n = 104)	–	–	–	–	–	–	–	31.0	–

Adapted from Kelly et al (1990).
[a] SF, San Francisco; [b] MACS, multicentre AIDS cohort study.

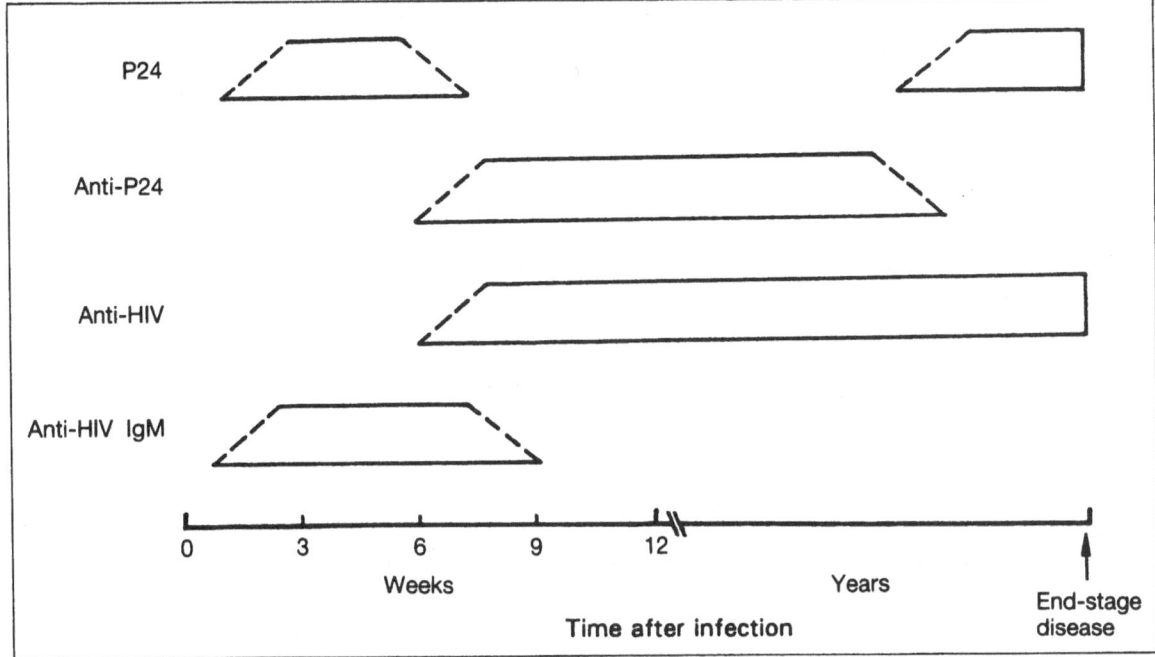

Figure 2.1. The development of antibodies in the serum to the core and surface proteins of the HIV virus.

remarkably similar, with an 11.5 per cent progression rate at 5 years and 18 per cent at 6 years (Eyster et al, 1987; Biggar, 1989; Kelly et al, 1990). This evidence suggests that the long-term progression to AIDS in haemophiliac and homosexual cohorts is very similar. Several studies have shown that there is a correlation between a more rapid progression to disease and increasing age at the time of infection.

In summary, the above discussion pertains to the progression to AIDS. If one includes other manifestations of symptomatic disease, that is CDC subgroups IVA–E, one can assume that around 50–60 per cent of patients infected with HIV will develop symptoms and signs of disease within 10 years of infection.

Natural History of Infection: Clinical Aspects

Acute infection with HIV may be accompanied by a transient, non-specific illness similar to glandular fever, which includes fever, malaise, myalgia, lymphadenopathy, pharyngitis and a rash. This has been reported as occurring in between 10–40 per cent of infected individuals

(Cooper et al, 1985). A transient acute meningo-encephalitis, myelopathy and polyneuropathy have also been described (Carne et al, 1985; Denning et al, 1987). The meningo-encephalitis may present with a prodrome of malaise, fever and personality change, progressing to severe headache, occasionally convulsions and various degrees of coma. Acute infection is accompanied by the development of antibodies in the serum to the core and surface proteins of the virus, usually in 2–6 weeks, although delayed seroconversions have been observed (Figure 2.1).

Chronic infection then ensues. This is asymptomatic in the early stages. Physical examination may show no abnormality, but about one-third of patients have persistent generalized lymphadenopathy (PGL) (nodes of 1 cm or more in diameter in two or more non-contiguous extra-inguinal sites, which cannot be explained by any other infection or condition). Many other patients have lesser degrees of lymphadenopathy. The nodes are symmetrical, mobile and non-tender. It was thought that asymptomatic patients with PGL had a more advanced infection than those without. However, prospective studies have shown that asymptomatic anti-HIV-positive patients with and without PGL progress to AIDS at the same rate (Polk et al, 1987). Biopsy is largely unrewarding in asymptomatic patients, revealing only non-specific follicular hyperplasia. If there are constitutional symptoms such as weight loss or fever, or the nodes are markedly asymmetrical, painful or rapidly enlarging, or if there is an extra-nodal mass or co-existent hilar nodes, then biopsy is essential. This will exclude tumour such as lymphoma or lymphopathic KS or an opportunistic infection such as mycobacteria. The commonest sites of lymphadenopathy are the cervical and axillary lymph nodes, and hilar lymphadenopathy is unusual.

Later in infection non-specific constitutional symptoms develop which may be intermittent or persistent. These include fever, with or without nightsweats, diarrhoea and weight loss (CDC subgroup IVA).

Patients may also be affected by a large number of 'minor' opportunistic infections or conditions that tend to affect the mucous membranes and skin (Table 2.6). Skin conditions encountered include a range of infections: dermatophytic fungal; viral (recalcitrant warts, molluscum contagiosum, herpes simplex and zoster); and bacterial (folliculitis, impetigo and furunculosis).

Table 2.6. Chronic manifestations of HIV infection in the skin/oropharynx

Skin
Viral
Herpes simplex (types 1 and 2) – anogenital
Varicella zoster
Human papilloma virus
Molluscum contagiosum

Fungi
Tinea cruris/pedis/corporis/other
Candidiasis – interdigital, perianal
Pityriasis versicolor

Bacterial
Folliculitis – mild, severe, acneiform
Impetigo

Other
Seborrhoeic dermatitis
Psoriasis

Oropharynx
Candidiasis
Hairy leucoplakia
Aphthous ulcers
Herpes simplex
Dental disease
   Periodontitis
   Abscesses
   Necrotizing ulcerative gingivitis

The spectrum of other skin disorders ranges from a mere dryness to severe seborrhoeic dermatitis. Hairy leucoplakia appears typically as greyish-white lesions on the lateral borders of the tongue. Its histological appearance is characteristic, with keratin projections resembling hairs, koilocytosis and mild atypia (Greenspan et al, 1985). Epstein–Barr virus (EBV DNA) has been demonstrated in these lesions by hybridization with specific probes, and there have been anecdotal reports of a response to acyclovir therapy. The collections of symptoms and signs which are often a prodrome to the development of major opportunistic infection or tumour are often referred to as AIDS-related complex (ARC).

Prognostic Laboratory Markers

Acute viraemia, detected by the presence of a core protein (p24), precedes the appearance of antibodies (IgG and IgM) to the whole virus. While the patient remains asymptomatic, high titres of antibodies to envelope and core proteins persist, and p24 remains detectable in only about 10–15 per cent of patients (Figure 2.1). As immune deficiency develops, the titre of antibody to p24 falls, but the titres of antibodies to the viral envelope proteins (gp41, 120 and 160) remain high. p24 antigenaemia recurs with time, and is found in 50–60 per cent of patients with symptomatic disease (Weber et al, 1987).

Prospective cohort studies have identified clinical and laboratory markers (Lange et al, 1989) which predict a more rapid progression to symptomatic disease (Table 2.7). Clinical features such as constitutional symptoms, oral candida, leucoplakia and herpes zoster skin rash have been well documented. Haematological abnormalities encountered include anaemia, a raised erythrocyte sedimentation rate and cytopenia. The lymphopenia is largely accounted for by progressive depletion of CD4-positive lymphocytes (Moss et al, 1988). The depletion and impaired

Table 2.7. Clinical and laboratory markers associated with increased risk of progression to AIDS

Clinical
Constitutional symptoms
Oral candidiasis
Oral hairy leucoplakia

Laboratory

Simple (late predictors)	*Sophisticated (early predictors)*
Anaemia (packed cell volume)	↓ CD4+ lymphocytes
Lymphopenia	↑ CD8+ lymphocytes
Neutropenia	↑ B_2 microglobulin
↑ Erythrocyte sedimentation rate	↑ p24 antigen
	↓ antibody to p24

function of this subset of T-lymphocytes seems to be the primary abnormality of immune dysfunction. The CD4 molecule, however, is also displayed at lower density by other cells such as monocytes, macrophages and some B-lymphocytes.

Tumours

Kaposi's Sarcoma

Kaposi's sarcoma is a presenting feature in approximately 20 per cent of patients, and is commoner in homosexual men than other groups at risk. The median survival is about 2–3 years, with some patients running a more rapid fulminant course and others remaining well for several years with minimal cutaneous disease. Death is usually caused by supervening life-threatening opportunistic infection. HIV-related KS is characterized by widespread skin, mucous membrane, visceral and lymph node involvement (see Chapter 5). Skin lesions are the most common presenting complaint. They appear as pink or red macules or violaceous plaques and nodules on the face, trunk or limbs. Early skin lesions may be difficult to differentiate from other benign skin conditions such as granulomas, bruises, naevi, dermatofibromas, secondary syphilis or lichen planus. Histologically, the tumours consist of spindle-shaped cells arranged in broad bands, with vascular slits and extravasation of erythrocytes between the cells (see Chapter 5).

The gastrointestinal tract is one of the commonest internal organs to be involved (Friedman et al, 1985). If upper and lower gastrointestinal endoscopies were performed at presentation, lesions would be demonstrated in about 40 per cent of patients. At postmortem they are present in more than 70 per cent. Lesions resemble the range seen in the skin, from small, flat lesions, not well demonstrated by contrast studies and only seen at endoscopy, to large nodular polypoid lesions. Endoscopic biopsies have a high false-negative rate, with only 23 per cent of suspicious lesions being confirmed histologically because of their predominant position deep in the submucosa. However, complications from involvement of the gut are unusual. Haemorrhage may occur but is rare. Nodules of KS also occur in the lungs. Chest X-ray appearances vary from confluent, irregular masses to interstitial nodularity. Computer tomography of the thorax may be useful in differential diagnosis. At bronchoscopy endobronchial lesions may be seen.

The diagnosis of KS in very early skin lesions may be extremely difficult, as little more may be seen than is seen regularly, that is, dilated vascular channels in the mid-dermis and a mild inflammatory cell infiltrate.

Lymphoma

Cancer Registry data in San Francisco and Los Angeles in 1983 indicated up to a 3-fold rise in high-grade lymphomas in young, never-married men, and the characteristics of 90 such cases in homosexual men were first reported by Ziegler et al (1984). The tumours are of B-cell origin and may present *de novo*, in a setting of prodromal lymphadenopathy, opportunistic infection or KS. The majority of patients present with an extra-nodal involvement, predominantly in the central nervous system (see Chapter 6), bone marrow and gut (Ziegler et al, 1984) (see Chapter 5). The diagnosis of lymphoma should also be considered in patients with weight loss, constitutional symptoms and anaemia. The mechanisms involved in the transition from the follicular hyperplasia of PGL with polyclonal B-cell activation, through to B-cell lymphoma and the role of HIV and EBV have yet to be fully determined.

Squamous carcinomas of the mouth and anorectum have also been described in homosexual men with HIV infection, and human papilloma virus may well play a part.

Opportunistic Infections

The organisms responsible for the opportunistic infections occurring in patients with AIDS are unusual pathogens. Most of the infections are due to reactivation of latent organisms in the host or, in some cases, ubiquitous organisms to which we are all continually exposed. The variation in prevalence of infections in different populations and regions of the world reflects the variation in endemic prevalence. The infections are often difficult to diagnose because conventional serological tests are unhelpful. Treatment often suppresses rather than eradicates the infection. Relapses are therefore common, and continuous treatment with drugs, which may cause side-effects, may be necessary.

Three main organ systems are affected – the respiratory system, the gastrointestinal tract and the nervous system. In addition, patients may present with a history of night sweats, chronic ill health, fevers, or weight loss.

Pulmonary Complications

Pneumocystis carinii pneumonia (PCP) is the commonest life-threatening opportunistic infection in patients who progress from chronic HIV infection to AIDS. The presentation is subacute, with malaise, fatigue, weight loss and shortness of breath often developing over several weeks. Typical symptoms include retrosternal or subcostal chest discomfort associated with increasing shortness of breath, a dry cough and fever. The chest X-ray at presentation may be normal or show bilateral, perihilar fine infiltrates. The arterial oxygen tension is usually depressed, and the carbon monoxide transfer factor, where available, is low and may be the earliest detectable abnormality. The diagnosis is confirmed by cytological examination of induced sputum, and if this is negative, by fibre-optic bronchoscopy, with bronchial lavage and transbronchial biopsy. At the same time, other causes of pneumonia or co-existent infection such as cytomegalovirus (CMV), mycobacteria and fungi may be excluded (Murray et al, 1984). CMV in the lung is more often a bystander than a pathogen in AIDS (Millar et al, 1990).

Pyogenic bacterial causes of pneumonia should always be considered, particularly as their presentation may be atypical. As many as 10 per cent of episodes of pneumonia in AIDS patients may be due to bacteria such as *Streptococcus pneumoniae*, *Haemophilus influenzae*, *Branhamella* and group B streptococci (Polsky et al, 1986). Radiological appearances may include diffuse infiltrates, as well as the more typical focal or lobar patterns. Another cause of diffuse abnormality is lymphocytic interstitial pneumonitis, first described in paediatric AIDS and now recognized increasingly in adults. The cause is still uncertain, but it is certainly associated with HIV infection.

Infection with *Mycobacterium tuberculosis* may also occur. It is particularly common in the developing world, and in black and hispanic drug users in the USA, where the annual attack rate in tuberculin-positive HIV-positive drug users may be as high as 7–8 per cent. A slowing of the downward trend of reported tuberculosis cases

has occurred in the United States *pari passu* with the HIV epidemic (Centers for Disease Control, 1986a). In studies of newly diagnosed cases of tuberculosis in Miami, Kinshasa (Zaire) and San Francisco, approximately 30 per cent of patients were found to be HIV-seropositive (Chaisson and Slutkin, 1989). It seems that pulmonary tuberculosis appears in chronic HIV infection earlier than the atypical mycobacterial infections that complicate the severe immune depression of AIDS. The former responds to conventional antituberculous therapy.

Gastrointestinal and Hepatic Complications (Table 2.8)

Retrosternal Discomfort and Dysphagia

Oral and oesophageal candidiasis is the commonest cause of dysphagia or retrosternal discomfort. Oral candidiasis alone does not fulfil the criteria for AIDS. Oesophageal infection is best shown by culture or biopsy at endoscopy, although plaques of *Candida albicans* can often be seen during a barium swallow. Ulceration may be focal or diffuse. Both CMV and herpes simplex virus (HSV) may cause a similar pattern of ulceration in the oesophagus (and also may affect the stomach and duodenum). It may be difficult to differentiate between them by barium

Table 2.8. Gastrointestinal complications of AIDS

Complications	Causes
Retrosternal discomfort and dysphagia	Candidiasis Cytomegalovirus Herpes simplex virus
Diarrhoea, weight loss and malabsorption	Unknown – enteropathy Cryptosporidiosis, *Isospora belli*, and microsporidial infection Cytomegalovirus/herpes simplex virus Enteric bacteria – *Salmonella*, *Campylobacter* Neoplasia
Hepatitis and cholestasis	Mycobacteria Cytomegalovirus Drug-induced *Cryptosporidium*
Perianal ulceration	Herpes simplex virus ? Cytomegalovirus
Neoplasia and miscellaneous	Kaposi's sarcoma Lymphoma Hairy leucoplakia Recalcitrant anorectal warts ? Squamous oral/anal carcinoma

studies. In most cases of patients presenting with oral candidiasis and oesophageal symptoms, a pragmatic approach is to treat empirically and only proceed to endoscopy should there be no response.

Diarrhoea, Malabsorption and Weight Loss

Diarrhoea is a common symptom in patients with chronic HIV infection, with or without other manifestations of AIDS. In many cases a cause is not found and symptomatic treatment is all there is to offer. Enteropathy with villous atrophy and malabsorption has been described (Kotler et al, 1984).

Cryptosporidium is a coccidian protozoal parasite, and probably the commonest pathogen isolated from patients with AIDS who have diarrhoea (Soave et al, 1984). It is the commonest of the protozol causes of diarrhoea, which also include *Isospora belli* and microsporidia. In immunocompetent human hosts, *Cryptosporidium* produces a transient diarrhoeal illness. In people infected with HIV it can cause transient, intermittent, or persistent diarrhoea, ranging from loose stools to watery diarrhoea, colic, and severe fluid and electrolyte loss. Oocysts (4–5 μm) can be found in stools. If direct smears of unconcentrated faecal samples stained with iodine or modified acid-fast stains fail to show the oocysts, the samples should be concentrated. The diagnosis should not be discounted without examining multiple specimens.

CMV and HSV can cause focal or diffuse ulceration of the gut, from the mouth to the anus. HSV most commonly causes mucocutaneous lesions at the upper and the lower ends of the gastrointestinal tract, whereas CMV infection may mimic inflammatory bowel disease (McGowan and Weller, 1990).

Atypical mycobacteria of the *M. avium–intracellulare* complex are ubiquitous organisms that have little virulence for the immunocompetent host. Disseminated infection of several organisms occurs in patients with AIDS. Gastrointestinal infection may be associated with fever, weight loss, diarrhoea and malabsorption. Diagnosis can be made by acid-fast staining of the stool or biopsy material, or both, or culture of blood or tissue. *M. tuberculosis* infection of the bowel does occur, but is less common. *Campylobacter* and *Salmonella* species infections may cause diarrhoea, but the latter more commonly presents as a fever of unknown origin with bacteraemia.

Hepatitis and Cholestasis

Hepatitis in patients with AIDS may present as fever, abdominal pain and hepatomegaly. An abnormal liver function test results, particularly a raised alkaline phosphatase activity, and needle-biopsy often shows granulomatous hepatitis, usually caused by atypical mycobacteria rather than *M. tuberculosis*. The herpes viruses may also occasionally cause hepatitis as part of a disseminated infection. When multiple drugs are being taken, drug-induced hepatitis must always be considered.

Acalculous cholecystis and cholangitis also occur with an endoscopic retrograde cholangiographic picture similar to that of primary sclerosing cholangitis, with strictures and dilatation of the biliary tree. *Cryptosporidium* and CMV have been shown or isolated either alone or together, and are implicated as a cause of this syndrome. (Dowsett et al, 1988).

Neurological Complications

These are discussed in Chapters 3 and 13.

Conclusions

In the developed world we are no longer observing the natural history of HIV infection because of the extended use of antivirals – particularly zidovudine – in symptomatic disease, and antimicrobial prophylaxis and early therapeutic intervention in asymptomatic infection on the basis of low CD4-positive lymphocyte counts.

The first multicentre randomized placebo-controlled trial of zidovudine in symptomatic HIV disease, which was terminated short of a planned 6 months' duration in September 1986, demonstrated a significant reduction in mortality and incidence of opportunistic infection, improvement in symptoms scores, significant increase in CD4-positive lymphocyte counts and an approximate 90 per cent reduction in p24 antigenaemia in patients remaining on full dose (Fischl et al, 1987). Zidovudine was licensed for severe HIV disease and released for use in Europe in the spring of 1987. Most physicians and patients were certain that the demonstrated short-term benefit in patients with severe HIV disease outweighed the cost of short-term toxicity and concerns about long-term efficacy and side-effects. Since then there has been a general consensus, based on survival studies using follow-up of the original trial and historical controls and clinical experience, that there has been an improvement in the probability of survival in AIDS. However, apart from the introduction of zidovudine, a number of factors have probably contributed to this. These include the extended use of antimicrobial prophylaxis, particularly for *Pneumocystis carinii* pneumonia, earlier diagnosis, and earlier and better therapy of some of the opportunistic infections. However, there is a consensus view that there is a window of efficacy for zidovudine in the long term, which may be associated with the emergence of resistant viral strains. Furthermore, the use of certain antimicrobial agents, particularly for primary and secondary prophylaxis and maintenance treatment, may provide cross-cover for other opportunistic infections. An example of this is fluconazole, used for treatment and maintenance therapy of oral and oesophageal candidiasis, which is also active against cryptococcus. In addition, there has been a considerable move towards primary prophylaxis for pneumocystis in asymptomatic patients with CD4 counts of less than 200, based on natural history studies in cohorts which show an increased risk for individuals with such counts (Centers for Disease Control, 1989). More recently, zidovudine has been licensed for use in asymptomatic individuals with low CD4 counts, based on the results of a large study which showed that the progression rate to AIDS and advanced AIDS-related complex in patients with baseline CD4 counts of less than 500 was approximately halved from 8 to 4 per 100 person years. However, the actual number of individuals who benefited was small, and the results short-term with a mean follow-up of only 55 weeks. Other trials in Europe and Australia are continuing in order to answer fully the long-term questions of efficacy and toxicity (Weller, 1989). Nevertheless, in the face of all these developments it is possible that we will observe over the coming years a change in the manifestations of end-stage disease, particularly in those areas that we are not controlling or suppressing with earlier intervention. It is highly likely that the various diseases of the nervous system will emerge as increasingly common complications which will pose an important challenge to therapy.

References

Barré-Sinoussi F, Cherman JC, Rey F et al. (1983) Isolation of T lymphotropic retrovirus from a patient at risk of acquired immune deficiency syndrome (AIDS). Science 220:868–70.

Biberfield G, Brown F, Esparza J et al. (1987) WHO Working Group on characterization of HIV-related retroviruses: criteria for characterization and proposal for a nomenclature system. AIDS 1:189–90.

Biggar RJ. (1989) Time-to-AIDS among 816 seroconvertors: Seroconvertors Working Group. V International Conference on AIDS, Montreal, Canada, June 1989 (Abstract).

Biggar RJ. (1990) International registry of seroconvertors, AIDS incubation in 1891 HIV seroconvertors from different exposure groups. AIDS 4:1059–66.

Carne CA, Tedder RS, Smith A et al. (1985) Acute encephalopathy coincident with seroconversion for anti-HTLV-III. Lancet ii:1206–8.

Centers for Disease Control. (1986a) Tuberculosis – United States 1985 and the possible impact of HTLV-III/LAV infection. MMWR 35:74–6.

Centers for Disease Control. (1986b) CDC Classification System for Human T Lymphotrophic Virus Type III – lymphadenopathy associated virus infections. MMWR 35:334–9.

Centers for Disease Control. (1987) Revision of the CDC surveillance case definition for acquired immunodeficiency syndrome. MMWR 36 (suppl).

Centers for Disease Control. (1989) Guidelines for prophylaxis against *pneumocystis carinii*. Pneumonia for persons infected with human immunodeficiency virus. MMWR 38:1–9.

Chaisson RE, Slutkin G. (1989) Tuberculosis and human immunodeficiency virus infection. J Infect Dis 159:96–9.

Cooper DA, Gold J, Maclean P et al. (1985) Acute AIDS retrovirus infection. Lancet i:537–40.

Denning DW, Anderson J, Rudge P et al. (1987) Acute myelopathy associated with primary infection with human immunodeficiency virus. Br Med J 294:143–4.

Dowsett JF, Miller R, Davidson R et al. (1988) Sclerosing cholangitis in acquired immunodeficiency syndrome. Scand J Gastroenterol 23:1267–74.

Eyster M, Gail M, Ballard J et al. (1987) Natural history of human immunodeficiency virus infections in haemophiliacs. Ann Intern Med 107:1–6.

Fischl MA, Rickman DD, Grieco MH et al. (1987) The efficacy of azidothymidine (AZT) in the treatment of patients with AIDS and AIDS-related complex. N Engl J Med 317:185–91.

Friedman-Kien A, Laubenstein L, Marmor M et al. (1981) Kaposi's sarcoma and pneumocystis among homosexual men – New York City and California. MMWR 30:305–8.

Friedman-Kien A, Laubenstein LJ, Rubinstein P et al. (1982) Disseminated Kaposi's sarcoma in homosexual men. Ann Intern Med 96:693–700.

Friedman SL, Wright TL, Atlman DF. (1985) Gastrointestinal Kaposi's sarcoma in patients with acquired immunodeficiency syndrome. Endoscopic and autopsy findings. Gastroenterology 89:102–8.

Gallo RC, Savin PS, Gelman EP et al. (1983) Isolation of human T cell leukaemia virus in acquired immune deficiency syndrome (AIDS). Science 220:865–7.

Giesecke J, Scalia-Tomba G, Beergland O et al. (1988) Progression to AIDS in haemophiliacs and blood transfusion recipients infected with human immunodeficiency virus. Br Med J 297:99–102.

Goedert JJ, Biggar RJ, Weiss SH et al. (1986) Three-year incidence of AIDS in five cohorts of HTLV-III infected risk group members. Science 231:992–5.

Gottlieb MS, Schanker HM, Fan PT et al. (1981a) Pneumocystis pneumonia – Los Angeles. MMWR 30:250–2.

Gottlieb MS, Schroff R, Schanker HM et al. (1981b) *Pneumocystis carinii* pneumonia and mucosal candidiasis in previously healthy homosexual men: evidence of a new acquired cellular immunodeficiency. N Engl J Med 305:1425–31.

Greenspan JS, Greenspan D, Lennette ET et al. (1985) Replication of Epstein–Barr virus within the epithelial cells of oral 'hairy' leucoplakia, an AIDS-associated lesion. N Engl J Med 313:1564–71.

Hessol NA, Rutherford GW, Lifson AR et al. (1988) The natural history of HIV infection in a cohort of homosexual and bisexual men: a decade of follow up. IV International Conference on AIDS, Stockholm, June 1988 (Abstract 4096).

Kelly G, Stanley B, Weller IVD. (1990) The natural history of human immunodeficiency virus infection: a five year study in a London cohort of homosexual men. Genitourin Med 66:238–43.

Kotler DP, Gaetz HP, Lange M et al. (1984) Enteropathy associated with the acquired immunodeficiency syndrome. Ann Intern Med 101:421–8.

Lange JMA, de Wolf F, Goudsmit J. (1989) Markers for progression in HIV infection. AIDS 3 (suppl 1):S153–S160.

Lee CA, Miller EJ, Griffiths PD et al. (1988) HIV disease in a cohort of 104 haemophiliacs. IV International Conference on AIDS, Stockholm, June 1988 (Abstract 7733).

McGowan I, Weller IVD. (1990) AIDS and the gut In: R Pounder, ed. Recent advances in gastroenterology no. 8. Churchill Livingstone: New York, 133–56.

Millar AB, Patou G, Miller RF et al. (1990) Cytomegalovirus in the lungs of patients with AIDS. Respiratory pathogen or passenger? Am Rev Respir Dis 141:1474–7.

Moss AR, Bacchetti P, Osmond D et al. (1988) Seropositivity for HIV and the development of AIDS or AIDS-related condition: 3 year follow-up of the San Francisco General Hospital cohort. Br Med J 296:744–50.

Munoz A, Wang MC, Good S et al. (1988) Estimation of the AIDS-free times after HIV-1 seroconversion. IV International Conference on AIDS, Stockholm, June 1988 (Abstract 4129).

Murray JF, Felton CP, Garay SM et al. (1984) Pulmonary complications of the acquired immunodeficiency syndrome. N Engl J Med 310:1682–8.

Phair J, Munoz A, Kingsley L et al. (1988) Incidence of AIDS in homosexual men developing HIV infection. IV International Conference on AIDS, Stockholm, June 1988 (Abstract 4093).

Polk BF, Fox R, Brookmeyer R et al. (1987) Predictors of the acquired immunodeficiency syndrome developing in a cohort of seropositive homosexual men. N Engl J Med 316:61–6.

Polsky B, Gold JWM, Whiimbey E et al. (1986) Bacterial pneumonia in patients with the acquired immunodeficiency syndrome. Ann Intern Med 104:38–41.

Schecter MT, Craib KJ, Willoughly B et al. (1988) Progression to AIDS in a cohort of homosexual men: results at 5 years. IV International Conference on AIDS, Stockholm, June 1988 (Abstract 4098).

Soave R, Danner RL, Honig CL et al. (1984) Cryptosporidiosis in homosexual men. Ann Intern Med 100:504–11.

Weber JN, Clapham PR, Weiss RA et al. (1987) Human immunodeficiency virus infection in two cohorts of homosexual men: neutralising sera and association of anti-gag antibody with prognosis. Lancet i:119–22.

Weller IVD. (1989) The treatment of asymptomatic HIV infection: lessons from the zidovudine experience. AIDS 3 Suppl 1:S215–S220.

WHO/CDC. (1990) Statistics from the World Health Organization and the Centers for Disease Control. AIDS 4:937–41.

Ziegler JL, Beckstead JH, Volberding P et al. (1984) Non-Hodgkin's lymphoma in 90 homosexual men: relation to generalised lymphadenopathy and the acquired immunodeficiency syndrome. N Engl J Med 311:565–70.

3 Neurological Complications of HIV Infection: Clinical Aspects

M.J.G. Harrison

The acquired immunodeficiency syndrome (AIDS), due to infection with the human immunodeficiency virus (HIV), produces clinical effects through patients' susceptibility to opportunistic infections and the development of tumours, such as lymphomas. It soon became clear that the nervous system was frequently involved (Snider et al, 1983). Severe and often fatal infections of the nervous system were experienced by those with immunodeficiency, and in 10 per cent of cases neurological problems were the first signs of the development of AIDS. As the epidemic due to HIV infection has grown, so too has the list of neurological manifestations (McArthur, 1987; Harrison and McAllister, 1991). The number of patients with AIDS-related neurological problems, as seen for example at the University of California group of hospitals in San Francisco, is doubling every year (Levy and Bredesen, 1988). To date (April 1991) there have been some 4228 cases of AIDS in the UK, of whom 50 per cent are dead.

Some of the neurological problems encountered could not be attributed to any known opportunistic pathogen or atypical tumour, raising the possibility that HIV was directly responsible for nervous system lesions (Table 3.1). The evidence that HIV could, indeed, be neurotropic as well as lymphotropic, was both virological and clinical. First, the virus is morphologically and genetically related to visna, a neurotropic and lymphotropic retrovirus which affects sheep. Second, infectious HIV can be recovered from cultures of cerebrospinal fluid (CSF), brain, and spinal cord (Ho et al, 1985). CSF samples in some patients also reveal unique oligoclonal IgG bands, and CSF levels of HIV-specific IgG are higher than in the serum, implying active central nervous system (CNS) infection. Also, CSF samples taken from asymptomatic individuals within 2 years of seroconversion show a cellular reaction, positive virus culture, or a related IgG rise (Appleman et al, 1988); again evidence of early low-grade infection of the CNS. Autopsies reveal abnormalities in 90 per cent, of which the commonest is an encephalitis with no overt cause other than HIV. The clinical evidence for a direct role of HIV in CNS disorders includes the appearance of meningitis or encephalitis at the time of seroconversion, and the fact that many patients with nervous system manifestations simply have no other explanation for their lesions. It is, of course, possible that new pathogens will be discovered, however.

Table 3.1. Effect of HIV on the nervous system

Lymphotropism	Opportunistic infections Lymphomas
Neurotropism	So-called direct effects of HIV, e.g. meningitis, myelopathy, encephalopathy

The immune deficiency produced by HIV infection makes patients with AIDS susceptible to infection by a variety of viruses, bacteria, fungi, and parasites, none of which is normally very pathogenic (Table 3.2). It also modifies the inflammatory response (Table 3.3). Thus, clinical signs of inflammation may be minimal, with no fever or meningism. Also, the CSF response may be muted with few cells involved. HIV infection impairs humoral as well as cellular immunity, so antibody production is defective. This means that antibody titres are of little use in the diagnosis of opportunistic infections. The tissue response may also be muted, and as a result a normal abscess wall may not develop, which, in turn, affects brain imaging characteristics. Multiple pathogens may cause simultaneous infections, complicating diagnosis and treatment. Finally, because relapse is common in the presence of immunosuppression, antimicrobial treatment usually has to be lifelong.

Table 3.2. Opportunistic pathogens in AIDS

Pathogen	Example
Parasites	*Toxoplasma*
Fungi	*Cryptococcus*
	Candida
Viruses	Cytomegalovirus
	Herpes zoster
	Herpes simplex
	JC virus
	(HIV)
Bacteria	*Mycobacterium avium–intracellulare complex*
	Mycobacterium tuberculosis
	Listeria monocytogenes

Table 3.3. Effects of immunosuppression on evidence of infection

Reduced incidence of fever and meningism
Muted CSF response
Impaired antibody response
Modified tissue response, e.g. cerebral abscess wall
Multiple infections
High risk of relapse

It is not yet clear whether the 'at risk' background affects the pattern of infectious complications. Black patients in New York and intravenous drug users appear to be more prone to the development of cryptococcal meningitis. It is also possible that the particular circumstances of infected haemophiliacs affect their risks of complications. Preliminary autopsy studies (Esiri et al, 1989; Lantos et al, 1989) suggest that opportunistic infections and encephalitis are less

common or less severe in haemophiliacs, but this may be an artefact of premature death from complications of the haemophilia.

Opportunistic infections and lymphomas complicate the late stages of HIV infection when immunosuppression supervenes. At earlier stages, conditions in which there is a prominent inflammatory response, such as an acute demyelinating neuropathy of Guillain–Barré type and polymyositis, are more likely. The timing of the different complications is set out in Table 3.4 and the rest of the discussion will follow a similar schema.

Table 3.4. Timing of commonest neurological complications

Stage of disease	Complication
Seroconversion	Meningitis
	Encephalitis
	Myelitis
Asymptomatic	CSF changes
	Meningitis
	Demyelinating neuropathy
	Myopathy/polymyositis
Asymptomatic, but immunosuppressed or ARC	Meningitis
	Mononeuritis multiplex
	Encephalopathy
	Myopathy
AIDS	Meningitis
	Myelopathy
	Axonal neuropathy
	Radiculopathy
	Encephalopathy
	Cerebrovascular disease
	Myopathy
ARC/AIDS	Opportunistic infections
	Abscesses
	Meningitis
	Encephalitis
	Progressive multifocal
	leukoencephalopathy
	Lymphomas

Seroconversion Illness

In most patients seroconversion passes unnoticed, but in some there is a mild influenza-like illness. More rarely, a glandular fever-like illness with fever, malaise, lymphadenopathy, and arthralgia occurs (Cooper et al, 1985). Neurological involvement is rare, but meningitis, encephalitis, myelitis and radiculitis have all been described (Carne et al, 1985). A mild meningitic illness, perhaps complicated by facial weakness, is the most common of these (McArthur, 1987). That it can even be subclinical is suggested by the CSF findings reported in asymptomatic HIV-positive individuals within a year or two of seroconversion (Appleman et al, 1988). No specific treatment is needed for these manifestations of CNS invasion, which are self-limiting.

Asymptomatic Period

HIV Meningitis

Patients may develop a meningitic illness (which appears to be due to HIV itself) at any stage from seroconversion to AIDS. Headache and meningism may be complicated by cranial nerve palsies (the 5th, 7th or 8th), and long-tract signs (McArthur, 1987). The CSF shows an increase in cells and protein. The illness is usually self-limiting, patients recovering in 1–4 weeks. However, it may run a lengthier course, with no more than persistent headache with a cellular CSF, or follow a relapsing and remitting pattern. All other causes of meningitis need to be considered, as HIV meningitis is a diagnosis of exclusion. The frequency of mildly abnormal CSF in asymptomatic HIV-positive individuals suggests that a low-grade meningeal inflammatory reaction is common, and this is supported by autopsy findings.

Peripheral Neuropathy

Early in the natural history of HIV infection (rarely, even at the time of seroconversion), and well before there are signs of immunosuppression, patients may develop an acute demyelinating neuropathy of Guillain–Barré type. The clinical picture is the familiar one of weakness of limb and facial muscles, minor sensory symptoms and loss of tendon reflexes, often accompanied by backache and limb pains. Nerve conduction studies and biopsies show that the predominant pathological change is demyelination. The clinical picture is slightly different from the classical post-infectious syndrome, in that the history may be rather longer, the autonomic nerves are usually spared, and recovery may be slower. In many, the condition looks more like the chronic form of idiopathic demyelinating neuropathy (Cornblath et al, 1987). The main difference concerns the CSF, which usually shows a pleocytosis, as well as a rise in protein. CSF IgG may be elevated and be oligoclonal. Severe cases may respond to plasmapheresis or steroids. There are thus several features suggesting an autoimmune process. Rare cases have, however, occurred during the AIDS-related complex stage (ARC) (Mishra et al, 1985), when this is less likely. Idiopathic cases of the Guillain–Barré syndrome are usually preceded by infection (Winer et al, 1988), for example, with *Campylobacter* or cytomegalovirus (CMV) (Dowling and Cook, 1981), both of which might occur in HIV-infected patients. In the cases described in ARC there might alternatively be a role for autoimmune clones of dysregulated T-cells.

Myopathy/Polymyositis

A progressive proximal muscle weakness, which on EMG and biopsy studies is revealed to be myopathic in nature, may occur at any stage, including the asymptomatic period. Muscle pain may be prominent, especially after exercise, and muscles may be tender on palpation (Dalakas et al, 1986). Serum creatine phosphokinase levels may or may not be elevated, and correlate poorly with the degree of weakness. Biopsies show muscle fibre degeneration, with little necrosis and inflammatory infiltrate (see Chapter 10). In the 25 per cent of cases with a more classical polymyositis on biopsy, there may be a good therapeutic response to corticosteroids, with relief of the symptoms and signs, and reversal of any muscle enzyme changes. The use of steroids in this group is not without risk, however, and may precipitate opportunistic infections. This clinical evidence suggests an autoimmune basis for the myopathy or polymyositis observed. However, HIV has been cultured from muscle, and viral antigen has been identified in the OKT4 lymphoid cells around muscle fibres, so a direct infection is also possible. If the patient is receiving antiviral treatment with zidovudine, a mitochondrial myopathy may develop that is clinically indistinguishable. Treatment may have to be stopped.

ARC/AIDS

Mononeuritis Multiplex

Immunosuppressed patients, usually with ARC, may develop mononeuritis multiplex. The clinical picture consists of a combination of individual nerve palsies, for example, an ulnar nerve palsy together with a foot drop, due to a common peroneal nerve lesion. Less commonly, cranial nerves are affected, or a plexopathy is seen, which is probably analogous. A vasculitis may be responsible and be detected on nerve biopsy. If it were not for the risk of adding to the immuno-paralysis, steroids would be the obvious treatment. The prognosis can be good, however, and some patients recover in a few weeks without steroids.

Myelopathy (see also Chapter 8)

Transient spinal cord symptoms and signs due to a subacute myelitis have occasionally been noted to occur at the time of seroconversion (Denning et al, 1987). Also, there have been rare examples of *Toxoplasma*, CMV, or HSV affecting the cord in AIDS (Jacobson and Mills, 1988; Jacobson et al, 1988). However, the most common finding in the spinal cord at autopsy, which is seen in as many as 40 per cent of AIDS patients, is a vacuolar myelopathy (Petito et al, 1985). This affects the posterior and lateral columns in the same pattern seen in subacute combined degeneration of the cord (SACD), but in the majority of cases B_{12} and folate levels are normal (see Chapter 8). The role of HIV is undecided. Ho et al (1985) isolated HIV from the CSF and cord in one case, and multinucleated giant cells, probably a sign of HIV infection, have been described (Maier et al, 1989). Other researchers have found no evidence of this or any other virus in affected cords. Although HTLV-I can cause cord damage, it does not appear to be responsible in AIDS.

The clinical counterpart of the autopsy findings is a paraparesis with sensory ataxia and incontinence in a patient with ARC or AIDS. One unusual feature, again reminiscent of SACD, is the tendency for lower limb reflexes not to be very brisk. Myelography or MRI of the cord is always needed to exclude compression of the cord, for example, by lymphoma. The CSF should also be cultured for HSV, CMV and varicella zoster virus (VZV). Most patients with the myelopathy show signs of dementia by the time cord features are obvious, and isolated involvement of the cord is rare in practice (McArthur, 1987). In many cases, the myelopathy is an autopsy finding in a severely demented patient in whom the relative role of the cord and brain in any long-tract signs that may be detected at the bedside is uncertain. It is not yet known whether zidovudine affects the signs of cord damage.

Peripheral Neuropathy

Estimates vary, but the peripheral nervous system is probably involved in 10–20 per cent of patients infected with HIV (McArthur, 1987). The commonest complication in patients with AIDS is an often painful axonal neuropathy. Patients complain of burning feet, and painful parasthesiae in the toes, and such glove and stocking sensory symptoms dominate even when weakness is also present. Examination reveals elevated sensory thresholds in the feet, often accompanied by absent ankle reflexes. A few cases show wasting of small foot muscles (Cornblath, 1988). Nerve conduction studies can be diagnostic with small or absent sensory action potentials, but they can be normal early on. There has been a report that zidovudine may influence the natural history of the neuropathy in some patients (Yarchoan et al, 1987), but this is difficult to judge since untreated it may be progressive or remain rather static. Sural nerve biopsies show axonal loss more often than demyelination, and at postmortem the picture is of a dying back neuropathy (Rance et al, 1988). Carbamazepine, clonazepam, tri-cyclic antidepressants, or even lumbar root nerve blocks, may be needed to control the limb pains. Autonomic neuropathy has occasionally been reported in AIDS, and may be present in 10 per cent of cases. It has probably been underdiagnosed (Cornblath, 1988). Rarely, a ganglioneuritis causes a sensory ataxia (Dalakas and Pezeshkpour, 1988).

The aetiology of the axonal type of neuropathy is unknown, but as it correlates with the presence of encephalopathy, it has been suggested that it may be a direct effect of HIV. In keeping with this, the virus has been cultured from the peripheral nerve (Ho et al, 1985), and IgM has been detected in a nerve biopsy. However, it is also possible that the neuropathy is related to that seen in advanced malignancy and in the critically ill. As many of these patients are on neurotoxic drugs like vincristine and metronidazole, the possibility of iatrogenic neuropathy also has to be considered in the differential diagnosis. Terminally, when the patient is cachectic, pressure palsies and nutritional neuropathies are possible. Metastatic Kaposi's sarcoma and lymphoma can infiltrate peripheral nerves.

Some patients describe odd areas of sensory disturbance, not attributable to a mononeuritis multiplex or a diffuse peripheral neuropathy (Rance et al, 1988). This seems to be analogous to chronic idiopathic demyelinating neuropathy, and often has a good prognosis (Lipkin et al, 1985).

A type of inflammatory polyradiculopathy, producing flaccid leg weakness, sacral sensory loss, and sphincter involvement with a striking CSF pleocytosis (polymorphs), has been attributed to CMV (Eidelberg et al, 1986a). Nerve conduction studies show proximal slowing and some denervation. The prognosis is variable, and the response to ganciclovir still controversial (see Chapter 10).

Encephalopathy

At first, complaints of poor memory and concentration were attributed to the psychological stresses associated with the diagnosis of AIDS. It soon became apparent, however, that patients with AIDS might develop a progressive dementia (Snider et al, 1983), and the postmortem studies revealed that patients frequently had evidence of an encephalitis (Snider et al, 1983; Jordan et al, 1985; Navia et al, 1987). At first, CMV was held responsible for the pathological changes, but it is now generally believed that HIV is, itself, the cause of the encephalitis. HIV has been identified in, and cultured from, the brains of patients with AIDS and encephalitis (Ho et al, 1985; Levy et al, 1985a). There were many drawbacks, however, to the simple concept that the encephalitis caused the dementia. For example, there is little virus replication in the brain and neurones are not directly affected (see Chapter 12). Not all demented patients have prominent encephalitis, and not all patients with encephalitis are demented in life. All patients with dementia attributed to HIV itself show laboratory evidence of immunological impairment, and it seems likely that although the virus seeds the CNS at the time of seroconversion, it only causes persistent or progressive changes in the cerebral substance when immunosuppression supervenes.

Navia et al (1986a), in a retrospective case note survey, found that 65 per cent of patients dying of AIDS had been thought to be demented. However, the true prevalence is proving to be more in the region of 10–20 per cent (see Chapter 7). Nevertheless, it still represents the commonest neurological complication (Snider et al, 1983; Koppel et al, 1985; Bruhn, 1987; McArthur, 1987). The incidence is dependent on the stage of the underlying disease. Although some patients are found to have dementia as the first manifestation of their immunosuppression and the CDC criteria for AIDS now include HIV encephalopathy (Levy and Bredesen, 1988), dementia is usually a late development in a sick patient.

Early on, friends and lovers report subtle changes in the patient's personality. Affected individuals show slowing of verbal and emotional responses. They then become apathetic and withdrawn with loss of libido, and a flat affect. They may be noted to be distractable or just said to be 'different'. The patients themselves are more aware of mental slowing and distressing forgetfulness. They lose the thread of conversations, need to make lists, and have difficulties with their daily affairs. Headaches, poor coordination, and seizures may also occur. Dysphasia and dyspraxia are late features when the dysfunction becomes global. Even when the changes are subtle, patients with HIV encephalopathy tend to be very sensitive to routine doses of sedatives, tranquillizers and analgesics.

The early picture of slowness, forgetfulness and apathy suggests that the dementia is of the 'subcortical' type (Tross et al, 1988) (check Chapters 6 and 7 for cortical lesions). Later cognitive changes are more striking, and eventually patients become mute and incontinent with seizures, myoclonus, grasp reflexes, hyperreflexia and extensor plantar responses. Terminally, paraplegia or decorticate posturing may be seen.

Soft signs, such as tremor and unsteadiness, have also been attributed to HIV encephalopathy (Navia et al, 1986a). The problem with this aspect of the 'AIDS dementia complex' is that such indefinite findings could also be due to a metabolic encephalopathy, and could thus be seen in febrile or systemically ill patients. Recently, abnormal eye movements have been demonstrated in patients with AIDS and have been claimed to correlate with the presence of dementia (Currie et al, 1988).

The differential diagnosis of mild mental changes includes depression, anxiety, bereavement reactions, drug side-effects, alcoholism, recreational drug use, and confusion due to coincidental infections. Dementia due to structural brain disease can be due to opportunistic infections of the CNS, and lymphomas. Detailed neuropsychological assessment and neurological investigation are clearly indicated.

EEGs show non-specific loss of faster rhythms and increase in slow activity. CT scans may be normal in the early stages or show atrophy. MR imaging reveals atrophy, but often also shows confluent areas of abnormal signal in the white matter, especially on T2-weighted images (see also Chapter 4). The CSF usually shows an elevated protein content. The diagnosis of HIV encephalopathy should only be made in this context, when all other possible diagnoses have been rigorously excluded. This will usually require both imaging and lumbar puncture, and occasionally even brain biopsy.

The possibility of subclinical changes has been much debated. To date, there is no firm evidence that asymptomatic patients show any increased incidence of cognitive or imaging abnormalities when compared with seronegative controls from similar 'at risk' backgrounds (Janssen et al, 1989; McArthur et al, 1989; Selnes et al, 1990).

Some have attributed the reduced prevalence of dementia in their recently-referred case load to the introduction and widespread use of zidovudine. The published trials are not yet convincing, however. Despite reducing mortality in one trial (Garcia et al, 1983), zidovudine failed to affect the results of psychological tests. The effect of inhibiting viral replication on the incidence and severity of encephalopathy is thus still *sub judice*.

Cerebrovascular Disease (see also Chapter 6)

Although a diagnosis of stroke is rarely made in AIDS patients, autopsy studies have shown that cerebral infarcts are not uncommon (Snider et al, 1983; Levy et al, 1985b). The possible causes include non-bacterial endocarditis as a source of embolism, and meningovascular syphilis (Berry et al, 1987). Others have a cerebral vasculitis which may be due to Herpes zoster, or perhaps to HIV itself (Eidelberg et al, 1986b; Yankner et al, 1986). The lupus anticoagulant has been detected in AIDS, and could cause a thrombotic tendency. Haemorrhage due to metastatic Kaposi's sarcoma or thrombocytopenic purpura is rare.

Myopathy

As noted already, a progressive proximal myopathic weakness can occur at any stage from asymptomatic to AIDS. Recently there have been reports that patients on long-term zidovudine may develop a myopathy or asymptomatic elevation of muscle enzymes (serum CK). The symptomatic myopathy may remit on stopping the drug (Besser and Weinberg 1988), but does

not always recur on rechallenge. Until further research clarifies the situation, it seems advisable to stop zidovudine in any patient who develops a proximal myopathic distribution of muscle weakness. Patients debilitated by other AIDS-related infections may have muscle weakness, which on biopsy shows non-specific type 2 fibre atrophy. Assessment is often complicated by cachectic weight loss and co-existent peripheral neuropathy.

Opportunistic Infections

Parasites

Toxoplasmosis

Toxoplasma gondii is an obligate intracellular parasite. The primary host is the cat, from which humans may acquire it by the faecal – oral route. It can also be caught by eating undercooked meat which contains tissue cysts from infected domestic animals, or rarely, by blood transfusion (Ruskin and Remington, 1976). Although the primary infection is usually asymptomatic, the organism forms cysts in all tissues. These are the source of reinfection in the immunocompromised host. In parts of the world where the exposure to toxoplasmosis is widespread, for example the southern USA and France, the risk of a toxoplasma encephalitis in AIDS may reach 12 per cent. In the UK, only 20 per cent of young adults have serological evidence of prior exposure, and the incidence of toxoplasma brain abscesses appears to be much lower. Overall, cerebral toxoplasmosis affects some 5–30 per cent of AIDS patients (Pitchenik et al, 1983; Snider et al, 1983; Luft et al, 1984; Wong et al, 1984), and is the commonest opportunistic infection in Europe.

Infection in the brain is usually multifocal. The clinical presentation may be focal or diffuse, or a combination of the two. Most patients present with personality change or mild confusion, combined with constitutional symptoms, and a progressive focal deficit. Headaches are reported in 60–70 per cent, and seizures in 25 per cent. Over 50 per cent have fever, and meningism is most unusual. The onset of symptoms can be sudden, mimicking a cerebrovascular accident, but is more usually progressive over a few days, or a week or two. The presence of headache and fever should be enough to suggest the diagnosis. The focal deficit is usually in the form of a hemiparesis, and is seen in two-thirds of cases. Toxoplasma abscesses are the commonest cause of a focal neurological deficit in AIDS.

CT or MRI scanning should precede any thoughts of a lumbar puncture, and usually reveal multiple focal mass lesions. The differential diagnosis of this clinical and radiological picture includes: abscesses due to other pathogens, such as *Candida*, *Aspergillus*, *Nocardia*, *Mucor*, *Cryptococcus*, *Mycobacterium*, or pyogenic organisms; or the changes of lymphoma, progressive multifocal leukoencephalopathy (PML), or infarction. Less commonly, toxoplasma causes a diffuse encephalitis (Gray et al, 1989).

Serological tests are often misleading in AIDS. IgM rarely rises, as this is a reactivation of latent infection and IgG titres may not show a diagnostic rise. There is hope that a diagnostic test can be based on IgA levels. At first, brain biopsy was recommended for all mass lesions because of the lack of specificity in CT or MRI appearances and the poor serological tests (Handler et al, 1983; Levy et al, 1985b). An alternative policy has been developed dependent on a trial of anti-toxoplasma therapy as a diagnostic test. This is based on the fact that a response can be expected in over 90 per cent of cases (Handler et al, 1983; Navia et al, 1986b). If the patient and the scan improve within 2 to 15 days, the diagnosis of *T. gondii* is considered established. If, instead, the patient and/or the scan lesions deteriorate, a biopsy is considered to exclude other treatable lesions. A biopsy is also considered if one CT/MRI lesion progresses, whilst others regress on anti-toxoplasma therapy. Single mass lesions are less likely to be due to toxoplasmosis, so biopsy is often indicated early on. Lymphoma is the usual finding.

Most patients, even those in coma, can make a good neurological recovery (Navia et al, 1986b; Leport et al, 1988). The prognosis is better than for any other cause of a cerebral mass lesion in an AIDS patient. Pyrimethamine and sulphadiazine are only active against the free trophozoites. They do not affect the encysted organism, so long-term treatment must continue to prevent relapse (Pinching, 1988). Long-term survival is now thought to be further improved if patients can tolerate zidovudine as well (Hermans et al, 1988). Survival for 6–18 months is common (Hermans et al, 1988; Holliman, 1988). It has been suggested that HIV-positive asymptomatic patients with serological evidence for earlier exposure to toxoplasma, should be treated prophylactically to protect them when immunodeficiency supervenes.

Fungi

Cryptococcal Meningitis

Cryptococcus neoformans is a ubiquitous budding encapsulated soil yeast, which gains access to the body via the respiratory tract (Pons et al, 1988). Blood-borne spread leads to a granulomatous meningitis, with or without clinically evident pulmonary or disseminated infection. Alternatively, and indeed in HIV-positive patients, there may be small cysts in the cerebral cortex. This is the most common CNS fungal infection in AIDS, and the second most common opportunistic infection. Cryptococcal meningitis or cryptococcomas are found in 5–10 per cent of AIDS patients (Dismukes, 1988). Black patients and intravenous drug users appear especially vulnerable.

The clinical presentation is usually with headache (75 per cent), fever (95 per cent) and malaise, and the diagnosis is likely to be missed if too much reliance is placed on the classical features of neck stiffness and photophobia, since they are present in only 30 per cent of cases or less (Pons et al, 1988; Zuger et al, 1986; Dismukes, 1988). Focal neurological signs due to cryptococcomas are picked up in only 10 per cent. Any feverish HIV-positive patient needs a lumbar puncture to exclude cryptococcal meningitis, unless some other cause of fever is already obvious. CT scanning should always precede lumbar puncture, however, because of the risk of silent mass lesions in AIDS (Dismukes, 1988). In cryptococcal meningitis imaging is usually normal. The diagnosis is further complicated by the tendency for the CSF to be normal (Zuger et al, 1986; Dismukes, 1988). India ink staining of fresh CSF shows yeasts in 70 per cent of patients, and serum or CSF cryptococcal antigen and culture are positive in 90–100 per cent (Koppel et al, 1985; Zuger et al, 1986; McArthur, 1987; Pons et al, 1988). Cryptococcal antigen determination in serum and CSF is thus the most useful investigation, and should be carried out on all samples from HIV-infected symptomatic patients (Eng et al, 1986). Serum testing may be a useful first step in the investigation of well patients with unexplained fever, but the CSF will usually be needed. Cryptococci may also be isolated from blood, sputum or bronchial washings, bone marrow, skin lesions and urine. A CSF cryptococcal antigen titre of over 1:10000 and a positive India ink smear indicate heavy infestation, and imply a worse prognosis (Zuger et al, 1986). The differential diagnosis includes tuberculomas and syphilitic and lymphomatous meningitis.

Treatment consists of a 6–8 week course of amphotericin B alone or with 5-flucytosine, giving a 70–85 per cent chance of a good clinical response. Successful treatment produces a striking fall in the titre of antigen in the CSF, and this is the best way to monitor therapy (Eng et al, 1986; Zuger et al, 1986). AIDS patients run a high risk of relapse and a high mortality in relapse (Zuger et al, 1986). Lifelong maintenance therapy, e.g. with weekly amphotericin B, is indicated. Two new imidazole derivatives, fluconazole and itraconazole, are being investigated as alternative treatments in the search for a less toxic regime. Fluconazole appears to be effective and adequate for maintenance, though slower to induce clinical change.

Candida

Oral and oesophageal candidiasis is common in ARC and AIDS. Involvement of the nervous system is rare (Levy and Bredesen, 1988), and is usually a terminal event associated with systemic infection. *Candida* brain abscesses have occasionally been diagnosed at biopsy or autopsy, the CT scan appearances being indistinguishable from those of other mass lesions in AIDS (Levy et al, 1983). Culture of a biopsy specimen is the best diagnostic procedure. The treatment of choice is probably amphotericin B.

Nocardia is another rare cause of brain abscess (Adair et al, 1987).

Viruses

Cytomegalovirus (CMV)

Serological evidence of prior exposure to CMV by the age of 40 years is found in 80 per cent of the population of the USA, and almost all AIDS patients can be shown to be excreting this virus. Also, CMV is readily reactivated by other infections, so its presence could merely reflect immunosuppression. Its significance is, therefore, difficult to determine. It is weakly lymphotropic and in vitro it enhances replication of HIV, so its presence could be synergistic with that of HIV (see Chapter 6). However, there are some cases in which CMV appears to have been the immediate and sole cause of an encephalitis, with fever and impaired conscious level. The diagnosis is difficult to make in life unless blood cultures are positive or a brain biopsy is carried out, when characteristic inclusion bodies may be seen. A brain stem encephalitis presenting with an internuclear ophthalmoplegia, for example, may be caused by CMV (Fuller et al, 1989). Similar inclusions have been reported in the spinal cord, and CMV may occasionally be the cause of a symptomatic myelitis (Dalakas and Pezeshkpour, 1986) (see also Chapter 8).

A slowly-evolving lesion in the cauda equina, with perineal sensory loss and sphincter disturbance together with a radicular pattern of muscle weakness and reflex loss in the legs, has recently been described and shown to be associated with CMV infection of lumbar and sacral roots (Tucker et al, 1985). The CSF often shows a marked pleocytosis (mostly polymorphs).

CMV is the proven cause of a frequent ocular infection (Dalakas and Pezeshkpour, 1986), which causes full thickness necrosis and haemorrhage of the retina, leading to blindness in untreated cases. Up to 30 per cent of AIDS patients develop evidence of this CMV retinopathy. Early on, fundoscopy reveals spreading coalescing white granular areas that become fluffy white exudates. The differential diagnosis at this stage includes a benign microvascular retinopathy with cotton wool spots, and microaneurysms that appear to be due to deposits of immune complexes. This is asymptomatic, non-progressive, and its fluffy exudates tend to regress spontaneously. By contrast, CMV lesions are progressive, and cause progressive, though painless, visual loss. The diagnosis must be made clinically, since it is impractical to obtain fluid or tissue from the eye for culture (Drew, 1988), though blood cultures are usually positive. Sudden deterioration in sight in an eye affected by CMV retinopathy can be due to retinal detachment. Most patients have AIDS before retinopathy is obvious, but a few have presented to ophthalmologists with no other manifestations of immunodeficiency (Dalakas and Pezeshkpour, 1986). Rarely, *Toxoplasma* or *Cryptococcus* causes intraocular infection. All patients should probably have regular ophthalmological examinations to detect early signs of CMV retinopathy, since it can be suppressed by the regular administration of ganciclovir (DHPG).

Herpes Zoster (Varicella Zoster Virus or VZV) (see Chapter 6)

Attacks of shingles with characteristic root pain and vesicular eruption affect 5–10 per cent of

HIV-infected individuals, due to the reactivation of VZV in dorsal root ganglion cells. Affected subjects are more likely to get a disseminated rash than are non-AIDS patients.

Rare cases of myelitis and of acute encephalitis have been described (Ryder et al, 1986; Dix and Bredesen, 1988).

Herpes Simplex (HSV)

HSV-1 or HSV-2 can cause an encephalitis, or more rarely, a myelitis or meningitic illness, in an AIDS patient (Tucker et al, 1985; Dix and Bredesen, 1988). There may be no cellular reaction in the CSF, and it is rare to culture the virus. In the encephalitis caused by HSV-1 (or exceptionally by HSV-2), the history usually begins with headache and fever with behavioural change, and then increasing drowsiness. Focal signs may be prominent. Paradoxically, the illness may be less acute in AIDS than in other anergic patients (Price et al, 1973), which suggests that local immune responses contribute to the tissue damage in the immunocompetent. CT scans and MRI may show changes in the temporal lobes. If HSV encephalitis or myelitis is suspected, the patient should have a course of intravenous acyclovir. HSV may have a synergistic role, since in vitro it can be shown to accelerate HIV replication.

Progressive Multifocal Leukoencephalopathy

Progressive multifocal leukoencephalopathy (PML) is a central demyelinating disease. It occurs almost exclusively in immunocompromised patients (Åstrom et al, 1958; Brooks and Walker, 1984), and appears to represent reactivation of a latent infection (Houff et al, 1988) of oligodendrocytes by a papovavirus (in the vast majority of cases the JC virus). In the USA, 70 per cent of the population show serological evidence of prior exposure to JC virus, compared to approximately 50 per cent in the UK. Progressive focal limb weakness is the commonest presenting complaint, followed by visual field defects, ataxia and dementia. CT scans show areas of low density without mass effect in the white matter, but MR images are more dramatic. The CSF is unhelpful, and brain biopsy is necessary for a definitive diagnosis (Schlitt et al, 1986; Berger et al, 1987) and the exclusion of CMV encephalitis, lymphoma and so on (see Chapter 4).

In AIDS, survival beyond 6 months is unusual, although occasional patients have survived 12–20 months with supportive care. PML is going to become common (Levy et al, 1988), and there is an urgent need for research into effective therapy.

Spirochaetes/Bacteria

Neurosyphilis

It has been suggested that HIV infection may alter the natural history of syphilis. An abnormally short interval between primary infection and meningovascular complications has been reported (Johns et al, 1987), and it has been suggested that retrobulbar neuritis and ophthalmoplegia may be particularly frequent in AIDS. This impression has yet to be confirmed, however. Both HIV meningitis and neurosyphilis can cause headache and cranial nerve palsies, and in either case the CSF may show a mild pleocytosis. A trial of anti-syphilitic therapy may be needed, as serological proof of the activity of lues can be difficult in the immunosuppressed patient (Centers for Disease Control, 1988). For example, the CSF VDRL test can be negative.

Mycobacteria

Atypical mycobacteria (organisms belonging to the Mycobacterium avium–intracellulare complex (MAI)) are the most common cause of systemic bacterial infection in AIDS patients. The infection is often a terminal manifestation, and MAI organisms can be detected in as many as 50 per cent of AIDS patients coming to autopsy (Armstrong et al, 1985). The clinical picture is non-specific, with fever and weight loss. MAI has been described as a rare cause of meningitis and of encephalopathy in AIDS patients (Zakouski et al, 1982; Armstrong et al, 1985). The diagnosis usually depends on identifying the organisms in the blood. It fails to respond to standard antituberculous medication (Young, 1988).

Tuberculous meningitis (Fischl et al, 1985) and tuberculomas due to M. tuberculosis (Bishburg et al, 1986) occur more frequently in AIDS patients than in the non-AIDS population. Abscess formation is more likely (Bishburg et al, 1986), so CT scans may show ring enhancing lesions. The clinical picture consists of headaches, seizures, or an altered mental state. The patient need not be febrile. The diagnosis may be difficult without biopsy, but blood cultures may be positive (Pitchenik et al, 1984). Standard quadruple therapy is indicated.

Listeria monocytogenes, which is a familiar pathogen in other immunosuppressed patients, is unexpectedly rare in AIDS, but can cause meningitis, encephalitis and microabscesses (Gould et al, 1986; Mascola et al, 1988).

Tumours

Lymphomas

AIDS is now the most common underlying cause for CNS lymphomas. They may be the first clinical manifestation of the immunodeficient state in 0.5 per cent of cases (Snider et al, 1983; Koppel et al, 1985; McArthur, 1987; Levy and Bredesen, 1988) and are second only to toxoplasmosis as a cause of intracranial mass lesions, eventually affecting up to 5 per cent of adults with AIDS. They are also the most common cause of a mass lesion in children with AIDS. CNS lymphomas are multicentric B-cell tumours.

More than two-thirds of patients present with non-specific problems, such as confusion, lethargy, and memory loss (So et al, 1986, 1988; McArthur, 1987; Hochberg and Miller, 1988). Around one-third have focal neurological deficit, which may progress rapidly, mimicking stroke or intracranial infection. Headache is usually a late feature, and is rarely the sole symptom. Seizures affect up to 20 per cent of patients.

It is impossible to be absolutely confident in differentiating lymphomas from *Toxoplasma* abscesses on CT or MRI, but a truly solitary lesion is more likely to be a lymphoma.

The CSF usually shows a moderate protein elevation and low glucose, with a modest mononuclear pleocytosis (Hochberg and Miller, 1988; So et al, 1988). It may be possible to demonstrate that apparently 'reactive' lymphocytes in the CSF are monoclonal and therefore neoplastic (Louriero et al, 1988).

The prognosis in AIDS is very poor, with median survival times of less than 6 months (So et al, 1986, 1988). However, there may be an initial response to radiotherapy (So et al, 1986, 1988; Louriero et al, 1988).

AIDS patients are also at risk of systemic non-Hodgkin's lymphoma. These aggressive B-cell tumours are likely to present at an advanced stage, with extranodal involvement (Italian Cooperative Group for AIDS-related Tumours,

1988; Knowles et al, 1988; Louriero et al, 1988). Up to two-thirds have CNS involvement, usually in the form of a lymphomatous meningitis with cranial nerve or spinal root involvement. Several CSF specimens may be needed before the diagnostic cells are seen. Alternatively, they may present with extradural spinal cord compression. Full remission can be achieved, but the prognosis overall is worse than in the non-AIDS patient. CNS prophylaxis should be considered in those who present with non-neurological disease, to prevent neurological relapse after systemic therapy (Louriero et al, 1988).

Metastatic Kaposi's sarcoma is extremely rare in the brain, but is a possible cause for cerebral haemorrhage, or a mass lesion.

Paediatric AIDS (see also Chapter 9)

Congenitally acquired HIV causes immunosuppression between 6 months and 2 years of age (Epstein et al, 1986). Opportunistic infections of the nervous system are, however, rare. Most neurologically affected children show signs of an encephalopathy (Epstein and Sharer, 1988), with lost milestones, ataxia and pyramidal signs. Older children develop language and other cognitive problems. The condition may appear to stabilize, but this is in fact only temporary and a stepwise deterioration is the rule. Retarded brain growth leads to acquired microcephaly (Pawha et al, 1986). CT scans may show atrophy or calcification, especially in the basal ganglia. The encephalopathy appears to be due to HIV itself (Sharer et al, 1986), and may affect as many as 50 per cent of children with AIDS or ARC (Epstein et al, 1986). An indistinguishable encephalopathy can be due to CMV, but this usually also causes extracerebral infection (Lechtenberg and Sher, 1988). A static encephalopathy, not accompanied by any evidence of active HIV infection in the brain, has been attributed to antenatal and perinatal problems. Myelopathy and neuropathy are rare.

Lymphoma, despite being rare, appears to be the commonest cause of a cerebral mass lesion in children with AIDS. Clinically, focal CNS problems are often superimposed on a background of intellectual impairment due to coincidental encephalopathy. Systemic lymphomas with CNS involvement have also been described (Epstein and Sharer, 1988; Epstein et al, 1988).

References

Adair JC, Beck AC, Apfelbaum RI et al. (1987) Nocardia cerebral abscess in the acquired immunodeficiency syndrome. Arch Neurol 44:548–50.

Appleman ME, Marshall DW, Brey RL et al. (1988) Cerebrospinal fluid abnormalities in patients without AIDS who are seropositive for the human immunodeficiency virus. J Infect Dis 158:193–8.

Armstrong D, Gold JWM, Dryjanski J et al. (1985) Treatment of infections in patients with the acquired immunodeficiency syndrome. Ann Intern Med 103:738–43.

Åstrom K-E, Mancall EL, Richardson EP. (1958) Progressive multifocal leukoencephalopathy: a hitherto unrecognised complication of chronic lymphatic leukaemia and Hodgkin's disease. Brain 81:93–111.

Berger JR, Kaszovitz B, Post MJD et al. (1987) Progressive multifocal leukoencephalopathy associated with human immunodeficiency virus infection. Ann Intern Med 107:78–87.

Berry CD, Hooton TM, Collier AC et al. (1987) Neurologic relapse after benzathine penicillin therapy for secondary syphilis in a patient with HIV infection. N Engl J Med 316:1587–9.

Besser P, Weinberg H. (1988) Severe polymyositis like syndrome associated with zidovudine therapy of AIDS and ARC. N Engl J Med 318:708.

Bishburg E, Sunderam G, Reichman LB et al. (1986) Central nervous system tuberculosis with the acquired immunodeficiency syndrome and its related complex. Ann Intern Med 105:210–13.

Brooks BR, Walker DL. (1984) Progressive multifocal leukoencephalopathy. Neurol Clin 2:299–313.

Bruhn P. (1987) AIDS and dementia: a quantitative neuropsychological study of unselected Danish patients. Acta Neurol Scand 76:443–7.

Carne CA, Smith A, Elkington SG et al. (1985) Acute encephalopathy coincident with seroconversion for anti HTLV-111. Lancet i:1206–8.

Centers for Disease Control, Division of Sexually Transmitted Diseases. (1988) Recommendations for diagnosing and treating syphilis in HIV-infected patients. MMWR 37:600–8.

Cooper DA, Gold J, MacLean P et al. (1985) Acute AIDS retrovirus infection. Definition of a clinical illness associated with seroconversion. Lancet i:537–40.

Cornblath DR. (1988) Treatment of the neuromuscular complications of human immunodeficiency virus infection. Ann Neurol 23 (suppl):s88–91.

Cornblath DR, McArthur JC, Kennedy PGE et al. (1987) Inflammatory demyelinating peripheral neuropathies associated with human T cell lymphotropic virus type III. Ann Neurol 21:32–40.

Currie J, Benson E, Ramsden B et al. (1988) Eye movement abnormalities as a predictor of the acquired immunodeficiency syndrome. Arch Neurol 45:949–53.

Dalakas MC, Pezeshkpour GH. (1988) Neuromuscular diseases associated with human immunodeficiency virus infection. Ann Neurol 23 (suppl):s38–48.

Dalakas MC, Pezeshkpour GH, Gravell M et al. (1986) Polymyositis associated with AIDS retrovirus. JAMA 256:2381–3.

Denning DW, Anderson J, Rudge P et al. (1987) Acute myelopathy associated with primary infection with human immunodeficiency virus. Br Med J 294:143–4.

Dismukes WE. (1988) Cryptococcal meningitis in patients with AIDS. J Infect Dis 157:624–8.

Dix RD, Bredesen DE. (1988) Opportunistic viral infections in acquired immunodeficiency syndrome. In: Rosenblum ML, Levy RM, Bredesen DE, ed. AIDS and the nervous system. Raven Press: New York, 221–61.

Dowling PC, Cook SD. (1981) Role of infection in Guillain–Barré syndrome: laboratory confirmation of Herpes viruses in 41 cases. Ann Neurol 9 (suppl):44–5.

Drew WL. (1988) Cytomegalovirus infection in patients with AIDS. J Infect Dis 158:449–56.

Eidelberg D, Sofrel A, Vogel H et al. (1986a) Progressive polyradiculoneuropathy in acquired immune deficiency syndrome. Neurology 36:912–16.

Eidelberg D, Sotrel A, Horoupian DS et al. (1986b) Thrombotic cerebral vasculopathy associated with Herpes zoster. Ann Neurol 19:7–14.

Eng RHK, Bishburg E, Smith SM et al. (1986) Cryptococcal infections in patients with acquired immune deficiency syndrome. Am J Med 81:19–23.

Epstein LG, Sharer LR. (1988) Neurology of human immunodeficiency virus infection in children. In: Rosenblum ML, Levy RM, Bredesen DE, ed. AIDS and the nervous system. Raven Press: New York, 79–101.

Epstein LG, Sharer LR, Oleske JM et al. (1986) Neurological manifestations of human immunodeficiency virus infection in children. Pediatrics 78:678–87.

Epstein LG, DiCarlo FJ, Joshi VV et al. (1988) Primary lymphoma of the central nervous system in children with acquired immunodeficiency syndrome. Pediatrics 82:355–63.

Esiri MM, Scaravilli F, Millard PR et al. (1989) Neuropathology of HIV infection in haemophiliacs: comparative necropsy study. Br Med J 299:1312–15.

Fischl MA, Pitchenik AE, Spira TJ. (1985) Tuberculous brain abscess and toxoplasma encephalitis in a patient with the acquired immunodeficiency syndrome. JAMA 253:3428–30.

Fuller GN, Guiloff RJ, Scaravilli F et al. (1989) Combined HIV–CMV encephalitis presenting with brainstem signs. J Neurol Neurosurg Psychiatry 52:975–9.

Garcia I, Fainstein V, Rios A et al. (1983) Nonbacterial thrombotic endocarditis in a male homosexual with Kaposi's sarcoma. Arch Intern Med 143:1243–4.

Gould IA, Belok LC, Handwerger S. (1986) Listeria monocytogenes: a rare cause of opportunistic infection in the acquired immunodeficiency syndrome (AIDS) and a new cause of meningitis in AIDS. A case report. AIDS Res Hum Retroviruses 2:231–4.

Gray F, Gherardi R, Wingate E et al. (1989) Diffuse 'encephalitic' cerebral toxoplasmosis in AIDS: report of four cases. J Neurol 236:273–7.

Handler MH, Ho V, Whelan M et al. (1983) Intracerebral toxoplasmosis in patients with acquired immune deficiency syndrome. J Neurosurg 59:994–1001.

Harrison MJG, McAllister R. (1991) Neurological complications of HIV infection. In: Lambert HP, ed. Handbook of infectious diseases. Edward Arnold: London, 343–60.

Hermans P, Magrez P, De Wit S et al. (1988) The benefit of zidovudine in the treatment of AIDS patients with cerebral toxoplasmosis. In: Program and abstracts of the IV International Conference on AIDS, Book 2; Abstract no. 3667, p. 183.

Ho D, Rota T, Schooley R et al. (1985) Isolation of HTLV-III from cerebrospinal fluid and neural tissues of patients with neurological syndromes related to the acquired immunodeficiency syndrome. N Engl J Med 313:1493–503.

Hochberg FH, Miller DC. (1988) Primary central nervous system lymphoma. J Neurosurg 68:835–53.

Holliman RE. (1988) Toxoplasmosis and the acquired immune deficiency syndrome. J Infect 16:193–7.

Houff SA, Major EO, Katz DA. (1988) Involvement of JC virus-infected mononuclear cells from the bone marrow and spleen in the pathogenesis of progressive multifocal leuco-encephalopathy. N Engl J Med 318:301–5.

Italian Cooperative Group for AIDS-related Tumours. (1988) Malignant lymphomas in patients with or at risk for AIDS in Italy. J Natl Cancer Inst 80:855–60.

Jacobson MA, Mills J. (1988) Serious cytomegalovirus disease in the acquired immunodeficiency syndrome (AIDS). Ann Intern Med 108:585–94.

Jacobson MA, Mills J, Rush J et al. (1988) Failure of antiviral therapy for acquired immunodeficiency syndrome related cytomegalovirus myelitis. Arch Neurol 45:1090–2.

Janssen RS, Saykin AJ, Cannon L et al. (1989) Neurological and neuropsychological manifestations of HIV-1 infection: association with AIDS-related complex but not asymptomatic HIV-1 infection. Ann Neurol 26:592–600.

Johns DR, Tierney M, Felsertein D. (1987) Alteration in the natural history of neurosyphilis by concurrent infection with the human immunodeficiency virus. N Engl J Med 316:1569–72.

Jordan BD, Navia BA, Petito CK et al. (1985) Neurological syndromes complicating AIDS. Front Radiat Ther Onc 19:82–7.

Knowles DM, Chamulak GA, Subar M et al. (1988) Lymphoid neoplasia associated with the acquired immunodeficiency syndrome (AIDS). Ann Intern Med 108:744–53.

Koppel BS, Wormser GP, Tuchman AJ et al. (1985) Central nervous system involvement in patients with acquired immune deficiency syndrome. Acta Neurol Scand 71:337–53.

Lantos PL, McLaughlin JE, Scholtz CL et al. (1989) Neuropathology of the brain in HIV infection. Lancet i:309–10.

Lechtenberg R, Sher JH. (1988) AIDS in the nervous system. Churchill Livingstone: New York.

Leport C, Raffi F, Mathieron S et al. (1988) Treatment of central nervous system toxoplasmosis with pyrimethamine/sulphadiazine combination in 35 patients with the acquired immunodeficiency syndrome. Am J Med 84:94–100.

Levy RM, Bredesen DE. (1988) Central nervous system dysfunction in acquired immunodeficiency syndrome. In: Rosenblum ML, Levy RM, Bredesen DE, ed. AIDS and the nervous system. Raven Press: New York, 29–63.

Levy RM, Pons VG, Rosenblum ML. (1983) Intracerebral mass lesions in the acquired immunodeficiency syndrome (AIDS). N Engl J Med 309:1454–5.

Levy JA, Shimabukuro J, Hollander H et al. (1985a) Isolation of AIDS associated retroviruses from cerebrospinal fluid and brain of patients with neurological symptoms. Lancet ii:586–8.

Levy RM, Bredesen DE, Rosenblum ML. (1985b) Neurological manifestations of the acquired immunodeficiency syndrome (AIDS): experience at UCSF and a review of the literature. J Neurosurg 62:475–95.

Levy RM, Janssen RS, Bush TJ et al. (1988) Neuroepidemiology of acquired immunodeficiency syndrome. In: Rosenblum ML, Levy RM, Bredesen DE, ed. AIDS and the nervous system. Raven Press: New York, 13–27.

Lipkin WI, Parry G, Kiprov D et al. (1985) Inflammatory neuropathy in homosexual men with lymphadenopathy. Neurology 35:1479–83.

Louriero CL, Gill PS, Meyer PR et al. (1988) Autopsy findings in AIDS-related lymphoma. Cancer 62:735–9.

Luft BJ, Brooks RG, Conley FK et al. (1984) Toxoplasmic encephalitis in patients with acquired immune deficiency syndrome. JAMA 257:913–17.

Maier H, Budka H, Lassmann H et al. (1989) Vacuolar myelopathy with multinucleate giant cells in the acquired immunodeficiency syndrome (AIDS). Light and electron microscopic distribution of human immunodeficiency virus (HIV) antigens. Acta Neuropathol 78:497–503.

Mascola L, Lieb L, Chiu J et al. (1988) Listeriosis: an uncommon opportunistic infection in patients with acquired immunodeficiency syndrome. Am J Med 84:162–4.

McArthur JC. (1987) Neurologic complications of AIDS. Medicine (Baltimore) 66:407–37.

McArthur JC, Cohen BA, Selnes OA et al. (1989) Low frequencies of neurological and neuropsychological abnormalities in otherwise healthy HIV-1 infected individuals: results from the multicentre AIDS cohort study. Ann Neurol 26:601–11.

Mishra BB, Sommers W, Koshi CL. (1985) Acute inflammatory demyelinating polyneuropathy in the acquired immune deficiency syndrome. Ann Neurol 18:131–2.

Navia BA, Jordan BD, Price RW. (1986a) The AIDS dementia complex: 1. Clinical features. Ann Neurol 19:517–24.

Navia BA, Petito CK, Gold JWM et al. (1986b) Cerebral toxoplasmosis complicating the acquired immune deficiency syndrome: clinical and neuropathological findings in 27 patients. Ann Neurol 19:224–38.

Navia BA, Cho E-S, Petito CK et al. (1987) The AIDS dementia complex: 2. Neuropathology. Ann Neurol 19:525–35.

Pawha S, Kaplan M, Fikrig S et al. (1986) Spectrum of human T-cell lymphotropic virus infection in children. JAMA 255:2299–305.

Petito CK, Navia BA, Cho E-S et al. (1985) Vacuolar myelopathy pathologically resembling subacute combined degeneration in patients with the acquired immunodeficiency syndrome. N Engl J Med 312:874–9.

Pinching AJ. (1988) Prophylactic and maintenance therapy for opportunist infections in AIDS. AIDS 2:335–43.

Pitchenik AE, Fischl MA, Dickinson GM. (1983) Opportunistic infections and Kaposi's sarcoma among Haitians: evidence of a new acquired immune deficiency state. Ann Intern Med 98:277–84.

Pitchenik AE, Cole C, Russel BW et al. (1984) Tuberculosis, atypical mycobacteriosis, and the acquired immunodeficiency syndrome among Haitian and non-Haitian patients in South Florida. Ann Intern Med 104:641–5.

Pons VG, Jacobs RA, Hollander H. (1988) Nonviral infections of the central nervous system in patients with acquired immunodeficiency syndrome. In: Rosenblum ML, Levy RM, Bredesen DE, ed. AIDS and the nervous system. Raven Press: New York, 263–83.

Price R, Chernik NL, Horta-Barbosa L et al. (1973) Herpes simplex encephalitis in an anergic patient. Am J Med 54:222–7.

Rance NE, McArthur JC, Cornblath DR et al. (1988) Gracile tract degeneration in patients with sensory neuropathy and AIDS. Neurology 38:265–71.

Ruskin J, Remington JS. (1976) Toxoplasmosis in the compromised host. Ann Intern Med 84:193–9.

Ryder JW, Croen D, Kleinschmidt-DeMasters BK et al. (1986) Progressive encephalitis 3 months after resolution of cutaneous zoster in a patient with AIDS. Ann Neurol 19:182–8.

Schlitt M, Morawetz RB, Bonnin J et al. (1986) Progressive multifocal leukoencephalopathy: three patients diagnosed by brain biopsy, with prolonged survival in two. Neurosurgery 18:407–14.

Selnes O, Miller EN, McArthur J et al. (1990) HIV-1 infection:

no evidence of cognitive decline during the asymptomatic stages. Neurology 40:204–8.

Sharer LR, Epstein LG, Cho E-S et al. (1986) Pathologic features of AIDS encephalopathy in children: evidence for LAV/HTLVIII infection in brain. Hum Pathol 17:271–84.

Snider WD, Simpson DM, Nielsen S et al. (1983) Neurological complications of acquired immune deficiency syndrome: analysis of 50 patients. Ann Neurol 14:403–18.

So YT, Beckstead JH, Davis RL. (1986) Primary central nervous system lymphoma in acquired immune deficiency syndrome: a clinical and pathological study. Ann Neurol 20:566–72.

So YT, Choucair A, Davis RL et al. (1988) Neoplasms of the central nervous system in acquired immunodeficiency syndrome. In: Rosenblum ML, Levy RM, Bredesen DE, ed. AIDS and the nervous system. Raven Press: New York, 285–300.

Tross S, Price RW, Navia B et al. (1988) Neuropsychological characterization of the AIDS dementia complex: a preliminary report. AIDS 2:81–8.

Tucker T, Dix RD, Davis KC et al. (1985) Cytomegalovirus and Herpes simplex virus ascending myelitis in a patient with acquired immune deficiency syndrome. Ann Neurol 18:74–9.

Winer JB, Hughes RAC, Anderson MJ et al. (1988) A prospective study of acute idiopathic neuropathy. (2) antecedent events. J Neurol Neurosurg Psychiatry 51:613–18.

Wong B, Gold JWM, Brown AE et al. (1984) Central nervous system toxoplasmosis in homosexual men and parenteral drug abusers. Ann Intern Med 100:36–42.

Yankner BA, Skolnik PR, Shoukimas GM et al. (1986) Cerebral granulomatous angiitis associated with isolation of human T-lymphotropic virus type III from the central nervous system. Ann Neurol 20:362–4.

Yarchoan R, Berg G, Brouwers P et al. (1987) Response of human immunodeficiency virus-associated neurological disease to 3-azido-3-deoxythymidine. Lancet i:132–5.

Young LS. (1988) *Mycobacterium avium* complex infection. J Infect Dis 157:863–7.

Zakowski P, Fliegel S, Berlin GW et al. (1982) Disseminated *Mycobacterium avium–intracellulare* infection in homosexual men dying of acquired immunodeficiency. JAMA 248:2980–2.

Zuger A, Louie E, Holzman RS et al. (1986) Cryptococcal disease in patients with the acquired immunodeficiency syndrome. Ann Intern Med 104:234–40.

4   Acquired Immune Deficiency: Neuroradiological Imaging

B. Kendall

The acquired immune deficiency syndrome (AIDS) retroviruses, HIV-1 and HIV-2, may be transmitted by sexual contact, injection of infected blood or blood products, and through the placenta of an infected mother to the fetus.

AIDS is widespread in Central Africa and Haiti. The disease should be suspected when appropriate neurological symptoms occur in immigrants from these countries, as well as in all homo- or bisexual males and their female partners, recipients of blood or blood products (especially haemophiliacs and addicts injecting drugs) and the children of mothers with AIDS. The incidence of superadded infection is lower in young children than in sexually-transmitted HIV infection. Differences in neurological manifestations may be related both to age-dependent metabolic requirements and to the lifestyles of the infected persons.

Pre-seroconversion

Following infection, there is a period of weeks or months before seroconversion. During this time, aseptic meningitis, sometimes associated with seizures or polymyositis, may occur due to invasion of the nervous system by the neurotropic retrovirus. Imaging of the CNS is not usually indicated at this stage; if performed it will generally be negative. However, at least one case of acute encephalopathy with necrotizing and demyelinating encephalitis predominantly involving the central hemispheric white matter of both cerebral hemispheres and the upper brainstem has been recorded in a 31-year-old woman. This was probably due to seronegative HIV infection (Jones et al, 1988).

Seroconversion to HIV-Positive

Just before, or at the time of seroconversion, there may be an autoimmune response. This may manifest as Guillain–Barré syndrome or as cranial nerve palsies. In such cases, immunoglobulins are present in the cerebrospinal fluid in two-thirds, and a pleocytosis is present in one-third, but no abnormality is shown in imaging of the CNS. A subacute HIV encephalitis causing the AIDS dementia complex, which usually manifests after other complications of AIDS are evident, may commence at this stage. It occurs as an isolated feature in 7–9 per cent of patients presenting in London, and reaches as high as 37 per cent in the United States (Levy and

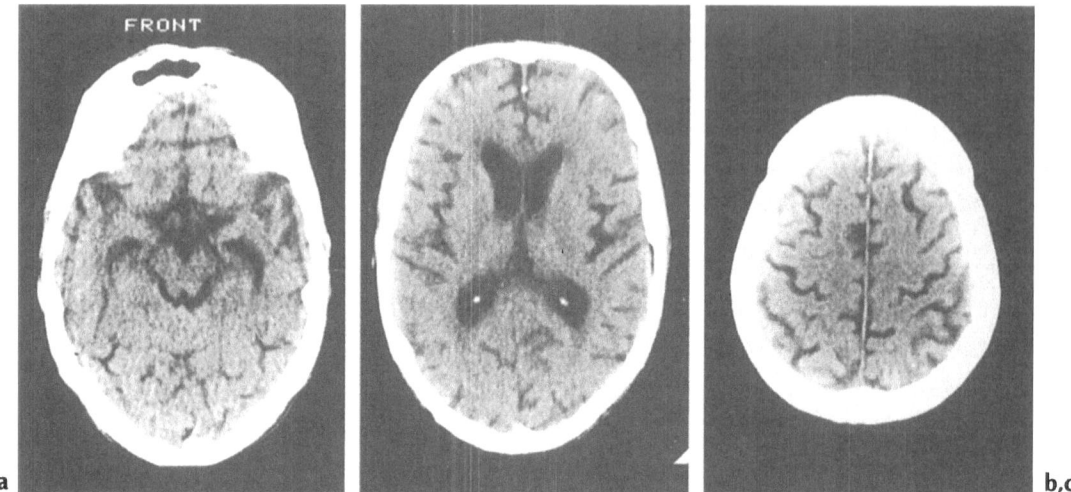

Figure 4.1a–c. 22-year-old HIV-positive homosexual male with progressive dementia. CT scans of the brain: **a** through the basal cisterns; **b** through the anterior horns of the lateral ventricles; **c** supraventricular. There is moderate dilatation of the lateral ventricles and the cerebral sulci. No other abnormality can be seen.

Bredesen, 1988). The clinico-pathological features of the AIDS dementia complex are described in Chapters 6 and 7.

AIDS

A diagnosis of full-blown AIDS is indicated when there is evidence of suppression of the immune system, followed by clinical manifestations of superadded infections, often by opportunistic organisms and/or by development of tumours. These include B-cell lymphomas and, outside the CNS, Kaposi's sarcomas, which are probably of infectious origin.

About 50 per cent of patients with AIDS and neurological symptoms show some abnormality on imaging. About two-thirds of these have enlarged lateral ventricles and/or wide cerebral sulci as the only manifestation (Figure 4.1). However, their significance should be interpreted with caution since they could be related to factors other than HIV, for example, excess alcohol intake, dehydration or drug abuse which may cause potentially reversible brain shrinkage, and previous head trauma or co-existent parenchymal neurosyphilis, cytomegalovirus (CMV), herpes simplex, *Toxoplasma*, or cryptococcal infections in which the atrophy may be arrested with treatment. Follow-up studies showing progressive

atrophy provide a more reliable indicator of encephalopathy (Figure 4.2). In the other third of patients showing abnormalities on imaging, there are focal lesions. Uncomplicated HIV infection may cause an ill-defined diffuse or multi-focal abnormality of the deep cerebral white matter (Figure 4.3), which is more frequent and more extensive on magnetic resonance imaging (MRI) (Figure 4.4) than on computed tomography (CT). Prognosis regarding the development of progressive neurological disease is better in patients with normal imaging, but, eventually, 40–50 per cent of HIV-positive patients will develop neurological symptoms (Berger et al, 1987; Levy et al, 1985), and 80 per cent of brains at autopsy will have evidence of reactive changes to the HIV and/or opportunistic infections (Jordan et al, 1985).

The brainstem and spinal cord lesions may cause progressive spastic paraparesis with ataxia. The lateral and posterior columns of the cord are principally affected, simulating subacute combined degeneration (Petito et al, 1985). Myelography is usually normal, but delayed post-myelogram CT may show abnormally increased density of contrast medium within the cord sub-stance, presumably retained in the vacuoles. T2-weighted MRI may reveal regions of high signal in the spinal cord, but in many proven cases no abnormality has been detected by imaging. The mononeuritis multiplex and peripheral neuro-pathy caused by the HIV are not associated with any abnormalities visible on imaging.

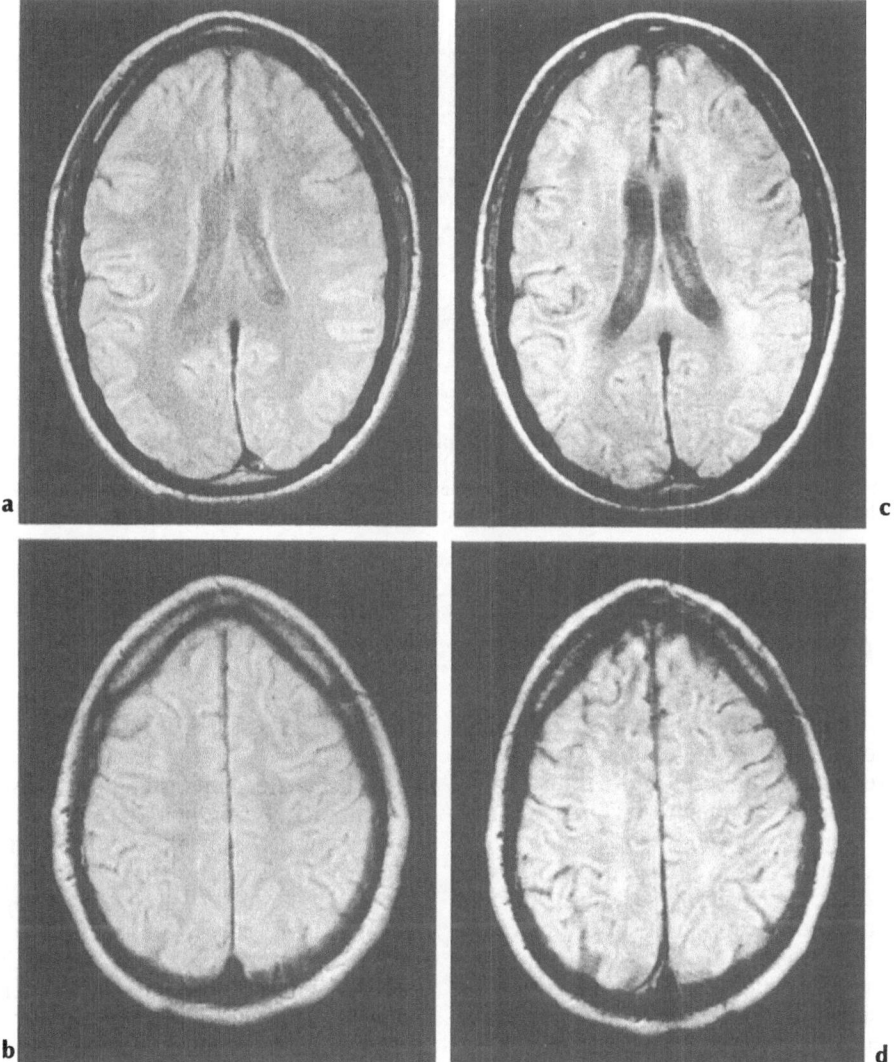

Figure 4.2a–d. 46-year-old HIV-positive homosexual male. At the time of the first MRI scan in March 1989 (**a,b**) he was asymptomatic. He remained physically well, but by the time of the second MRI scan performed in October 1990 (**c,d**) there was evidence of cognitive dysfunction. **a,b** MRI proton-density-weighted images showing no abnormality. **c,d** Similar sections, although there is now moderate diffuse widening of cerebral sulci and slight increase in the size of the lateral ventricles. There is also high signal in the deep cerebral white matter due to AIDS encephalopathy with progressive atrophy.

Cytomegalovirus Infection

Of the superadded viral infections, cytomegalovirus (CMV) is the most common, and has been recorded in 17–26 per cent of cases at autopsy (Anders et al, 1986). It causes a variable degree of involvement of both grey and white matter, sometimes with viral inclusions in cortical glial nodules. There may also be radiculomyelitis. Imaging signs are usually indistinguishable from HIV infection alone. The superinfection may rarely present clinically with acute alteration in mental state, as a focal brainstem lesion, or with visual impairment due to optic neuritis. More frequently, in as many as 30 per cent of cases, it causes an exudative retinitis with recurrent retinal detachment. Concomitant ventriculitis is frequent (Wiley et al, 1986), which on imaging shows as high signal in the periventricular white matter on T2-weighted MRI, occasionally with enhancement of the ventricular margins. In extreme cases, periaqueductal encephalitis may lead to partial obstruction sufficient to cause hydrocephalus (Vinters, 1989).

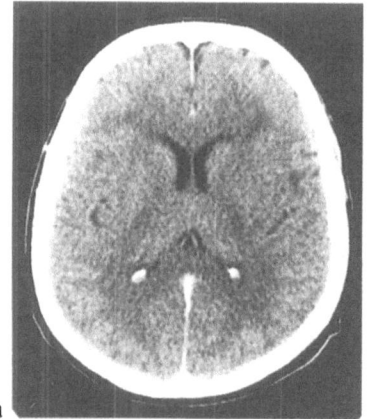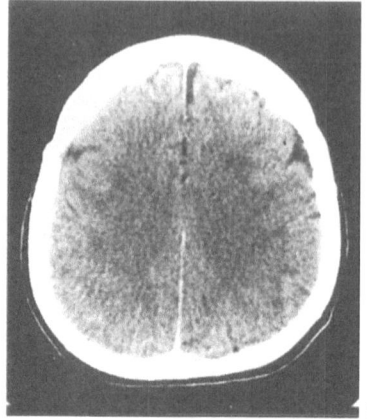

Figure 4.3a,b. 27-year-old homosexual male with AIDS. Complained of headaches and lethargy. There were no abnormalities on neurological examination. Enhanced CT sections: **a** at level of anterior horns; **b** supraventricular. There are multiple very small low-density lesions in the white matter of both cerebral hemispheres. There was no evidence of superadded opportunistic infection, and the lesions were unchanged on a further CT scan made after an interval of 5 months.

Progressive Multifocal Leukoencephalopathy

Progressive multifocal leukoencephalopathy (PML) due to infection by the papovavirus occurs much less frequently (2–7 per cent) than CMV, but produces more characteristic imaging abnormalities. The virus invades oligodendrocytes and there is secondary demyelination and necrosis of white matter, particularly in the centrum semiovale and subcortical regions of the cerebral hemispheres. Extension of the lesion into the intracortical myelin, with sparing of the cortex, is particularly suggestive of PML in the appropriate clinical setting.

On imaging, there are well-defined lesions which may be single, but are more usually multiple and asymmetrical, and are found mainly in the white matter and particularly in the parieto-occipital lobes. In AIDS patients, however, involvement of grey matter has been described (Ledoux et al, 1989) and the extent and location of the disease in white matter tends to be more diffuse with infratentorial lesions occurring in about 10 per cent of cases. These lesions are well-defined, with little oedema or mass effect, and they commonly, but by no means invariably, enhance after injection of an intravenous contrast medium (Figure 4.5). The lesions are usually evident as low densities on CT, but T2-weighted MRI, being much more sensitive, shows more extensive and often more numerous high-signal foci (Figure 4.6) (Krupp et al, 1985). The disease is inexorably progressive with increasing signs

appropriate to the focal involvement. The diagnosis is confirmed by histology.

Varicella Zoster Infection

Varicella zoster infection may cause multifocal encephalitis and/or myelitis (Ryder et al, 1986), which may be associated with direct invasion of blood vessels by the virus causing a vasculitis (Linnemann and Alvira, 1980; Morgello et al, 1988). In such cases, there is non-specific multifocal white matter high signal on T2-weighted MRI sequences. However, the incidence in AIDS is low (2 per cent or less).

Herpes zoster ophthalmicus may be the presenting feature of AIDS. The infection may spread from the trigeminal ganglion to involve the internal carotid artery, and lesions resembling cerebral infarcts may ensue (Figure 4.7).

Herpes Simplex Infection

In HIV-positive patients prior to the development of AIDS, lesions due to Herpes simplex are similar to those affecting HIV-negative patients. The temporal and inferior frontal lobes are usually involved, with hypodense swelling sometimes with foci of haemorrhagic change. The basal ganglia are spared and there is usually striking demarcation of the insulae. In manifest AIDS, herpes encephalitis tends to be more diffuse, but milder. The swelling is less marked and the symptoms are not as devastating, and

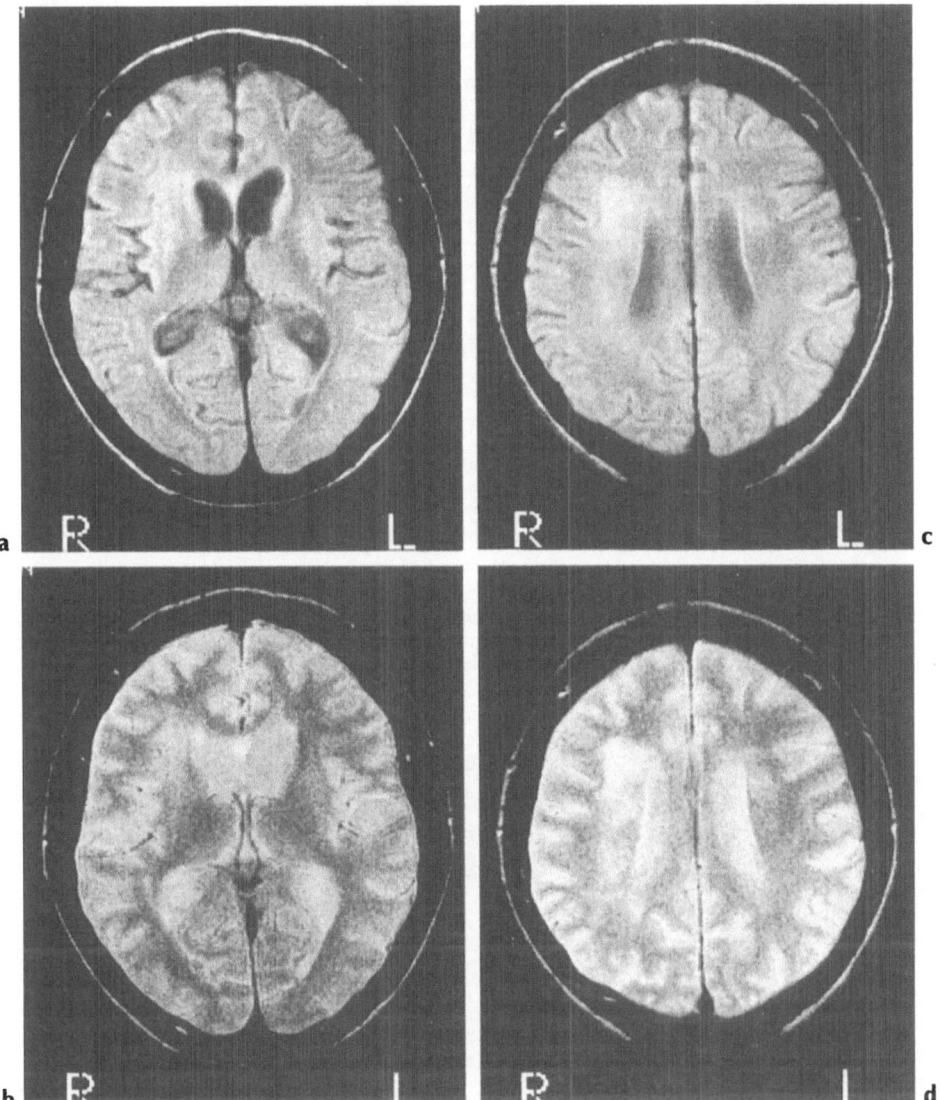

Figure 4.4a–d. 30-year-old homosexual male with AIDS and progressive dementia. CT had revealed atrophy but no other abnormality. MRI: **a,c** proton-density-weighted images; **b,d** T2-weighted images; **a,b** at the level of the anterior horns; **c,d** at the level of the roofs of the lateral ventricles. There is high signal with ill-defined margins in the white matter of both cerebral hemispheres, more extensive on the right. There is mild cerebral atrophy.

since the incidence of Herpes simplex is only slightly increased in AIDS, the diagnosis may not be suspected clinically.

Toxoplasmosis

Of the non-viral infective agents, *Toxoplasma gondii* is most frequent. *Toxoplasma* infection is commonly the primary presentation of AIDS (Snyder, 1989). The usual clinical features are of focal neurological signs related to the site of the inflammatory mass, plus confusion and/or lethargy. Incidence varies with that of latent infection in particular communities, which may be assessed from the frequency of congenital infection – approximately 4 per cent in London and 15 per cent in the USA, for example.

The organisms may be free within the brain substance, or contained within cysts. They may cause multiple abscesses or, much less frequently, a diffuse micronodular encephalitis. Histologically, vasculitis and necrosis are prominent features. The appearances on imaging reflect

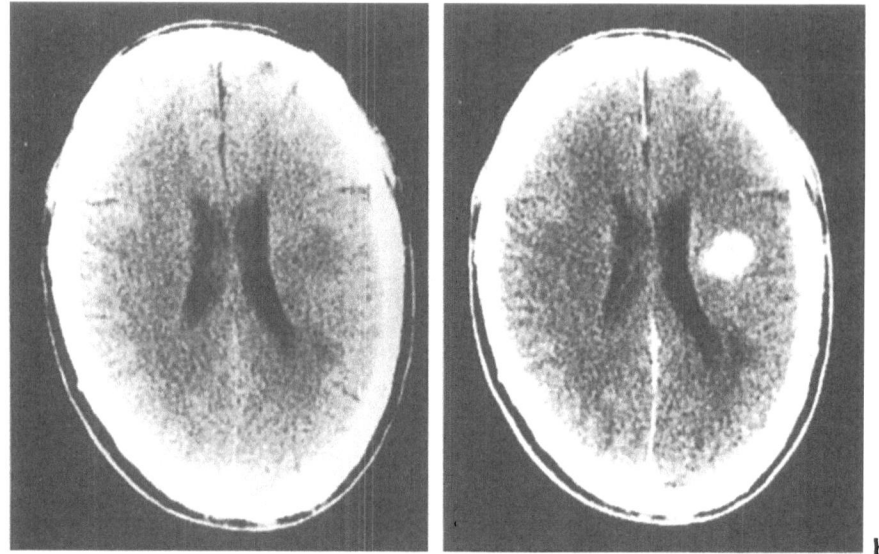

Figure 4.5a,b. 30-year-old homosexual male with progressive hemiparesis due to AIDS plus progressive multifocal leuko-encephalopathy. CT scan. **a** Plain. There are low-density lesions without mass effect in the left posterior frontal and parietal white matter. **b** Similar level after injection of intravenous contrast medium. The posterior frontal lesion enhances, but the left parietal lesion shows no enhancement. Biopsy of the frontal lesion confirmed progressive multifocal leukoencephalopathy.

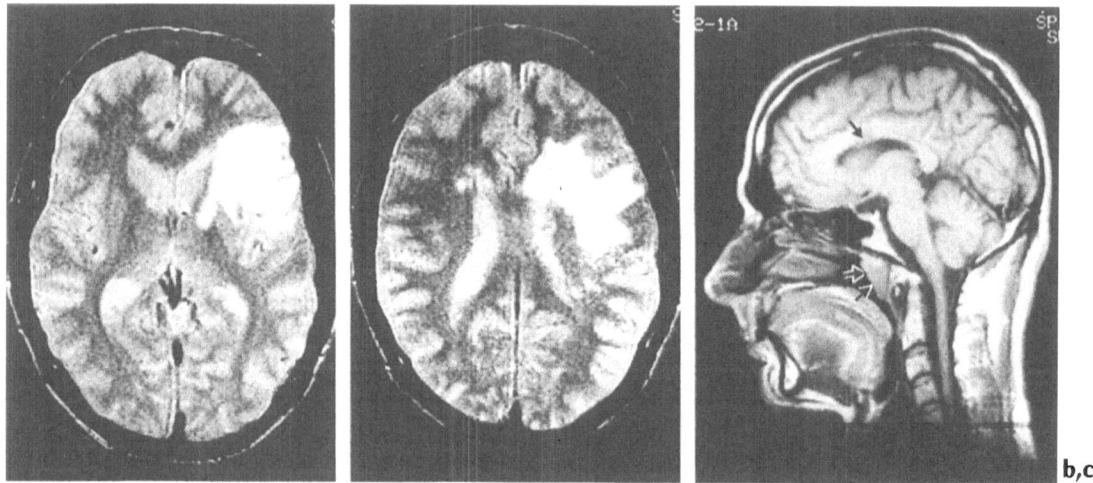

Figure 4.6a–c. AIDS plus progressive multifocal leukoencephalopathy. MRI axial T2-weighted sections: **a** through the anterior horn; **b** slightly higher. A large high-signal lesion without mass effect involves mainly white matter of the left frontal lobe, but extends into the cortex and deep grey matter. There are further small high-signal lesions in the right parietal and left frontal white matter. **c** MRI sagittal T1-weighted paramedian section. The large lesion extends into the body of the corpus callosum causing slightly diminished signal return. The adenoids are enlarged (*arrows*).

the pathology, revealing multifocal, or less commonly, single masses with surrounding oedema. They are characteristically sited in the deep grey matter (60 per cent), and less often in the cortex. They may fail to enhance, but usually show nodular (Figure 4.8), or more commonly

thin-wall ring enhancement (Figure 4.9). In the cortex they may cause gyral enhancement with underlying oedema. Such lesions are generally treated presumptively with anti-*Toxoplasma* therapy, to which clinical and imaging response generally occurs quickly, almost always within

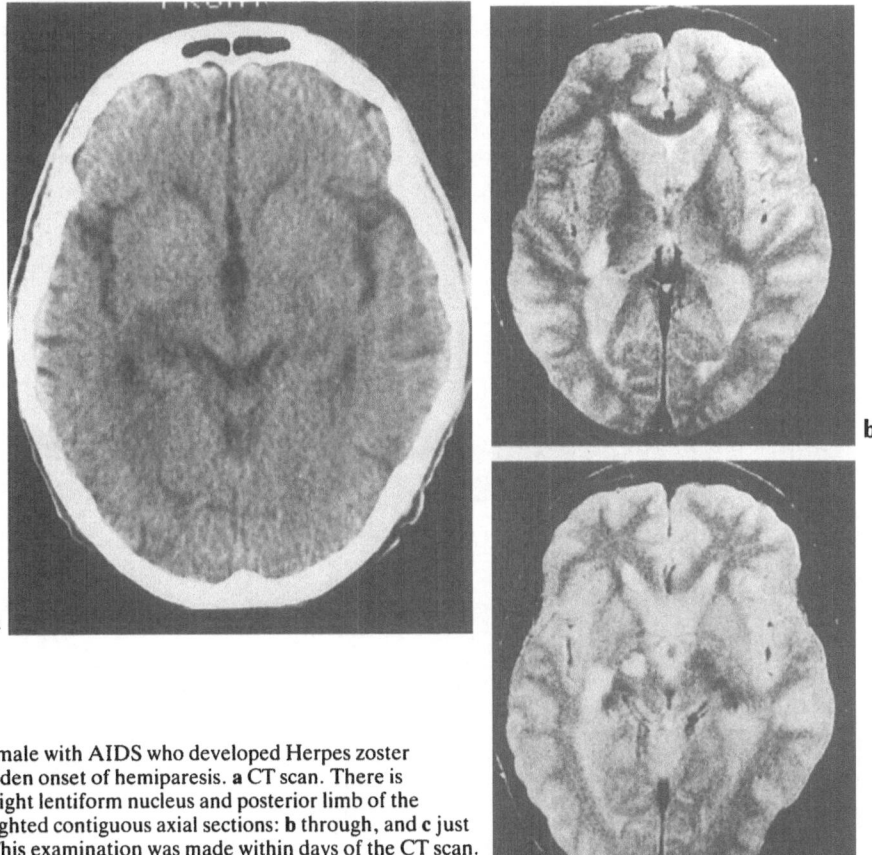

Figure 4.7a–c. Homosexual male with AIDS who developed Herpes zoster ophthalmicus followed by sudden onset of hemiparesis. **a** CT scan. There is ill-defined low density in the right lentiform nucleus and posterior limb of the internal capsule. MRI T2-weighted contiguous axial sections: **b** through, and **c** just below, the internal capsule. This examination was made within days of the CT scan. The lesion causes well-defined high signal, which is much more obvious and more extensive than is shown on CT.

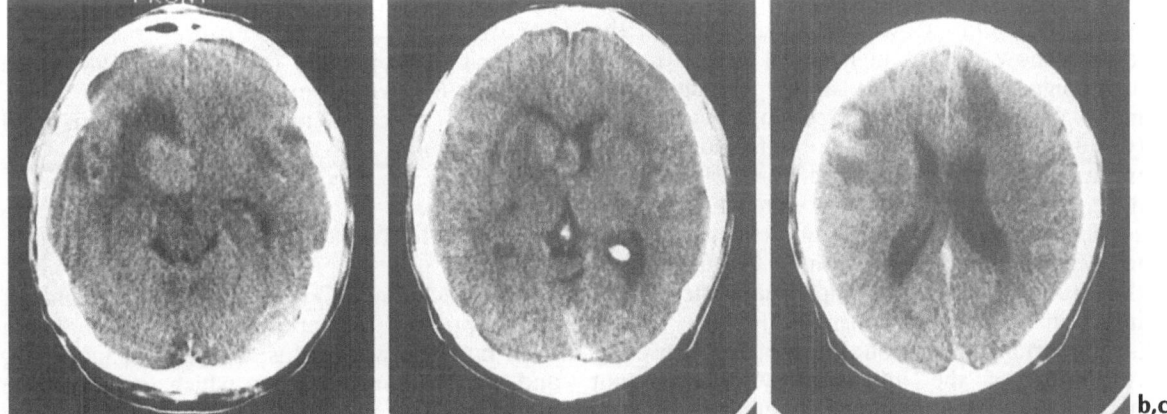

Figure 4.8a–c. AIDS plus toxoplasmosis. Enhanced CT scan, axial sections: **a** through the temporal horns; **b** through the frontal horns; **c** through the bodies of the lateral ventricles. There are homogeneously enhancing masses with adjacent oedema involving the head of the right caudate nucleus and adjacent inferior frontal lobe, and the right frontal convexity cortex. There is also oedema in the left frontal lobe adjacent to a further lesion which is not included on the section.

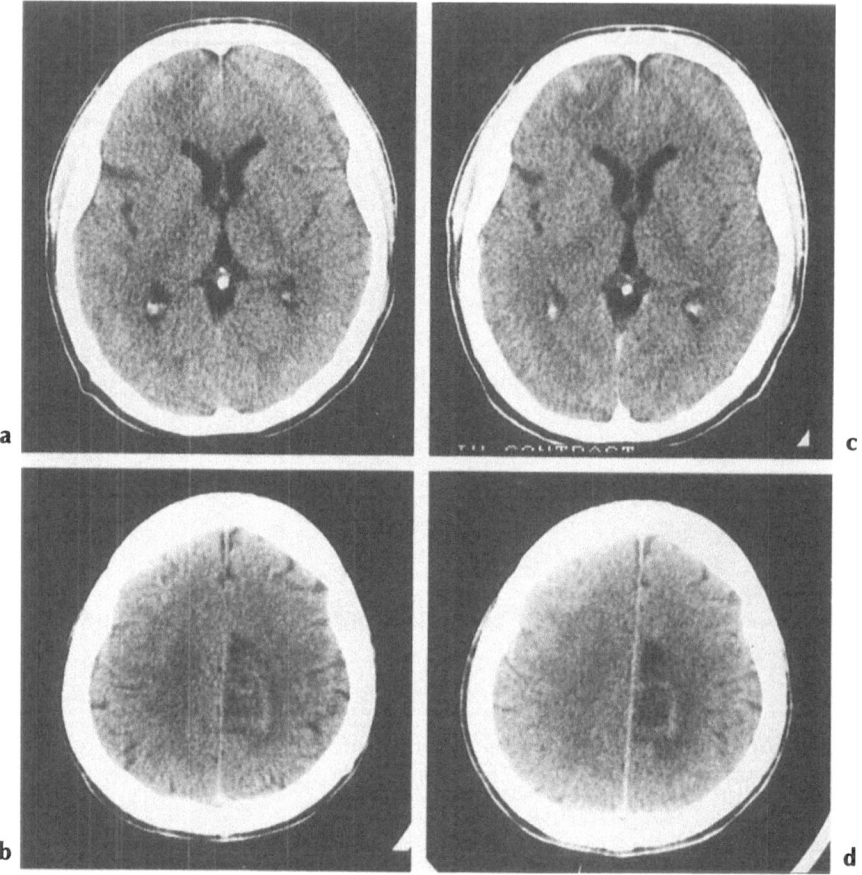

Figure 4.9a–d. AIDS plus toxoplasmosis. Plain CT scan, axial sections: **a** through the frontal horns; **b** through the supraventricular region; **c,d** similar sections after injection of intravenous contrast medium. There are mixed density lesions in the right frontal and left parietal regions, which show irregular ring enhancement with surrounding oedema.

4 weeks, though it may be slower. The *Toxoplasma* abscesses are of low signal on T1-weighted sequences. On T2-weighted sequences, the lesions show central high signal within a thin lower-signal ring from the capsule with surrounding high-signal oedema. Occasionally *ab initio* (Chaudhari et al, 1989), but more frequently after treatment with sulphonomide, the central signal may be high on both T1- and T2-weighted sequences due to haemorrhagic change. Failure to respond to treatment within a few days raises the strong possibility of lymphoma, or less likely infection with tuberculosis or fungus, and is an important indication for biopsy. However, resolution of the inflammatory changes not uncommonly takes months (Sze et al, 1987) or is incomplete, and some residual gliosis remains. Acute relapse after cessation of therapy is frequent (Figure 4.10) and, though response may occur again when therapy is reinstated, some

authorities consider that treatment should be lifelong (Cohn et al, 1989).

Toxoplasma may also cause diffuse non-enhancing encephalitis (Gray et al, 1989) (see Chapter 6), which may simulate HIV alone or CMV, or cause meningitis with meningeal enhancement.

Fungal Infections

Fungi, including *Cryptococcus* and *Candida*, rarely present as parenchymal masses. These form nodules, with diffuse or ring enhancement, and surrounding oedema which may not be marked. There may be accompanying meningitis with enhancement on CT and MRI. Complications include hydrocephalus associated with meningitis, and involvement of blood vessels resulting in infarction.

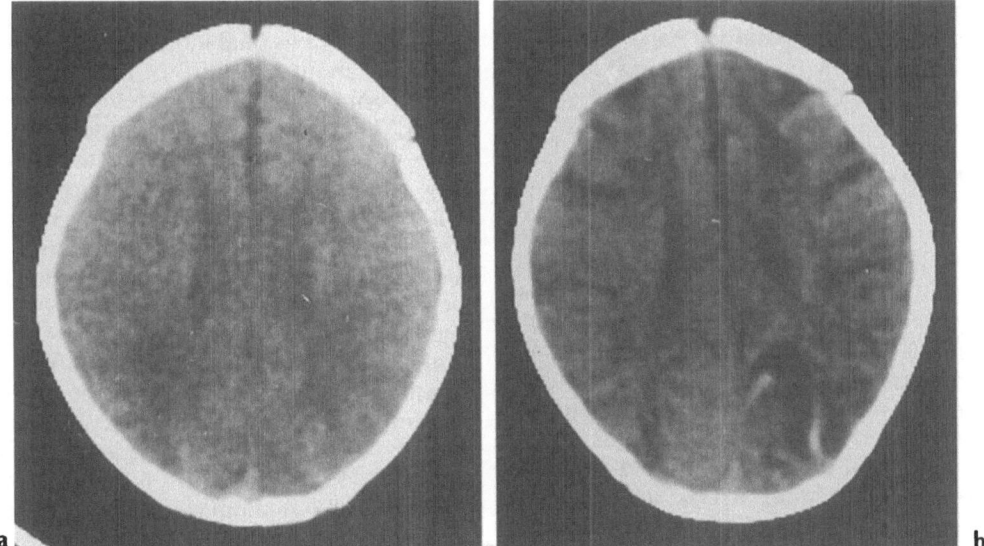

Figure 4.10a,b. AIDS plus toxoplasmosis. **a** CT scan made at the end of sulfonamide therapy. There is ill-defined low density in the parietal white matter. No other abnormality is present. **b** CT scan made 10 days later after clinical relapse. There is a large mixed-density lesion in the right parietal lobe, which responded to further anti-*Toxoplasma* therapy.

Cryptococcal Infection

Cryptococcal infection is the commonest mycosis seen in AIDS, occurring in 2–5 per cent of patients (Zuger et al, 1986), and in 25 per cent of these it is the presenting feature. It most typically causes subacute ependymitis and meningitis, with extension into the adjacent brain. However, imaging is usually negative or shows only atrophy or hydrocephalus for quite long periods after the onset of meningeal infection at a time when the patient is very ill with altered mental state,

meningitic signs and fever. Cryptococcal masses occur in 11–25 per cent of patients (Long et al, 1980; Popovich et al, 1990). These may be nodular or ring enhancing, or low-density non-enhancing lesions on CT (Figure 4.11). They are due to gelatinous masses of organisms and are usually parenchymal, but rarely they occur within, and distend parts of, the ventricular system. MRI is useful in revealing them at an earlier stage and locating them more exactly. They appear as fluid-like masses, which can be distinguished from infarcts, but MRI does not

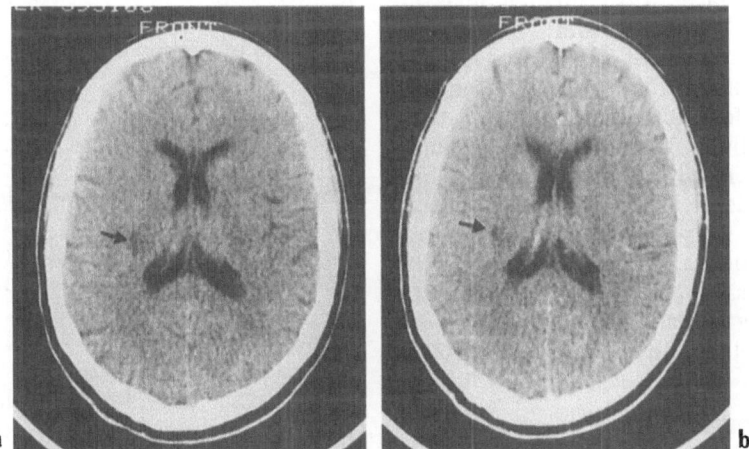

Figure 4.11a,b. AIDS plus cryptococcal infection. CT scan, axial sections: **a** plain; **b** after injection of intravenous contrast medium. There is a small low-density non-enhancing lesion in the right corona radiata, which proved to be a cryptococcal granuloma at autopsy (*arrows*).

allow a specific diagnosis from most of the other infections or from lymphoma (Popovich et al, 1990). Diagnosis is by CSF cryptococcal antigen, or by culture from CSF.

Other Infections

Infection by other organisms is much less frequent, but is important because they are often responsive to appropriate antibiotic treatment. Tuberculosis and atypical mycobacteria (Greene et al, 1982), *Aspergillus fumigatus*, *Listeria monocytogenes*, *Nocardia asteroides*, *Histoplasma capsulatum*, *Candida albicans* and meningo-vascular syphilis have all been described.

Since cell- rather than humoral-mediated responses are affected in AIDS, the incidence of pyogenic bacterial infections should not theoretically be increased. However, abscess formation due to *E. coli* has been recorded.

Tuberculosis

Though intracranial tuberculomas are an uncommon complication of AIDS, they are relatively more frequent in groups at higher risk for tuberculosis, such as Afro-Caribbeans and Africans. They appear to be more frequent in intravenous drug abusers than in patients with sexually-acquired AIDS. Indolent tuberculomas when small are usually solid homogeneously-enhancing nodules of high density on CT and high signal on T2-weighted MRI. Larger tuberculomas with central caseation show ring enhancement, and occasionally, central calcification may cause a target-like appearance. Most tuberculomas are surrounded by oedema. Acute tuberculous abscesses may be single or multiple, with thin-wall ring enhancement. They simulate, and are usually mistaken for, toxoplasmosis.

The manifestations of granulomatous meningitis with enhancement of tissue within the basal cisterns, though identical with that of cryptococcal infection, should suggest tuberculosis in appropriate ethnic groups. There may be infarcts, usually in the basal ganglia, due to involvement of vessels in the granulation tissue and/or hydrocephalus due to obstruction of the basal cisterns.

B-Cell Lymphomas

B-cell lymphomas occur in about 6 per cent of AIDS patients, the primary form being found in approximately 2–5 per cent and representing at least a six-fold increase in immune-suppressed compared with normal individuals (Burstein et al, 1963). In about one-third of these cases, neurological symptoms are the presenting feature of AIDS. The tumours are multifocal in about 50 per cent of cases and frequently involve the deep grey matter. They present with confusion and lethargy (So et al, 1986). However, they may arise anywhere, including the posterior fossa in about 13 per cent of cases (Jack et al, 1986) and the spinal cord, causing the appropriate focal signs in these instances. The tumours are of high grade, and patients usually survive for only a few months. On CT they may form the more typical iso- or hyperdense homogeneously-enhancing masses, often with little surrounding oedema (Poon et al, 1989) (Figure 4.12), but in AIDS, primary CNS lymphomas tend to undergo central necrosis and are of low density with ring enhancement (Gill et al, 1985). Most such lesions are indistinguishable from toxoplasma, until they fail to respond to appropriate treatment for the latter, or biopsy is performed. Primary lymphoma may also present as a diffuse lesion, involving the ependyma or white matter. The former causes an iso- or hyperdense enhancing periventricular rim, and the latter diffuse deep white matter low density on CT and high signal on MRI.

About 20 per cent of systemic non-Hodgkin's lymphomas in AIDS patients have metastases involving the CNS (Ziegler et al, 1984). These usually present as leptomeningeal or ependymal spread, or as extra-axial masses which are often flat and more apparent when imaged in profile.

Metastases to the CNS from Kaposi's sarcoma are very rare (Levy et al, 1984). They are prone to haemorrhage and should be considered likely if an haemorrhagic intra-axial mass is shown in the presence of a known Kaposi's sarcoma.

Strokes

Cerebral infarcts are no more frequent in autopsy series of AIDS patients than in age-matched controls, and the spectrum of cerebrovascular disease is in general similar (Berger et al, 1990). Some of them arise from terminal emboli detached from the heart valves in marasmic acute non-bacterial endocarditis, or from disseminated intravascular coagulopathy. They may also complicate dehydration or be associated with arterial involvement by granulomatous meningitis, meningovascular syphilis (Holland et al, 1986), Herpes zoster and lymphoma. However, the HIV itself can cause vasculitis, which is a much

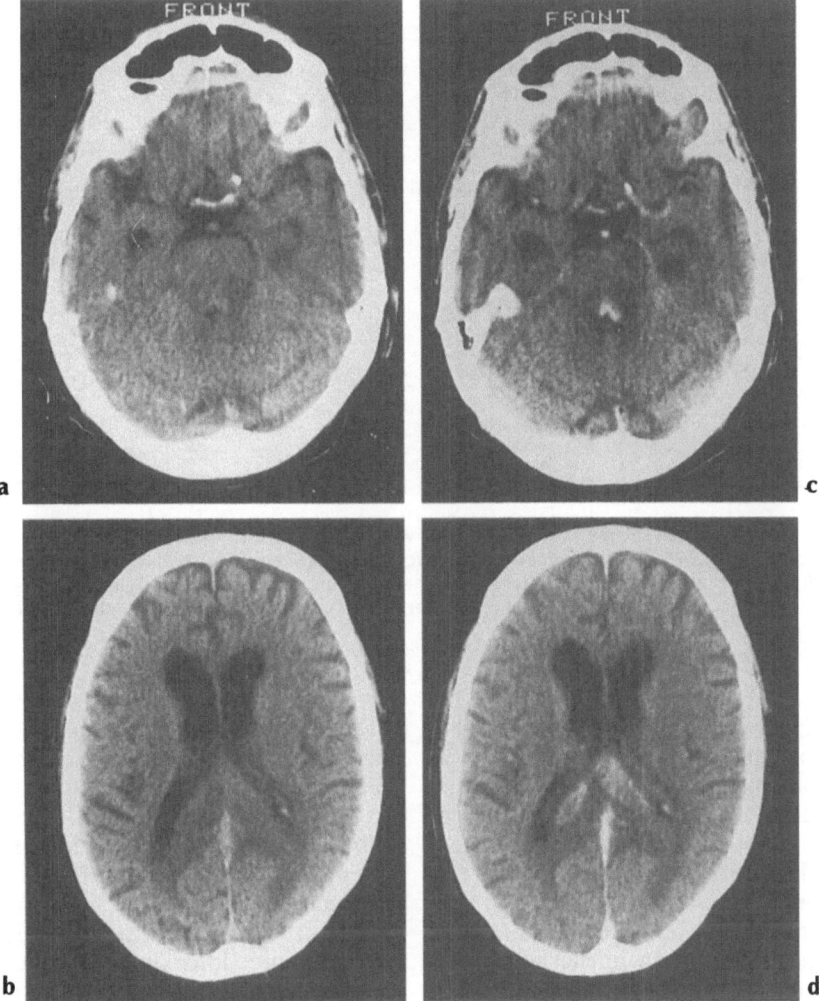

Figure 4.12a–d. AIDS plus B-cell lymphoma (biopsy proof). Plain CT scan, axial sections: **a** through the brainstem; **b** through the splenium of the corpus callosum. **c,d** Similar levels after injection of intravenous contrast medium. There are mixed-density masses in the brainstem and in the splenium of the corpus callosum. The higher-density components enhance; the lower-density components are probably due to oedema.

commoner cause of cerebral ischaemia or infarction (Figure 4.13) than in non-AIDS patients.

Thrombocytopenia is not infrequent in AIDS (Levy et al, 1985) and, rarely, spontaneous cerebral haemorrhage due to disorder of coagulation may occur during the course of, or even be the presenting feature of, AIDS. There is no clear evidence that HIV infection increases the high risk of intracranial haemorrhage in haemophiliacs due to their deficiencies in clotting factors. However, about one-third of haemophiliacs die of intracranial haemorrhage and many more of the diseases associated with blood transfusion, so that a smaller percentage of haemophiliacs manifest full-blown AIDS at autopsy than is found in

sexually-transmitted and other groups (Esiri et al, 1989).

Haemorrhage may also complicate cerebral angiitis, mycotic aneurysm or abuse of certain drugs, including cocaine and amphetamines. It may also occur within neoplastic and inflammatory lesions, or from an incidental aneurysm or angiomatous malformation (Snider et al, 1983).

Paediatric AIDS

In AIDS acquired congenitally or early in life, there is failure to thrive often associated with

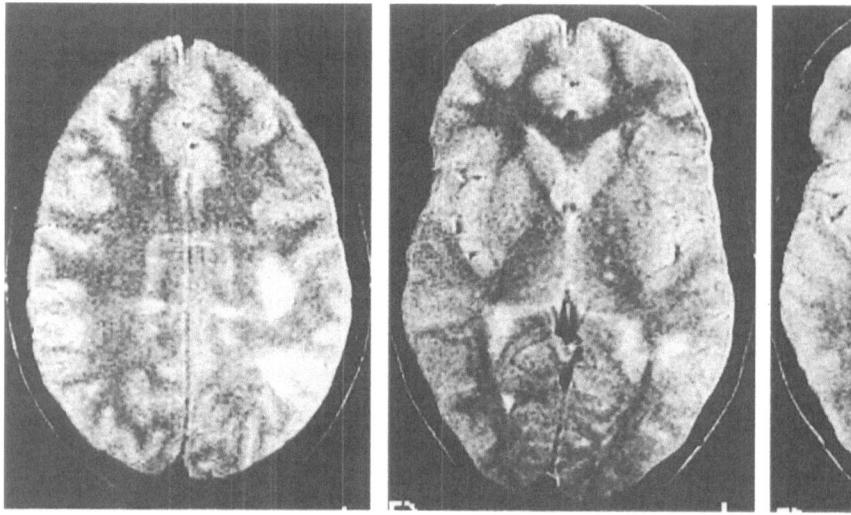

a   b,c

Figure 4.13a–c. 31-year-old female intravenous drug abuser with AIDS. History of acute neurological deficits. Cardiolipin antibody level elevated. MRI T2-weighted sequence: **a** above the lateral ventricles; **b** through the frontal horns and third ventricle; **c** through the inferior frontal and occipital lobes. There are large and small high-signal lesions in the cerebral white matter, and a larger infarct involving both white and grey matter posteriorly in the left temporal lobe extending into the occipital and parietal lobes.

hepatosplenomegaly and lymphadenopathy. CNS involvement was first described by Shaw et al (1985) who emphasized the rarity of opportunistic infections. Other complications, such as lymphomas, are reported by Epstein et al (1988). The clinical and pathological presentation of this group of patients is described in Chapter 9. Calcification, described by Sharer et al (1986), is shown by CT in the lentiform nuclei and in the white matter of the cerebral hemispheres in about 20 per cent of cases (Figure 4.14) and by T2-weighted MRI as an abnormal signal in the hemispheric white matter in about 15 per cent of cases. These changes may well reflect competitive inhibition of pterin metabolism due to production of neopterin by infected macrophages. Such changes are not diagnostic of congenital AIDS, but they should suggest the possibility of the diagnosis in appropriate clinical circumstances. In older children with acquired AIDS, secondary infections are more frequent.

Response to Therapy

Although there is evidence that zidovudine decreases the rate of replication of the virus and may delay the progression of AIDS, reversal of atrophy does not occur.

The importance of imaging and close monitoring of response to anti-*Toxoplasma* therapy has been mentioned in relation to presumptive *Toxoplasma* superinfection. Occasionally, sulphonamides cause depression of platelets sufficient to produce haemorrhagic changes within *Toxoplasma* lesions.

The presence of multiple simultaneous or consecutive infections or neoplastic lesions of the central nervous system is a not-infrequent occurrence in AIDS. The possibility should be suggested if appearances typical of different pathologies are noted, or if some lesions persist or advance while others are regressing on therapy.

Imaging Methods

In the investigation of suspected diseases of the brain and spinal cord, MRI is the least invasive and most sensitive modality (Bradley et al, 1984). The parenchymal lesions complicating HIV infection are all shown earlier and better using MRI than any other imaging method (Sze et al, 1987). The meningeal and ependymal lesions also are best revealed by enhancement with gadolinium-DTPA, which is more sensitive than iodine-enhanced CT. Intra-axial tumours which are not visible on CT are also commonly revealed by

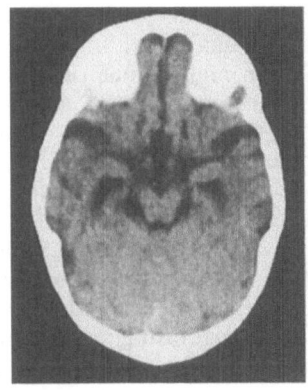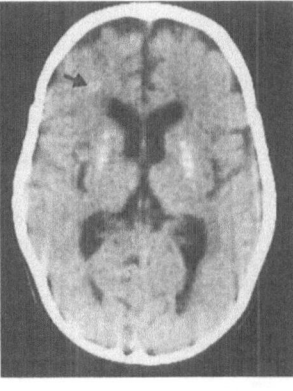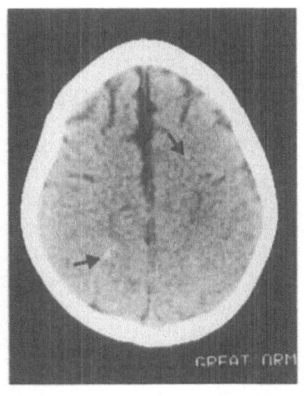

Figure 4.14a–c. 15-month-old child with congenital AIDS and retardation. Plain CT scan, axial sections: **a** through the temporal horns; **b** through the frontal horns; **c** supraventricular. The lateral ventricles are enlarged with prominence of the subarachnoid spaces. There is nodular calcification in the lentiform nuclei and less-dense foci of calcification in the hemispheric white matter (*arrows*).

MRI. A diagnosis based on the features of a single lesion shown on CT may be modified by demonstration of multiple lesions on MRI and consideration of their distribution (Drayer, 1987; Brandt-Zawadzki, 1988).

Infarcts are shown at an earlier stage (Unger et al, 1987) and haemorrhage can still be recognized on MRI at a time when the pathognomonic density on CT has resolved (Gomori et al, 1985). In the spine also, extradural masses are revealed non-invasively by MRI, with similar accuracy to myelography plus CT (Modic, 1986). Leptomeningeal and intradural tumours, and inflammatory changes are generally well shown with gadolinium enhancement (Sze et al, 1988), and intramedullary lesions can also be recognized in the absence of swelling or atrophy (Miller et al, 1987).

Currently, the severe limitation of MRI is its limited availability, so that despite being less effective and/or more invasive, CT and myelography are often used as primary studies.

Imaging Abnormalities

Intracranial Lesions

Atrophy

The size of the intracranial CSF spaces is equally well shown by CT and MRI. Variations in size within the upper normal range are difficult to assess subjectively and 'atrophy of minor degree' is as frequently diagnosed in HIV-negative homo-sexual men used as controls as in HIV-positive and AIDS patients. Reversible enlargement of CSF spaces may occur with dehydration, malnutrition, alcohol abuse, and on drug therapy, particularly with steroids. Generalized atrophy is non-specific: it may follow all types of brain injury including head trauma and cerebral ischaemia, as well as being due to the AIDS dementia complex alone and with superadded infection particularly with CMV (Fig. 4.1). For all these reasons atrophy occurring in isolation should not be equated with involvement of the brain by AIDS.

White Matter Disease

Low density without mass effect may be shown by CT, but MRI is much more sensitive in revealing white matter disease (Sze et al, 1987) as an increased signal on T2-weighted sequences. Ill-defined diffuse patchy signal increase in the hemisphere white matter, internal capsules, cerebellum and/or brainstem occurs due to AIDS alone. The presence of similar changes in age-matched HIV-negative controls (McArthur et al, 1990) indicates that they cannot be considered to be unequivocal evidence of AIDS *per se*. The changes in the controls did not progress with time and this may be useful as a means of distinction from significant lesions.

Superadded CMV infection and ischaemic damage may give similar appearances also. Well-defined lesions, particularly with an outer border extending into the intracortical cores of myelin, suggest PML. When grey matter is involved in addition, the rarer Varicella zoster encephalitis may also be considered.

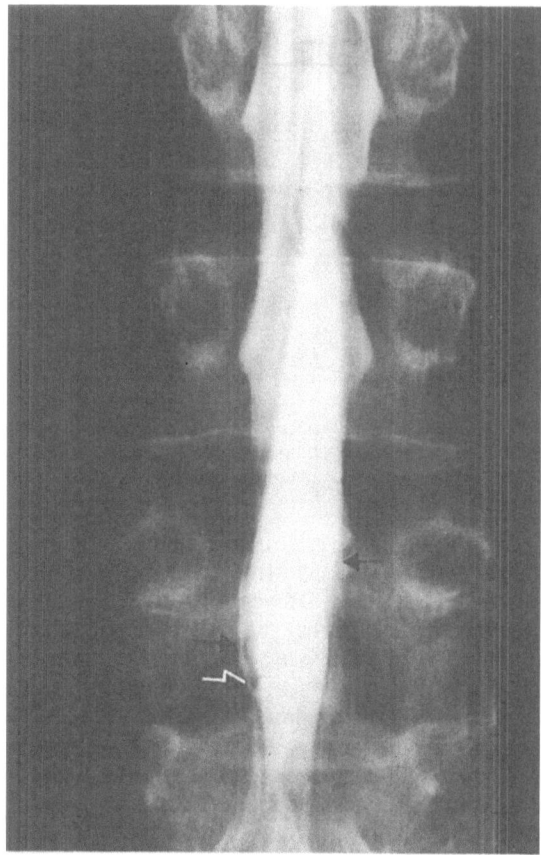

Figure 4.15. HIV-positive homosexual male presenting with meningitis and cauda equina syndrome. Iohexol radiculogram. There is irregularity of the theca and small nodules are present on the nerve roots (*arrow*). Biopsy confirmed CMV meningitis.

Mass lesions causing CT density or MR signal change, the latter being a more sensitive indicator (Scroth et al, 1987), occur with the necrotizing encephalitis of toxoplasmosis, with other secondary infections which cause granulomas or abscesses, with acute infarcts, and with lymphoma deposits or metastases. Differential diagnosis is sometimes suggested by combined clinical and imaging features against the background of statistics indicating a more than 50 per cent chance of infection with *Toxoplasma* and 25 per cent chance of lymphoma in adults, but response to specific therapy and/or biopsy is usually necessary for definitive diagnosis.

Leptomeningeal and/or Ependymal Disease

Although leptomeningeal and/or ependymal disease are commonly present on histopathology,

imaging is often negative or non-specific by showing hydrocephalus alone. Subependymal low density on CT or high signal on T2-weighted MRI, and enhancement in the meninges and/ or ependyma (more sensitively shown with gadolinium-enhanced MRI than by CT), are most frequently found in TB, toxoplasmosis, CMV infection and cryptococcal meningitis, and in lymphoma. However, they occur much more rarely with HIV, syphilis, and other fungal and atypical mycobacterial meningitides. Infarcts in the distribution of perforating arteries may occur in all these conditions.

Myelopathy

Myelopathy may be difficult to recognize clinically in the presence of encephalopathy and/or neuropathies. It is most commonly due to infection of the spinal cord, but it may also be caused by epidural masses compressing the cord.

As mentioned previously, MRI is the study of choice since as well as being non-invasive it will reveal not only extrinsic masses and their effects in the spinal theca and cord, but also any distortion and/or swelling of the cord, and/or abnormality of the signal returning from it. However, extrinsic masses may be adequately revealed by plain X-rays or CT, and their effects on the theca and cord will be consistently shown by myelography or, to better effect, by computed myelography (CM). In acute myelopathy, cord swelling may be present and later atrophy may be evident, both of which can be shown by myelography plus CT. Minor irregularity of the subarachnoid surfaces may be shown by myelography in CMV (Figure 4.15) and *Candida* meningitis. Lesions affecting only the internal structure of the cord, including the common vacuolar myelopathy, will not be shown by myelography and/or CM unless delayed computed myelography is also performed. There may be excessive retention of contrast medium in the microcystic spaces within the affected spinal cord, but this can only be distinguished from normal contrast retention when it is marked and localized, and therefore in advanced cases.

High signal from the cord in T2-weighted sequences is the most reliable indicator of myelitis, which may be due to HIV alone (Figure 4.16) or to Varicella zoster infection, CMV infection or toxoplasmosis. Similar appearances may be caused by lymphoma, and gadolinium enhancement occurs in both, though additional

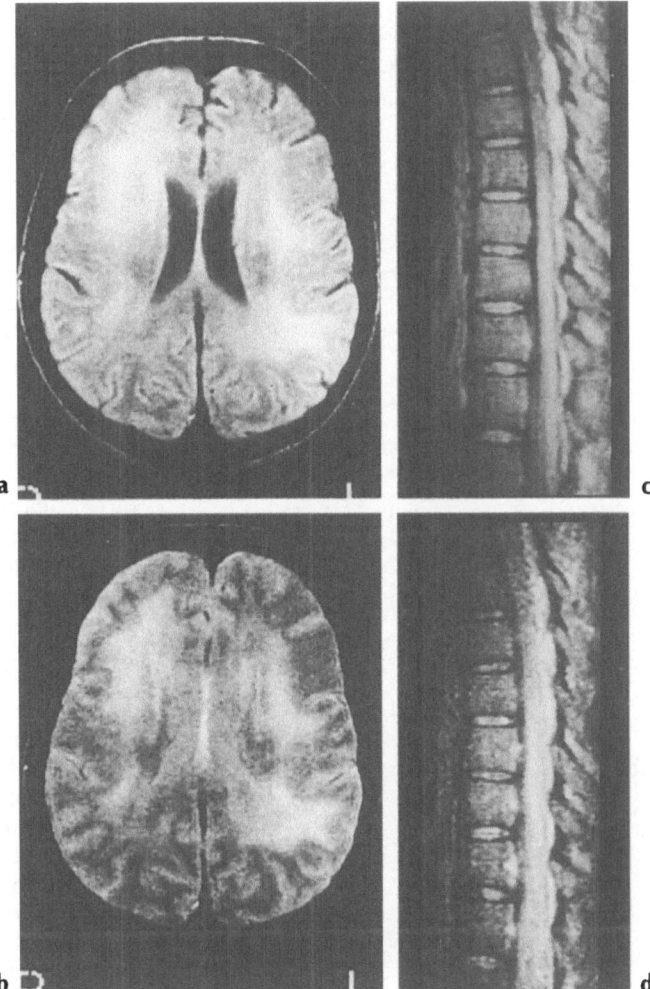

Figure 4.16a–d. 37-year-old HIV-positive homosexual male presenting with a spastic right leg. This progressed to spastic paraparesis and the patient became increasingly demented. MRI scan of the brain: **a** proton-density-weighted; **b** T2-weighted images. There is extensive abnormal high signal in the cerebral white matter. The biopsy was consistent with HIV encephalitis. MRI scan of the dorsal spine: **c** proton-density-weighted; **d** T2-weighted images. There is abnormal slightly increased signal returned from the mid-dorsal spinal cord consistent with HIV myelitis.

meningeal enhancement favours an inflammatory process.

Compressive lesions related to AIDS may be inflammatory or neoplastic. The former are usually epidural abscesses or granulomas in which the spinal structures may also be involved. The latter are usually B-cell lymphomas extending from affected vertebrae or from retroperitoneal or mediastinal lymphoid masses, in which case the extension may not involve bone, but may pass through the intervertebral foramina into the spinal canal. The degree of thecal compression is usually sufficient to obstruct partly or completely the subarachnoid space. However, presentation may occur at an earlier stage, when encroachment may show only subtle changes on imaging. Also the disease may be present over many segments. For these reasons, it is most easily assessed by MRI. Kaposi's sarcoma and squamous cell car-

cinoma of the nasopharynx, which is also slightly more common in AIDS patients (Silverman et al, 1986), may also extend into the spinal canal and compress the spinal cord. Similar considerations apply to these conditions.

Head and Neck Disease

Head and neck disease is common in ARC and AIDS. In adult patients presenting for cerebral imaging at a time when the clinical diagnosis of AIDS is unsuspected, the presence of enlarged tonsils and adenoids (Figure 4.6) noted as an incidental finding may alert the radiologist to the possibility of the diagnosis. Large masses may be the reason for referral, to determine the extent and elucidate the possibility of involvement of the spinal canal or cranial cavity.

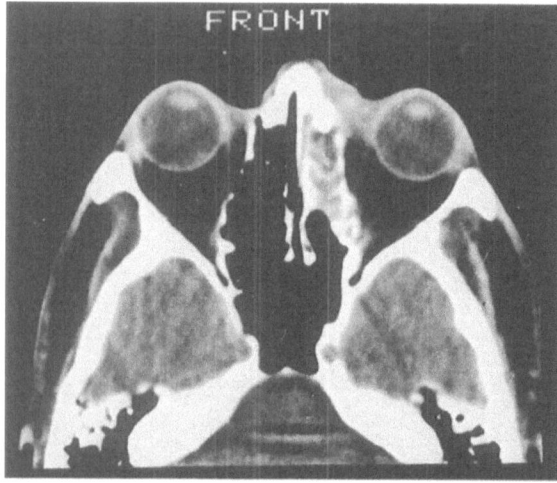

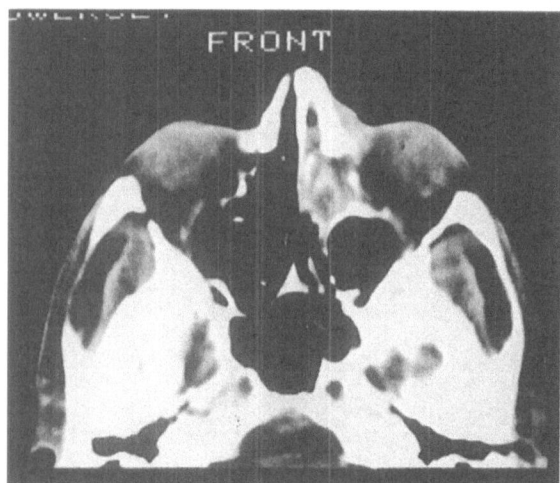

Figure 4.17a,b. HIV-positive homosexual male presenting with swelling over the left side of the nose and medial canthus. **a,b** Contiguous CT section through the ethmoid sinuses and orbits. The left ethmoid sinuses are filled with soft tissue which is contiguous with a mass lying medially in the prefascial compartment of the left orbit and over the left nasal bone. There is no evidence of bone destruction.

centres of low density on CT and of high signal on T2-weighted MRI. In mycobacterial infections, there is much less cellulitis and the fluid-filled lymph node abscesses have moderately thick, but well-defined, capsules.

Kaposi's sarcoma (Marcusen and Sooy, 1985) commonly affects the mouth or pharynx in AIDS. It tends to form nodular, non-ulcerating masses bulging into the mucosa (Emery et al, 1986), plus enlarged lymph nodes with well-defined margins and low-density centres.

B-cell lymphoma may involve any of the lymphoid tissues. It commonly causes large nodal masses, which are homogeneous in structure and tend to remain well defined without tissue plane infiltration. Tonsils, adenoids and lymphoid tissue in salivary glands may be similarly involved, causing homogeneous enlargement of the affected structures. Carcinoma (Figure 4.17) is an uncommon complicating lesion causing soft tissue masses in the nasopharynx, parotid glands, and rarely, in sinuses, where it progresses to bone destruction. It usually responds to deep X-ray therapy. Unilateral and bilateral parotid enlargement may be caused by benign cystic lymphoepithelial lesions. These are shown as low-density well-defined masses on CT, confirmed by fine-needle aspiration and removed by superficial parotidectomy (Finfer et al, 1988).

Unfortunately, even with MRI which is the most sensitive imaging modality, primary HIV encephalitis (Post et al, 1988) and superadded CMV infection, particularly when the brainstem and cerebellum are involved, frequently remain undetected until autopsy. This is even more apparent with the vacuolar myelopathy caused by HIV infection, which commonly remains undetected unless the very latest MRI technology is available.

In the lymphadenopathy of ARC, the nodes are multiple, generally under 2 cm in diameter, and are homogeneous without necrosis or surrounding inflammatory reaction causing obscuration of fat planes. The tonsils and adenoids are also usually enlarged and mucosal thickening is frequent in the paranasal sinuses.

In cellulitis, which is usually of bacterial origin, the inflammatory process extends through and obliterates fatty tissue planes between muscles, vessels and other structures, and infiltrates into subcutaneous fat. When necrosis occurs, forming bacterial abscesses, these have fluid-filled necrotic

References

Anders KH, Guerra WF, Tomiyasu U et al. (1986) The neuropathology of AIDS. UCLA experience and review. Am J Pathol 124:537–58.

Berger JR, Moskowitz L, Fischl M et al. (1987) Neurologic disease as the presenting manifestation of acquired immunodeficiency syndrome. South Med J 80:683–6.

Berger JR, Harris JO, Gregorios J et al. (1990) Cerebrovascular disease in AIDS: a case-control study. AIDS 4: 239–44.

Bradley WG, Waluch V, Yadley RA et al. (1984) Comparison of CT and MR in 400 patients with suspected disease of the brain and spinal cord. Radiology 152:695–702.

Brandt-Zawadzki M. (1988) MR imaging of the brain.

Radiology 166:1–10.

Burstein SD, Kernohan JW, Uihlein A. (1963) Neoplasms of the reticuloendothelial system of the brain. Cancer 16: 289–305.

Chaudhari AB, Singh A, Jindal S et al. (1989) Haemorrhage in cerebral toxoplasmosis—a report on a patient with the acquired immunodeficiency syndrome. S Afr Med J 76: 272–4.

Cohn JA, McMeeking A, Cohen W et al. (1989) Evaluation of the policy of empiric treatment of suspected Toxoplasma encephalitis in patients with the acquired immunodeficiency syndrome. Am J Med 86:521–7.

Drayer BP. (1987) Supratentorial tumours: MRI strategies. MRI Decisions 1:309.

Emery CD, Wall SD, Federie MP et al. (1986) Pharyngeal Kaposi's sarcoma in patients with AIDS. AJR 147:919–22.

Epstein LG, DiCarlo J, Joshi VV et al. (1988) Primary lymphoma of the central nervous system in children with acquired immunodeficiency syndrome. Pediatrics 82: 355–63.

Esiri MM, Scaravilli F, Millard PR et al. (1989) Neuro-pathology of HIV infection in haemophiliacs: comparative necropsy study. Br Med J 299:1312–15.

Finfer MD, Schinella RA, Rothstein SG et al. (1988) Cystic parotid lesions in patients at risk for the acquired immuno-deficiency syndrome. Arch Otolaryngol Head Neck Surg 114:1290–4.

Gill PS, Levine AM, Meyer PR et al. (1985) Primary central nervous system lymphoma in homosexual men. Clinical, immunologic and pathologic features. Am J Med 78:742–8.

Gomori JM, Grossman RI, Goldberg HI et al. (1985) Intra-cranial hematomas: imaging by highfield MR. Radiology 157:87–93.

Gray F, Gherard R, Wingate E et al. (1989) Diffuse 'en-cephalitic' cerebral toxoplasmosis in AIDS. Report of four cases. J Neurol 236:273–7.

Greene JB, Sidhu GS, Lewin S et al. (1982) Mycobacterium avium intracellulare, a cause of disseminated life-threatening infection in homosexuals and drug abusers. Ann Intern Med 97:539–46.

Holland BA, Perrett LV, Mills CM. (1986) Meningovascular syphilis: CT and MR findings. Radiology 158:439–42.

Jack CR Jr, Reese DF, Scheithauer BW. (1986) Radiographic findings in 32 cases of primary CNS lymphomas. AJR 146:271–6.

Jones HR Jr, Ho DD, Forgacs P et al. (1988) Acute fulminat-ing fatal leukoencephalopathy as the only manifestation of human immunodeficiency virus infection. Ann Neurol 23: 519–22.

Jordan BD, Navia BA, Petito C et al. (1985) Neurological syndromes complicating AIDS. Front Radiat Ther Oncol 19:82–7.

Krupp LB, Lipton RB, Swerdlow ML et al. (1985) Progressive multifocal leukoencephalopathy: clinical and radiographic features. Ann Neurol 87:344–9.

Ledoux S, Libman I, Robert F et al. (1989) Progressive multifocal leukoencephalopathy with gray matter involve-ment. Can J Neurol Sci 16:200–2.

Levy RM, Pons VG, Rosenblum ML. (1984) Central nervous system mass lesions in the acquired immunodeficiency syn-drome (AIDS). J Neurosurg 61:9–16.

Levy RM, Bredesen DE, Rosenblum ML. (1985) Neuro-logical manifestations of the acquired immunodeficiency syndrome (AIDS): experience at UCSF and review of the literature. J Neurosurg 62:475–95.

Levy RM, Bredesen DE. (1988) Central nervous system dysfunction in the acquired immunodeficiency syndrome. In: Rosenblum MC, Levy RM, Bredesen DE, eds. AIDS

and the nervous system. Raven Press: New York, 29–63.

Linnemann CC, Alvira MM. (1980) Pathogenesis of varicella-zoster angiitis in the CNS. Arch Neurol 37:239–40.

Long JA Jr, Herdt JR, Di Chiro G et al. (1980) Cerebral mass lesions in torulosis demonstrated by computed tomography. J Comput Assist Tomogr 4:766–9.

Marcusen DC, Sooy CD. (1985) Otolaryngologic and head and neck manifestations of acquired immunodeficiency syndrome (AIDS). Laryngoscope 95:401–5.

McArthur JC, Kumar AJ, Johnson DW et al. (1990) The Multicenter AIDS Cohort Study. J AIDS 3:252–60.

Miller DH, McDonald WI, Blumhardt LD et al. (1987) Magnetic resonance imaging in isolated noncompressive spinal cord syndromes. Ann Neurol 22:714–23.

Modic MT. (1986) Magnetic resonance of the musculoskeletal system and spine. In: Budinger TF, Margulis AR, eds. Magnetic resonance imaging and spectroscopy. A primer. Society of Magnetic Resonance in Medicine: San Francisco, 144–64.

Morgello S, Block GA, Price RW et al. (1988) Varicella-zoster virus leucoencephalitis and cerebral vasculopathy. Arch Pathol Lab Med 112:173–7.

Petito CL, Navia BA, Cho ES et al. (1985) Vacuolar myelo-pathy pathologically resembling subacute combined de-generation in patients with the acquired immunodeficiency syndrome. N Engl J Med 312:874–9.

Poon T, Matosoo I, Tchertkoff V et al. (1989) CT features of primary cerebral lymphoma in AIDS and non-AIDS patients. J Comput Assist Tomogr 13:6–9.

Popovich MJ, Arthur RH, Helmer E. (1990) CT of intra-cranial cryptococcosis. AJNR 11:139–42.

Post MJ, Tate JG, Quencer RM et al. (1988) CT, MR, and pathology in HIV encephalitis and meningitis. AJR 151: 373–80.

Ryder JW, Croen K, Kleinschmidt-DeMasters BK et al. (1986) Progressive dementia encephalitis three months after resolution of cutaneous zoster in a patient with AIDS. Ann Neurol 19:182–8.

Scroth G, Kretzschmar K, Gawehn J et al. (1987) Advantage of magnetic resonance imaging in the diagnosis of cerebral infections. Neuroradiology 29:120–6.

Sharer LR, Epstein LG, Cho ES et al. (1986) Pathologic features of AIDS encephalopathy in children: evidence for LAV/HTLV-III infection of brain. Hum Pathol 17:271–84.

Shaw GM, Harper ME, Hahn BH et al. (1985) HTLV-III infection in brains of children and adults with AIDS encephalopathy. Science 227:177–82.

Silverman S, Migliorati CA, Lozada-Nur F et al. (1986) Oral findings in people with or at high risk for AIDS: a study of 375 homosexual males. J Am Acad Dermatol 112:187–92.

Snider WD, Simpson DM, Nielsen S et al. (1983) Neurological complications of acquired immune deficiency syndrome: analysis of 50 patients. Ann Neurol 14:403–18.

Snyder HS. (1989) CNS toxoplasmosis as the initial presen-tation of the acquired immunodeficiency syndrome. Am J Emerg Med 7:588–92.

So YT, Beckstead JH, Davis RL. (1986) Primary central nervous system lymphoma in acquired immune deficiency syndrome. Ann Neurol 20:566–72.

Sze G, Brandt-Zawadzki MN, Norman D et al. (1987) The neuroradiology of AIDS. Semin Radiol 22:42–53.

Sze G, Abramson A, Krol G et al. (1988) Gadolinium-DTPA in the evaluation of intradural extramedullary spinal diseases. AJR 150:911–21.

Unger EC, Gado MH, Fulling K et al. (1987) Acute cerebral infarction in monkeys: an experimental study using MR imaging. Radiology 162:789–95.

Vinters HV. (1989) AIDS, cytomegalovirus, and the brainstem. Ann Neurol 25:311–12.

Wiley CA, Schrier RD, Denaro FJ et al. (1986) Localization of cytomegalovirus proteins and genome during fulminant central nervous system infection in an AIDS patient. J Neuropathol Exp Neurol 45:127–39.

Ziegler JL, Beckstead JA, Volberding PA et al. (1984) Non-Hodgkin's lymphoma in 90 homosexual men. Relation to generalized lymphadenopathy and the acquired immunodeficiency syndrome. N Engl J Med 311:565–70.

Zuger A, Louie E, Holzman RS et al. (1986) Cryptococcal disease in patients with the acquired immunodeficiency syndrome. Diagnostic features and outcome of treatment. Ann Intern Med 104:234–40.

5 General Pathology

J.N. Harcourt-Webster

Introduction

The affinity of the human immunodeficiency virus (HIV) for the CD4 receptor brings about severe depletion of T4 lymphocytes. As a consequence of the ensuing severe deficit in cell-mediated immunity, AIDS patients become extremely sensitive to infectious agents and may also develop certain neoplasms in fulminant forms previously uncommon or unknown. Most of the infections are by opportunist organisms which include viruses (predominantly cytomegalovirus, but also Herpes simplex and zoster), bacteria (*Mycobacterium avium–intracellulare* and sometimes *M. tuberculosis*), fungi (*Candida albicans, Cryptococcus neoformans*) and protozoa (*Toxoplasma gondii, Pneumocystis carinii*). Autopsy studies have revealed that virtually every organ or system may be affected; in addition to opportunistic infection, the nervous system may also develop a unique form of encephalitis (see Chapter 6) as a direct effect of HIV. A characteristic feature of this infection is the coexistence in the same patient, indeed in the same organ, of lesions due to more than one agent.

A knowledge of the pathologies which may at some stage coexist with the involvement of the nervous system is essential for the neuropathologist. Extraneural illnesses may be responsible for the death of a patient before any neurological involvement becomes apparent or during a neurological illness. On occasion, however, the latter may be produced by an extraneural ailment through, for example, spread of infection or anoxia. This chapter is an attempt to present the neuropathologist with the pattern of HIV-related non-CNS changes, some of which may be associated with involvement of that system.

Lymphoreticular System

HIV-1 has a particular affinity for cells expressing the CD4 receptor. As a consequence, cells carrying this receptor – helper/inducer T-lymphocytes, monocytes, some macrophages, and plasmacytoid T-cells – (Funke et al, 1987; Spickett and Dalgleish, 1988), decrease in number following the infection. Plasmacytoid T-cells, surrounded by T-lymphocytes, are found in the paracortex of lymphoid tissue. Monocyte-macrophage cells are a widespread family which include Langerhans cells of the skin and represent the route of dissemination through the host (Gartner et al, 1986; Popovic and Gartner, 1987). Many immunological functions depend on the central inducing role of T4 lymphocytes. With the disruption of the system brought about by HIV-1, the cell-mediated immunity is severely impaired. In addition to the effects on defence mechanisms against organisms, immune surveillance is also compromised with the development of frequently fulminating malignancies (Kaposi's sarcoma and malignant lymphomas: Levy and Ziegler, 1983; Ziegler et al, 1984b; Marchevsky et al, 1985; Kaplan et al, 1987b).

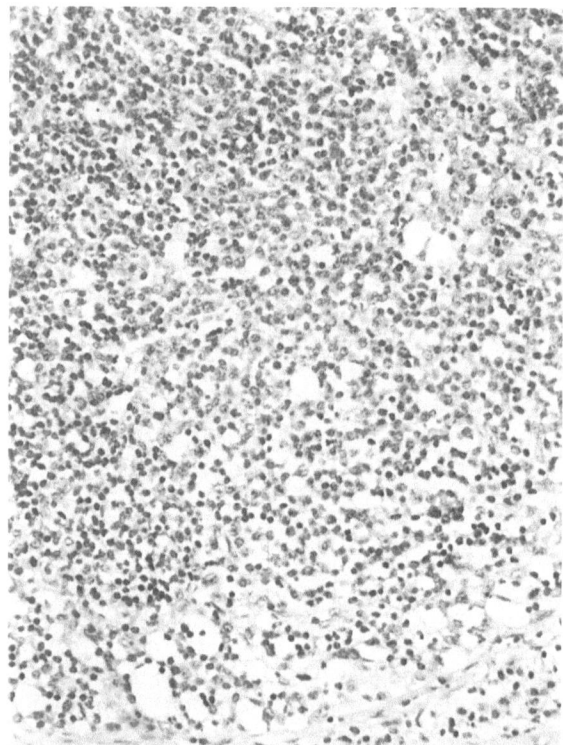

Figure 5.1. Lymph node with paracortical lymphocyte hyperplasia, but loss of follicles; early fibrosis of capsule at bottom.

ing HIV infection PGL is an early and relatively mild manifestation, but it is part of a spectrum with characteristic, although non-specific, histological patterns (Ioachim et al, 1983; Metroka et al, 1983). Three such pictures are recognized and for each there are close correlations with the clinical and immunological status of patients so infected (Ewing et al, 1985).

With type I the nodes may be greatly enlarged; they retain their capsules and resemble an acute lymphadenitis of virus origin. The hyperplasia affects all areas but is florid in the follicles, which may coalesce. Germinal centres are numerous, large and very active throughout all areas of the node with many mitoses, tingible body macrophages phagocytosing nuclear debris and extensive cytolysis. Disruption of the dendritic cell network is considerable and the small lymphocytes of the mantle may be very attenuated, sometimes absent; these latter cells may impinge upon or enter the germinal centres, a picture known as 'follicular lysis'. A conspicuous finding is the apparent increase in vascularity, with branching, thickened walls and swollen endothelium. The paracortex especially is affected and may include many small haemorrhages with aggregates of transformed monocytoid B-lymphocytes adjacent to the blood vessels and sinuses (Ioachim, 1990a). Amidst these cells are frequently increased numbers of neutrophil granulocytes, sometimes eosinophils, and lesser numbers of plasma cells, mast cells, small lymphocytes and immunoblasts. In both germinal centres and the paracortex there may be scattered, multinucleated giant cells similar to those seen in virus diseases in general and especially the Warthin–Finkeldy type, a feature of measles (Kjeldsberg and Kim, 1981); they may be HIV-infected helper/inducer T-cells (O'Hara, 1989).

These lymph node changes can be seen in various degrees within reactive nodes in the absence of HIV infection; therefore they are by no means pathognomonic of PGL. However, they occur most frequently in the latter, whilst follicle cell lysis, dermatopathic changes and granulomas are seen with about the same frequency in the two conditions.

Type II, or the transition stage, is characterized by the effacement or near effacement of germinal centres with partial follicle regression and progressive paracortical hyperplasia (Figure 5.1). This occurs in about 50–65 per cent of patients with type I over a period of 2 years or more, although the rate can be very variable (Ewing et al, 1985; Turner et al, 1987). With this there is vascular proliferation and sinus cell hyperplasia.

After the first exposure to the virus some patients may develop an acute illness similar to infectious mononucleosis which may last for up to about 2 weeks (see Chapter 2); others show no symptoms whatsoever and the vast majority are unaware of any constitutional change (Kaplan et al, 1988). Whereas some patients from both groups remain further unaffected, others develop an early manifestation known as the syndrome of persistent generalized lymphadenopathy (PGL), this being persistent lymphadenopathy involving two or more extra-inguinal sites for 3 months or more in the absence of any other diagnosable causes of lymphadenopathy (Centers for Disease Control, 1982; Meyer et al, 1984; see Chapter 2).

Lymph Nodes

The lymph nodes are clearly in a unique situation and are ideal for removal in patients suspected as being at risk from AIDS (Ewing et al, 1985). With AIDS, examination of a lymph node may both determine the origin of the lymphadenopathy and indicate the cause of a systemic disease. Follow-

Capsular thickening due to fibrosis as well as some of the latter in the medulla are frequent and venules are both conspicuous and frequent with adventitial thickening and intramural lymphocytes. Within the paracortex the cellular composition varies within a node and from node to node. Sometimes in this stage, however, particular features are conspicuous and are such a constant finding that these variations may be described separately:

Follicular hyperplasia with hypervascularity, in which proliferating vessels are mainly paracortical with some being hyalinized and extending into follicles. It is this variant that may be associated with Kaposi's sarcoma in the same or an adjacent node, although occasionally the sarcoma is seen in the type I stage (Harris, 1984; Brynes and Gill, 1990).

Angioimmunoblastic-lymphadenopathy-like, in which there are effacement of the follicular architecture, multiple branching blood vessels with delicate hyaline fibrosis of their walls, sometimes erythrophagocytosis, and proliferating immunoblasts and plasma cells (Blumenfeld and Beckstead, 1983; Brynes and Gill, 1990). This pattern, however, is separated from true angioimmunoblastic lymphadenopathy by the comparatively reduced number of proliferating immunoblasts and plasma cells and the absence of the characteristic PAS-positive interstitial material (Marche et al, 1984).

A *Castleman's-disease-like* pattern is reported also (Dickson et al, 1985; Lowenthal et al, 1987), where the follicles are atrophic and hyalinized with conspicuous, thick hyaline-walled blood vessels enclosing and running into the atrophic germinal centres; additionally the mantle zone cells are concentrically arranged.

The type III or involution pattern is fairly uniform by comparison with the earlier stages; uncomplicated lymph nodes are severely atrophic and difficult to locate (Figure 5.2). The node capsule may be thickened and fibrotic, whilst blood vessels are plentiful, often congested and lined by flat endothelium, but intramural lymphocytes are rare. Lymphocyte depletion is massive throughout and follicles are absent, or if present scanty and small. Immunoblasts and plasma cells are plentiful and set in a background of fibrosis. An almost ever-present feature is sinus cell proliferation (Reichert et al, 1983), sometimes severe, amongst which may be occasional multinucleate cells (Ewing et al, 1985); erythrophagocytosis is more frequent than the latter (Reichert et al, 1983).

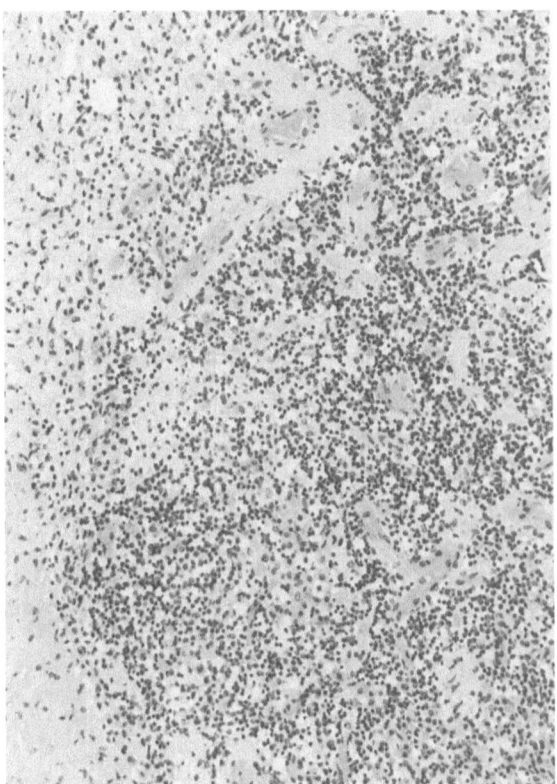

Figure 5.2. Atrophic lymph node with severe depletion of lymphocytes and absent follicles but an increase of histiocytes; fibrosis affects both the capsule and parenchyma, the latter including focal sclerosis.

Following initial infection of lymph nodes by HIV the various architectural changes described above may ensue and in some patients the three patterns of change follow sequentially (Turner et al, 1987). Progression is unidirectional and never reverses, but the type II and III patterns are not necessarily to be found in each infected patient, however long histological study is maintained. The initial lymphadenopathy is due to cortical and paracortical hyperplasia; if the overall lymph node reaction balances viral replication and consequent destruction of helper T-lymphocytes the condition of the patient may stabilize. When the lymph nodes are unable to develop a strong hyperplastic response, or cannot maintain a balance between hyperplasia and virus replication, the overall destruction of helper T-lymphocytes may result in immunological collapse and uncontrolled replication of the virus. Accompanying these changes is depletion of B-lymphocytes with shrinkage of the nodes, and it is the loss of the germinal centre with the evolution

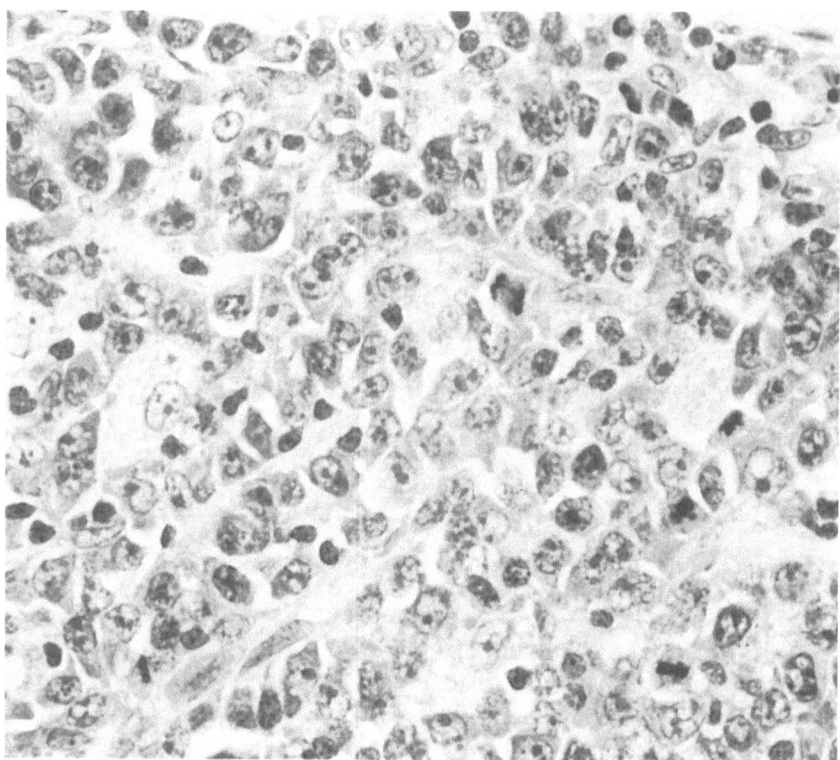

Figure 5.3. Poorly differentiated, diffuse non-Hodgkin's malignant lymphoma with intermediate-sized cells predominating, many mitoses and dispersed tingible body macrophages.

to the type II pattern which is a point of no return. In parallel the CD4 subset of T-cells declines to very low levels with inversion of the helper-inducer T/suppressor-cytotoxic T ratio. For patients with a helper T-cell count of less than $200/mm^3$ the cumulative incidence rate for AIDS reaches 84 per cent (Kaplan et al, 1988).

In the latter stage of pure HIV infection and in AIDS the lymph nodes are commonly afflicted by a variety of changes superimposed on the involution pattern (Kaplan et al, 1987b; Ioachim, 1990a). A range of infectious agents may be found and sudden node enlargement is clinical suspicion for a malignant lymphoma or Kaposi's sarcoma: with both, behaviour is often atypical and aggressive by comparison with these same neoplasms in non-AIDS patients (Ioachim, 1990b).

Neoplasms

Neoplasms arising in AIDS and those complicating non-AIDS immunosuppressive disorders are similar in type, frequency and behaviour; furthermore, they may occur in unusual sites with, for example, malignant lymphomas occurring in atypical extranodal locations. Kaposi's sarcoma may present almost anywhere, sometimes in a most subtle way making the differential diagnosis from other vascular, or semivascular, lesions difficult. The occurrence of both these neoplasms in increased frequency, by comparison with all non-suppressed patients, is explained by the common denominator of immunodeficiency; the depleted surveillance and defence mechanisms are subject to what is now a near-overwhelming influence of numerous infectious agents, some of which, singly or in combination, lead to neoplastic transformation and development.

Malignant Lymphoma

Malignant lymphoma occurs in patients with AIDS as it does in the general population, but with a significantly increased incidence (Kaplan et al, 1987b; Harnly et al, 1988; Kristal et al, 1988;

Knowles and Chadburn, 1990). Non-Hodgkin's lymphoma (NHL) is the type which predominates, and constitutes the second most frequent neoplasm associated with AIDS; in those age groups where the frequency of HIV infection is highest the incidence of NHL is increased up to fivefold (Harnly et al, 1988). Whilst Hodgkin's disease is seen (Alfonso et al, 1988), the incidence is much less than that of NHL (Kaplan et al, 1987b; Brynes and Gill, 1990), in sharp contrast with the relative frequencies amongst an age-matched non-HIV-infected population (Knowles et al, 1988). Of the AIDS patients with NHL about 80 per cent are homosexual or bisexual men, most of the remainder being intravenous drug abusers, but there is an increasing number of women so affected.

The majority of patients with AIDS-associated NHL have widely disseminated disease, including much extranodal involvement. Sites include the gastrointestinal tract, liver, bone marrow and central nervous system, but on occasion most organs may be implicated (Ioachim and Cooper, 1986; Knowles et al, 1988), localized lymph node disease being limited to around one third of all cases.

Both Hodgkin's and NHL associated with AIDS are frequently of high-grade type. With NHL most of the cases are within the combined poorly differentiated lymphocytic lymphomas of Burkitt type and the usually more common immunoblastic sarcomas, both aggressive in nature. The former have an almost monotonous picture of intermediate-sized, polygonal cells with round non-cleaved nuclei, small clearly defined nucleoli, a fine chromatin meshwork, frequent mitoses and a narrow rim of often pyroninophilic cytoplasm (Figure 5.3). Tingible body macrophages are scattered through this infiltrate creating a starry-sky appearance; many show active cell lysis and phagocytosis. Extensive zones of necrosis are a common finding in all high-grade malignant lymphomas which, almost without exception, are of B-cell type.

The immunoblastic sarcomas are made up of predominantly large lymphoid cells with large, sometimes pleomorphic nuclei, one or two readily seen eosinophilic nucleoli, dispersed chromatin and numerous, often bizarre mitoses (Figure 5.4). Cytoplasm is abundant with pyroninophilia and sometimes basophilia, as well as occasionally plasmacytoid differentiation. Zones of necrosis are again frequent.

Occasional examples of high-grade NHL have only a patchy, sometimes unconvincing, starry-sky pattern with most of the lymphoid cells of

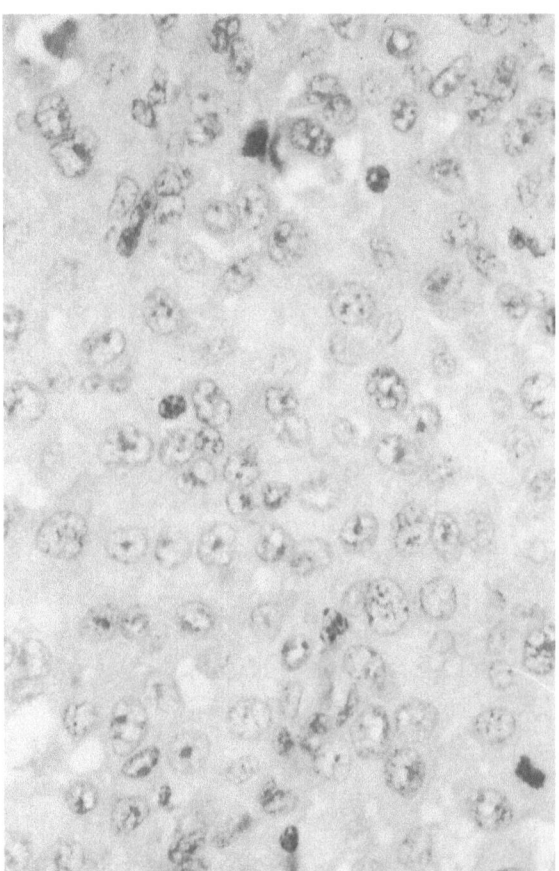

Figure 5.4. Immunoblastic non-Hodgkin's malignant lymphoma of B-cell type with vesicular nuclei, conspicuous nucleoli and many mitoses.

no more than intermediate size (Figure 5.5), although again cell necrosis is frequent and mitoses readily found. These unusual features suggest classification as an undifferentiated or poorly differentiated lymphocytic lymphoma, but a tendency to plasmacytoid differentiation and the periodic presence of typical immunoblast-like cells is a contraindicator. Categorization is well nigh impossible and, as with other processes seen in AIDS, these lymphomas are probably an occasional aberration of the more usual immunoblastic sarcomas or poorly differentiated lymphocytic lymphomas, and have similarities with some lymphomas reported in immunodeficiency after transplantation (Frizzera et al, 1981).

In all, the high-grade lymphomas make up about 70 per cent of those seen with AIDS, the remainder being of lesser grade (Knowles and Chadburn, 1990); the latter are nearly always diffuse lymphomas of follicle centre cell origin, the cells being large and cleaved or non-cleaved,

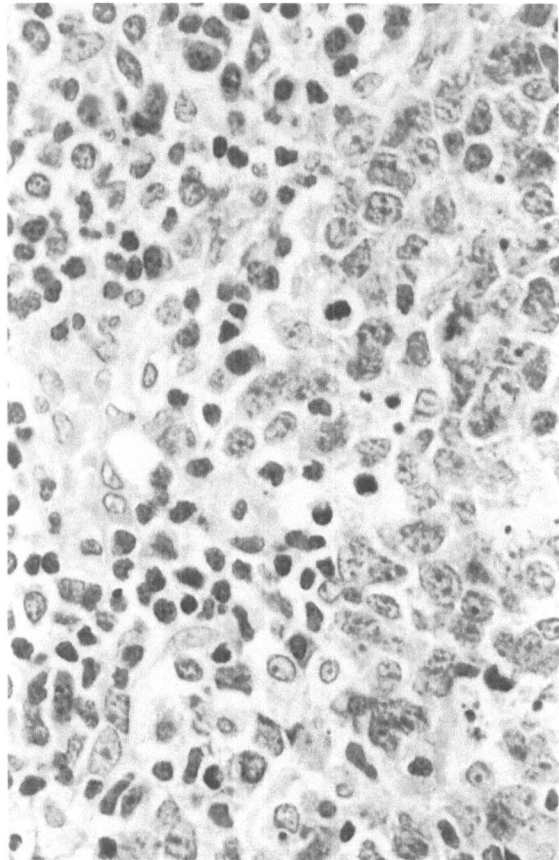

Figure 5.5. Mixed, in part poorly differentiated, intermediate and small-sized non-Hodgkin's malignant lymphoma.

or small and cleaved. Occasionally nodular lymphoma and other similarly low-grade neoplasms are seen but, as with the occasional T-cell lymphoma (Presant et al, 1987; Longacre et al, 1989), their relationship to HIV infection is doubtful.

When malignant lymphomas are diagnosed in AIDS almost 70 per cent of patients have stage III–IV clinical disease (Ziegler et al, 1984a; Ioachim et al, 1985; Di Carlo et al, 1986). Extranodal spread ultimately affects up to 90 per cent of cases, and particularly involves the bone marrow, gastrointestinal tract, liver and central nervous system. Amongst sites not usually associated with lymphoma, the myocardium, anorectum and oropharynx are well recognized for their involvement in HIV infection (Ioachim et al, 1985, 1987; Silverman et al, 1986; Gill et al, 1987).

Hodgkin's Disease. Amongst the general population, and allowing for sex and age variation,

Hodgkin's disease is much more frequent than NHL; with HIV infection and AIDS this is sharply reversed (Temple and Andes, 1986; Levine 1987; Knowles et al, 1988; Hamilton-Dutoit et al, 1991). The natural course of Hodgkin's disease is significantly altered in AIDS and, as with all malignant lymphomas associated with HIV, such a diagnosis represents an ominous progression especially amongst homosexual men; hitherto only a few cases have been reported in intravenous drug abusers and the other at-risk categories, but recently a high incidence in drug abusers has been reported (Anonymous 1988; Andrieu et al, 1988; Tirelli et al, 1989; Serrano et al, 1990). In the experience of the author lymphoma in affected women follows closely the trends seen in men.

At initial presentation up to 90 per cent of patients are in stage III–IV and there is frequent involvement of multiple groups of lymph nodes and/or extranodal sites such as bone marrow, liver, skin and soft tissue (Ioachim et al, 1985; Knowles et al, 1988; Tirelli et al, 1989; Serrano et al, 1990). Splenic involvement is less common than usual with this lymphoma and mediastinal and hilar lymph nodes are often unaffected (Knowles et al, 1988).

Structurally the neoplasms are of mixed cellularity or nodular sclerosing types with the usual cell composition including Reed–Sternberg cells; the Epstein–Barr virus (EBV) genome is to be found in the latter and the reactive lymphocytes (Guarner et al, 1990). Lymphocytes not surprisingly are depleted, sometimes unusually severely, and when analysed CD4 T-cells (helper/inducer) are few whilst the CD8 (suppressor/cytotoxic) subset is clearly dominant. However, B-cells remain polyclonal as in all patients with Hodgkin's disease. Rarely Hodgkin's disease is reported after treatment of NHL (Senaldi et al, 1990), and similarly both lymphomas can occur synchronously (Gallagher and Meschter, 1990).

Kaposi's Sarcoma

Kaposi's sarcoma involving lymph nodes is commonly found as a cause of lymphadenopathy both in surgical pathology and, especially, at necropsy, where the incidence can reach 70 per cent (Guarda et al, 1984). However, the frequency is thought to be falling, although both sexes and all ages are clearly affected. Within a lymph node involvement usually starts in the subcapsular sinus, followed by infiltration and later penetration of the capsule. Spreading along paracortical and medullary sinuses the lymphoid tissues are replaced by plump spindle cells proliferating in solid clusters or, when more fibroblast-like, forming slit-like spaces often filled with blood. Sometimes there are additional vascular channels lined by plump endothelial cells which may themselves form clusters. Most cells of all types have large, somewhat ovoid nuclei and conspicuous nucleoli, and there are varied numbers of mitoses, occasionally bizarre. However, cytological changes in early lesions may be minimal, whilst well-established florid lesions may resemble angiosarcoma.

Extravasated erythrocytes and variable amounts of haemosiderin are frequent, especially in those established lesions which have taken over much, if not the whole, of the lymph node. Lymphocytes are always few whilst plasma cells may be plentiful. Occasionally Kaposi's sarcoma may show focal fibrosis and, at low magnification, simulate nodular sclerosing Hodgkin's disease (Schofield et al, 1989). Similarly highly vascularized angioimmunoblastic lymphadenopathy can cause confusion initially (see the section on skin below).

Other Neoplasms

A variety of other neoplasms, haemopoietic and non-haemopoietic, are described in association with HIV infection. Malignant lymphoma of T-cell type and both acute and chronic leukaemias including significant numbers of T-cells are described, as are occasional patients with plasmacytoma and multiple myelomatosis (Knowles and Chadburn, 1990). Most other malignancies are seen, sometimes with a suspected increased incidence, as with both anorectal and oral carcinomas and germ cell tumours; there is no evidence, however, to link their cause directly with HIV infection, nor is there conclusive proof that immunodeficiency induced by the latter plays any significant role.

Opportunistic Infections of Lymph Nodes

Lymph nodes are commonly sites for opportunistic infections in AIDS, usually with lymphadenopathy (Brynes and Gill, 1990; Ioachim, 1990a; Strand, 1990). Infections are often multiple and sometimes are found in association with neoplasms (cf. skin). Amongst the organisms are the Mycobacteria; with *M. tuberculosis* (MTb) any tissue may be involved, with the usual features including necrosis, granulomas, Langhans' giant

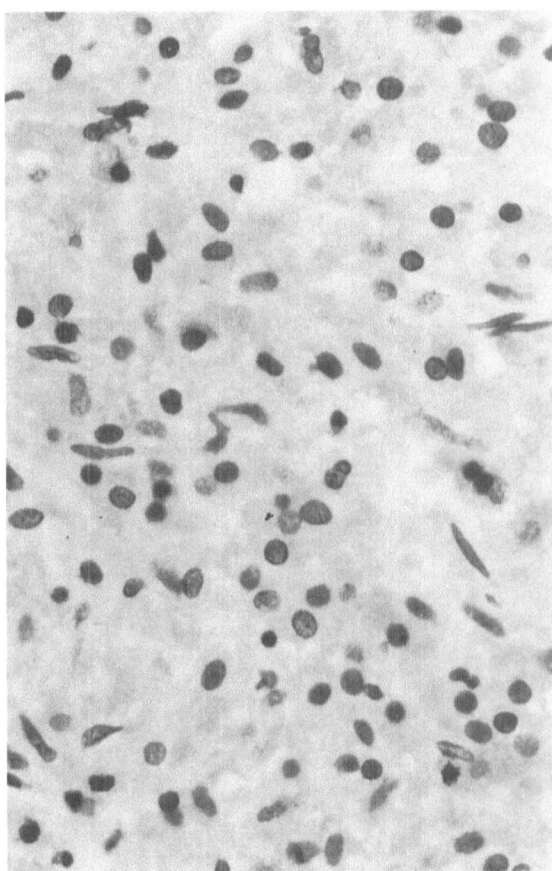

Figure 5.6. Lymph node with a diffuse and heavy infiltrate of macrophages almost completely replacing the normal structure. Acid–alcohol fast staining would show phagocytic cells to teem with Mycobacteria, identifiable as *M. avium–intracellulare*.

cells, fibrosis and calcification. With increasing immunodeficiency, however, the tissue reactions may be substantially modified; atypical lesions are seen, dissemination occurs, and with AIDS there is often anergy. By contrast the atypical forms such as *M. avium–intracellulare* (MAI) include a group of closely related organisms which are frequently seen in AIDS patients; they are a common cause of bacteraemia (Strand, 1990), and both in life and at necropsy are widely disseminated (Miller-Catchpole et al, 1989), but significant pulmonary involvement is unusual. In lymph nodes there is almost always a near-total replacement of the cortex and much of the medulla by clusters and sheets of foamy histiocytes usually nearly packed with the lipid-coated acid–alcohol fast bacilli (Figure 5.6); these are shorter and plumper than MTb and stain additionally with PAS and Gomori's methenamine

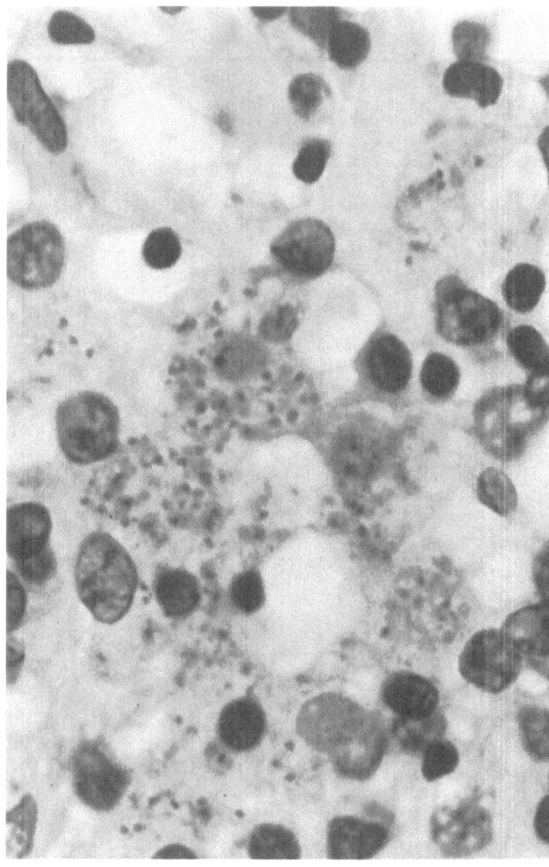

Figure 5.7. Lymph node with structure distorted by many macrophages filled by *Leishmania donovani*.

silver stains. In patients with AIDS necrosis with MAI is most unusual and a granulomatous reaction, if ever seen, is barely recognizable (Farhi et al, 1986). The clinical symptoms and signs are often non-specific and the pathogenicity of MAI is certainly open to debate.

Lymph nodes may become infected by a variety of other organisms, but these are usually secondary involvement during dissemination of an already recognized complication of HIV infection. Fungi do this and the yeast form of *Cryptococcus neoformans* with its mucicarmine-staining, thick mucoid capsule may be seen replacing much of the node with swollen histiocytes and mucin. Occasionally coccidioidomycosis, toxoplasmosis, histoplasmosis, leishmaniasis (Figure 5.7) and pneumocystis are recorded; each may be accompanied by a restricted granulomatous response with a few giant cells, but usually the reaction is restricted to clusters of epithelioid-like histiocytes. Diagnosis, essential for successful treatment, depends upon the same procedures as

elsewhere, namely specific monoclonal markers where available and microbiology both for confirmation and drug sensitivity.

Spleen

The spleen undergoes changes reflecting the pathology elsewhere in the lymphoid system (Brynes and Gill, 1990). When removed surgically, this is often for severe and uncontrollable immune thrombocytopenic purpura (ITP) (Costello et al, 1986). The lymphoid tissue shows great atrophy of T- and B-cell zones, with total loss of germinal centres; the follicles are reduced to a central arteriole cuffed by a pale zone in which plasma cells predominate. In the pulp, erythrophagocytosis and haemosiderin are common, with occasionally extramedullary haematopoiesis (Reichert et al, 1983).

Bone Marrow

From the early stages of HIV infection throughout the span of AIDS, the bone marrow shows great variability in cellularity, terminating in exhaustion, whilst reticulin may be increased (Ioachim, 1990a). The final result is ineffective haemopoiesis, due to direct and indirect effects from HIV, made worse by infections and drugs (Zon et al, 1987). Haematological side effects have been dramatically exacerbated by the extensive use of zidovudine, repeated transfusions being needed to combat severe aplasia (Costello et al, 1987; Richman et al, 1987).

In the marrow, the balance between the precursors is often changed, myeloid hyperplasia outstripping that of the erythroid series (Ioachim, 1990a). Megakaryocytes are often increased and tend to cluster. Immature and atypical forms of all lines tend to increase, with an absolute drop in mature forms. Megaloblastic erythroid changes may occur and are due either to myelodysplasia or to concurrent treatment with co-trimoxazole (Costello, 1988). Dysplasia is common amongst all types, with bizarre cells and nuclei occasionally occurring. Increased plasma cells reflect the frequent hypergammaglobulinaemia and presence of infections, whilst eosinophilia is associated with the immunological disturbances (Ioachim et al, 1983).

A reticulin stain on most AIDS marrows shows at least some increase with a delicate meshwork (Zon and Groupman, 1988), especially around granulomas and, in up to a third of such marrows,

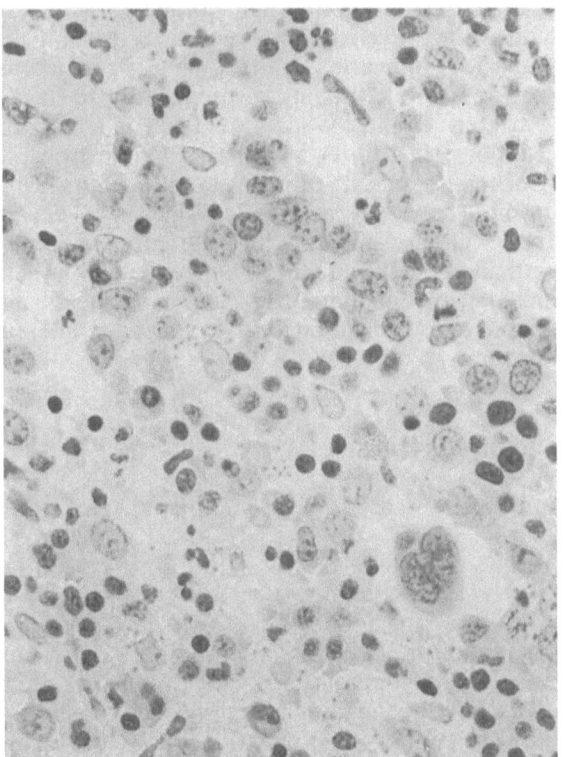

Figure 5.8. Bone marrow smear showing Leishman–Donovan bodies within and sometimes outside histiocytes.

lymphoid aggregates (Osborne et al, 1984; Spivak et al, 1984). The latter found in no particular position probably reflect virus infections; sometimes the cytology simulates lymphoma, but surveillance shows lymphoma does not always occur.

Characteristic granulomas with giant cells and necrosis are rarely seen, if ever; much more frequent, and an almost certain manifestation of opportunistic infection, are macrophages, sometimes singly, more often in clusters, infiltrating the marrow. A foamy appearance suggests Mycobacteria, especially MAI, which can be further elucidated by appropriate special stains (see above). Other intracellular organisms may be seen at high magnification (Heyman and Rasmussen, 1987; Namiki et al, 1987) (Figure 5.8). Appropriate special stains assist in identification of cryptococci, *Histoplasma*, *Leishmania*, *Pneumocystis* and *Toxoplasma*, whilst increasingly available specific monoclonal antibodies enable immunocytochemistry to confirm many diagnoses.

Microbiological culture can be a useful aid with both bone marrow aspiration and biopsy (Bishburg et al, 1986; Strand, 1990), the results

sometimes being surprisingly helpful, as with MAI (Castella et al, 1985). Such overall investigations can be rewarding, even in the absence of suggestive histological findings.

Hodgkin's disease and NHL may involve the bone marrow in AIDS and that may be the primary site (Ziegler et al, 1984a; Ioachim et al, 1990); certainly marrow involvement needs to be excluded in management staging. The author is unaware of Kaposi's sarcoma being recorded in bone marrow.

Pigment is almost entirely restricted to haemosiderin, which usually is increased due to a reticuloendothelial iron block and presents as focal deposits; this is probably due to repeated episodes of infection and haemolysis. Melanin is rare, there being no increased incidence of malignant melanoma in AIDS.

Other Sites

At other sites where lymphoreticular tissue is found, such as around the pharynx and in the salivary glands (Ioachim et al, 1988), the potential changes are the same for this tissue elsewhere.

Skin

Cutaneous pathology, which frequently accompanies HIV infection (Fisher and Warner, 1987; Kaplan et al, 1987a; Matis et al, 1987; Valle, 1987), may be the first manifestation of the infection or just another outward reflection of the disorganized immunosuppression in these patients. Skin disorders may still appear just as they do in the population as a whole, but as immunocompetence declines their course may at least be varied. Common skin disorders may follow an unusually florid or disseminated course with unusual patterns. Clinical features may change, with different lesions showing similar appearances. Similarly non-neoplastic cutaneous lesions may be mistaken for Kaposi's sarcoma, even after microscopy. The role of adequate skin sampling and of both histopathology and microbiology cannot be overestimated, and biopsy of skin is as frequent, perhaps more so, as that of the respiratory tract (Kory et al, 1987).

Inflammatory infiltrates with abundant plasma cells or individually necrotic keratinocytes in the epidermis, or the finding of infectious agents that rarely cause skin pathology, suggest the possibility of HIV seropositivity (Cockerell, 1986), especially in an at-risk patient of either sex. A correlation between the frequency of skin changes and the degree of immunosuppression often exists (Kaplan et al, 1987a; Triana et al, 1987), with a similar relationship involving unusual presentations, both clinical and structural. An identical situation undoubtedly exists between the frequency, and possibly severity, of cutaneous infective disorders and the seriousness of the HIV infection as measured by the falling helper T-cell count; most cutaneous disorders appear when the latter drops below 100 cells/mm^3 (Kaplan et al, 1987a). Further complications include finding several skin disorders coexisting, or a particular abnormality with more than one possible aetiological agent (Penneys and Hicks, 1985; Kwan and Kaufman, 1986; Boudreau et al, 1988; Smith et al, 1989).

Kaposi's sarcoma, whether a single lesion or a generalized change affecting many areas and possibly associated with internal organ involvement, is amongst the Centers for Disease Control (CDC) criteria for recognition of AIDS (Friedman-Kien et al, 1981); the same applies for disseminated infection with *Histoplasma capsulatum* and *Mycobacterium avium–intracellulare*, in both of which skin involvement may be a feature (Centers for Disease Control, 1985a, 1987).

Non-infectious Disorders

Non-infectious skin disorders include a macular rash often seen with acute HIV infection (Cooper et al, 1985), and a papular eruption (James et al, 1986); both occur anywhere, but especially on exposed areas. The basic pathology is a perivascular lymphocyte infiltrate in the dermis with lesser numbers of other chronic inflammatory cells (Sindrup et al, 1987), the latter eruption distinguished by acanthosis and parakeratosis.

A common cutaneous manifestation and early sign of HIV infection is a seborrhoeic-like dermatitis affecting the usual sites, but unusually involving the trunk and extremities (Mathes and Douglas, 1985; Soepreno et al, 1986). This HIV-related form is distinguished from the usual seborrhoeic dermatitis by histopathology. The more severe changes are closely associated with declining cell immunity and an increasingly poor prognosis, some features suggesting a graft-versus-host reaction. Other non-infectious conditions include alopecia, seen in a high proportion of younger patients with AIDS; there are also

changes in the skin and hair normally associated with ageing, such as frontal recession, thinning on the crown and vertex of the scalp and premature greying, this often being severe in the late stages (Farthing et al, 1986). Also seen are drug eruptions (Gordin et al, 1984), papular urticaria (Cockerell, 1989), psoriasis and Reiter's syndrome (Duvic et al, 1987; Winchester et al, 1987). Other skin conditions occur sometimes with increased frequency, and especially when linked with nutritional deficiency (Penneys and Hicks, 1985), but their histopathology is altered but rarely.

Infectious Diseases

Infectious diseases are frequent and widespread.

Bacterial Infections

Bacteria, especially *Staphylococcus aureus* but occasionally diphtheroids and *Staphylococcus albus*, produce inflammatory changes, unusually florid in the HIV-infected; the beard and flexion sites of the neck are particularly affected (Muhlemann et al, 1986). Subcutaneous abscesses are common, particularly in the perianal area, with folliculitis the basis of all lesions (Kaplan et al, 1987a), there being acute inflammation or a granulomatous response to a disrupted follicle. Elsewhere, multiple pruritic papules and nodules may occur on the trunk and extremities; often there is a plug or 'grain' of bacteria in the dermis with surrounding mixed inflammation. This is called botryomycosis (Patterson et al, 1987), and organisms involved are usually Gram-positive. Other organisms occasionally found in such cutaneous inflammatory lesions include *Pseudomonas* and *Haemophilus influenzae* (Kaplan et al, 1987a).

Mycobacterial Infections

Mycobacteria, both *Mycobacterium tuberculosis* and atypical forms such as *M. avium–intracellulare*, are occasionally seen in the skin (Penneys and Hicks, 1985; Kaplan et al, 1987a), but rarely is there a recognizable granulomatous reaction. Both forms, but especially the atypical, are almost always part of a disseminated process; further, sometimes the organisms are identifiable amidst other pathology such as Kaposi's sarcoma (Freed et al, 1987). Histopathology varies from a suppurative focus to a diffuse infiltrate of histio-cytes, atypical bacilli being seen with PAS and acid–alcohol fast stains, usually in vast numbers. However, occasionally there is the typical granu-lomatous change with scanty organisms as seen in immunocompetent patients, AIDS always threatening if not producing the unexpected.

Syphilis

Although the classic picture of secondary syphilis may be seen in the HIV-infected, there can be an unusually swift progression from the primary to the tertiary stage with neurosyphilis of unusual severity (Berry et al, 1987; Johns et al, 1987). Despite numerous *Treponema pallidum* some-times being seen there are often abnormal immune responses such that seronegativity is not uncommon (Radolf and Kaplan, 1988), although intense seropositivity can occur also. Micro-scopically there may be a superficial, perivascular, very sparse plasma cell infiltrate, or the infiltrate may be very heavy, include histiocytes, be lichenoid and involve the full thickness of the skin. Spirochaetes usually teem and there is no correlation between their numbers and the strength of the cellular changes. Currently mono-clonal antibodies for *T. pallidum* are assisting diagnosis at an early stage (Cockerell, 1990).

Candidiasis

Candida albicans infection of the mouth is fre-quent (see the section on the alimentary system below). On the other hand, cutaneous candidiasis is uncommon, but there may be focal hyperplasia of the squamous epithelium with hyphae and spores on the surface amidst keratin debris, some-times acutely inflamed. However, despite full penetration there may be no inflammation and the features mimic leukoplakia. There is an increased incidence of vaginal candidiasis in HIV-infected women, and in children so affected a florid napkin-type rash (Kaplan et al, 1987a), both of which will be seen more frequently with the rising incidence of the heterosexual spread of AIDS. Nowadays the finding of *Candida* spp. should arouse the suspicion of HIV seropositivity in the at-risk.

Cryptococcal Infection

Cryptococcal infection of the skin may be part of a dissemination from a focus elsewhere. The lesions may be hypopigmented papules and resemble molluscum contagiosum (Rico and Penneys, 1985), although a herpetiform pattern is more common (Borton and Wintroub, 1984). In the dermis, inflammation is slight, but there are numerous pale staining areas which correspond to the mucopolysaccharide-laden capsules sur-rounding the mainly extracellular organisms, phagocytosis being inhibited by the capsule and immunosuppression. PAS and more specifically Southgate's mucicarmine stains show up the capsule whilst silver impregnation (e.g. Masson Fontana) blackens the organism wall.

Histoplasmosis

Histoplasma duboisii, the cause of African histo-plasmosis, mainly involves the skin; elsewhere in the world, however, *H. capsulatum* is the agent and only reaches the skin with immunosup-pression; in HIV infection cutaneous involvement is always part of disseminated histoplasmosis, the lung being the portal of entry. There is a maculopapular rash or acneiform picture (Hazelhurst and Vismer, 1985; Kalter et al, 1985), microscopy showing multiple macrophages in the dermis with no more than a small ac-companiment of other inflammatory cells. Occasionally giant cells and even an early granuloma develop and surface ulceration can occur. Careful scrutiny of H&E preparations identifies both intrahistiocytic and free organisms, the fungus being haloed by retraction of the cyto-plasm from the cell envelope. PAS is of limited value, but Grocott's methenamine silver is highly successful in identification (Figure 5.9); however, differentiation from *Leishmania* and *Pneumocystis carinii* requires a monoclonal immunocyto-chemical marker or contained microbiological culture.

Leishmaniasis

Leishmaniasis is another parasitic infection asso-ciated with HIV infection predominantly in tropical and subtropical regions. There are numerous species of *Leishmania*, identical in morphology but separable by isoenzyme profiles and DNA probes of the nucleus (Lucas, 1989). Purely skin lesions are uncommon in AIDS, if occurring at all, and such involvement is almost certainly a manifestation of wide dissemination in visceral leishmaniasis; the cytoplasm of the lymphoreticular system and particularly that of

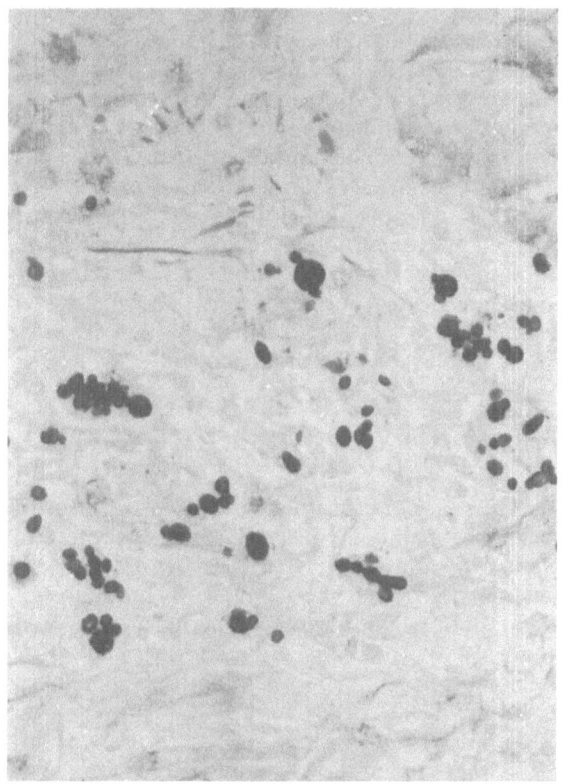

Figure 5.9. Skin with numerous *Histoplasma capsulatum* stained uniformly black amidst intradermal chronic inflammation; some show budding and a few lie within the tip of a rete peg on right (Grocott).

histiocytes in bone marrow is the favourite site for Leishman–Donovan bodies. Granulomas are not found.

Leishmania as amastigotes can occur in the macrophages of the dermis and are readily seen with Leishman or Giemsa stains, but they have to be distinguished from *Histoplasma*, *Toxoplasma gondii*, *Trypanosoma cruzi*, *Microspora* and *Pneumocystis carinii*. This can be achieved using appropriate monoclonal antibodies (see above), but where species-specific sera are not available a less specific antibody may have to be used in conjunction with serology and microbiological culture to finalize diagnosis.

Dermatophytosis

Dermatophytosis and other superficial fungal infections are fairly frequent in HIV-infected patients and often more florid than in the immunocompetent (Matis et al, 1987); despite treatment they tend to recur. Histopathology

ranges from a very sparse inflammatory cell infiltrate to a relatively dense mixed infiltrate in the dermis, with neutrophil granulocytes amidst the parakeratosis which overlies foci of spongiosis (Radolf and Kaplan, 1988). It is a wise precaution with skin showing these changes always to search with fungus stains.

Psoriasiform Dermatitis

Psoriasiform dermatitis may be associated with numerous mites of scabies (*Sarcoptes scabiei*) in the cornified layer (Sadick et al, 1986; Drabick et al, 1987), but other parasitic infections are rare despite HIV infection. However, *Pneumocystis carinii* (Coulman et al, 1987), *Acanthamoeba castellani* and *Demodex folliculorum* have been reported, the latter two in acute necrotising dermatitis and eosinophilic pustular folliculitis respectively (Harawi and Kurban, 1989).

Toxoplasmosis

Toxoplasma gondii has been reported amidst a non-specific inflammatory infiltrate in the dermis (Hirschmann and Chu, 1988), but absolute diagnosis is near impossible without carefully controlled immunocytochemistry.

Viral Diseases

Viral diseases are common in HIV-infected patients. Both Herpes simplex and zoster produce intra-epidermal vesicles with associated acantholytic, multinucleated giant cells and an underlying dense, mixed, chronic inflammation of the dermis (Cockerell, 1990). However, secondary bacterial infection is frequent and the underlying virus may only be detected by culture or microscopy.

Condyloma Acuminatum

Condyloma acuminatum in HIV infection is frequent and often large and florid (Cockerell, 1989); not infrequently the cytological changes are atypical and may indicate a diagnosis of Bowenoid papulosis (Croxson et al, 1984), with occasional direct transformation to Bowen's disease or a full-blown squamous cell carcinoma. Similarly the verrucas may be increased in number and unresponsive to treatment.

Molluscum Contagiosum

Molluscum contagiosum is also seen more frequently and sometimes reaches large size and
ulcerates, the genitalia, face and scalp being particularly affected (Kaplan et al, 1987a; Katzman
et al, 1987; Friedman-Kien and Cockerell, 1988),
whilst atypical changes can distort the clinical
picture. Inflammation is usually slight, but the
intracytoplasmic inclusion bodies may reach very
large size; occasionally the inflammation is dense
with atypical lymphocytes and lymphoma may be
suspected.

Cytomegalovirus Infections

Cytomegalovirus (CMV) infections of the skin
present as papular nodules or ulcers usually close
to mucocutaneous junctions and in patients with
underlying widely disseminated disease (Niedt
and Schinella, 1985); characteristically there is a
dense, mixed inflammatory cell infiltrate not only
extending deep into the dermis (Cockerell, 1989)
(Figure 5.10) but possibly also involving underlying muscle. The usual sites for the CMV inclusion bodies are fibroblasts, endothelial and
histiocytic cells and muscle fibres; epithelial cell
involvement is most unusual.

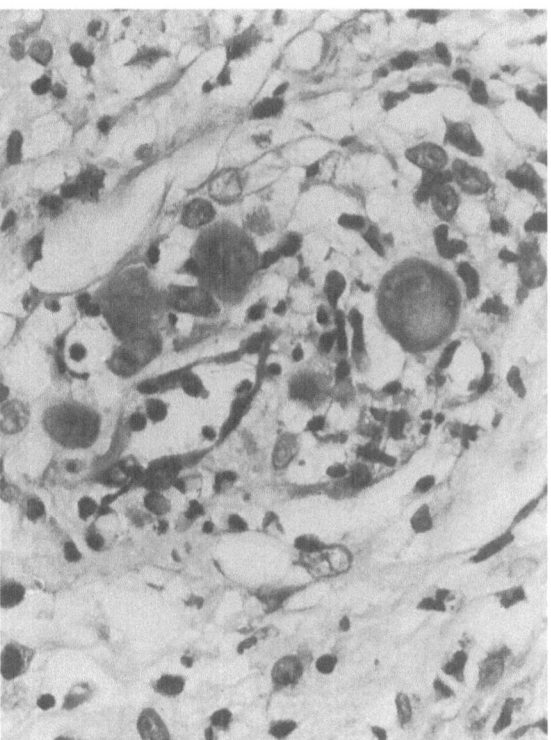

Figure 5.10. Lip with chronically inflamed subepithelial connective tissues; multiple typical intranuclear and intracytoplasmic inclusion bodies of cytomegalovirus affecting some
fibroblasts and histiocytes.

Bacillary Epithelioid Angiomatosis

Bacillary epithelioid angiomatosis (bacillary
angiomatosis of AIDS), first described in the skin
and lymph nodes of HIV-infected patients
(Cockerell et al, 1987; Knobler et al, 1988), is
now recognized in visceral organs and is clearly
linked with immunodeficiency (Knobler et al,
1988), although reported with immunocompetence (Cockerell et al, 1990). The lesions
present as solitary cutaneous polyps or subcutaneous nodules, but occasionally there are
more diffuse, purple, vascular patches not unlike
Kaposi's sarcoma. Microscopically the deep
dermis and subcutaneous tissues include a
dense collection of plump endothelial and
cuboidal histiocytic cells, sometimes with necrosis,
whilst superficially there is often a part necrotic,
unusually cellular, pyogenic granulomatous
picture. Granular eosinophilic material is often
seen interstitially and a Warthin–Starry stain
shows this to be argyrophilic bacteria (Le Boit
et al, 1988), confirmed by electron microscopy.
Recently the bacteria have been identified as a
hitherto uncharacterized *Rickettsia*-like organism

closely related to *R. quintana* (Relman et al,
1990).

Neoplasms

The most common neoplasm in AIDS is Kaposi's
sarcoma, a classic and diagnostic feature of the
condition (Friedman-Kien et al, 1981). Others
(basal cell and squamous carcinoma and non-
Hodgkin's malignant lymphoma) are seen
periodically; all tend to pursue a more aggressive
course than is usual, but there seems to be no
greater incidence than with immunocompetent
patients of similar age and origin. This contrasts
with their increased frequency in those immunosuppressed following organ transplantation.

Kaposi's Sarcoma

During the 1970s an association was seen between
Kaposi's sarcoma (Kaposi, 1872) and iatrogenic
immunodeficiency states; with the emergence
of AIDS there was a dramatic change and a

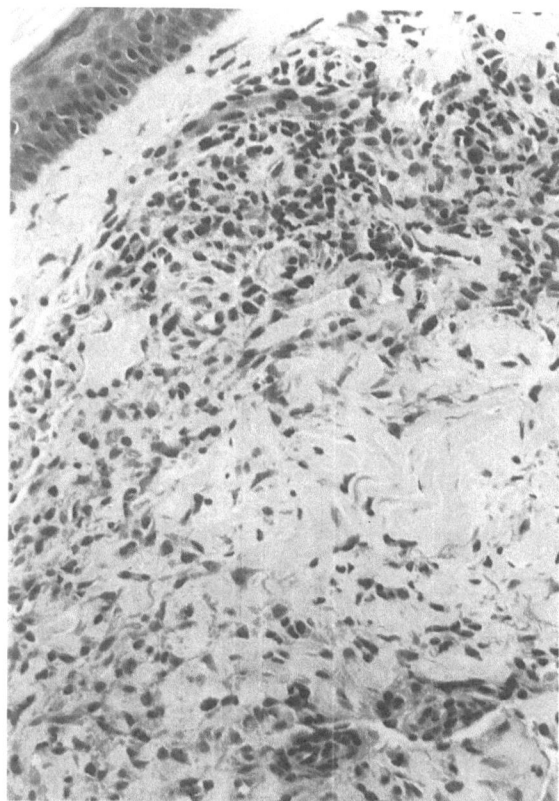

Figure 5.11. Skin with upper and mid-dermal Kaposi's sarcoma including solid clusters of cells, slit-like spaces dissecting collagen and slight chronic inflammation.

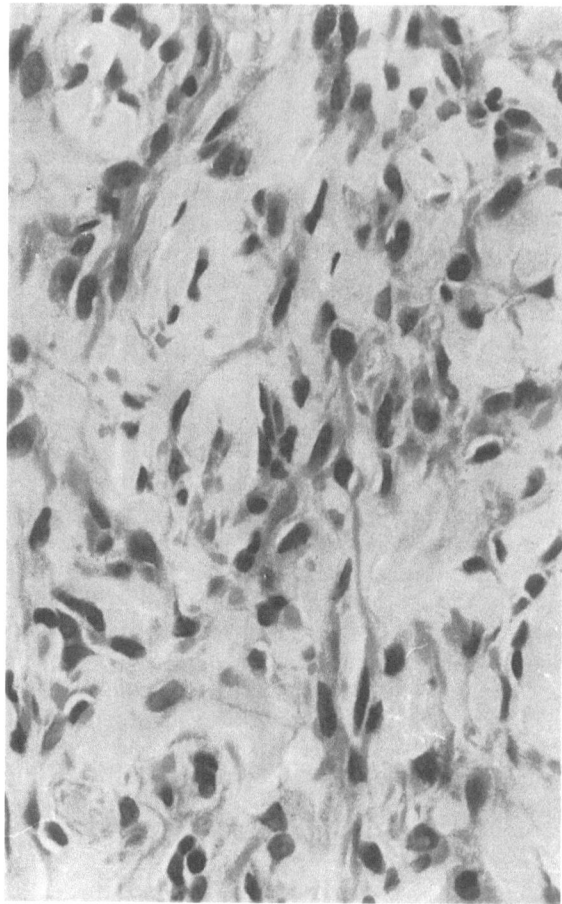

Figure 5.12. Kaposi's sarcoma amidst collagen, many cells lining vascular slits dissecting the latter and showing varied atypia, whilst red cells have extravasated. Perls' stain would show a fine deposit of haemosiderin.

generalized, often fulminant form reached epidemic proportions.

This sarcoma involves blood vessels. Dorfman (1962) suggested an origin from lymphatics, but controversy still persists as to whether the predominating 'endothelial cell' of Kaposi's sarcoma is of blood vessel (Rutgers et al, 1986; Scully et al, 1988) or lymphatic origin (Jones et al, 1986; Dorfman, 1988).

Kaposi's sarcoma can be seen anywhere in the skin. The lesions present as variously sized patches, plaques or nodules of deep red-brown, changing to purple-grey or brown with increasing duration. Histological features include angiomatous, inflammatory and mixed patterns whilst pleomorphic and spindle cell types are recognized (Templeton, 1981; Gottleib and Ackerman, 1982; Schwartz et al, 1983; Kalengayi and Kashala, 1984; Moskowitz et al, 1985). The features showing wide variation include the degree of cellularity and vascular space formation, differences in cell and nuclear atypia, numbers of mitoses, the pre-

sence of intracellular eosinophilic globules, extravasation of erythrocytes, haemosiderin deposition, and the usually accompanying though restricted chronic inflammatory infiltrate, mainly plasma cells (Gottleib and Ackerman, 1982) (Figures 5.11 and 5.12). Microscopy is all-important for the diagnosis of Kaposi's sarcoma, which is a criterion for applying the label of AIDS in HIV infection. The differential diagnoses include capillary haemangioma, melanocytic naevus, benign fibrous histiocytoma, angiosarcoma and bacillary epithelioid angiomatosis (Knobler et al, 1988; Schofield et al, 1989), whilst large and frequent deposits of haemosiderin, as distinct from a fine scattering, make the diagnosis unlikely.

Epidemic, AIDS-related Kaposi's sarcoma can be distinguished from the classic form (Ziegler

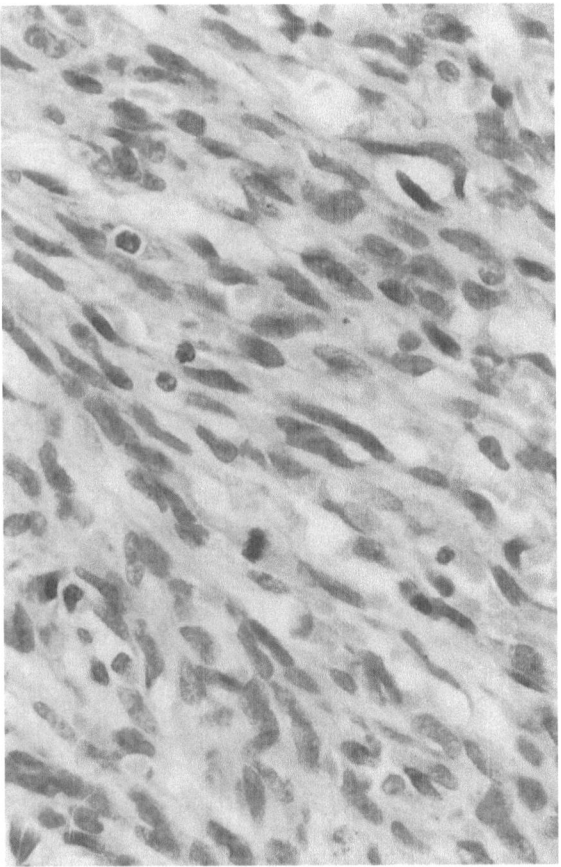

Figure 5.13. Highly cellular, mixed spindle and polygonal cell Kaposi's sarcoma with conspicuous nuclear atypia, several mitoses and a few vascular slits.

arborizing channels, but also narrow slit-like spaces not otherwise easily seen. Cell and nuclear atypia with mitoses are usually present (Figure 5.13), but usually very much less severe than in most angiosarcomas.

Well-established older lesions show fewer vascular spaces although some channels of varied size remain. Spindle cells are numerous and spread through the dermis, and may involve the subcutaneous fat. Whilst some of these cells ramify as slender columns, others surround vascular slits, a reticulin preparation showing up not only the latter but also a meshwork of fibres, although mature collagen is infrequent. At this stage atypia and mitoses are clearly visible, but only rarely are severe, whilst a Perls' stain shows up a fine dust of haemosiderin, variable in density but sometimes fairly heavy; the pigment can always be found using levels.

Extravasation of red cells is frequent; whilst close to the atypical vessels in the early stages, later on they are throughout the lesion, being clearly seen in the stroma and sometimes within spindle cells. A frequent and diagnostically helpful feature is intracytoplasmic, eosinophilic globules usually smaller than an erythrocyte (Blumenfeld et al, 1985); they are easily seen with H&E and stain bright red with phloxine tartrazine. Special stains, immunocytochemistry and electron microscopy give variable, often conflicting results, and debate as to the nature of these globules continues (Aziz et al, 1985). Ultrastructural evidence indicates that they are most likely to be giant lysosomes or fragments of altered erythrocytes.

Certain microscopic features point to a diagnosis of Kaposi's sarcoma: a proliferation of slit-like vascular spaces amongst and around apparently normal vessels and nerves; angiomatoid foci lined by 'hobnail' type endothelium and surrounded by spindle cells; dissection of adjacent stroma and collagen by spindle cells and vascular slits and spaces (Francis et al, 1986) (Figure 5.14). In the early stages, however, the features are subtle and, if there is doubt, more levels or even additional material may be required; if clarification is not obtained the diagnosis may be suggested but not made, for Kaposi's sarcoma is a criterion of AIDS in HIV infection. Immunocytochemistry and electron microscopy are of little help in clarifying such a situation. The basic structural pattern may be altered only a little, but such distortions can be due to additional infections which require exclusion (Penneys and Hicks, 1985; Smith et al, 1989) (Figure 5.15).

et al, 1984b). With HIV infection the early lesion may be the merest hint of an abnormal endothelial cell proliferation in the upper dermis, this often surrounding the adnexa, whilst cytological changes are minimal. At this stage there is frequently a chronic inflammatory cell infiltrate with plasma cells, histiocytes and lymphocytes in that order of declining numbers; the former are more frequent than in most of the differential diagnoses. Occasionally the features mimic granulation tissue, but often they are of lesser degree and inflammation is less than in classic non-HIV-associated Kaposi's sarcoma of similar development (Santucci et al, 1988). An architectural change is the formation of thin-walled, sometimes dilated, anastomosing vascular channels lined by flat or plump endothelial cells; the latter may push into the lumina of the former as well as forming clusters between the vessels. A Gordon and Sweets' reticulin stain shows not only the

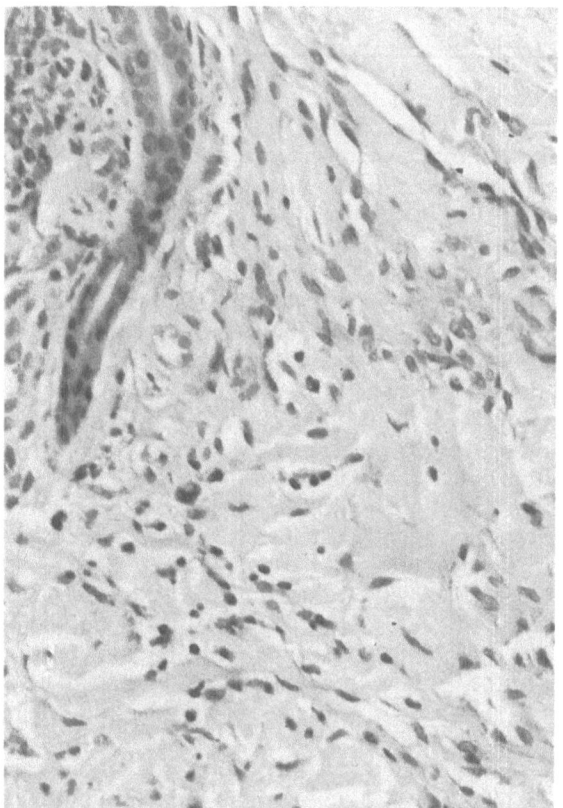

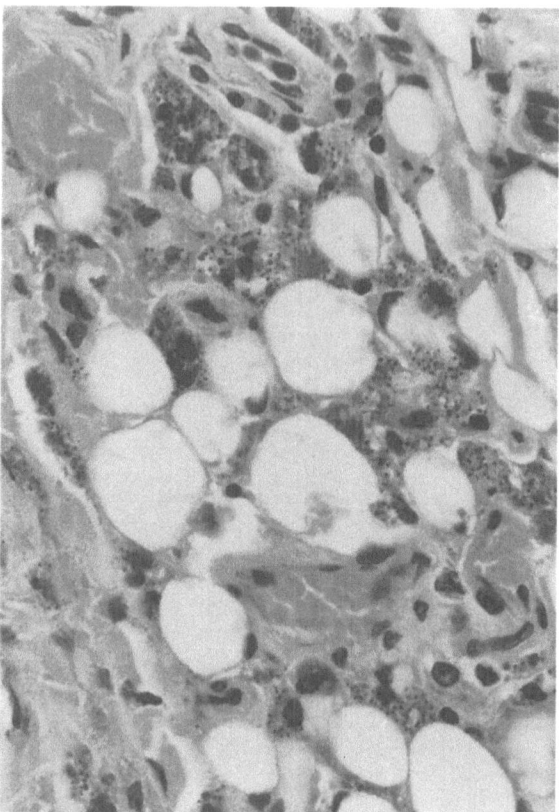

Figure 5.14. Deep dermal Kaposi's sarcoma with investment of sweat glands and again a vascular pattern.

Figure 5.15. Kaposi's sarcoma with copious haemosiderin amidst dermal collagen and subcutaneous fat; additionally numerous Leishman–Donovan bodies lie both within phagocytic cells and loosely in the stroma.

Respiratory System

Pulmonary complications are frequent in patients with AIDS, over 60 per cent presenting with pneumonia, whilst respiratory failure is a common cause of death, manifested by severe refractory hypoxia (Kovacs et al, 1984). The vast majority of cases, around 65 per cent (Murray et al, 1984; Marchevsky et al, 1985), are due to the protozoon/fungus *Pneumocystis carinii*. However, about 25 per cent of the respiratory disorders occurring in AIDS are due to infections other than *Pneumocystis* (Murray et al, 1984; Purdy et al, 1986). These include cytomegalovirus, *Mycobacterium*, both atypical and *M. tuberculosis*, *Legionella*, pneumococci, cryptococci and other fungi. About 5–10 per cent of the pneumonias are mixed coexisting infections by two or more of these organisms, this often explaining treatment failures when only one of the agents has been identified. Neoplasms can occur, with or without accompanying infection, and are often a postmortem finding.

Pneumocystis carinii Pneumonia

Until 1981 *Pneumocystis carinii* infection was restricted to a pneumonia which originated during World War II as an epidemic amongst debilitated, premature or underweight infants (Spencer, 1977). *Pneumocystis carinii* pneumonia (PCP) later became associated with patients suffering from malignant lymphomas and leukaemias as well as those on immunosuppressive therapy or on long-term steroid or cytotoxic therapy (Peters and Prakash, 1987). With the appearance of AIDS the incidence of PCP has dramatically increased, especially when the CD4 subset count falls below $200/mm^3$ (Masur et al, 1989).

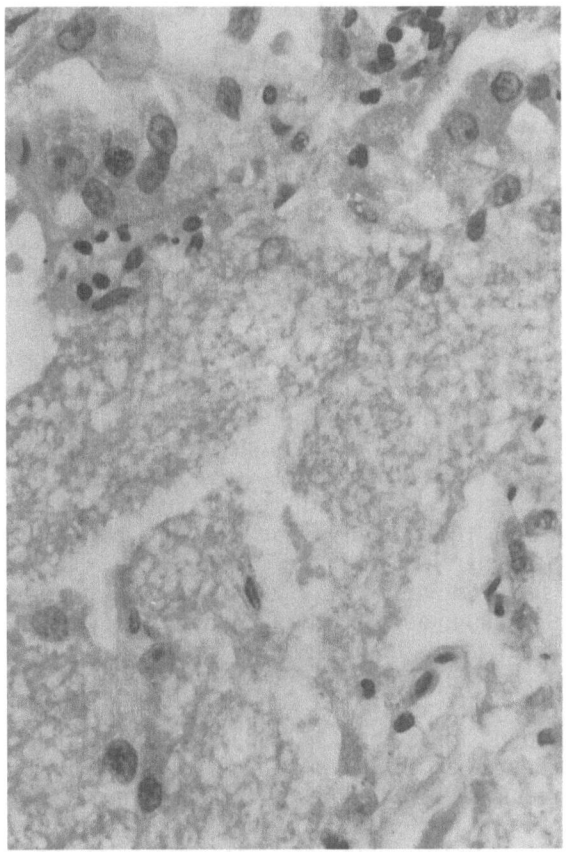

Figure 5.16. *Pneumocystis carinii* pneumonia with the characteristic foamy exudate within alveoli; there is minimal oedema and only slight, non-specific inflammation.

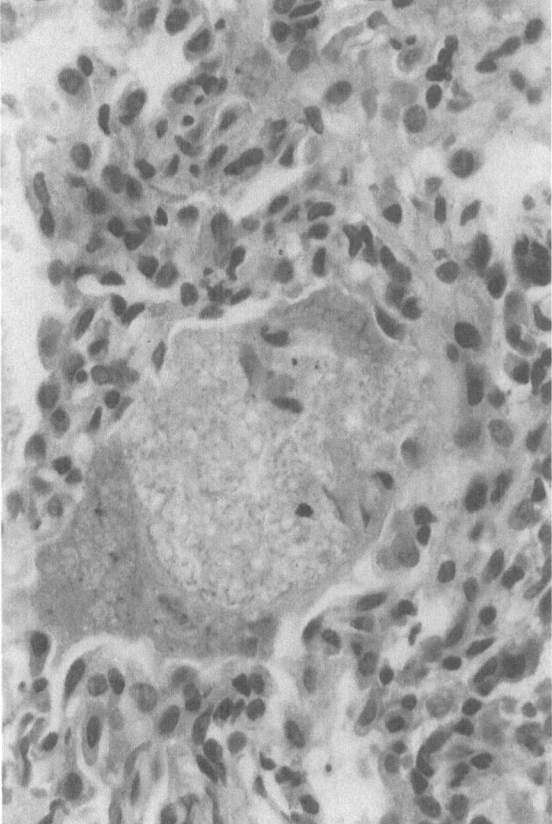

Figure 5.17. Foamy exudate in an alveolus surrounded by chronic inflammation; there is some hyaline membrane, a little haemorrhage and early interstitial fibrosis. In such circumstances the organisms from within the walled-off exudate are difficult to find in sputum or bronchial lavage.

Pneumocystis carinii is a unicellular, opportunistic organism sometimes regarded as one of the Sporozoa family of protozoa; others, however, consider the organism to be a fungus, but doubt remains (Edman et al, 1988; Mackenzie, 1990). The organism is probably acquired by the airborne route and exists saprophytically in the lungs, activation and replication arising with developing immunodeficiency.

Within the lung the pneumocysts are found in the alveoli as cysts each containing several sporozoites; the latter have fine surface projections by which they anchor to the alveolar wall, hence their infrequency in sputum. In the alveolar space sporozoites liberated from the cyst mature into trophozoites which pass through early, intermediate and late stages before becoming a mature cyst. Electron microscopy shows trophozoites to be attached largely to type I pneumocytes and more rarely to those of type II (Long et al, 1986).

Microscopically inflammation is usually minimal even with the most widespread and severe change; the cardinal feature is the eosinophilic, 'frothy', intra-alveolar, fibrinous exudate within which are characteristic basophilic dots, but few cells (Figure 5.16). Infiltrating cells, usually mononuclear, are few and the interstitium is involved to an even lesser degree. Foci of intra-alveolar haemorrhage, hyaline membrane and mild interstitial fibrosis are sometimes seen (Figure 5.17). Whereas recurrent infections in patients after transplantation or other non-HIV-associated immunodeficiency may be associated with epithelioid granulomas, multinucleated giant cells, numerous intra-alveolar macrophages and widespread fibrosis, these are not features of PCP in AIDS.

Replicating pneumocysts appear to increase directly the permeability of the alveolar capillary membrane with intra-alveolar oedema and sub-

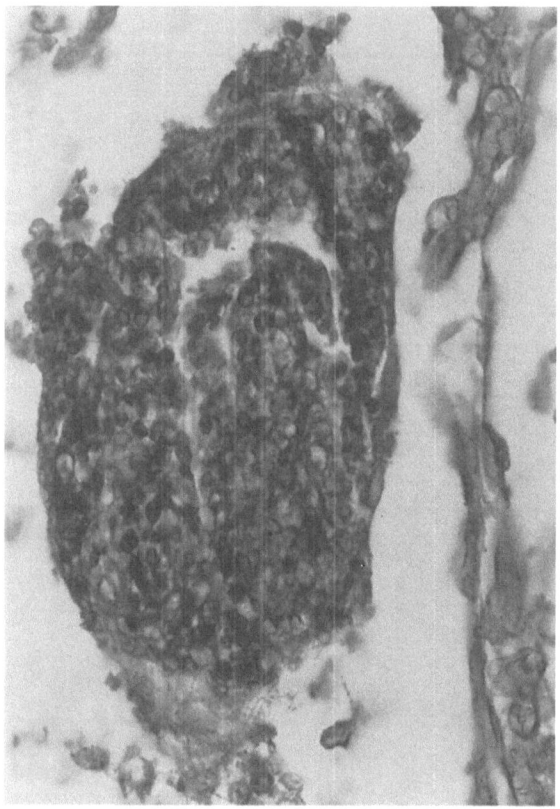

Figure 5.18. Alveoli filled with exudate and stained by Grocott's method to show up the walls of the many empty cysts, just a few containing the protozoa.

sequently damage to the type I pneumocytes alone, all factors impairing oxygenization.

Grossly early involvement is accompanied by patchy consolidation in the affected lung; later usually both are heavy with diffuse, pinkish consolidation resembling lobular pancreatic tissue, although appearances may be modified by accompanying pathology, seen in significant numbers of patients.

Extrapulmonary (lymph nodes, spleen, kidneys, liver (Radin et al, 1990), bone marrow (Heyman and Rasmussen, 1987), thyroid (Drucker et al, 1990), skin (Davey et al, 1989)) spread of *Pneumocystis carinii* is rarely recorded and the route is probably via the bloodstream.

Definitive diagnosis relies on the identification of the organism in tissue biopsy, bronchoalveolar lavage or induced sputum (Hopewell, 1988; Strand, 1990; Midgley et al, 1991). The cysts are identified by Grocott's methenamine silver stain (Figure 5.18) but with any overstaining erythrocytes and leucocytes may be mistaken for parasites.

Cytomegalovirus Pneumonitis

Cytomegalovirus (CMV) pneumonitis is the second most common respiratory infection in AIDS (Klatt and Shibata, 1988; Travis et al, 1989) and is frequently part of disseminated CMV disease; often other infections are present also. Within the host cell, CMV provokes increased protein synthesis with enlargement of the cell and the formation of viral particles, creating an easily seen inclusion within the nucleus, separated from its membrane by a halo. In the lung both intranuclear and intracytoplasmic forms of inclusion body can be seen in both alveolar and endothelial cells as well as histiocytes. The pneumonitis is characteristically interstitial, focal or extensive, the frequent former modes accounting for the poor results when attempting diagnosis by bronchoscopy (Nash and Fligiel, 1984a; Niedt and Schinella, 1985). Most commonly the only change is a few scattered CMV-infected cells with almost no inflammatory reaction (Figure 5.19), although occasionally there are accompanying large zones of acute inflammation with some necrosis and rarely haemorrhage, with or without alveolar septal thickening. Significant inflammatory changes and inclusion bodies are not seen in the pleura (Klatt and Shibata, 1988; Ioachim, 1990a).

Confirmation of the diagnosis relies on immunocytochemistry with anti-CMV antibodies and in situ hybridization, as described in Chapter 6.

Other Types of Pneumonitis

Herpes virus pneumonitis is very rare. *Toxoplasma gondii* is also rarely a cause of pneumonitis, despite its frequency as an agent for necrotizing encephalitis (Murray et al, 1984; Marchevsky et al, 1985) (see Chapter 6). The occasional report of significant infection shows a mild chronic interstitial inflammation with proliferation of the alveolar lining cells, the trophozoites being found in the intra-alveolar exudate and associated macrophages. Grossly there are multiple, often confluent red and grey zones of consolidation, but certainly no specific change. The presence of the parasite can only be detected by careful microscopy with immunocytochemical confirmation, be it as a pure or, more likely, a mixed infection with other organisms. Diagnosis during life is highly unusual.

Cryptosporidium is another parasite infrequently found in the lung, on the surface of the respiratory epithelium. It is almost always a com-

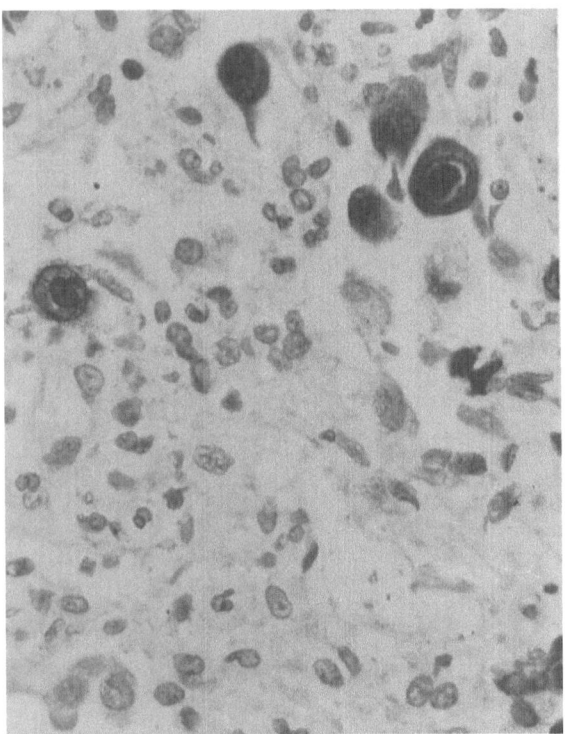

Figure 5.19. Cytomegalovirus pneumonitis: a specific monoclonal marker showing several inclusion bodies, mostly intranuclear, in the walls and lumina of alveoli, with some oedema and minimal chronic inflammatory cells.

plication of intestinal infection where it is common (see the section on the alimentary system below).

Fungal pneumonitis almost always occurs as part of a mixed infection; primary involvement of the lung is most unusual (Ioachim, 1990a). *Candida*, usually *C. albicans*, characteristically appears as hyphae and spores in micro- and larger abscesses as well as on the surface and in the wall of respiratory passages. *Aspergillus* and *Cryptococcus neoformans* occasionally occur; the former is associated with an acute necrotizing pneumonitis. With the latter organism, interstitial oedema and inflammation are almost non-existent as with the meninges (see Chapter 6), but the organisms with the mucoid capsule (see the section on skin above) may be seen amidst the intra-alveolar froth. *Histoplasma capsulatum*, always acquired by the airborne route, may cause a pneumonitis if the inoculum is large or if there is underlying lung pathology, but often the infection is asymptomatic. Most organisms are within macrophages and often in clusters as are the latter, usually amidst a mild interstitial inflam-

mation which itself is diffuse, but granulomas are not seen (Wheat et al, 1985). Other fungi rarely seen in the United Kingdom and Europe may occur as the AIDS population increases and air travel brings contaminated persons from afar. Coccidioidomycosis, blastomycosis and zygomycosis are thus possibilities.

Mycobacterium tuberculosis in AIDS has a higher frequency (up to 3 times) in extrapulmonary sites than in the non-immunosuppressed and atypical presentations in the lungs are frequent (Maguire et al, 1987; Manoff et al, 1987). *Mycobacterium avium–intracellulare* (MAI) are the most common organisms, and the most common cause of bacteraemia in AIDS (Wong et al, 1985; Hawkins et al, 1986; Strand, 1990). Primary or predominant pulmonary involvement is unusual (O'Brien, 1988). Whilst poorly defined granulomas are described in the lung the inflammatory reaction is usually slight (Amberson et al, 1985), the significant finding being sheets of macrophages many of which are teeming with organisms. The organisms are demonstrable not only as acid–alcohol fast but by PAS and sometimes by silver precipitation, as well as by appropriate culture, the latter being essential for the management of bacteraemia (Strand, 1990).

Amongst other infective causes of pulmonary pathology, and frequently producing a lobar-type inflammatory response, are *Pneumococcus* and *Haemophilus influenzae*. *Legionella pneumophila* amongst many other bacteria does occur, but none of them are other than occasional. Another suggested cause is the mycoplasmas; with *M. incognitus* changes range from none to fulminant necrosis (Lo et al, 1989), and such infections may be on the increase.

Lymphoid interstitial pneumonia in AIDS is usually restricted to children, sometimes in association with hypergammaglobulinaemia, but there are occasional reports in adults, often in association with a variety of immunological disturbances (Saldana et al, 1983; Guillon et al, 1988; Ioachim, 1990a). Throughout much of the lung there is a diffuse, polyclonal, chronic inflammatory type B-cell infiltrate which is thought to be reactive in character on the basis of immunocytochemical studies. Varied numbers of plasma cells, including occasional Russell bodies, are included and nodular aggregates of lymphoid cells, sometimes with germinal centre formation, may occur. Again fibrosis of the bronchial submucosa and capillary hyperplasia within the lymphoid nodules can occur, but neither true granulomas nor destruction of bronchi are seen.

Neoplasms

Kaposi's sarcoma is fairly frequent in the respiratory tract and may present as a pure pulmonary disorder; however, much more commonly the involvement, and especially that of the lungs, is part of a widely disseminated process (Nash and Fligiel, 1984b; Garay et al, 1987). The microscopic changes are basically those described for Kaposi's sarcoma in the skin (see above). They may occur anywhere in the respiratory tract and, when in the lung, from hilum to the pleura, though rarely the parietal layer; even without infiltration the mesothelial cells of the pleura often show severe reactive changes which can be a helpful diagnostic pointer. This sarcoma in the respiratory system accounts for much morbidity and mortality in AIDS and may be associated with massive intrapulmonary haemorrhage, occasionally into the pleural space but only very rarely into the respiratory passages. Involvement of respiratory passage walls, including the mucosa, is frequent; the larynx and epiglottis are affected but less often. Necropsy on patients with Kaposi's sarcoma shows respiratory involvement in up to 50 per cent, and with or without accompanying infection it is a frequent cause of death (Meduri et al, 1986).

Lymphoma is rare and involvement of the respiratory tract is part of extranodal dissemination; a primary site in the lung is most unlikely. The structure follows the patterns seen elsewhere (see the section on lymph nodes above).

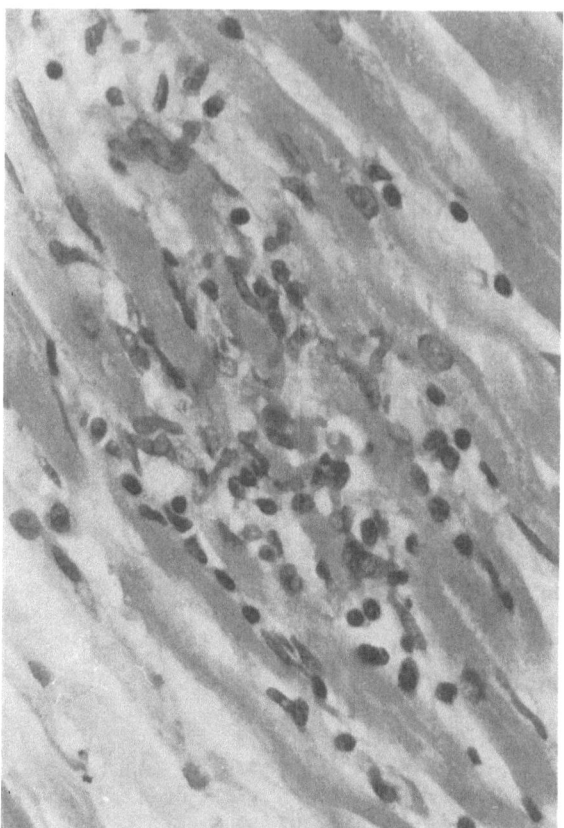

Figure 5.20. Myocarditis in AIDS with a typical small focus of interstitial lymphocytes showing many more than five cells per high-power field, myocytolysis and a few degenerate fibres.

Cardiovascular System

Since the epidemic of AIDS began, cardiac morbidity and mortality have been clearly recognized, although the rates for all patients with the syndrome (6–7 per cent for the former, 1–6 per cent for the latter) have changed little (Anderson and Virmani, 1990a). In the USA, as the number of people with AIDS increases, several thousand additional cases of heart disease are predicted each year for the foreseeable future (Corallo et al, 1988; Levy et al, 1989), with cardiac complications attributable to AIDS causing significant numbers of deaths per annum. Study of hearts from such patients by the author indicates that the role of cardiac pathology in causing such complications is inadequately recognized.

Various groups of cardiac dysfunction are known (Himelman et al, 1989; Anderson and

Virmani, 1990a): cardiac tamponade due to pericardial effusion or haemorrhage, dilated cardiomyopathy, ventricular tachycardia, chronic cardiac failure, and both infectious and non-infectious endocarditis with thrombosis and possible embolism (Anderson and Virmani, 1990b). Most of these pathologies are found incidentally at necropsy even though most such series report around 50 per cent cardiac involvement (Cammarosono and Lewis, 1985; Anderson et al, 1988; Lewis, 1989). The consistently most frequent finding is a chronic myocarditis (Roldan et al, 1987; Baroldi et al, 1988; Reilly et al, 1988; Anderson et al, 1988), defined as at least a few foci of chronic inflammation with five or more lymphocytes in the interstitium (Anderson et al, 1988) (Figure 5.20). Degeneration of muscle fibres is usually, but not always, present. For most patients a viral aetiology is postulated, and particularly so where myocarditis is associated with biventricular dilatation (Roldan et al, 1987; Baroldi et al, 1988; Levy et al, 1988; Reilly et al,

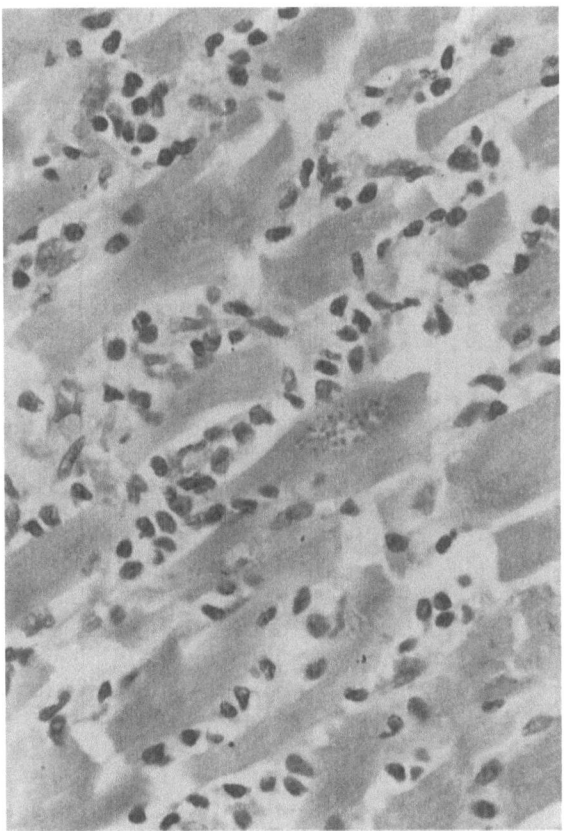

Figure 5.21. Myocarditis in AIDS with *Toxoplasma gondii* bradyzoites packed inside a pseudocyst forming within a muscle fibre, itself surrounded by a diffuse, mainly lymphocyte infiltrate.

1988); occasionally a variety of micro-organisms are found (Anderson et al, 1988). A likely explanation for many of these myocarditises, their paucity reflecting immunosuppression, is that HIV inflicts damage on the muscle fibres, either directly or by releasing cytotoxic substances, when replicating in the interstitial inflammatory cells – the 'innocent bystander hypothesis' (Anderson et al, 1988).

Opportunistic changes include reactivation of quiescent infection (nocardial (Holtz et al, 1985), mycobacterial and cryptococcal myocarditis and pericarditis (Anderson et al, 1988)). *Mycobacterium tuberculosis* (pericardium), as distinct from *M. avium–intracellulare*, and the protozoon *Toxoplasma gondii* (myocardium) cause cell damage and, despite the underlying immunodeficiency, there is often an undoubted although small inflammatory cell response as usually applies at the other sites of similar infection (Figure 5.21).

Kaposi's sarcoma occurs (Silver et al, 1984; Cammarosano and Lewis, 1985; Lewis, 1989), particularly in the pericardium, but also in the myocardium and always with involvement elsewhere. Usually an incidental finding at necropsy, clinical suspicion may be raised by tamponade or chronic cardiac failure in a patient in whom Kaposi's sarcoma has already been diagnosed elsewhere (Roldan et al, 1987). Myocardial involvement by malignant lymphoma (B-cell type) occasionally occurs with cardiac dysfunction (Anderson and Virmani, 1990b); usually of high-grade, large cell type it contrasts with the mainly pericardial, usually lesser-grade lymphomas of non-AIDS patients (Balasubramanyam et al, 1986).

Other pathological conditions may be found, such as ischaemic changes, endocarditis with vegetations and tertiary syphilis, but these are incidental and appear totally unrelated to the immunosuppression underlying the other changes.

Blood vessels are not generally involved in any specific pathology associated with AIDS, but there are occasional reports of unexplained changes in different sites; cerebromeningeal vascular changes occur (see Chapter 6), whilst significant arterial changes occur in the synovium, the narrowing being associated with premature fibrosis (Dalton et al, 1990). With disseminated Kaposi's sarcoma the adventitia of arteries is often involved, either as the origin or by spread from nearby.

Alimentary System

The gastrointestinal tract and associated structures are the site from which multitudinous symptoms and signs can arise in AIDS patients. These include diarrhoea, often intractable, and swallowing difficulty, and are frequently associated with malabsorption, wasting and weight loss. The chronic diarrhoea and weight loss, now recognized as two of the three major criteria of AIDS in developing regions (WHO 1986, Widy-Wirski et al, 1988), are so frequent and debilitating within Central Africa that AIDS is known as 'slim disease' or enteropathic AIDS (Lucas, 1987; Sewankambo et al, 1987).

The range of pathology underlying gastrointestinal disease falls into three categories: *opportunistic infections*, following the development of immunosuppression; the latter may also promote *malignancy* (Kaposi's sarcoma, lymphoma and possibly a variety of carcinomas);

thirdly, *enteropathy* and a loosely defined malabsorption syndrome, probably by direct injury to the mucosa by HIV.

Within the gastrointestinal tract and associated structures the various pathologies can present as discolorations, thickenings, nodules or ulcers visible to the eye directly, or more often through an appropriate endoscope. Following biopsy, special stains and, less frequently, immunohistochemistry, electron microscopy or special microbiological procedures may be required.

Mouth and Pharynx

Mouth and pharyngeal involvement is frequent and consists of infections and neoplasms, these lesions often being seen first by dentists and hygienists.

Candidiasis

Infection of the oral and oesophageal mucosa with *Candida* spp., usually *C. albicans*, is frequent (up to 90 per cent of the patients) (Malebranche et al, 1983); it arises at any stage and sometimes predates diagnosis of the syndrome (Klein et al, 1984). It may occur in isolation, but its incidence is boosted by other pathologies. The lesions are seen as thick white patches, sometimes becoming confluent. Microscopy shows the yeasts as spores and entangled pseudohyphae amidst a pseudomembrane made up of cell debris with squames and at least some inflammatory exudate. The elements of the fungus can be readily highlighted by PAS and Grocott preparations.

Hairy Leukoplakia

Hairy leukoplakia (HL), or condyloma planum, localized on the lateral borders and/or under surface of the tongue, usually presents as poorly defined, firmly adherent, white excrescences with a folded or hair-like surface (Eversole et al, 1986; Greenspan and Greenspan, 1988), but may be flat. Once restricted to homosexual HIV-positive males, HL has nowadays a more extended association which includes all the other risk groups (Centers for Disease Control, 1985b; Greenspan et al, 1985, 1986a, 1987; Eversole et al, 1986; de Maubeuge et al, 1986; Rindum et al, 1987; Schiodt et al, 1987; Ficarra et al, 1988; Greenspan and Greenspan, 1988), and is recognized as a diagnostic criterion by the Centers for Disease

Control (1985b). Changes are limited to the squamous epithelium, which shows acanthosis, parakeratosis and either surface corrugations or slender, bent, hair-like projections, although occasionally they are flat. Beneath this there are numerous koilocytotic cells with clear cytoplasm and large nuclei which sometimes include inclusion bodies. Inflammation is minimal and non-specific.

The finding of intranuclear inclusions suggests a cytopathic effect of virus infection (Fowler et al, 1989). Whilst there is no evidence of HIV or its products, various techniques have, however, shown human papilloma virus (HPV) and Epstein–Barr virus (EBV) antigens (Greenspan et al, 1985; Morgan et al, 1989), as well as EBV virus particles, although evidence for HPV particles is conflicting (Ficarra et al, 1988; Kanas et al, 1988). Virions of Herpes simplex have been seen ultrastructurally (Belton and Eversole, 1986). The effect of HIV on the cell is probably to pave the way for these other influences. Secondary *Candida* infection is common (Greenspan et al, 1984), and needs to be recognized by microscopy or antifungal therapy.

Herpes Simplex Infection

Herpes simplex virus (HSV) causes large, shallow, painful ulcers in up to 10 per cent of patients with AIDS (Barr and Torosian, 1986; Silverman et al, 1986; Phelan et al, 1987). The normal self-limiting nature of the lesions is radically altered in this syndrome, in which there is intractable persistence, often with additional involvement of the genital and perianal areas (Raufman, 1988).

The infection presents as multiple, small fluid-filled vesicles with central umbilication and a pale yellow, slender rim. On rupture, the clean-cut linear ulcers have raised granular margins and a haemorrhagic base; coalescence may occur to form densely inflamed erosions. Specific microscopic features are ballooned, multinucleated giant cells with moulded nuclei, many having intranuclear eosinophilic inclusions with or without surrounding halo. The youngest such bodies may be basophilic and fill the nucleus; at all stages confirmation of diagnosis requires surface markers.

Other Infections

Other infections include cytomegalovirus, fungi, such as *Histoplasma*, and *Mycobacterium avium–*

intracellulare (MAI) (Volpe et al, 1985); in AIDS these are always part of a systemic process and their identification is almost certainly diagnostic. The lesions are usually non-granulomatous, inflammatory nodules on the lips and in the mouth. Often they ulcerate and diagnosis depends upon recognition of the underlying organism(s). Rarely other organisms, including Enterobacteria, occur (Greenspan et al, 1986b).

Other unusual changes of uncertain origin include non-healing, intensely painful ulcers, probably a form of aphthous ulceration restricted to AIDS. They show no evidence of a viral origin. Acute necrotizing gingivitis and an aggressive periodontal disease occur also (Malebranche et al, 1983; Greenspan et al, 1986b; Winkler et al, 1986, 1987).

Neoplasms

Neoplasms may arise on the lips and in the mouth.

Kaposi's sarcoma occurs as single or multiple patches or nodules, purple or red, and of very varied size; these are localized predominantly to the hard palate and the mucosa adjacent to the upper molar teeth and may be part of a wide dissemination, but can be isolated (Lumerman et al, 1988). The microscopic features are described elsewhere (see the section on skin above).

Malignant lymphoma (non-Hodgkin's type) of the oral cavity can occur anywhere (Silverman et al, 1986), with the same structures as elsewhere; however, an additional site is localization to the salivary gland lymph nodes (Ziegler et al, 1984a; Ioachim et al, 1988), which can also be the site of Kaposi's sarcoma (Puterman and Goldstein, 1983), and often the benign lymphadenopathies (PGL) (Ryan et al, 1985; Ioachim et al, 1988). The involvement of the salivary gland lymph nodes in this infection highlights the role of the oral portal of entry in the aetiology of HIV-related disease and of saliva as a source of virus spread.

With the exception of squamous cell carcinoma there is no change in the incidence of other neoplasms compared with their frequency in previously healthy and immunocompetent patients. Although oral squamous cell carcinoma is the most frequent of this group recorded in AIDS (Conant et al, 1982; Silverman et al, 1986; Kaplan et al, 1987b; Overly and Jakubek, 1987) there is no evidence that HIV plays a direct role in its cause, although it may do so indirectly (Anonymous, 1985; Bradbeer, 1987; Levine, 1987).

Oesophagus and Stomach

Lesions of the oesophagus and stomach manifest as dysphagia, odynophagia, nausea, vomiting, retrosternal pain and sometimes bleeding. Diagnostic radiology is rarely of help and cytology smear and histology are essential for diagnosis (Connolly et al, 1989a). Association with similar pathology in the oral cavity and even elsewhere is fairly common and important in management.

Candidiasis

Candidiasis is included in the criteria of recognition for the syndrome and is almost always associated with oral involvement and occasional spread to the stomach. The fairly high incidence of oesophageal candidiasis is sometimes particularly so associated, as in Africa and Haiti (Malebranche et al, 1983; Piot et al, 1984), and it is the likely cause of oesophageal symptoms in those with oral infections (Gelb and Miller, 1986; Tavitian et al, 1986; Raufman, 1988; Porro et al, 1989). The rare further spread of candidiasis is attributed to the retention of neutrophil granulocyte formation in HIV infection which restricts movement of the fungus (Gillim et al, 1989). Gastric candidiasis often overlies other pathology such as lymphoma (Porro et al, 1989).

The distal segment of the oesophagus is the most frequently affected and involvement is most often circumferential, although there may be a single small ulcer only (Bier et al, 1985). Occasionally the whole mucosa is involved with a pseudomembrane overlying haemorrhagic plaques and nodules that are sometimes ulcerated. Microscopy shows the same findings as in the oral cavity; pseudohyphae may nearly penetrate the wall. The occasional secondary infection must not be forgotten.

Herpes Simplex and Cytomegalovirus Infections

Herpes simplex and cytomegalovirus (CMV) present with the same clinical picture as *Candida*, whilst the naked eye appearances are as in the perioral region; Herpes is rarely if ever seen beyond the cardia, although CMV in the gastric mucosa does occasionally occur. Endoscopically

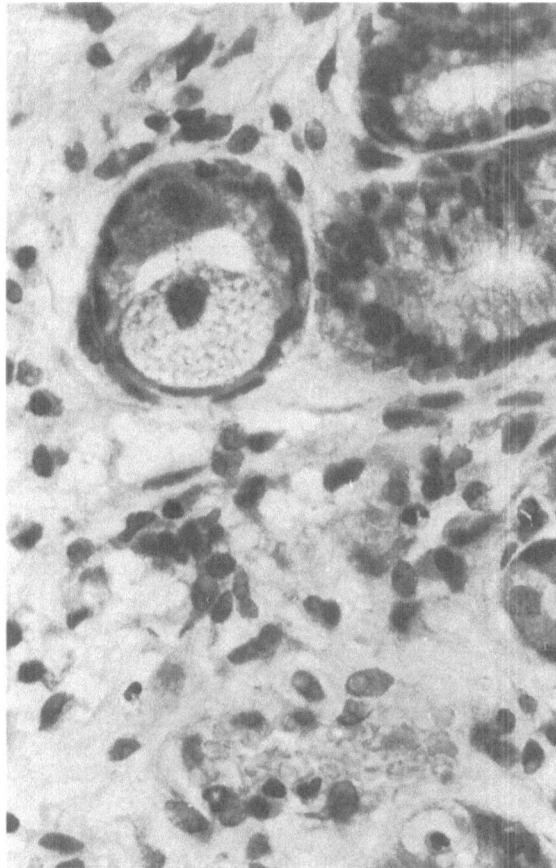

Figure 5.22. Gastric mucosal glands from close to the cardia with inclusion bodies of cytomegalovirus; one is intranuclear as yet without a halo, another shows many virions in the swollen cytoplasm. No significant inflammation.

the fluid-filled vesicles or clean-cut linear ulcers of Herpes may be distinguished from the often longer and deeper ulcers of CMV (Villar et al, 1984), which sometimes simulate oesophageal carcinoma. In the stomach there is punctate ulceration of a patchy non-specific gastritis, but the occasional swelling may mimic Kaposi's sarcoma or lymphoma (Elta et al, 1986).

With herpetic ulcers, diagnosis relies on the typical multinucleate giant cells and the intranuclear inclusion bodies. In CMV infection variations in the degree of inflammation and necrosis are common, both between different patients and between nearby ulcers in the same patient. All types of inflammatory cell may be present but not granulomas, and with severe changes perforation may occur. The diagnostic feature is the finding of enlarged glandular epithelial, histiocytic or endothelial cells with an intranuclear inclusion highlighted by a halo and/

or a part eosinophilic intracytoplasmic body (Figure 5.22). These changes are rarely seen in the squamous epithelium. At necropsy similar changes, often severe, are to be seen together in oesophagus and stomach as well as elsewhere.

Other Infections

Other infections occurring occasionally, but requiring exclusion until the true explanation for the symptoms is found, include cryptosporidiosis and *Mycobacterium avium–intracellulare* (MAI), usually in association with similar involvement of the intestinal tract.

Neoplasms

Neoplasms at both sites are represented by Kaposi's sarcoma and the less frequent malignant lymphoma, both of which are usually part of multifocal presentation, and prognosis is poor. Kaposi's sarcoma occurs as single or multiple, variously sized plaques or nodules, involving up to all layers of both organs with intramural bleeding, but only rarely ulceration, the commonest site for that being the cardia (Figure 5.23). Coexistence with other pathologies occasionally happens. Malignant lymphoma (NHL) occurs at the cardia and in the stomach arising from the intrinsic lymphoid tissue, but not elsewhere unless by spread from local lymph nodes, which may cause problems by distortion alone.

A few patients presenting with fever, malaise, a skin rash and dysphagia may show a few, discrete, superficial ulcers or, occasionally, more widespread aphthous ulcers which usually accompany similar lesions in the mouth. No organisms or inclusion bodies are found. Sometimes the mucosa is normal. However, electron microscopy has shown HIV-like viral particles and on that basis an HIV oesophagitis has been postulated (Rabeneck et al, 1986). *Campylobacter* (*Helicobacter*) *pylori* is frequent with oesophagitis, gastritis and peptic ulceration, but incidence, distribution and number are not changed in AIDS.

Intestinal Tract

The intestinal tract is both a digestive and an immunological organ. Potential pathogens and foreign antigens abound within the intestine, but, in the immunocompetent, the body is protected, largely by the combined activities of the lympho-

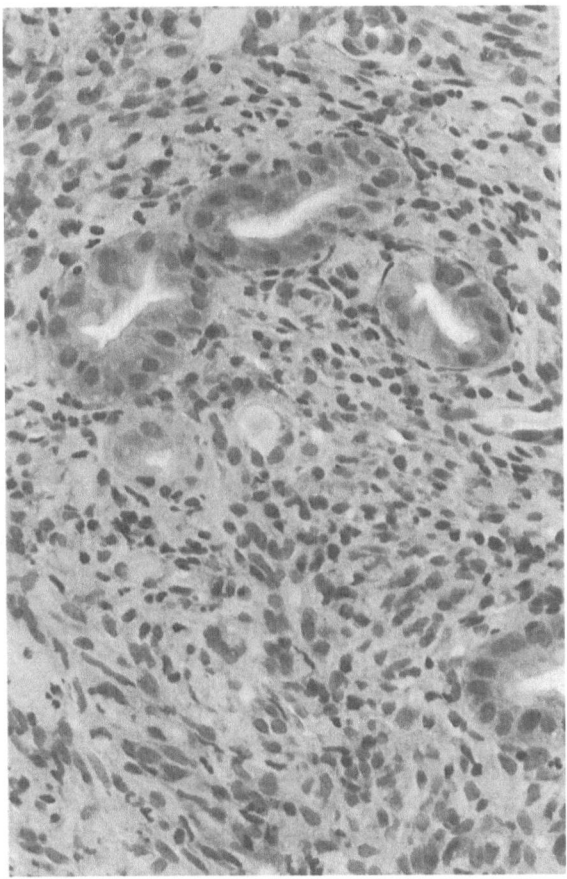

Figure 5.23. Kaposi's sarcoma distorting the lamina propria at the cardia of the stomach.

cytes amidst the epithelium and lamina propria and by the components of both the lymphoid follicles and Peyer's patches. The initial exposure of the intestine to new antigens involves the M-cells of the epithelium; these transfer both soluble and insoluble antigens to the macrophages found within the lymphoid aggregates of the lamina propria. After passage through these macrophages, antigens are made available to the CD4 T-cells which proliferate and induce plasma cells to produce IgA, the main antibody found in the mucosa. Other T-cells exercising their influential role in this immunological system cause further proliferation and differentiation of B-cells within the various lymphoid nodules, and of plasma cells during their migration to the mesenteric lymph nodes; similar changes involve any antigen-activated T-cells in the immunocompetent.

Following HIV infection the CD4/CD8 ratio diminishes in both epithelium and lamina propria of the intestine as happens in the blood and lymph nodes (Rodgers et al, 1986; Weber and Dobbins, 1986; Ellakany et al, 1987). Furthermore the IgA-producing plasma cells in the mucosa of the small intestine of patients with AIDS are diminished despite a normal or even increased level of IgA in their serum (Kotler and Scholes, 1984). This cell depletion parallels the deficiency of secretory IgA in some non-HIV patients who show an increased incidence of intestinal infections. It is believed that this immunoglobulin functions as a blocking antibody that prevents a variety of antigenic agents including bacteria, parasites and toxins from attacking the intestinal tract (Hodges and Wright, 1982).

Since the surviving (CD8) lymphocytes are of the natural killer type (Elson, 1985), they play a significant role in protecting the epithelial cells against virus infections and by virtue of their cytotoxic properties can trigger immunological injury or apoptosis, sometimes seen in the crypts of intestinal and rectal mucosa (Kotler et al, 1986) (Figure 5.24). Their cytotoxic role is probably accentuated by the reversal of the CD4/CD8 ratio (Weber and Dobbins, 1986), apoptosis becoming significant with evidence suggesting that cells so affected are infected with HIV (Nelson et al, 1988).

In AIDS the CD8 cells and their functions are undiminished so that the protective mechanism against virus infections is intact and this may explain the infrequency of viral (mostly CMV and Herpes simplex) diarrhoeas, although it is quite possible that many enteroviruses are not recognized. The limitations of the epithelium surveillance may be reflected in the fact that whilst neoplasms of the epithelium are not increased in incidence with AIDS, those arising in the submucosa, mainly Kaposi's sarcoma and malignant lymphoma, are common and of undoubtedly increased frequency.

Although intestinal disease in AIDS can usually be related to identifiable opportunistic agents, in some patients no appropriate pathogens are recognized and in yet others eradication of identified pathogens does not lead to full clinical recovery (Gillin et al, 1985). Various studies in HIV-positive and AIDS patients have shown functional and structural changes (Kotler et al, 1984; Ullrich et al, 1989) in the absence of identifiable pathogens. The combined findings indicate that by itself HIV can cause limited mucosal atrophy with impaired maturation of enterocytes sufficient to cause malabsorption; the malnutrition arising directly or indirectly will itself cause a deterioration in the patient's clinical course, especially as a result of deficiencies in

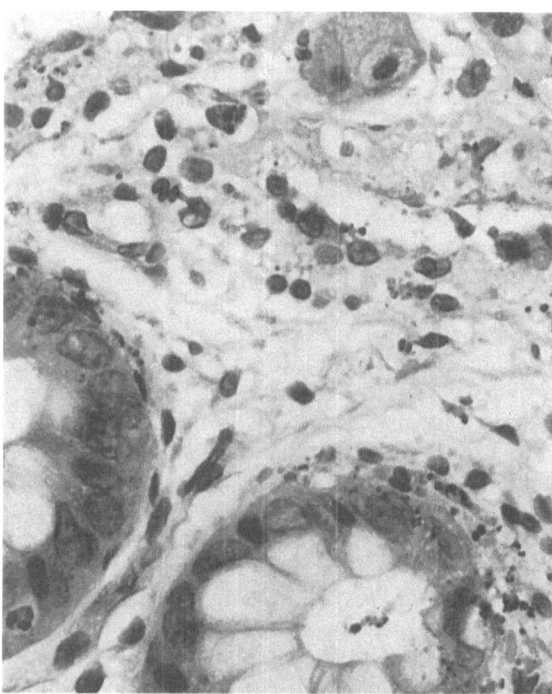

Figure 5.24. Colonic crypts with apoptosis, some epithelial cells undergoing disintegration with apoptotic bodies amidst the cell debris, but only minimal inflammation. Cytomegalovirus inclusion bodies are seen in the stroma.

protein and base metal intake, for elements such as zinc are essential for maintaining normal cell-mediated immunity (Cunningham-Rundles et al, 1983; Gillin et al, 1985). Similarly total T-lymphocyte and CD4 counts as well as immuno-globulin production are adversely affected (Chandra, 1983).

Cytomegalovirus Infection

Cytomegalovirus potentially involves not only the colon but also the liver, gall bladder, bile ducts and pancreas, with the oral cavity and anus the least common sites. Many patients with AIDS or in at-risk groups first present with symptoms related to this system and diarrhoea is frequent, with CMV the most common of the infections (Laughon et al, 1988). These infections are thought to indicate reactivation of latent disease, and their ever-widening spread (Gold, 1985). Serology studies show many immunocompetent people to have had prior exposure to CMV and this is particularly so amongst patients with AIDS. The estimated incidence of CMV infection

in any of the systems in patients with AIDS is 10 per cent (Centers for Disease Control, 1986), but there are variations in some series, especially necropsy studies, where disseminated CMV infection may reach close to 100 per cent (Mobley et al, 1985).

The colon, and especially the ascending colon, often shows the largest number of inclusion bodies in the entire alimentary system (Hinnant et al, 1986; Rotterdam, 1987). However, involvement is often patchy. The appearances of the infection range from a pseudomembranous colitis with near-white plaques and nodules to erosions, usually multiple, or shallow ulcers resembling inflammatory bowel disease, that are seen in up to 50 per cent of patients (Gertler et al, 1983; Hinnant et al, 1986; Klatt and Shibata, 1988); there may be haemorrhagic pseudotumour-like lesions resembling Kaposi's sarcoma (Meiselman et al, 1985). The latter may cause bleeding and perforation can occur, particularly in the appendix. CMV can occur without obvious changes (Klatt and Shibata, 1988; Dieterich and Rahmin, 1991), but be associated with malabsorption or protein-losing enteropathy.

Microscopically, changes range from minimal, non-specific chronic inflammation in the lamina propria to severe, highly active, widespread inflammation accompanied by necrosis, haemorrhage and rupture. The diagnostic intranuclear and intracytoplasmic inclusions vary both in number and in size, the former usually being larger (Figure 5.24). They can be seen in the epithelial cells especially in the stomach and upper small intestine, but endothelial cells of small vessels, histiocytes and possibly pericytes may also be involved, particularly in the colon, whereas in the connective tissue typical owl's eye cells (Cowdry's type A) are few (Figures 5.25 and 5.26). Additionally irregular granular inclusions within the cytoplasm may also be present. Immunocytochemistry shows up involvement of apparently normal cells close to those with typical inclusions (Francis et al, 1989) (Figure 5.27). Typical inclusion bodies are seen in the lamina muscularis, but more often infected cells are elongated with poorly defined, deeply basophilic smudged nuclei without a halo.

The involvement of endothelial cells is not associated with an active vasculitis and only rarely with a thrombus. It is likely that CMV infection of muscle cells is of significance in causing necrosis and perforation (Fernandes et al, 1986; Tatum et al, 1989).

There is a link between the severity of the disease and the total number of infected cells, but

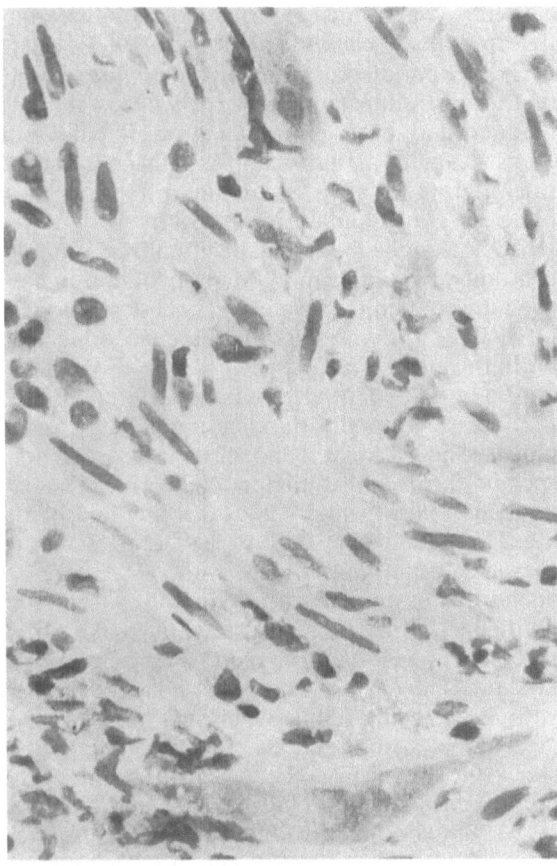

Figure 5.25. Connective tissue of colonic wall with cytomegalovirus inclusion bodies; several are in form of lengthy 'smudged' nuclei within tapering cytoplasm of cells, probably muscular.

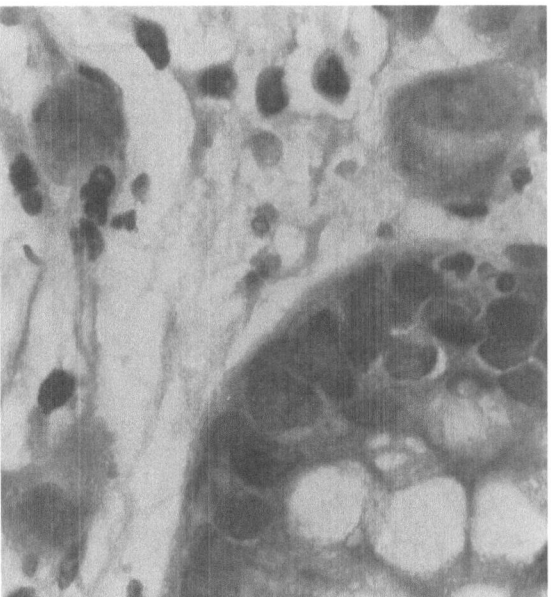

Figure 5.26. Cytomegalovirus inclusion bodies in a histiocyte of the lamina propria and two capillary endothelial cells.

exceptions do occur; irrespective of these factors CMV infection of the intestine is serious in that further dissemination soon follows, readily seen at necropsy.

Cryptosporidiosis

Cryptosporidium is a coccidian protozoon prevalent in bovine animals. It can be transmitted to man (Schultz, 1983), usually through contaminated food or water, though person to person transmission is likely especially in day-care centres (Wolfson et al, 1985; Janoff and Reller, 1987), with hospital cross-infection and between homosexuals, where the incidence of infection is increased (Webber and Philip, 1983; Soave et al, 1984). In the immunocompetent, infection has a low incidence (2 per cent) and manifests as 'traveller's diarrhoea' only. This is a self-limiting,

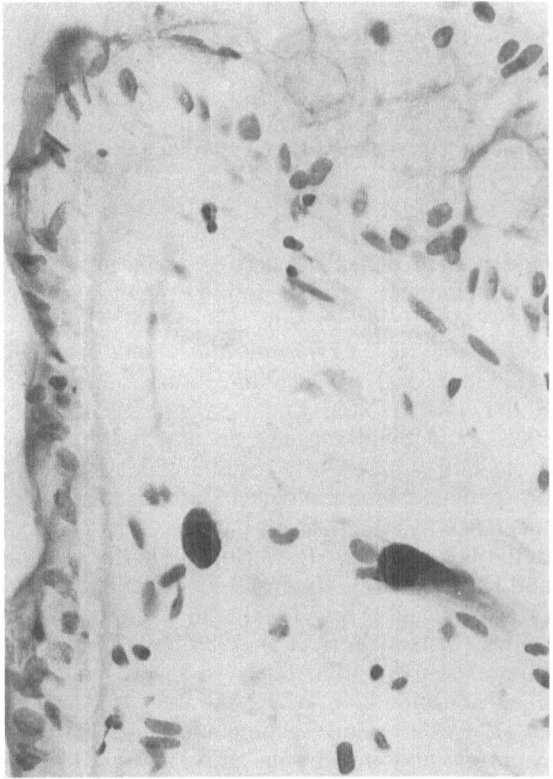

Figure 5.27. Immunocytochemical staining of colonic mucosa using a monoclonal antibody and showing several inclusion bodies not otherwise readily detectable.

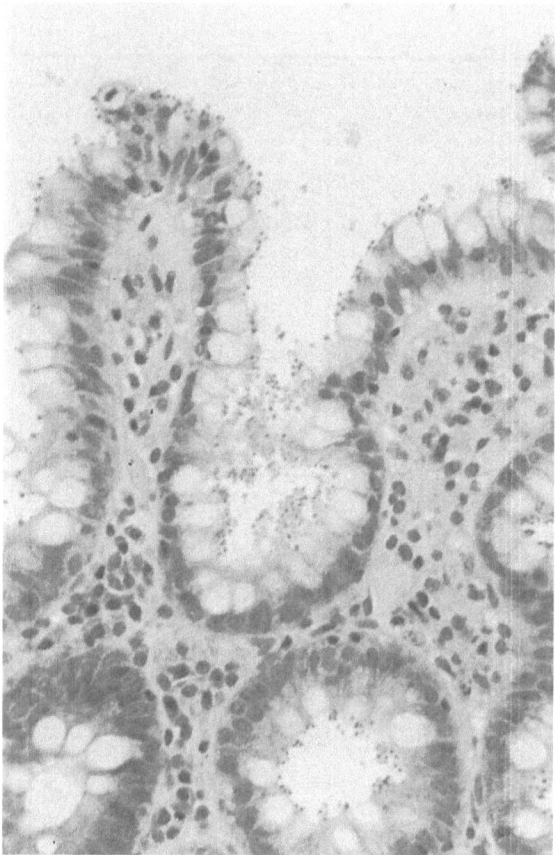

Figure 5.28. Cryptosporidial infestation of minimally inflamed colon, many organisms at the brush border of both surface and crypt epithelium, with a few apparently free in the lumen.

mild watery diarrhoea, lasting no more than 2 weeks, and no toxin is found. Infectious diarrhoeas may be due to virus or idiopathic (Dupont, 1985); however, *Cryptosporidium* may cause up to 13 per cent of such infections (Janoff and Reller, 1987; Holten-Andersen et al, 1984; Cook, 1988), and children may be affected too, the protozoon subsequently remaining dormant until the patient's immunological status is disturbed. Incidence in HIV-positive patients is 11 per cent in the UK and USA (Dworkin et al, 1985; Navin and Hardy, 1987), about 40–45 per cent in Haiti (Malebranche et al, 1983) and even higher in Africa (Colebunders et al, 1988). With immunosuppression the diarrhoea is severe, choleralike, non-bloody and very watery (Casemore et al, 1985a), and accompanied by anorexia, vomiting, abdominal pain and cramps with low-grade fever (Berkowitz, 1985; Casemore et al, 1985a). The duration is protracted with a high rate of fatality from deteriorating malnutrition

caused by water and electrolyte loss and no effective treatment (Navin and Juranek, 1984).

Cryptosporidiosis can involve the whole length of the gastrointestinal tract, as well as the gall bladder and both extra- and intrahepatic bile ducts (Kahn et al, 1987), but it is the small intestine and especially the jejunum that is most affected (Navin and Juranek, 1984).

Oocysts in the faeces can be visualized either by a modified (cold) Ziehl–Neelsen or auramine–rhodamine stain (Garcia et al, 1983; Casemore et al, 1985b), or by immunofluorescence, which can be both sensitive and specific with a monoclonal antibody (Garcia et al, 1987; Loose et al, 1989).

Biopsies from the duodenojejunal and rectal mucosa show a slight to moderate increase in non-specific chronic inflammatory cells; with the small intestine there is mild villous atrophy and blunting. The diagnostic feature is the presence of the cryptosporidia, which with H&E appear anchored to the brush border and outside the epithelial cell as pale blue, slightly ovoid structures up to 4–5 µm (Figure 5.28); they are Giemsa- and Gram-positive, but not invariably so with PAS, whilst unlike in faeces they are not acid-fast in tissue sections. Electron microscopy shows the organisms to be extracytoplasmic at the apical poles of the cells (Marcial and Madara, 1986), but contained within a membrane formed from the attenuation or fusion of host membranes (Lefkowitch et al, 1984; Dobbins and Weinstein, 1985). Cryptosporidial infection of extraintestinal tissues further complicates eradication and contributes to the chronicity of the infection. The gall bladder and bile duct system are infected in up to 10 per cent of patients (Marcial and Madara, 1986; Margulis et al, 1986; Soave and Johnson, 1988) and may act as a reservoir perpetuating the infection (Soave and Johnson, 1988). Rupture of the gall bladder may ensue, especially when associated with a CMV cholecystitis (Blumberg et al, 1984). Pulmonary cryptosporidiosis probably follows aspiration of gastroenteric contents in patients with AIDS (Brady et al, 1984; Ma et al, 1984b; Hojlyng and Jensen, 1988).

Giardiasis

Giardia lamblia, a flagellate protozoon, can be found in association with cryptosporidiosis, although giardiasis is not a criterion for the Centers for Disease Control's definition of AIDS. Both organisms may contaminate water (Issac-Renton et al, 1987) and be jointly spread, whilst *Giardia* has a high frequency in all homosexuals

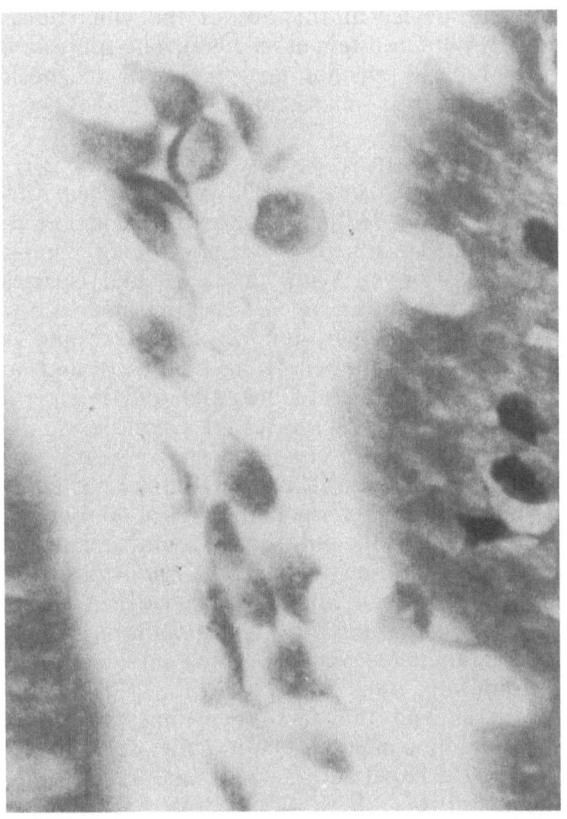

Figure 5.29. *Giardia lamblia* in the lumen of the duodenum with two attached to mucosa by their ventral adhesive disc.

(Connolly et al, 1989b). Whilst the protozoon may be found in the intestinal tract of HIV sero-positives with diarrhoea, there is no evidence that there is any change in incidence when compared with the immunocompetent; furthermore, giardiasis does not change its pattern of behaviour in any way with AIDS. This duodenojejunal infection is often asymptomatic and at most elicits a mild chronic inflammatory reaction, although some villous atrophy may occur. The pear-shaped trophozoites are seen close to and between the villi, some being attached by their broad end to enterocytes (Figure 5.29).

Mycobacterium avium–intracellulare Infection

Infection of the intestine by *Myobacterium avium–intracellulare* (MAI) usually reflects disseminated disease (Hawkins et al, 1986; Gold and Armstrong, 1989). The precise responsibility for

the symptoms is difficult to place as MAI often occurs late in the individual patient's history, coexisting with fungal, protozoal or viral infections. MAI may be found in faeces in the absence of symptoms and such patients may survive for long periods with no evidence of organ abnormality. Once the latter occurs, however, patients with AIDS decline rapidly (Hawkins et al, 1986), but MAI is rarely, if ever, the cause of death; the enormous numbers of bacilli seen with acid–alcohol fast and PAS stains (cf. *M. tuberculosis*) indicate the organisms to be of very low or relatively low individual pathogenicity.

The most significant changes within the intestinal tract involve the mucosa, particularly that of the small intestine; colonic and gastric involvement rarely occur. Affected segments of mucosa are pallid, faintly yellow and granular with the villi both widened and stunted, creating a picture similar to that of subtotal atrophy, interspersed with normal and intermediate areas. Affected villi are intensely infiltrated with swollen, foamy histiocytes, the fairly granular cytoplasm being filled by organisms (Roth et al, 1985; Gray and Rabeneck, 1989) (Figure 5.30). Granulocytes, lymphocytes and plasma cells are virtually non-existent and giant cells, granulomas and necrosis are rare (Klatt et al, 1987); their presence makes the exclusion of coexistent *M. tuberculosis* prudent.

The overall clinical (Gillin et al, 1983; Hawkins et al, 1986; Gray and Rabeneck, 1989) and pathological changes closely resemble Whipple's disease but, in the latter, the bacillus, whilst PAS positive, is acid–alcohol fast negative although Gram-positive. *Corynebacterium equi* has been associated with a similar Whipple-like presentation and pathology (Wang et al, 1986) and may need exclusion in patients with atypical features.

MAI in the late stages often involves lymph nodes, which may enlarge considerably (see the section on lymphoreticular tissue above), creating obstruction to the viscera, although the author has not seen MAI cause perforation or peritonitis.

Other Protozoal Infections

Microsporidia are a ubiquitous group of intracellular parasitic protozoa of which over 80 are recognized (Nichols, 1986). The organism has until recently not been described in man, but has now been reported in the immunocompromised (Dobbins and Weinstein, 1985; Modigliani et al, 1985). The incidence in patients with chronic

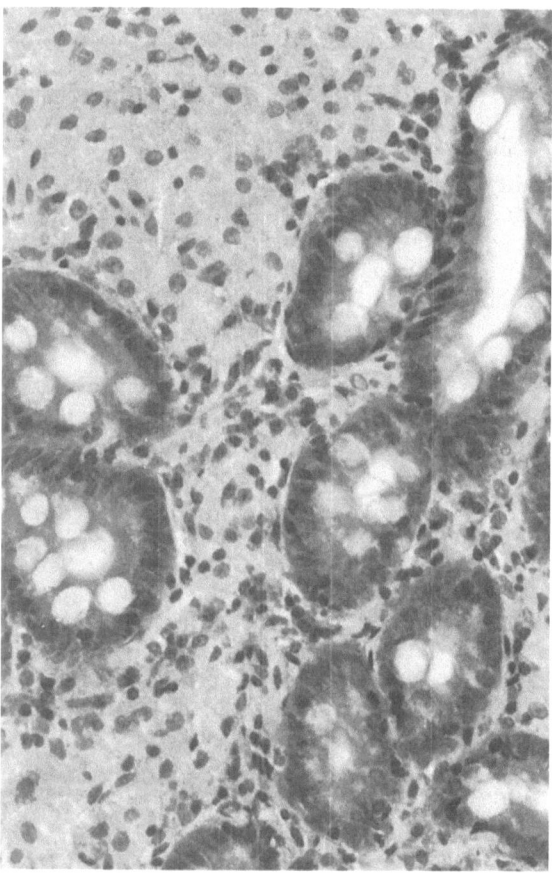

Figure 5.30. *Mycobacterium avium–intracellulare* infection of the small intestine with separation of glands by faintly granular histiocytes, each teeming with acid–alcohol fast bacilli.

diarrhoea and malabsorption appears to be increasing as detection of the tiny spores (less than 2 μm) in duodenal and jejunal mucosa improves (Sun, 1988; Orenstein et al, 1990; Efftinck Schattenkerk et al, 1991).

Another common organism in these patients is *Enterocytozoon bieneusi*, first identified by electron microscopy in a 29-year-old Haitian man with multiple infections and undoubted AIDS (Desportes et al, 1985). Their coiled filament within the spore uncoils after ingestion and the infective sporoplasm enters the enterocytes of the jejunum, preferentially of the duodenum (Orenstein et al, 1990), and develops in direct contact with the cytoplasm of the epithelial cell. No significant tissue response occurs. The role of the organisms in the pathogenesis of diarrhoea and its sequelae remains unclear. The organisms can be identified in thin, resin-embedded sections using H&E, Giemsa and Brown–Brenn stains as having tiny refractile bodies above the nucleus in

the enterocytes at the tips of the villi (Lucas et al, 1989; Orenstein et al, 1990). The spores are variably acid–alcohol fast. Effective diagnosis needs electron microscopy (Cali and Owen, 1988).

Another of these parasites, *E. cuniculi*, is reported in patients with AIDS as causing hepatitis (Terada et al, 1987) and peritonitis (Zender et al, 1989), the histological changes being similar to those above. With this particular species identification again needs electron necroscopy (Anonymous, 1990), but serological testing is available and antibodies have been found in homosexual men (Bergquist et al, 1984).

Isospora belli, a species of Coccidia closely related to *Cryptosporidium*, is known world-wide as an intestinal parasite of many animals including domestic pets and farm animals (Brandborg et al, 1970). *I. belli* is the only species found in man and its features are very similar to *Cryptosporidium*, being larger than Microsporidia and, in tissue, similar to *Toxoplasma*. This organism, spread by contaminated water or food and causing a self-limiting, mild watery diarrhoea, was found in faeces and the small intestinal mucosa in the immunocompetent before the epidemic of AIDS (De Hovitz et al, 1986; Sun, 1988). Since the onset of AIDS, homosexual transmission has been recognized (Ma et al, 1984a), and the prevalence of *I. belli* as a human pathogen has increased (De Hovitz et al, 1986; Navin and Hardy, 1987; Sun, 1988). Overall the clinical picture closely follows that of cryptosporidiosis, but there is a vital difference in that isospora infection is linked with a tissue and peripheral blood eosinophilia and, if recognized, can be treated (Pape et al, 1989), although recurrences do occur. Diagnosis is usually based on examination of faeces (see *Cryptosporidium* above), repeated concentration techniques being advisable. The oocysts infect the enterocytes of the small intestine and PAS, Giemsa or H&E with Alcian blue may show parasites in various stages of development within the cell cytoplasm, making some cells appear degenerate. Histological changes include varied villous atrophy with crypt hyperplasia, and up to moderately severe, non-specific chronic inflammation with eosinophilia, but no giant cells or granulomas. The degree of inflammatory response and the intracellular position of the oocysts distinguishes *I. belli* (up to 20 μm) from *Cryptosporidium* (up to 5 μm) in tissue. Although largely restricted to the small intestine, intestinal and mesenteric lymph nodes may be involved, with small granulomas (Restrepo et al, 1987).

Spirochaetosis

Spirochaetosis is an exacerbation of the presence of spirochaetes as commensals in the anorectal region of some normal people (Nielsen et al, 1983), with an increased incidence (up to 30 per cent) in immunocompetent homosexuals (Surawicz et al, 1987). It is usually restricted to the intestinal mucosa distal to the ileocaecal valve and can include the appendix, but in many patients, with and without AIDS, this treatable finding appears to be of minor significance (Lee et al, 1971; Ruane et al, 1989), although mucoid diarrhoea and bleeding may occur (Harland and Lee, 1967; Douglas and Crucioli, 1981).

No naked-eye or endoscopic changes are recognized and microscopically there is no significant inflammation or specific cell change, whilst the clinical and pathological significance of spirochaetosis is uncertain. Diagnosis depends on the H&E finding of a continuous or interrupted hazy, thick basophilic band along the brush border of the colonic and rectal surface epithelium; the vertically aligned spirochaetes are confirmed by PAS or Warthin–Starry stains.

Other Infections

Other infections include those organisms that periodically occur in hitherto normal immunocompetent persons but are increased in incidence in seronegative homosexuals and much more frequent in the HIV seropositive (Cone et al, 1986; Rolston et al, 1986; Shanson, 1989). This group includes *Salmonella* spp. and *Shigella* spp., *Entamoeba* (Sperber and Schleupner, 1987; Simor et al, 1989; Strand, 1990), and the parasites *Trichuris* and *Strongyloides stercoralis*.

The nematode *Strongyloides stercoralis* is frequent in warm wet climates, where it affects up to 85 per cent of the population (Meyers et al, 1976), but is rare elsewhere and especially in Northern Europe. Human carriers may remain infected indefinitely, strongyloidiasis developing many years after original contact (Hakim and Genta, 1986). In AIDS patients there may be direct entry as by homosexual contact, or reactivation of clinically dormant disease. So far and for whatever reasons hyperinfection with this nematode in AIDS is reported rarely (Harcourt-Webster et al, 1991) and, although included originally, strongyloidiasis was deleted by the Centers for Disease Control from the opportunistic infections associated with AIDS (Centers for Disease Control, 1987).

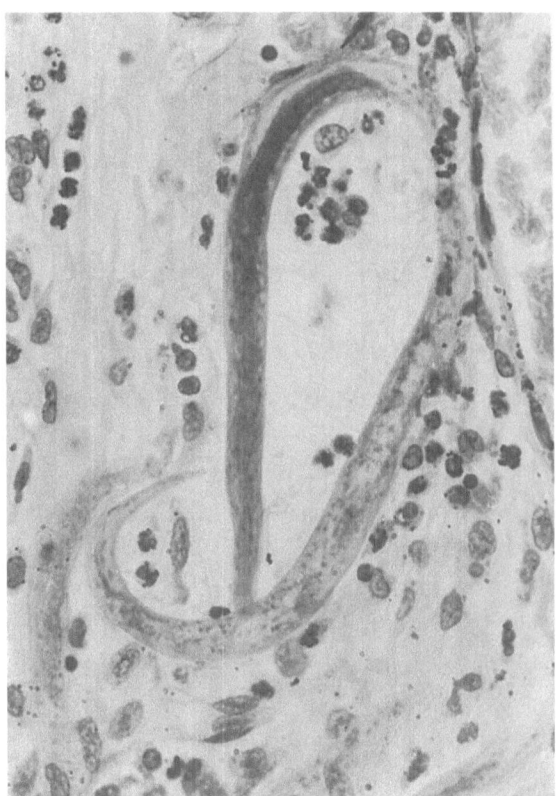

Figure 5.31. Filariform larva of *Strongyloides stercoralis* with its long, thin body, tapering tail and long oesophagus within a lymphatic alongside a vascular sinusoid in the jejunal wall.

The pathophysiology of the infection is well documented (Meyers et al, 1976; Grove, 1989); extra-intestinal strongyloidiasis is almost certainly fatal unless recognized early and treated effectively. Diagnosis depends upon identification of the larvae in faeces, although lighter infection or intermittent excretion may necessitate search of body fluids and aspirates as well as upper intestinal mucosa (Figure 5.31). ELISA and indirect immunofluorescence for the detection of IgG antibodies against the parasitic antigen may be the best future option (Hakim and Genta, 1986). When the larvae die a foreign-body-type giant cell granuloma is likely to develop and, with or without larval debris, remains as a tombstone (Figure 5.32).

On occasion the clinical picture associated with the 'other infections' is labelled the 'gay bowel syndrome'. However, sometimes the chronic diarrhoea, malabsorption and other symptoms persist but no pathogens or other cause are identified and with endoscopy of the small and large intestine no abnormality or only a

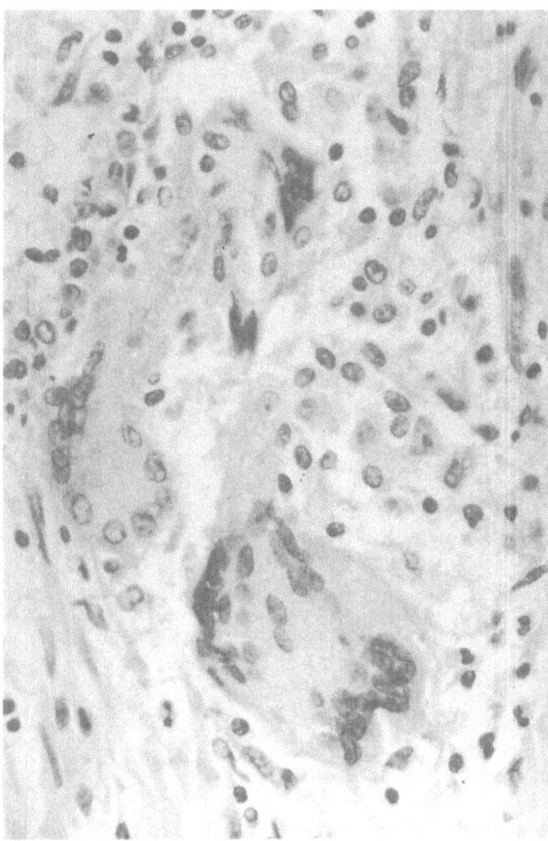

Figure 5.32. A 'tombstone' of *Strongyloides stercoralis* made up of foreign-body-type giant cells around the remains of a larva in the connective tissues of the intestine.

mild flattening of the villous pattern is revealed. With microscopy a varied degree of villous shortening and crypt hyperplasia is seen (Kotler et al, 1984; Dworkin et al, 1985; Gillin et al, 1985), with an accompanying mild, non-specific non-granulomatous chronic inflammation, but insufficiently severe to suggest coeliac disease. In the colon and rectum the inflammatory picture is similar, and may suggest quiescent ulcerative colitis or chronic graft-versus-host disease. Recently adenoviruses have been detected in colonic mucosa, adding further to the complexities of AIDS (Janoff et al, 1991).

Neoplasms

The pattern of gastrointestinal symptoms so far attributed to the various infections may be due to Kaposi's sarcoma, but with the additional effects of tissue haemorrhage (Krown, 1987; Steis and

Longo, 1988). However, the sarcoma may be asymptomatic, its presence only becoming clear at necropsy, where an incidence of nearly 50 per cent may be reached. Involvement of the gastro-intestinal tract alone is rare, but the lesions, usually multiple, can occur anywhere including within the appendicular wall. Structurally the sarcoma is usually centred in the submucosa and is as seen elsewhere with no specific system features.

Malignant lymphoma, almost always non-Hodgkin's lymphoma, is common in the gastro-intestinal tract, which is the most frequent of all the extranodal sites; most lymphomas occur in the small intestine (Niedt and Schinella, 1985). Microscopically all cytological and structural varieties may occur (Ioachim et al, 1985).

Hepatopancreatic System

Hepatic dysfunctions occur frequently, not only in patients with AIDS but sometimes also in those who are still only HIV seropositive.

Liver

Liver pathology is frequent, very few people having a liver of normal appearance (Orenstein et al, 1985; Kahn et al, 1986; Schneiderman et al, 1987). There is a wide range of changes but the effect of immunosuppression again modifies many so that particular patterns are rarely followed. This is well seen with granulomatous hepatitis, which can occur in up to 50 per cent of needle biopsies but is much less frequent in necropsies, where the consequences of HIV infection are maximized. Such granulomas vary in number, but can be numerous and almost always consist of small clusters of loosely arranged histiocytes with few, if any, other inflammatory cells; giant cells and necrosis are both rare. Most granulomas are associated with mycobacteria and fungi, the most frequent being *M. avium–intracellulare* (MAI) (up to 30 per cent of cases) whilst *M. tuberculosis*, *Candida* sp., *Cryptococcus*, *Histoplasma* and *Pneumocystis carinii* are much less frequent, all liver infections being part of disseminated disease (Reichert et al, 1983; Orenstein et al, 1985; Kahn et al, 1986; Schneiderman et al, 1987). Where such infection is severe, fungi in the liver may produce microabscesses.

Very often, in place of true granulomas, microscopy shows small numbers of macrophages mixed with a few other inflammatory cells, sometimes including neutrophil granulocytes; these collections are usually seen in the portal tracts and intralobular zones. All these appearances can be deceptive, especially at necropsy; large numbers of organisms may be identified in histiocytes, less often in Kupffer cells, and sometimes in hepatocytes.

MAI is quite commonly found in association with fatty droplet change, the latter often explained by a high alcohol intake or dietary imbalance. Overall the total incidence of liver involvement by this organism may reach 60 per cent in patients with disseminated infection (Schneiderman et al, 1987), although with some necropsy series there are even higher frequencies (Klatt et al, 1987).

Granulomatous hepatitis cannot always be linked with an identifiable organism and in some instances significant numbers of eosinophil granulocytes are found in the infiltrate, especially in the portal tracts. With such patients clinico-pathological correlation often suggests an adverse drug reaction, sometimes a sulphonamide, although other therapies, potentially hepatotoxic, may be the cause. On other occasions polarized light may show up birefringent crystals, a finding which may be seen in the porta hepatis lymph nodes also, such talc granulomas being long known in intravenous drug abusers.

Disseminated virus infections may involve the liver. Cytomegalovirus (CMV) hepatitis can occur in about 30 per cent of such patients (Reichert et al, 1983; Guarda et al, 1984) and may rarely be associated with granulomas, but virus infections usually are associated with little or no inflammatory response. Diagnosis depends on finding the characteristic CMV inclusion bodies, which may involve bile duct epithelium, endothelial cells, hepatocytes or Kupffer cells. Herpes simplex, occasionally seen in the liver at necropsy (Godwin and Felix, 1987), may be found with coagulative necrosis and a few acute inflammatory cells, but again diagnosis hinges on finding the inclusion bodies.

Most patients who are HIV seropositive have serological evidence of hepatitis B (HB), although they may have cleared the virus (Lebovics et al, 1985; Schneiderman, 1988); many also give a history of previous jaundice or hepatitis (Dworkin et al, 1987). Indeed, HB is almost certainly the commonest infection complicating these patients. Nevertheless, the carrier rate for the B virus in both the HIV-infected and in those with AIDS is similar (Lebovics et al, 1985; Schneiderman et al, 1987), and not greatly different from those who are otherwise healthy.

Histological evidence of virus hepatitis is found in up to 25 per cent of patients with AIDS (Waisman et al, 1987), whilst methods for the detection of HB surface antigen in the cytoplasm and the core antigen in the nuclei will often confirm the suspected and, additionally, sometimes detect unknown infections.

The liver may be involved by Kaposi's sarcoma and less frequently malignant lymphoma, with both as part of disseminated disease.

Nevertheless, a miscellany of changes may be seen. Other non-specific changes include cholestasis, sinusoidal dilatation sometimes with congestion, proliferating Kupffer cells, steatosis, occasional hepatocyte degeneration or necrosis and focal inflammatory cell infiltration. On the other hand, no increase in non-B virus hepatitis is reported. These findings occur in various combinations and may present as a non-specific reactive hepatitis. Furthermore, in patients with lymphadenopathy hepatic changes may reflect obstruction at or adjacent to the porta hepatis.

Gall Bladder and Extrahepatic Bile Ducts

The gall bladder and extrahepatic bile ducts are affected by the same opportunistic infections as occur in the gastrointestinal tract, cholecystitis and cholangitis often arising by their spread, probably retrograde. *Cryptosporidium* is a common finding also in the absence of significant inflammation (see pp. 79–80). Severe, sometimes necrotising, cholecystitis with cholangitis may be attributed to CMV. Occasionally the organism responsible is *Campylobacter fetus* (Costel et al, 1984; Wheeler and Gregg, 1986).

Although gall stones are rarely found with AIDS, sludged cellular debris in the ducts with cholestasis can occur and is attributable to opportunistic infections. These changes may culminate with strictures. Biliary obstruction occasionally follows extrinsic pressure from lymphadenopathy due to infection such as MAI, Kaposi's sarcoma or malignant lymphoma.

Pancreas

Pancreatic changes in AIDS include infection, inflammation and neoplasms, and necropsy studies have indicated about half such patients may be so affected (Brivet et al, 1987; Dowell et al, 1990). Opportunistic infections account for about a third of the latter, CMV being the commonest and almost always part of disseminated disease; usually the inclusion bodies are found in both the acinar epithelium and the islets of Langerhans, but inflammation is almost non-existent. A few such cases have been described in conjunction with acute pancreatitis (Iwasaki et al, 1987; Zazzo et al, 1987), which may fol- ·low disruption of virus-infected acinar cells or ischaemia from similarly injured endothelium.

Another reported cause of pancreatitis is iatrogenic following pentamidine or co-trimoxazole treatment for infection elsewhere (Salmeron et al, 1986; Brivet et al, 1987). However, the cause of the pancreatitis can remain equivocal, although in its absence organisms including mycobacteria and cryptococci have occurred, whilst Kaposi's sarcoma and malignant lymphoma may be seen.

Anorectum

Anorectal tissues are the site of the same infections with the same changes as are found in the intestinal tract and skin, but the possibility of venereal organisms such as *Chlamydia trachomatis*, *Neisseria gonorrhoeae* and *Treponema pallidum* must never be overlooked. In homosexuals the incidence, chronicity and aggressiveness are all increased, with further deterioration up to and including AIDS (Wexner et al, 1986). The tissues are congested and friable with frequent ulceration; abscesses are common and fissures, fistulae and strictures commonly occur. The complexities of pathology dictate a combined microbiological and histopathological approach to diagnosis on all occasions.

Kaposi's sarcoma is far from rare at this site and needs to be distinguished from granulation tissue, traumatic or infective in origin, and haemorrhoids. On the other hand, other neoplasms are common and an AIDS-associated lymphoma is particularly localized in this area (Ioachim et al, 1987). Both men and women who are HIV seropositive have an increased frequency of sometimes exuberant anal and genital warts which may cover a wide area of squamous epithelium. Structurally these variants of condyloma acuminatum show the usual changes. Immunocytochemistry and in situ hybridization show the presence of human papillomavirus (HPV) (Lauwers et al, 1989), which has long been associated in the female with anogenital warts and is now closely linked with squamous cell carcinoma arising in the female genital tract (Editorial, 1985; Howley, 1986; zur Hausen, 1987).

Anal and perianal epithelium behaves similarly to that of the cervix uteri and vagina in that it

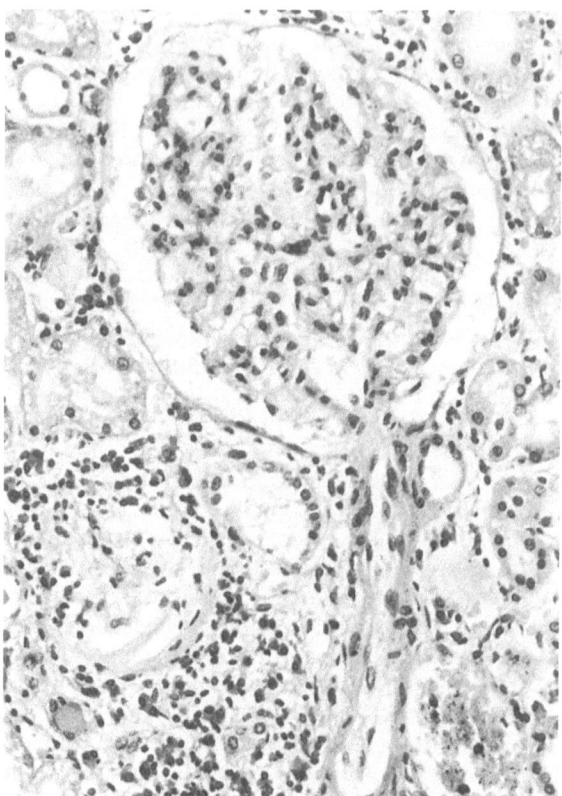

Figure 5.33. Focal early sclerosis of the glomerular tuft with adhesions between the capillaries and the capsule, hyaline change in the arteriole and mild chronic inflammation of the interstitium with a few tubular casts.

may undergo dysplasia. Intraepithelial neoplasia grades I to III has increased incidence in AIDS.

HPV can be detected in anogenital carcinomas (Howley, 1986; zur Hausen, 1987; Gal et al, 1987), which may be squamous or basaloid and unusually aggressive in behaviour. Despite this, recurrence and prognosis are determined by the extent of spread and depth of infiltration rather than the cytological features.

The role of the HIV is not clear-cut; anorectal carcinoma is more frequent in homosexuals than heterosexuals (Daling et al, 1987), and this was shown before the advent of HIV infection. Both men and women who practise anoreceptive intercourse have an increased incidence of these neoplasms (Daling et al, 1987; Wells et al, 1988), and it seems highly likely that these carcinomas are associated with sexual practices. The two viruses may well act synergistically in the development of these lesions; this is particularly likely with carcinoma, where the development may be facilitated by the increasing immunosuppression arising in the HIV seropositive.

Genito-urinary System

Study of the genito-urinary system, especially at necropsy, shows the components to be involved infrequently by any of the infections or neoplasms associated with HIV-seropositivity or AIDS (Welch et al, 1984; Niedt and Schinella, 1985). In the experience of the author, whilst occasional opportunistic infections and neoplasms occur, involvement of the genito-urinary system is unusual, even with widespread and severe disease elsewhere. Carcinoma occurs occasionally; nevertheless, as with many other infrequent findings amongst which are haemolytic uraemic syndrome, thrombotic thrombocytopenic purpura and fungal pyelonephritis, this is almost certainly a coincident occurrence and there is no evidence to suggest a linkage with HIV infection. Certainly any suggestions that HIV-seropositivity directly increases the incidence, age for age, of any of the usual genito-urinary infections or neoplasms is not supported by currently available evidence.

Renal impairment of varied degree is reported in up to 33 per cent of patients with AIDS (Gardenswartz et al, 1984; Rao et al, 1987), but a specific aetiology usually presenting as acute tubular necrosis is found in only about one third of those cases. This is usually secondary to drug toxicity, allergic reactions or ischaemia, all of which are fairly common in these patients; its frequency and end results are similar to those seen in patients with equally severe, but different, illnesses (Rao et al, 1987).

The remaining two thirds of these patients present with proteinuria, with or without azotaemia, and move on to end-stage kidney disease known as AIDS nephropathy; occasionally renal dysfunction is the presenting symptom for HIV infection (Alpers et al, 1988; Cohen and Nast, 1988), but in almost all these patients the pathology, wherever found, is that of an aggressive form of focal and segmental glomerulosclerosis (Pardo et al, 1984; Rao et al, 1984, 1987; Bourgoigne et al, 1988; Cohen and Nast, 1988; Rao and Friedman, 1989; Soni et al, 1989) (Figure 5.33). Infrequently glomerulonephritis of membranous or membranoproliferative type is encountered (Soni et al, 1989).

With HIV-associated glomerulopathy the early changes are enlarged visceral epithelial cells, both in size and in number, within the capillary loops with intracytoplasmic coarse vacuoles, increased mesangial matrix and the underlying capillaries normal or collapsed (Cohen and Nast, 1988). In the advanced stage of segmental and global

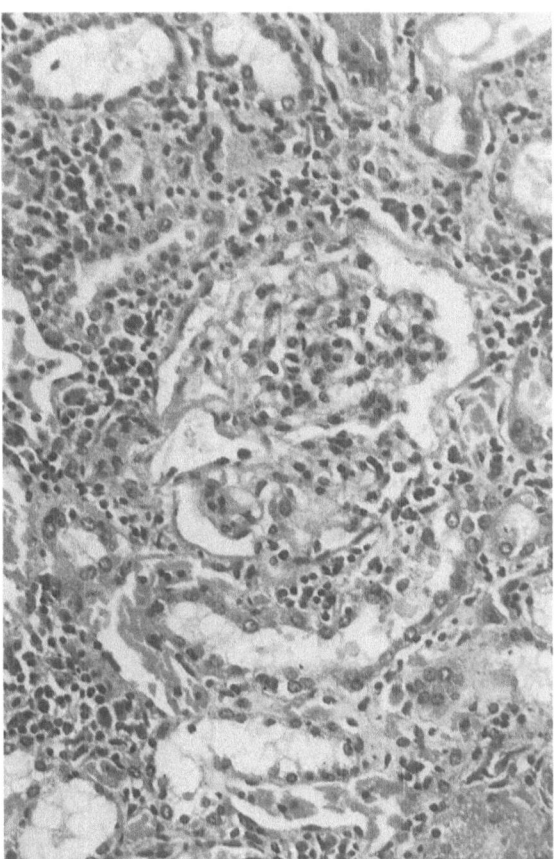

Figure 5.34. A shrinking glomerular tuft with capsular adhesions, varied dilatation of the tubules, some atrophic, casts and an unusually heavy interstitial inflammatory infiltrate. Multinucleate cells are present in the lining of several tubules.

sclerosis there are many atrophic glomeruli and a surrounding cuff of vacuolated, large visceral epithelial cells (Pardo et al, 1987); then a split between the latter cells and the basement membrane appears and often the urinary space enlarges. A further conspicuous feature is widespread tubular change associated with various stages of acute necrosis and multiple protein reabsorption droplets in the proximal segment; tubule dilatation with intraluminal, large, pale staining, PAS-negative amorphous casts throughout the kidney (microcystic dilatation) can occur. This is sometimes accompanied by multinucleate cells, usually tubule epithelium in origin, morphologically similar to those seen elsewhere in HIV infection where the virus can be detected (Figure 5.34). There is no more than a sparse chronic inflammatory cell infiltrate of the interstitium whilst fibrosis and arterial change are absent. Hypertension is rarely seen, despite deteriorating renal function.

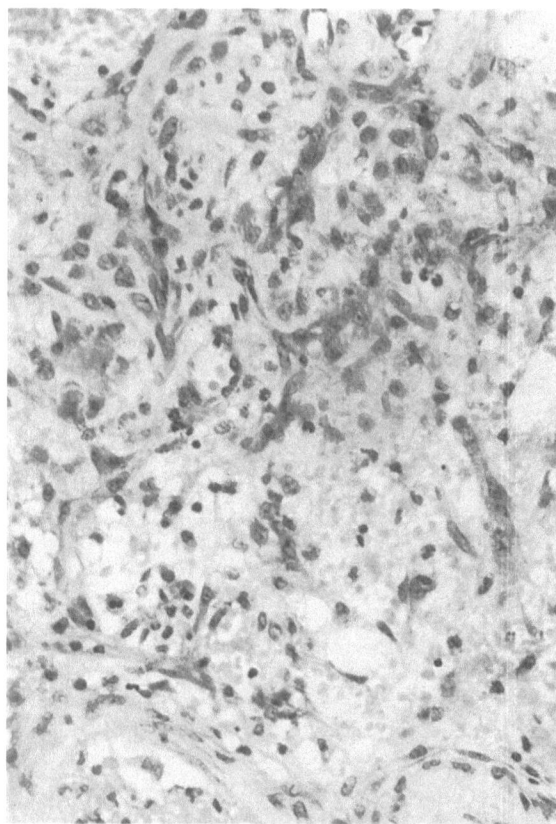

Figure 5.35. Testis with loss of spermatogenesis, thickening of the tubule basement membranes and Kaposi's sarcoma infiltrating the interstitium.

Electron microscopy of glomeruli identifies fusion of epithelial foot processes, villous transformation of those cells and their separation from the basement membrane. The finding of numerous tubuloreticular inclusions in the endothelium of glomerular and peritubular capillaries with other changes is almost a hallmark of HIV infection (Alpers et al, 1988); this supports additional evidence (Chander et al, 1987; Cohen et al, 1989) suggesting that HIV-associated nephropathy is virus-induced.

HIV infection of the testis occurs (Lecatsas et al, 1985; Ehrlich et al, 1989; Da Silva et al, 1990), and maturation arrest, thickening of the seminiferous tubule basement membranes and interstitial fibrosis are the major abnormalities in AIDS (Dalton and Harcourt-Webster, 1991). The latter two findings sometimes progress to hyalinization with loss of reticulin in both structures. Inflammation is usually present, but no more than moderate, whilst opportunistic infections are rare. The epididymis often appears normal although obstructive features are sometimes found in the duct (Dalton and Harcourt-Webster, 1991).

Neoplasms, including Kaposi's sarcoma (Rogers and Klatt, 1988; Dalton and Harcourt-Webster, 1991), of both testis and epididymis are uncommon (Figure 5.35). The bladder and prostate are often mildly chronically inflamed and sometimes fibrotic, but other pathology is rare. The pattern of change affecting the female genitalia has yet to be established.

Endocrine System

With *Cryptococcus*, cytomegalovirus (CMV), *Pneumocystis carinii* and *Toxoplasma gondii* infections of the brain, the pituitary may become involved by direct spread (Ferreiro and Vinters, 1988; Sano et al, 1989); occasionally, in the absence of such evidence, spread appears to be blood-borne. Evidence of disturbed pituitary function in life is rare although pan-hypopituitarism is described in toxoplasmosis (Milligan et al, 1984) (see also Chapter 6).

CMV inclusion bodies and cryptococci in the thyroid are reported at necropsy (Welch et al, 1984; Machac et al, 1985; Frank et al, 1987), as is Kaposi's sarcoma (Welch et al, 1984), and there are reports of successfully treated *Pneumocystis carinii* thyroiditis (Gallant et al, 1988; Drucker et al, 1990). Dysfunction is uncommon, but alterations in thyroid function tests occur (Poretsky et al, 1990).

Several necropsy series of AIDS patients show CMV to be fairly frequent in the adrenal glands with a high incidence (up to 75 per cent) of involvement amongst patients (Tapper et al, 1984; Welch et al, 1984; Klatt and Shibata, 1988; Pulakhandam and Dincsoy, 1990; Tomita et al, 1990) (Figure 5.36). Acid–alcohol fast bacilli consistent with atypical mycobacteria, cryptococci, *Pneumocystis carinii* and Kaposi's sarcoma occur also (Glasgow et al, 1985; Radin et al, 1990), although much less frequently. Evidence of dysfunction in adrenals so affected is variable (Aron, 1989; Merenich et al, 1990; Poretsky et al, 1990; Pulakhandam and Dincsoy, 1990), but sometimes significant. Many glands with CMV infection show up to 70 per cent necrosis, there being close correlation between their respective extents, whilst irrespective of any other changes the adrenals in AIDS not surprisingly show undoubted lipid depletion (Glasgow et al, 1985).

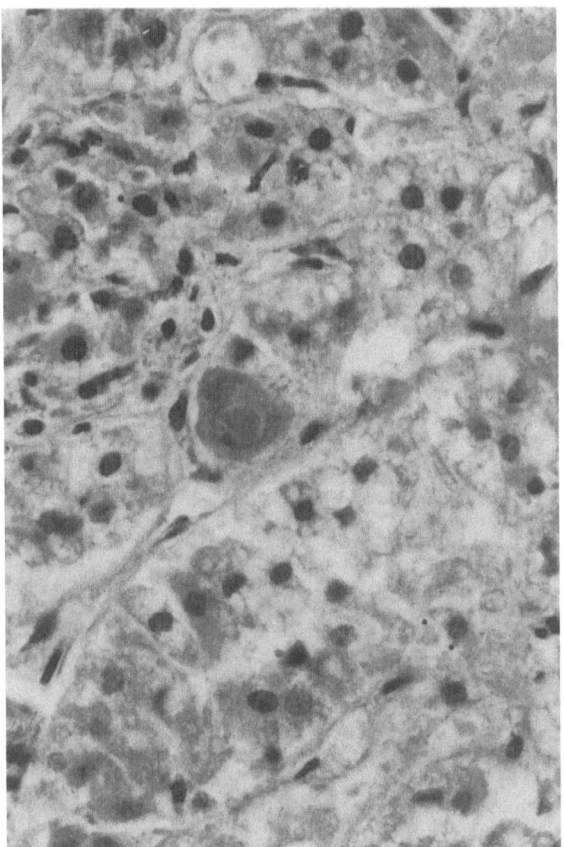

Figure 5.36. Cytomegalovirus inclusion body with minimal chronic inflammation amidst adrenal cortex.

Acknowledgements. The author would like to express thanks to Mr Reinhard Wentz and Miss Carrie Searle, Assistant Librarians, Westminster Medical School; Miss Frances Sellins, Senior Photographer, Westminster Hospital Medical Illustration; Mrs Kate Mathers and Mrs Alicia Martinez, Personal Secretaries, for typing the manuscript; Miss J. Midgley, Principal Microbiologist, for technical advice: and Mr A.J. Hall, Senior Chief MLSO, for technical assistance.

References

Alfonso PG, Sañundo EF, Carretero JM et al. (1988) Hodgkin's disease in HIV-infected patients. Biomed Pharmacother 42:321–5.

Alpers CE, Harawi S, Rennke HG. (1988) Focal glomerulosclerosis with tubuloreticular inclusions: possible predictive value for acquired immunodeficiency syndrome. Am J Kidney Dis 12:240–2.

Amberson JB, Dicarlo EF, Metroka CE et al. (1985) Diagnostic pathology in AIDS. Surgical pathology and cytology experience with 67 patients. Arch Pathol Lab Med 109: 345–51.

Anderson DW, Virmani R. (1990a) Emerging pattern of heart disease in human immunodeficiency virus infection. Hum Pathol 21:253–9.

Anderson DW, Virmani R. (1990b) Cardiac pathology of HIV disease. In: Joshi VV, ed. Pathology of AIDS and other manifestations of HIV infection. Igaku-Shoin: Tokyo, 165–85.

Anderson DW, Virmani R, Reilly JM et al. (1988) Prevalent myocarditis at necropsy in the acquired immunodeficiency syndrome. J Am Coll Cardiol 11:792–9.

Andrieu JM, Toledano M, Raphael M et al. (1988) HIV-related haematological neoplasias in France. Recent Results Cancer Res 112:46–53.

Anonymous. (1985) Genital warts, human papillomaviruses, and cervical cancer (Editorial). Lancet 2:1045–6.

Anonymous. (1988) Malignant lymphomas in patients with or at risk for AIDS in Italy. Italian Cooperative Group for AIDS-Related Tumors. J Natl Cancer Inst 80:855–60.

Anonymous. (1990) Microsporidian keratoconjunctivitis in patients with AIDS. MMWR 39:188–9.

Aron DC. (1989) Endocrine complications of the acquired immunodeficiency syndrome. Arch Intern Med 149:330–3.

Aziz DC, Srolovitz HD, Brisson ML et al. (1985) Characterisation of eosinophilic hyaline bodies in Kaposi's sarcoma (abstract). Lab Invest 52:4.

Balasubramanyam A, Waxman M, Kazal HL et al. (1986) Malignant lymphoma of the heart in acquired immune deficiency syndrome. Chest 90:243–6.

Baroldi G, Corallo S, Moroni M et al. (1988) Focal lymphocytic myocarditis in acquired immunodeficiency syndrome (AIDS): a correlative morphologic and clinical study in 26 consecutive fatal cases. J Am Coll Cardiol 12:463–9.

Barr CE, Torosian JP. (1986) Oral manifestations in inpatients with AIDS or AIDS-related complex. Lancet 2:288.

Belton CM, Eversole LR. (1986) Oral hairy leukoplakia: ultrastructural features. J Oral Pathol 15:493–9.

Bergquist R, Morfeldt-Mansoon L, Pehrson PO et al. (1984) Antibody against *Encephalitozoon cuniculi* in Swedish homosexual men. Scand J Infect Dis 16:389–91.

Berkowitz CD. (1985) AIDS and parasitic infections, including *Pneumocystis carinii* and cryptosporidiosis. Pediatr Clin North Am 32:933–52.

Berry CD, Hooton TM, Collier AC et al. (1987) Neurologic relapse after benzathine penicillin therapy for secondary syphilis in a patient with HIV infection. N Engl J Med 316:1587–9.

Bier SJ, Keller RJ, Krivisky BA et al. (1985) Oesophageal moniliasis: a new radiographic presentation. Am J Gastroenterol 80:734–7.

Bishburg E, Eng RHK, Smith SM et al. (1986) Yield diseases in patients with acquired immunodeficiency syndrome. J Clin Microbiol 24:312–14.

Blumberg RS, Kelsey P, Perrone T et al. (1984) Cytomegalovirus- and cryptosporidium-associated acalculous gangrenous cholecystitis. Am J Med 76:1118–23.

Blumenfeld W, Beckstead JH. (1983) Angioimmunoblastic lymphadenopathy with dysproteinemia in homosexual men with acquired immune deficiency syndrome. Arch Pathol Lab Med 107:567–9.

Blumenfeld W, Egbert BM, Sagebiel WR. (1985) Differential diagnosis of Kaposi's sarcoma. Arch Pathol Lab Med 109: 123–7.

Borton LK, Wintroub BU. (1984) Disseminated cryptococcosis

presenting as herpetiform lesions in a homosexual man with acquired immunodeficiency syndrome. J Am Acad Dermatol 10:387–90.

Boudreau S, Hines HC, Hood AF. (1988) Dermal abscesses with *Staph. aureus*, CMV and acid-fast bacilli in a patient with AIDS. J Cutan Pathol 15:53–7.

Bourgoignie JJ, Meneses R, Ortiz C et al. (1988) The clinical spectrum of renal disease associated with human immunodeficiency virus. Am J Kidney Dis 12:131–7.

Bradbeer C. (1987) Is infection with HIV a risk factor for cervical intraepithelial neoplasia? Lancet 2:1277–8.

Brady EM, Margois ML, Korzeniowski OM. (1984) Pulmonary cryptosporidiosis in acquired immune deficiency syndrome. JAMA 252:89–90.

Brandborg LL, Goldberg SB, Breidenbach WC. (1970) Human coccidiosis: a possible cause of malabsorption. The life cycle in small-bowel mucosal biopsies as a diagnostic feature. N Engl J Med 283:1306–13.

Brivet F, Coffin B, Bedossa et al. (1987) Pancreatic lesions in AIDS (letter). Lancet 2:570–1.

Brynes KB, Gill PS. (1990) Clinical characteristics, immunologic abnormalities, and haematopathology of HIV infection. In: Joshi VV, ed. Pathology of AIDS and other manifestations of HIV infection. Igaku-Shoin: Tokyo, 21–41.

Cali A, Owen RL. (1988) Microsporidiosis. In: Balows A, Hausler WJ, Lenette EH, eds. The laboratory diagnosis of infectious diseases: principles and practice, vol 1, Bacterial, mycotic and parasitic diseases. Springer: New York, 929–50.

Cammarosano C, Lewis W. (1985) Cardiac lesions in acquired immune deficiency syndrome (AIDS). J Am Coll Cardiol 5:703–6.

Casemore DP, Sands RL, Curry A. (1985a) Cryptosporidium species: a 'new' human pathogen. J Clin Pathol 38:1321–36.

Casemore DP, Armstrong M, Sands RL. (1985b) Laboratory diagnosis of cryptosporidiosis. J Clin Pathol 38:1337–41.

Castella A, Croxson TS, Mildvan D et al. (1985) The bone marrow in AIDS: a histologic, hematologic and microbiologic study. Am J Clin Pathol 84:425–32.

Centers for Disease Control. (1982) Persistent generalized lymphadenopathy among homosexual males. MMWR 31:249–67.

Centers for Disease Control. (1985a) Revision of the case definition of acquired immunodeficiency syndrome for national reporting: United States. MMWR 34:373–5.

Centers for Disease Control. (1985b) Oral viral lesion (hairy leukoplakia) associated with acquired immunodeficiency syndrome. MMWR 34:549–50.

Centers for Disease Control. (1986) Update: acquired immunodeficiency syndrome: Europe. MMWR 35:35–46.

Centers for Disease Control. (1987) Revision of the CDC surveillance case definition for acquired immunodeficiency syndrome. MMWR 36:1S–15S.

Chander P, Soni A, Suri A et al. (1987) Renal ultrastructural markers in AIDS-associated nephropathy. Am J Pathol 126:513–26.

Chandra RK. (1983) Nutrition, immunity and infection: present knowledge and future directions. Lancet 1:688–91.

Cockerell CJ. (1986) The dermatopathology of acquired immunodeficiency syndrome (AIDS). Semin Dermatol 5:316–29.

Cockerell CJ. (1989) Other mucocutaneous conditions in HIV-infection. In: Friedman-Kien AE, ed. A color atlas of AIDS. Saunders: Philadelphia.

Cockerell CJ. (1990) Pathology of the skin in patients with human immunodeficiency virus infection. In: Joshi VV, ed. Pathology of AIDS and other manifestations of HIV infection. Igaku-Shoin: Tokyo, 209–25.

Cockerell CJ, Whitlow MA, Webster GF et al. (1987) Epithelioid angiomatosis: a distinct vascular disorder in patients with the acquired immunodeficiency syndrome or AIDS-related complex. Lancet 2:654–6.

Cockerell CJ, Bergstresser PR, Myrie-Williams C et al. (1990) Bacillary epithelioid angiomatosis occurring in an immunocompetent individual. Arch Dermatol 126: 787–90.

Cohen AH, Nast CC. (1988) HIV-associated nephropathy. A unique combined glomerular, tubular and interstitial lesion. Mod Pathol 1:87–97.

Cohen AH, Sun NCJ, Shapshak P et al. (1989) Demonstration of human immunodeficiency virus in renal epithelium in HIV-associated nephropathy. Mod Pathol 2:125–8.

Colebunders R, Lusakumuni K, Nelson AM et al. (1988) Persistent diarrhoea in Zairian AIDS patients: an endoscopic and histological study. Gut 29:1687–91.

Conant MA, Volberding P, Fletcher V et al. (1982) Squamous cell carcinoma in sexual partner of Kaposi's sarcoma patient. Lancet 1:286.

Cone LA, Woodard DR, Potts BE et al. (1986) An update on the acquired immunodeficiency syndrome (AIDS). Associated disorders of the alimentary tract. Dis Colon Rectum 29:60–4.

Connolly GM, Hawkins D, Harcourt-Webster JN et al. (1989a) Oesophageal symptoms: their causes, treatment, and prognosis in patients with the acquired immunodeficiency syndrome. Gut 30:1033–9.

Connolly GM, Shanson D, Hawkins DA et al. (1989b) Noncryptosporidial diarrhoea in human immunodeficiency virus (HIV) infected patients. Gut 30:195–200.

Cook GC. (1988) Small-intestinal coccidiosis: an emergent clinical problem. J Infect 16:213–19.

Cooper DA, Gold J, MacLean P et al. (1985) Acute AIDS retrovirus infection: definition of a clinical illness associated with seroconversion. Lancet 1:537–40.

Corallo S, Mutinelli MR, Moroni M et al. (1988) Echocardiography detects myocardial damage in AIDS: prospective study in 102 patients. Eur Heart J 9:887–92.

Costel EE, Wheeler AP, Gregg CR. (1984) *Campylobacter fetus* ssp. *fetus* cholecystitis and relapsing bacteremia in a patient with acquired immunodeficiency syndrome. South Med J 77:927–8.

Costello C. (1988) Haematological abnormalities in human immunodeficiency virus (HIV) infection. J Clin Pathol 41: 711–15.

Costello C, Treacy M, Lai L. (1986) Treatment of immune thrombocytopenic purpura in homosexual men. Scand J Haematol 36:507–10.

Costello C, Mir N, Luckit J. (1987) The blood transfusion service and zidovudine treatment for AIDS. Br Med J 295:1486.

Coulman CU, Greene I, Archibald RWR. (1987) Cutaneous pneumocystosis. Ann Intern Med 106:396–8.

Croxson T, Chaborn AB, Rorat E et al. (1984) Intra-epithelial carcinoma of the anus in homosexual men. Dis Colon Rectum 27:325–30.

Cunningham-Rundles C, Cunningham-Rundles S, Iwata T et al. (1983) Zinc deficiency, depressed thymic hormones and T lymphocyte dysfunction in patients with hypogammaglobulinemia. Clin Immunol Immunopathol 21:387–96.

Daling JR, Weiss NS, Hislop TG et al. (1987) Sexual practices, sexually transmitted diseases, and the incidence of anal cancer. N Engl J Med 317:973–7.

Dalton ADA, Harcourt-Webster JN. (1991) The histopathology of the testis and epididymis in AIDS: a post-mortem study. J Pathol 163:47–52.

Dalton ADA, Harcourt-Webster JN, Keat ACS. (1990) Synovium in AIDS: a post-mortem study. Br Med J 300: 1239–40.

Da Silva M, Shevchuck MM, Cronin WJ et al. (1990) Detection of HIV-related protein in testes and prostates of patients with AIDS. Am J Clin Pathol 93:196–201.

Davey RT, Margolis D, Kleiner D et al. (1989) Digital necrosis and disseminated Pneumocystis carinii infection after aerosolized pentamidine prophylaxis. Ann Intern Med 111:681–2.

De Hovitz JA, Pape JW, Boncy M et al. (1986) Clinical manifestations and therapy of Isospora belli infection in patients with the acquired immunodeficiency syndrome. N Engl J Med 315:87–90.

de Maubeuge J, Ledoux M, Feremans W et al. (1986) Oral 'hairy' leukoplakia in an African AIDS patient. J Cutan Pathol 13:235–41.

Desportes I, Le Charpentier Y, Galian A et al. (1985) Occurrence of a new microsporidan: Enterocytozoon bieneusi, n.g., n.sp., in the enterocytes of a human patient with AIDS. J Protozool 32:250–4.

Di Carlo EF, Amberson JB, Metroka CE et al. (1986) Malignant lymphomas and the acquired immunodeficiency syndrome. Evaluation of 30 cases using a working formulation. Arch Pathol Lab Med 110:1012–16.

Dickson D, Ben-Ezra JM, Reed J et al. (1985) Multicentric giant lymph node hyperplasia, Kaposi's sarcoma and lymphoma. Arch Pathol Lab Med 109:1013–18.

Dieterich DT, Rahmin M.(1991) Cytomegalovirus colitis in acquired immunodeficiency syndrome (AIDS): presentation in 44 patients and a review of the literature. J AIDS 4 (Suppl 1):29–35.

Dobbins WO 3rd, Weinstein WM. (1985) Electron microscopy of the intestine and rectum in acquired immunodeficiency syndrome. Gastroenterology 88:738–49.

Dorfman RF. (1962) Kaposi's sarcoma, the contribution of enzyme histochemistry to the identification of cell types. Acta Unio Int Contra Cancrum 18:464–76.

Dorfman RF. (1988) Kaposi's sarcoma: evidence supporting its origin from the lymphatic system. Lymphology 21: 45–52.

Douglas JG, Crucioli V. (1981) Spirochaetosis: a remedial cause of diarrhoea and rectal bleeding? Br Med J 283:1362.

Dowell SF, Moore GW, Hutchings GM. (1990) The spectrum of pancreatic pathology in patients with AIDS. Mod Pathol 3:49–53.

Drabick JJ, Lupton GP, Tomtkins K et al. (1987) Crusted scabies in HIV infection (letter). J Am Acad Dermatol 17:142.

Drucker DJ, Bailey D, Rotstein L. (1990) Thyroiditis as the presenting manifestation of disseminated extrapulmonary Pneumocystis carinii infection. J Clin Endocrinol Metab 71:1663–5.

Dupont HL. (1985) Cryptosporidiosis and the healthy host. N Engl J Med 312:1319–20.

Duvic M, Johnson TM, Rapini RP et al. (1987) Acquired immunodeficiency syndrome associated psoriasis and Reiter's syndrome. Arch Dermatol 123:1622–32.

Dworkin B, Wormser GP, Rosenthal WS et al. (1985) Gastrointestinal manifestations of the acquired immunodeficiency syndrome: a review of 22 cases. Am J Gastroenterol 80: 774–8.

Dworkin BM, Stahl RE, Giardina MA et al. (1987) The liver in acquired immune deficiency syndrome: emphasis on

patients with intravenous drug abuse. Am J Gastroenterol 82:231–6.

Editorial. (1985) Genital warts, human papillomaviruses, and cervical cancer. Lancet 2:1045–6.

Edman JC, Kovacs JA, Masur H et al. (1988) Ribosomal RNA sequences shows Pneumocystis carinii to be a member of the fungi. Nature 334:519–22.

Efftinck Schattenkerk JK, Van Gool T, Van Ketel RJ et al. (1991) Clinical significance of small-intestinal microsporidiosis in HIV-I-infected-individuals. Lancet 337:895–8.

Ehrilich GD, Davey FR, Kirshner JJ et al. (1989) A polyclonal CD4+ and CD8+ lymphocytosis in a patient doubly infected with HTLV-1 and HIV-1: a clinical and molecular analysis. Am J Hematol 30:128–39.

Ellakany S, Whiteside TL, Schade RR et al. (1987) Analysis of intestinal lymphocyte subpopulations in patients with acquired immunodeficiency syndrome (AIDS) and AIDS-related complex. Am J Clin Pathol 87:356–64.

Elson CO. (1985) Induction and control of the gastrointestinal immune system. Scand J Gastroenterol 114 (Suppl 1): 1–15.

Elta G, Turnage R, Eckhauser FE et al. (1986) A submucosal antral mass caused by cytomegalovirus infection in a patient with acquired immunodeficiency syndrome. Am J Gastroenterol 81:714–17.

Eversole LR, Jacobsen P, Stone CE et al. (1986) Oral condyloma planus (hairy leukoplakia) among homosexual men: a clinicopathologic study of thirty-six cases. Oral Surg Oral Med Oral Pathol 61:249–55.

Ewing EP Jr, Chandler FW, Spira TJ et al. (1985) Primary lymph node pathology in AIDS and AIDS-related lymphadenopathy. Arch Pathol Lab Med 109:977–81.

Farhi DC, Mason UG 3rd, Horsburgh CR. (1986) Pathologic findings in disseminated Mycobacterium avium intracellulare infection: a report of 11 cases. Am J Clin Pathol 85:67–72.

Farthing CF, Brown SE, Staughton RCD et al. (1986) A colour atlas of AIDS. Wolfe Medical Publications: London.

Fernandes B, Brunton J, Koven I. (1986) Ileal perforation due to cytomegaloviral enteritis. Can J Surg 29:453–6.

Ferreiro J, Vinters HV. (1988) Pathology of the pituitary gland in patients with the acquired immune deficiency syndrome AIDS. J Pathol 20:211–15.

Ficarra G, Barone R, Gaglioti D et al. (1988) Oral hairy leukoplakia among HIV-positive intravenous drug abusers: a clinicopathologic and ultrastructural study. Oral Surg Oral Med Oral Pathol 65:421–6.

Fisher BK, Warner LC. (1987) Cutaneous manifestations of acquired immunodeficiency syndrome (AIDS). Int J Dermatol 26:615–30.

Fowler CB, Reed KD, Brannon RB. (1989) Intranuclear inclusions correlate with the ultrastructural detection of herpes-type virions in oral hairy leukoplakia. Am J Surg Pathol 13:114–19.

Francis ND, Parkin JM, Weber J et al. (1986) Kaposi's sarcoma in acquired immune deficiency syndrome (AIDS). J Clin Pathol 39:469–74.

Francis ND, Boylston AW, Roberts AHG et al. (1989) Cytomegalovirus infection in gastrointestinal tracts of patients infected with HIV-1 or AIDS. J Clin Pathol 42: 1055–64.

Frank TS, Livolsi VA, Connor AM. (1987) Cytomegalovirus infection of the thyroid in immunocompromised adults. Yale J Biol Med 60:1–8.

Freed TA, Perrez NK, Chen V et al. (1987) Cutaneous mycobacteriosis: occurrence and significance in two patients with AIDS. Arch Dermatol 123:1601–3.

Friedman-Kien AE, Cockerell CJ. (1988) Management of skin infections in patients with HIV infection. In: Leoung GS, ed. Opportunistic infections in patients with acquired immunodeficiency syndrome. Marcel Dekker: New York.

Friedman-Kien AE, Laubstein L, Marmor M et al. (1981) Centers for Disease Control. Kaposi's sarcoma and *Pneumocystis* pneumonia among homosexual men: New York City and California. MMWR 30:305–8.

Frizzera G, Hanto DW, Gajl-Peczalska KJ et al. (1981) Polymorphic diffuse B-cell hyperplasias and lymphomas in renal transplant recipients. Cancer Res 41:4262–79.

Funke I, Hahn A, Rieber EP et al. (1987) The cellular receptor (CD4) of the human immunodeficiency virus is expressed on neurons and glial cells in human brain. J Exp Med 165:1230–5.

Gal AA, Meyer PR, Taylor CR. (1987) Papillomavirus antigens in anorectal condyloma and carcinoma in homosexual men. JAMA 257:337–40.

Gallagher JG, Meschter SC. (1990) Synchronous Hodgkin's disease and non-Hodgkin's lymphoma in an adult with acquired immunodeficiency syndrome (AIDS). Cancer Detect Prev 14:395–401.

Gallant JE, Enriquez RE, Cohen KL et al. (1988) *Pneumocystis carinii* thyroiditis. Am J Med 84:303–6.

Garay S, Belenko M, Fazzini E et al. (1987) Pulmonary manifestations of Kaposi's sarcoma. Chest 91:39–43.

Garcia LS, Bruckner DA, Brewer TC et al. (1983) Techniques for the recovery and identification of *Cryptosporidium* oocysts from stool specimens. J Clin Microbiol 18:185–90.

Garcia LS, Brewer TC, Bruckner DA. (1987) Fluorescence detection of *Cryptosporidium* oocysts in human faecal specimens by using monoclonal antibodies. J Clin Microbiol 25:119–21.

Gardenswartz MH, Lerner CW, Seligson GR et al. (1984) Renal disease in patients with AIDS: a clinicopathologic study. Clin Nephrol 21:197–204.

Gartner S, Markovits P, Markovitz DM et al. (1986) The role of mononuclear phagocytes in HTLV-III/LAV infection. Science 233:215–19.

Gelb A, Miller S. (1986) AIDS and gastroenterology. Am J Gastroenterol 81:619–22.

Gertler SL, Pressman J, Price P et al. (1983) Gastrointestinal cytomegalovirus infection in a homosexual man with severe acquired immunodeficiency syndrome. Gastroenterology 85:1403–6.

Gill PS, Chandraratna PA, Meyer PR et al. (1987) Malignant lymphoma: cardiac involvement at initial presentation. J Clin Oncol 5:216–24.

Gillim SE, Harawi SJ, Gottlieb LS. (1989) The digestive system and liver. In: Harawi SJ, O'Hara CJ, eds. Pathology and pathophysiology of AIDS and HIV-related diseases. Chapman and Hall: London, 327–51.

Gillin JS, Urmacher C, West R et al. (1983) Disseminated *Mycobacterium avium–intracellulare* infection in acquired immunodeficiency syndrome mimicking Whipple's disease. Gastroenterology 85:1187–91.

Gillin JS, Shike M, Alcock N et al. (1985) Malabsorption and mucosal abnormalities of the small intestine in the acquired immunodeficiency syndrome. Ann Intern Med 102:619–22.

Glasgow BJ, Steinsapir KD, Anders K et al. (1985) Adrenal pathology in the acquired immune deficiency syndrome. Am J Clin Pathol 84:594–7.

Godwin TA, Felix JC. (1987) Histopathology of the liver in AIDS: an autopsy study. Lab Invest 56:27A.

Gold JW. (1985) Clinical spectrum of infections in patients with HTLV-III-associated diseases. Cancer Res 45 (Suppl 9):4652s–4654s.

Gold JW, Armstrong D. (1989) Opportunistic infections in AIDS patients. In: Ma P, Armstrong D, eds. The acquired immune deficiency syndrome and infections of homosexual men. Butterworth: Stoneham, Mass, 325–35.

Gordin FM, Simon GL, Wofsy CB et al. (1984) Adverse reactions to trimethoprimsulfamethoxazole in patients with the acquired immunodeficiency syndrome. Ann Intern Med 100:495.

Gottleib GJ, Ackerman AB. (1982) Kaposi's sarcoma: an extensively disseminated form in young homosexuals. Hum Pathol 13:882–92.

Gray JR, Rabeneck L. (1989) Atypical mycobacterial infection of the gastrointestinal tract in AIDS patients. Am J Gastroenterol 84:1521–4.

Greenspan D, Greenspan JS. (1988) The oral clinical features of HIV infection. Gastroenterol Clin North Am 17:535–43.

Greenspan D, Greenspan JS, Conant M et al. (1984) Oral 'hairy' leukoplakia in male homosexuals: evidence of association with both papillomavirus and a herpes-group virus. Lancet 2:831–4.

Greenspan JS, Greenspan D, Lennette ET et al. (1985) Replication of Epstein–Barr virus within the epithelial cells of oral 'hairy' leukoplakia, an AIDS-associated lesion. N Engl J Med 313:1564–71.

Greenspan D, Hollander H, Friedman-Kien A et al. (1986a) Oral hairy leukoplakia in two women, a haemophiliac and a tranfusion recipient. Lancet 2:978–9.

Greenspan D, Greenspan JS, Pindborg JJ et al. (1986b) AIDS and the dental team. Munksgaard: Copenhagen, 35–75.

Greenspan D, Greenspan JS, Hearst NG et al. (1987) Relation of oral hairy leukoplakia to infection with the human immunodeficiency virus and the risk of developing AIDS. J Infect Dis 155:475–81.

Grove DI. (1989) Clinical manifestations. In: Grove DI, ed. Strongyloidiasis: a major roundworm infection of man. Taylor and Francis: London, 155–73.

Guarda LA, Luna MA, Smith JL et al. (1984) Acquired immunodeficiency syndrome: postmortem findings. Am J Clin Pathol 81:549–57.

Guarner J, Del Rio C, Hendrix L et al. (1990) Composite Hodgkin's and non-Hodgkin's lymphoma in a patient with acquired immune deficiency syndrome. In-situ demonstration of Epstein–Barr virus. Cancer 66:796–800.

Guillon JM, Autran B, Denis M et al. (1988) Human immunodeficiency virus related lymphocytic alveolitis. Chest 94:1264–70.

Hakim SZ, Genta RM. (1986) Fatal disseminated strongyloidiasis in a Vietnam war veteran. Arch Pathol Lab Med 110:809–12.

Hamilton-Dutoit SJ, Pallesen G, Franzmann MB et al. (1991) AIDS-related lymphoma. Histopathology, immunophenotype, and association with Epstein–Barr virus as demonstrated by in situ nucleic acid hybridization. Am J Pathol 138:149–63.

Harawi SJ, Kurban AK. (1989) The skin. In: Harawi SJ, O'Hara CJ, eds. Pathology and pathophysiology of AIDS and HIV-related diseases. Chapman and Hall: London, 269–99.

Harcourt-Webster JN, Scaravilli F, Darwish AH. (1991) *Strongyloides stercoralis* hyperinfection in an HIV positive patient. J Clin Pathol 44:346–8.

Harland WA, Lee FD. (1967) Intestinal spirochaetosis. Br Med J 3:718–19.

Harnly ME, Swan SH, Holly EA et al. (1988) Temporal trends in the incidence of non-Hodgkin's lymphoma and selected malignancies in a population with a high incidence of acquired immunodeficiency syndrome (AIDS). Am J Epidemiol 128:261–7.

Harris NL. (1984) Hypervascular follicular hyperplasia and Kaposi's sarcoma in patients at risk for AIDS. N Engl J Med 310:462–3.

Hawkins CC, Gold JW, Whimbey E et al. (1986) *Mycobacterium avium* complex infections in patients with the acquired immunodeficiency syndrome. Ann Intern Med 105:184–8.

Hazelhurst JA, Vismer HF. (1985) Histoplasmosis presenting with unusual skin lesions in AIDS. Br J Haematol 113: 345–8.

Heyman MR, Rasmussen P. (1987) *Pneumocystis carinii* involvement of the bone marrow in acquired immunodeficiency syndrome. Am J Clin Pathol 87:780–3.

Himelman RB, Chung WS, Chernoff DN et al. (1989) Cardiac manifestations of human immunodeficiency virus infection: a two-dimensional echocardiographic study. J Am Coll Cardiol 13:1030–6.

Hinnant KL, Rotterdam HZ, Bell ET et al. (1986) Cytomegalovirus infection of the alimentary tract: a clinico-pathological correlation. Am J Gastroenterol 81:944–50.

Hirschmann JV, Chu AC. (1988) Skin lesions with disseminated toxoplasmosis in a patient with acquired immunodeficiency syndrome. Arch Dermatol 124:1446–7.

Hodges JR, Wright R. (1982) Normal immune responses in the gut and liver. Clin Sci 63:339–47.

Hojlyng N, Jensen BN. (1988) Respiratory cryptosporidiosis in HIV-positive patients. Lancet 1:590–4.

Holten-Andersen W, Gerstoft J, Henriksen SA et al. (1984) Prevalence of *Cryptosporidium* among patients with acute enteric infection. J Infect 9:277–82.

Holtz HA, Lavery DP, Kapila R. (1985) *Actinomycetales* infection in the acquired immunodeficiency syndrome. Ann Intern Med 102:203–5.

Hopewell PC. (1988) *Pneumocystis carinii* pneumonia: diagnosis. J Infect Dis 157:1115–19.

Howley PM. (1986) On human papillomaviruses. N Engl J Med 315:1089–90.

Ioachim HL. (1990a) Biopsy diagnosis in human immunodeficiency virus infection and acquired immunodeficiency syndrome. Arch Pathol Lab Med 114:284–94.

Ioachim HL. (1990b) The opportunistic tumours of immune deficiency. Adv Cancer Res 54:301–17.

Ioachim HL, Cooper MC. (1986) Lymphomas of AIDS. Lancet 1:96.

Ioachim HL, Lerner CW, Tapper ML. (1983) The lymphoid lesions associated with the acquired immunodeficiency syndrome. Am J Surg Pathol 7:543–53.

Ioachim HL, Cooper MC, Hellman GC. (1985) Lymphomas in men at high risk for acquired immune deficiency syndrome (AIDS): a study of 21 cases. Cancer 56:2831–42.

Ioachim HL, Weinstein MA, Robbins RD et al. (1987) Primary ano-rectal lymphoma: a new manifestation of the acquired immune deficiency syndrome (AIDS). Cancer 60:1449–53.

Ioachim HL, Ryan JR, Blaugrund SM. (1988) Salivary gland lymph nodes: the site of lymphadenopathies and lymphomas associated with human immunodeficiency virus infection. Arch Pathol Lab Med 112:1224–8.

Ioachim HL, Cronin W, Roy M et al. (1990) Persistent lymphadenopathies in people at high risk for HIV infection: clinicopathologic correlations and long term follow-up in 79 cases. Am J Clin Pathol 93:208–18.

Issac-Renton JL, Fogel D, Stibbs HH et al. (1987) *Giardia* and *Cryptosporidium* in drinking water (letter). Lancet 1:973–4.

Iwasaki T, Tashiro A, Satodate R et al. (1987) Acute pancreatitis with cytomegalovirus infection. Acta Pathol Jpn 37:1661–8.

James WD, Redfield RR, Lypton GP. (1986) A papular eruption associated with HTLV III disease. J Am Acad Dermatol 14:561–4.

Janoff EN, Reller LB. (1987) *Cryptosporidium* species, a protean protozoan. J Clin Microbiol 25:967–75.

Janoff EN, Orenstein JM, Manischewitz JF et al. (1991) Adenovirus colitis in the acquired immunodeficiency syndrome. Gastroenterology 100:976–9.

Johns DR, Tierney M, Felsenstein D. (1987) Alteration of the natural history of neurosyphilis by concurrent infection with the human immunodeficiency virus. N Engl J Med 316: 1569–72.

Jones RR, Spaull J, Spry C et al. (1986) Histogenesis of Kaposi's sarcoma in patients with and without acquired immunodeficiency syndrome (AIDS). J Clin Pathol 39: 742–9.

Kahn SA, Saltzman BR, Klein RS et al. (1986) Hepatic disorders in the acquired immune deficiency syndrome: a clinical and pathologic study. Am J Gastroenterol 81: 1145–8.

Kahn DG, Garfinkle JM, Klonoff DC et al. (1987) Cryptosporidial and cytomegaloviral hepatitis and cholecystitis. Arch Pathol Lab Med 111:879–81.

Kalengayi MM, Kashala L. (1984) Clinicopathologic features of Kaposi's sarcoma in Zaire. IARC Scientific Publications 63:559–82.

Kalter DC, Tschen JA, Kilma M. (1985) Maculopapular rash in a patient with acquired immunodeficiency syndrome. Arch Dermatol 121:1455–6.

Kanas RJ, Abrams AM, Jensen JL et al. (1988) Oral hairy leukoplakia: ultrastructural observations. Oral Surg Oral Med Oral Pathol 65:333–8.

Kaplan MH, Sadick N, McNutt S et al. (1987a) Dermatologic findings and manifestations of acquired immunodeficiency syndrome (AIDS). J Am Acad Dermatol 16:485–506.

Kaplan MH, Susin M, Pahwa SG et al. (1987b) Neoplastic complications of HTLV-III infection. Lymphomas and solid tumours. Am J Med 82:389–96.

Kaplan JE, Spira TJ, Fishbein DB et al. (1988) A six-year follow-up of HIV-infected homosexual men with lymphadenopathy. Evidence for an increased risk for developing AIDS after the third year of lymphadenopathy. JAMA 260:2694–7.

Kaposi M. (1872) Idiopathisches multiples Pigment-sarkom der Haut. Arch Dermatol Syph 4:265–73.

Katzman M, Carey JT, Elmets CA. (1987) Molluscum contagiosum and AIDS: clinical and immunological details of two cases. Br J Dermatol 116:131–8.

Kjeldsberg CR, Kim H. (1981) Polykaryocytes resembling Warthin–Finkeldey giant cells in reactive and neoplastic lymphoid disorders. Hum Pathol 12:267–72.

Klatt EC, Shibata D. (1988) Cytomegalovirus infection in the acquired immunodeficiency syndrome. Clinical and autopsy findings. Arch Pathol Lab Med 112:540–4.

Klatt EC, Jensen DF, Meyer PR. (1987) Pathology of *Mycobacterium avium–intracellulare* infection in acquired immunodeficiency syndrome. Hum Pathol 18:709–14.

Klein RS, Harris CA, Small CB et al. (1984) Oral candidiasis in high risk patients as the initial manifestation of the acquired immunodeficiency syndrome. N Engl J Med 311: 354–8.

Knobler EH, Silvers DN, Fine KC et al. (1988) Unique vascular skin lesions associated with HIV. JAMA 260: 524–7.

Knowles DM, Chadburn A. (1990) The neoplasms associated with AIDS. In: Joshi VV, ed. Pathology of AIDS and other manifestations of HIV infection. Igaku-Shoin: Tokyo, 83–120.

Knowles DM, Chamulak GA, Subar M et al. (1988) Lymphoid

neoplasia associated with the acquired immunodeficiency syndrome (AIDS). The New York University Medical Center experience with 105 patients (1981–1986). Ann Intern Med 108:744–53.

Kory WP, Rico MJ, Gould E et al. (1987) Dermatologic findings in patients with acquired immunodeficiency syndrome (AIDS). South Med J 80:1529–32.

Kotler DP, Scholes JV. (1984) Altered intestinal plasma cell immunoglobulins in the acquired immunodeficiency syndrome. Gastroenterology 86:1144.

Kotler DP, Gaetz HP, Lange M et al. (1984) Enteropathy associated with the acquired immunodeficiency syndrome. Ann Intern Med 101:421–8.

Kotler DP, Weaver SC, Terzakis JA. (1986) Ultrastructural features of epithelial cell degeneration in rectal crypts of patients with AIDS. Am J Surg Pathol 10:531–8.

Kovacs JA, Hiemenz JW, Macher AM et al. (1984) Pneumocystis carinii pneumonia: a comparison between patients with the acquired immunodeficiency syndrome and patients with other immunodeficiencies. Ann Intern Med 100:663–71.

Kristal AR, Nasca PC, Burnett WS et al. (1988) Changes in the epidemiology of non-Hodgkin's lymphoma associated with epidemic human immunodeficiency virus (HIV) infection. Am J Epidemiol 128:711–18.

Krown SE. (1987) Neoplasia in AIDS. Bull NY Acad Med 63:679–91.

Kwan TH, Kaufman HW. (1986) Acid-fast bacilli with cytomegalovirus and herpesvirus inclusions in the skin of an AIDS patient. Am J Clin Pathol 85:236–8.

Laughon BE, Druckman DA, Vernon A et al. (1988) Prevalence of enteric pathogens in homosexual men with and without acquired immunodeficiency syndrome, Gastroenterology 94:984–93.

Lauwers G, Weinstein MA, Sohn N et al. (1989) Condyloma, Bowen's disease and squamous cell carcinoma in homosexual men. Mod Pathol 2:51A.

Le Boit PE, Berger TG, Egbert BM et al. (1988) Epithelioid haemangioma-like vascular proliferation in AIDS: manifestation of cat scratch disease bacillus infection? Lancet 1:960–3.

Lebovics E, Thung SN, Schaffner F et al. (1985) The liver in the acquired immunodeficiency syndrome: a clinical and histologic study. Hepatology 5:293–8.

Lecatsas G, Houff S, Macher A et al. (1985) Retrovirus-like particles in salivary glands, prostate and testes of AIDS patients. Proc Soc Exp Med 178:653–5.

Lee FD, Kraszewski A, Gordon J et al. (1971) Intestinal spirochaetosis. Gut 12:126–33.

Lefkowitch JH, Krumholz S, Fen-Chen K et al. (1984) Cryptosporidiosis of the human small intestine: a light and electron microscopic study. Hum Pathol 15:746–52.

Levine AM. (1987) Non-Hodgkin's lymphomas and other malignancies in the acquired immune deficiency syndrome. Semin Oncol 14 (Suppl 3):34–9.

Levy JA, Ziegler JL. (1983) Acquired immunodeficiency syndrome is an opportunistic infection and Kaposi's sarcoma results from secondary immune stimulation. Lancet 2:78–81.

Levy WS, Varghese J, Anderson DW et al. (1988) Myocarditis diagnosed by endomyocardial biopsy in human immunodeficiency virus infection with cardiac dysfunction. Am J Cardiol 62:658–9.

Levy WS, Simon GL, Rios JC et al. (1989) Prevalence of cardiac abnormalities in human immunodeficiency virus infection. Am J Cardiol 63:86–9.

Lewis W. (1989) AIDS: cardiac findings from 115 autopsies. Prog Cardiovasc Dis 32:207–15.

Lo SC, Dawson MS, Wong DM et al. (1989) Identification of Mycoplasma incognitus infection in patients with AIDS: an immunohistochemical, in situ hybridisation and ultrastructural study. Am J Trop Med Hyg 41:601–16.

Long EG, Smith JS, Meier JL. (1986) Attachment of PC to rat pneumocytes. Lab Invest 54:609–15.

Longacre TA, Foucar K, Koster F et al. (1989) Atypical cutaneous lymphoproliferative disorder resembling mycosis fungoides in AIDS. Report of a case with concurrent Kaposi's sarcoma. Am J Dermatopathol 11:451–6.

Loose JH, Sedergran DJ, Cooper HS. (1989) Identification of Cryptosporidium in paraffin embedded tissue sections with the use of a monoclonal antibody. Am J Clin Pathol 91: 206–9.

Lowenthal DA, Filippa DA, Richardson ME et al. (1987) Generalised lymphadenopathy with morphologic features of Castleman's disease in an HIV-positive man. Cancer 60: 2454–8.

Lucas S. (1987) Africa's AIDS problem. Br Med J 295:49.

Lucas SB. (1989) Aspects of infectious disease. In: Anthony PP, MacSween RNM, eds. Recent advances in histopathology. Churchill Livingstone: Edinburgh, 281–302.

Lucas SB, Papadaki L, Conlon C et al. (1989) Diagnosis of intestinal microsporidiosis in patients with AIDS (letter). J Clin Pathol 42:885–7.

Lumerman H, Freedman PD, Kerpel SM et al. (1988) Oral Kaposi's sarcoma: a clinicopathologic study of 23 homosexual and bisexual men from New York metropolitan area. Oral Surg Oral Med Oral Pathol 65:711–16.

Ma P, Kaufman D, Montana J. (1984a) Isospora belli diarrheal infection in homosexual men. Aids Res 1:327–38.

Ma P, Villanueva TG, Kaufman D et al. (1984b) Respiratory cryptosporidiosis in the acquired immune deficiency syndrome. JAMA 252:1298–301.

Machac J, Nejatheim M, Goldsmith SJ. (1985) Gallium-67 citrate uptake in cryptococcal thyroiditis in a homosexual male. J Nucl Med Allied Sci 29:283–5.

Mackenzie DWR. (1990) Pneumocystis carinii: A nomadic taxon. In: Mycoses in AIDS patients. Plenum Press: New York, 55–63.

Maguire GP, Delorenzo LJ, Brown RB et al. (1987) Case report: endobronchial tuberculosis simulating bronchogenic carcinoma in a patient with acquired immunodeficiency syndrome. Am J Med Sci 294:42–4.

Malebranche R, Arnoux E, Guerin JM et al. (1983) Acquired immunodeficiency syndrome with severe gastrointestinal manifestations in Haiti. Lancet 2:873–8.

Manoff SB, Stoneburner RL, Milberg JA et al. (1987) An old disease meets a new disease: tuberculosis and acquired immune deficiency syndrome in New York City (Abstract). Third International Conference on AIDS, Washington, DC, 69.

Marche C, Kernbaum S, Saimot AG et al. (1984) Histopathological study of lymph nodes in patients with lymphadenopathy or acquired immune deficiency syndrome. Eur J Clin Microbiol 3:75–6.

Marchevsky A, Rosen MJ, Chrystal G et al. (1985) Pulmonary complications of the acquired immunodeficiency syndrome: a clinicopathologic study of 70 cases. Hum Pathol 16: 659–70.

Marcial MA, Madara JL. (1986) Cryptosporidium: cellular localisation, structural analysis of absorptive cell–parasite membrane–membrane interactions in guinea pigs, and suggestion of protozoan transport by M cells. Gastroenterology 90:583–94.

Margulis SJ, Honig CL, Soave R et al. (1986) Biliary tract obstruction in the acquired immunodeficiency syndrome. Ann Intern Med 105:207–10.

Masur H, Ognibene FP, Yarchoan R et al. (1989) CD4 counts as predictors of opportunistic pneumonias in HIV infection. Ann Intern Med 111:223–31.

Mathes BM, Douglas MC. (1985) Seborrhoeic dermatitis in patients with acquired immunodeficiency syndrome (AIDS). J Am Acad Dermatol 13:947–51.

Matis WL, Triana A, Shapiro RS et al. (1987) Dermatologic findings associated with HIV infection. J Am Acad Dermatol 17:746–51.

Meduri GU, Stover DE, Lee M et al. (1986) Pulmonary Kaposi's sarcoma in AIDS: clinical, radiographic, and pathologic manifestations. Am J Med 81:11–18.

Meiselman MS, Cello JP, Margarettan W. (1985) Cytomegalovirus colitis: report of the clinical, endoscopic, and pathologic findings in two patients with the acquired immune deficiency syndrome. Gastroenterology 88:171–5.

Merenich JA, McDermott MT, Asp AA et al. (1990) Evidence of endocrine involvement early in the course of human immunodeficiency virus infection. J Clin Endocrinol Metab 70:566–71.

Metroka CE, Cunningham-Rundles S, Pollack MS et al. (1983) Generalised lymphadenopathy in homosexual men. Ann Intern Med 99:585–91.

Meyer PR, Yanagihara ET, Parker JW et al. (1984) A distinctive follicular hyperplasia in the acquired immune deficiency syndrome and the AIDS-related complex. A prelymphomatous state for B cell lymphomas? Haematol Oncol 2:319–47.

Meyers WM, Connor DH, Neafie RC. (1976) Strongyloidiasis. In: Binford CH, Connor DH, eds. Pathology of tropical and extraordinary diseases. Armed Forces Institute of Pathology: Washington, DC, 428–32.

Midgley J, Parsons PA, Shanson DC et al. (1991) Monoclonal immunofluorescence compared with silver stain for investigating Pneumocystis carinii pneumonia. J Clin Pathol 44:75–6.

Miller-Catchpole R, Variakojis D, Anastasi J et al. (1989) The Chicago AIDS autopsy study: opportunistic infections, neoplasms, and findings from selected organ systems with a comparison to national data. Chicago Associated Pathologists. Mod Pathol 2:277–94.

Milligan SA, Katz MS, Craven PC et al. (1984) Toxoplasmosis presenting as panhypopituitarism in a patient with the acquired immune deficiency syndrome. Am J Med 77:760–4.

Mobley K, Rotterdam HZ, Lerner CW et al. (1985) Autopsy findings in the acquired immunodeficiency syndrome. Pathol Annu 20:45–65.

Modigliani R, Bories C, Le Charpentier et al. (1985) Diarrhoea and malabsorption in acquired immune deficiency syndrome: a study of four cases with special emphasis on opportunistic protozoan infestations. Gut 26:179–87.

Morgan J, Herbst JS, Resnick L et al. (1989) Comparison of scrapings and biopsies of oral hairy leukoplakia for the detection of Epstein–Barr virus utilising DNA in-situ hybridisation. Mod Pathol 2:64A.

Moskowitz LB, Hensley GT, Gould EW et al. (1985) Frequency and anatomic distribution of lymphadenopathic Kaposi's sarcoma in AIDS: an autopsy series. Hum Pathol 16:447–56.

Muhlemann MF, Anderson MG, Paradinas FJ et al. (1986) Early warning skin signs in AIDS and persistent generalised lymphadenopathy. Br J Dermatol 114:419–24.

Murray JF, Felton CP, Garay SM et al. (1984) Pulmonary complications of the acquired immunodeficiency syndrome: report of a National Heart, Lung and Blood Institute Workshop. N Engl J Med 310:1682–8.

Namiki TS, Boone DC, Meyer PR. (1987) A comparison of bone marrow findings in patients with acquired immunodeficiency syndrome (AIDS) and AIDS-related conditions. Haematol Oncol 5:99–106.

Nash G, Fligiel S. (1984a) Pathologic features of the lung in AIDS: an autopsy study of 17 homosexual males. Am J Clin Pathol 81:6–12.

Nash G, Fligiel S. (1984b) Kaposi's sarcoma presenting as pulmonary disease in AIDS: diagnosed by lung biopsy. Hum Pathol 15:999–1001.

Navin TR, Hardy AM. (1987) Cryptosporidiosis in patients with AIDS (letter). J Infect Dis 155:150.

Navin TR, Juranek DD. (1984) Cryptosporidiosis: clinical, epidemiologic and parasitologic review. Rev Infect Dis 6:313–27.

Nelson JA, Wiley CA, Reynolds-Kohler C et al. (1988) Human immunodeficiency virus detected in bowel epithelium from patients with gastrointestinal symptoms. Lancet 1:259–62.

Nichols G. (1986) Microsporidiosis in AIDS. Communicable Dis Rev 86:3–4.

Niedt GW, Schinella RA. (1985) Acquired immunodeficiency syndrome. Clinicopathologic study of 56 autopsies. Arch Pathol Lab Med 109:727–34.

Nielsen RH, Orholm M, Pedersen JO et al. (1983) Colorectal spirochetosis: clinical significance of the infestation. Gastroenterology 85:62–7.

O'Brien RJ. (1988) Pulmonary disease due to Mycobacterium avium complex. Semin Resp Med 9:492–7.

O'Hara CJ. (1989) The lymphoid and haematopoietic systems. In: Harawi SJ, O'Hara CJ, eds. Pathology and pathophysiology of AIDS and HIV-related diseases. Chapman and Hall: London, 135–83.

Orenstein JM, Chiang J, Steinberg W et al. (1990) Intestinal microsporidiosis as a cause of diarrhoea in human immunodeficiency virus-infected patients: a report of 20 cases. Hum Pathol 21:475–81.

Orenstein MS, Tavitian A, Yonk B et al. (1985) Granulomatous involvement of the liver in patients with AIDS. Gut 26:1220–5.

Osborne BM, Guarda LA, Butler JJ. (1984) Bone marrow biopsies in patients with the acquired immunodeficiency syndrome. Hum Pathol 15:1048–53.

Overly WL, Jakubek DJ. (1987) Multiple squamous cell carcinomas and human immunodeficiency virus infection. Ann Intern Med 106:334.

Pape JW, Verdier RI, Johnson WD Jr. (1989) Treatment and prophylaxis of Isospora belli infection in patients with the acquired immunodeficiency syndrome. N Engl J Med 320:1044–7.

Pardo V, Aldana M, Colton RM et al. (1984) Glomerular lesions in the acquired immunodeficiency syndrome. Ann Intern Med 101:429–34.

Pardo V, Meneses R, Ossa L et al. (1987) AIDS-related glomerulopathy: occurrence in specific risk groups. Kidney Int 31:1167–73.

Patterson JW, Kitces EN, Neafie RC. (1987) Cutaneous botryomycosis with AIDS. J Am Acad Dermatol 16:238–42.

Penneys NS, Hicks B. (1985) Unusual cutaneous lesions associated with AIDS. J Am Acad Dermatol 13:845–52.

Peters SG, Prakash UBS. (1987) PCP review of 53 cases. Am J Med 82:73–8.

Phelan JA, Saltzman BR, Friedland GH et al. (1987) Oral findings in patients with the acquired immunodeficiency syndrome. Oral Surg Oral Med Oral Pathol 64:50–6.

Piot P, Quinn TC, Taelman H et al. (1984) Acquired immunodeficiency syndrome in a heterosexual population in Zaire. Lancet 2:65–9.

Popovic M, Gartner S. (1987) Isolation of HIV-I from monocytes but not T lymphocytes. Lancet 2:916.

Poretsky L, Maran A, Zumoff B. (1990) Endocrinologic and metabolic manifestations of the acquired immunodeficiency syndrome. Mount Sinai J Med 57:236–41.

Porro GB, Parente F, Cernuschi M. (1989) The diagnosis of oesophageal candidiasis in patients with acquired immune deficiency syndrome: is endoscopy always necessary? Am J Gastroenterol 84:143–6.

Presant CA, Gala K, Wiseman C et al. (1987) Human immunodeficiency virus-associated T-cell lymphoblastic lymphoma in AIDS. Cancer 60:1459–61.

Pulakhandam U, Dincsoy HP. (1990) Cytomegaloviral adrenalitis and adrenal insufficiency in AIDS. Am J Clin Pathol 93:651–6.

Purdy LJ, Colby TV, Yousem SA et al. (1986) Pulmonary Kaposi's sarcoma: premortem histologic diagnosis. Am J Surg Pathol 10:301–11.

Puterman M, Goldstein J. (1983) Primary lymph nodal Kaposi's sarcoma of the parotid gland. Head Neck Surg 5:535–8.

Rabeneck L, Boyko WJ, McLean DM et al. (1986) Unusual oesophageal ulcers containing enveloped virus-like particles in homosexual men. Gastroenterology 90:1882–9.

Radin DR, Baker EL, Klatt EC et al. (1990) Visceral and nodal calcification in patients with AIDS: related *Pneumocystis carinii* infection. AJR 154:27–31.

Radolf JD, Kaplan RP. (1988) Unusual manifestations of secondary syphilis and abnormal humoral immune response to *Treponema pallidum* antigens in a homosexual man with asymptomatic HIV infection. J Am Acad Dermatol 18:423–8.

Rao TKS, Friedman EA. (1989) AIDS (HIV)-associated nephropathy: does it exist? An in-depth review. Am J Nephrol 9:441–53.

Rao TKS, Filippone EJ, Nicastri AD et al. (1984) Associated focal and segmental glomerulosclerosis in the acquired immunodeficiency syndrome: N Engl J Med 310:669–73.

Rao TKS, Friedman EA, Nicastri AD. (1987) The types of renal disease in the acquired immunodeficiency syndrome. N Engl J Med 316:1062–8.

Raufman JP. (1988) Odynophagia/dysphagia in AIDS. Gastroenterol Clin North Am 17:599–614.

Reichert CM, O'Leary TJ, Levens DL et al. (1983) Autopsy pathology in the acquired immune deficiency syndrome. Am J Pathol 112:357–82.

Reilly JM, Cunnion RE, Anderson DW et al. (1988) Frequency of myocarditis, left ventricular dysfunction and ventricular tachycardia in the acquired immune deficiency syndrome. Am J Cardiol 62:789–93.

Relman DA, Loutit JS, Schmidt TM et al. (1990) The agent of bacillary angiomatosis: an approach to the identification of uncultured pathogens. N Engl J Med 323:1573–80.

Restrepo C, Macher AM, Radany EH. (1987) Disseminated extraintestinal isosporiasis in a patient with acquired immune deficiency syndrome. Am J Clin Pathol 87:536–42.

Richman DD, Fischl M, Grieco MH et al. (1987) The toxicity of azidothymidine (AZT) in the treatment of patients with AIDS and AIDS-related complex: a double-blind, placebo-controlled trial. N Engl J Med 317:192–7.

Rico NJ, Penneys NS. (1985) Cutaneous cryptococcus resembling molluscum contagiosum in a patient with AIDS. Arch Dermatol 121:901–2.

Rindum JL, Schiodt M, Pindborg JJ et al. (1987) Oral hairy leukoplakia in three haemophiliacs with human immunodeficiency virus infection. Oral Surg Oral Med Oral Pathol 63:437–40.

Rodgers VD, Fassett R, Kagnoff MF. (1986) Abnormalities in intestinal mucosal T cells in homosexual populations including those with the lymphadenopathy syndrome and acquired immunodeficiency syndrome. Gastroenterology 90:552–8.

Rogers C, Klatt EC. (1988) Pathology of the testis in acquired immunodeficiency syndrome. Histopathology 12:659–65.

Roldan EO, Moskowitz L, Hensley GT. (1987) Pathology of the heart in acquired immunodeficiency syndrome. Arch Pathol Lab Med 111:943–6.

Rolston KVI, Hoy J, Mansell PWA. (1986) Diarrhea caused by 'non-pathogenic amoebae' in patients with AIDS. N Engl J Med 315:192.

Roth RI, Owen RL, Keren DF et al. (1985) Intestinal infection with *Mycobacterium avium* in acquired immune deficiency syndrome (AIDS): histological and clinical comparison with Whipple's disease. Dig Dis Sci 30:497–504.

Rotterdam H. (1987) Tissue diagnosis of selected AIDS-related opportunistic infections. Am J Surg Pathol 11 (Suppl 1):3–15.

Ruane PJ, Nakata MM, Reinhardt JF et al. (1989) Spirochete-like organisms in the human gastrointestianl tract. Rev Infect Dis 11:184–96.

Rutgers JL, Wieczorek R, Bonett F et al. (1986) The expression of endothelial cell surface antigens by AIDS-associated Kaposi's sarcoma: evidence of a vascular endothelial cell origin. Am J Pathol 122:493–9.

Ryan JR, Ioachim HL, Marmer J et al. (1985) Acquired immune deficiency syndrome-related lymphadenopathies presenting in the salivary gland lymph nodes. Arch Otolaryngol 111:554–6.

Sadick N, Kaplan MH, Pahwa SG et al. (1986) Unusual features of scabies complicating HTLV III infection. J Am Acad Dermatol 15:482–6.

Saldana MJ, Mones J, Buck BE. (1983) Lymphoid interstitial pneumonia in Haitian residents of Florida. Chest 84:347.

Salmeron S, Petitpretz P, Katlama C et al. (1986) Pentamidine and pancreatitis. Ann Intern Med 105:140–1.

Sano T, Kovacs K, Scheithauer BW, Rosenblum MK et al. (1989) Pituitary pathology in acquired immunodeficiency syndrome. Arch Pathol Lab Med 113:1066–70.

Santucci M, Pimpinelli N, Moretti S et al. (1988) Classic and immunodeficiency-associated Kaposi's sarcoma. Arch Pathol Lab Med 112:1214–20.

Schiodt M, Greenspan D, Daniels T et al. (1987) Clinical and histologic spectrum of oral hairy leukoplakia. Oral Surg Oral Med Oral Pathol 64:716–20.

Schneiderman DJ. (1988) Hepatobiliary abnormalities in AIDS. Gastroenterol Clin North Am 17:615–30.

Schneiderman DJ, Arensen DM, Cello JP et al. (1987) Hepatic disease in patients with the acquired immune deficiency syndrome (AIDS). Hepatology 7:925–30.

Schofield JB, Lindley RP, Harcourt-Webster JN. (1989) Biopsy pathology of HIV infection: experience at St Stephen's Hospital, London. Histopathology 14:277–88.

Schultz MG. (1983) Emerging zoonoses. N Engl J Med 308:1285–6.

Schwartz RA, Kardashian JF, McNutt NS et al. (1983) Cutaneous angiosarcoma resembling anaplastic Kaposi's sarcoma in a homosexual man. Cancer 51:721–6.

Scully PA, Steinman HK, Kennedy C et al. (1988) AIDS-related Kaposi's sarcoma displays differential expression of endothelial surface antigens. Am J Pathol 130:244–51.

Senaldi E, Lee MH, Toth I et al. (1990) Hodgkin's disease after non-Hodgkin's malignant lymphoma in acquired immune deficiency syndrome. Cancer 66:960–4.

Serrano M, Bellas C, Campo E et al. (1990) Hodgkin's disease in patients with antibodies to human immunodeficiency virus. A study of 22 patients. Cancer 65:2248–54.

Sewankambo N, Mugerwa RD, Goodgame R et al. (1987) Enteropathic AIDS in Uganda: an endoscopic, histological and microbiological study. AIDS 1:9–13.

Shanson DC. (1989) AIDS and other diseases caused by retroviruses. In: Shanson DC, ed. Microbiology in clinical practice, 2nd edn. Wright: London, 478–506.

Silver MA, Macher AM, Reichert CM et al. (1984) Cardiac involvement by Kaposi's sarcoma in acquired immune deficiency syndrome (AIDS). Am J Cardiol 53:983–5.

Silverman S Jr, Migliorati CA, Lozada-Nur F et al. (1986) Oral findings in people with or at high risk for AIDS: a study of 375 homosexual males. J Am Dent Assoc 112: 187–92.

Simor AE, Poon R, Boreczyk A. (1989) Chronic *Shigella flexneri* infection preceding development of acquired immunodeficiency syndrome. J Clin Microbiol 27:353–5.

Sindrup JH, Lisby G, Weismann K et al. (1987) Skin manifestations in AIDS, HIV infection and ARC. Int J Dermatol 26:267–72.

Smith D, Gazzard B, Lindley RP et al. (1989) Visceral leishmaniasis (kala azar) in a patient with AIDS. AIDS 3:41–3.

Soave R, Johnson WD. (1988) *Cryptosporidium and Isospora belli* infections. J Infect Dis 157:225–9.

Soave R, Danner RL, Honig LC et al. (1984) Cryptosporidiosis in homosexual men. Ann Intern Med 100:504–11.

Soepreno FF, Schinella RA, Cockerell CJ et al. (1986) Seborrhoeic-like dermatitis of acquired immunodeficiency syndrome. J Am Acad Dermatol 14:1020.

Soni A, Agarwal A, Chander P et al. (1989) Evidence for an HIV-related nephropathy: a clinico-pathologic study. Clin Nephrol 31:12–17.

Spencer H. (1977) Pneumocystis pneumonia. In: Spencer H, ed. Pathology of the lung. Pergamon Press: Oxford, 331–6.

Sperber SJ, Schleupner CJ. (1987) Salmonellosis during infection with human immunodeficiency virus. Rev Infect Dis 9:925–34.

Spickett GP, Dalgleish AG. (1988) Cellular immunology of HIV-infection. Clin Exp Immunol 71:1–7.

Spivak JL, Bender BS, Quinn TC. (1984) Haematologic abnormalities in the acquired immune deficiency syndrome. Am J Med 77:224–8.

Steis RG, Longo DL. (1988) Clinical, biologic and therapeutic aspects of malignancies associated with the acquired immunodeficiency syndrome: part 1. Ann Allergy 60: 310–14.

Strand CL. (1990) Role of the microbiology laboratory in the diagnosis of opportunistic infections in persons infected with human immunodeficiency virus. Arch Pathol Lab Med 114:277–83.

Sun T. (1988) Opportunistic parasitic infections in patients with acquired immunodeficiency syndrome. Pathol Annu 23:1–32.

Surawicz CM, Roberts PL, Rompalo A et al. (1987) Intestinal spirochetosis in homosexual men. Am J Med 82:587–92.

Tapper ML, Rotterdam HZ, Lerner CW et al. (1984) Adrenal necrosis in the acquired immunodeficiency syndrome. Ann Intern Med 100:239–41.

Tatum ET, Sun PC, Cohn DL. (1989) Cytomegalovirus vasculitis and colon perforation in a patient with the acquired immunodeficiency syndrome. J Pathol 21:235–8.

Tavitian A, Raufman J-P, Rosenthal LE. (1986) Oral candidiasis as a marker for esophageal candidiasis in the acquired immunodeficiency syndrome. Ann Intern Med 104:54–5.

Temple JJ, Andes WA. (1986) AIDS and Hodgkin's disease. Lancet 2:454–5.

Templeton AC. (1981) Kaposi's sarcoma. Pathol Annu 2: 315–36.

Terada S, Reddy KR, Jeffers LJ et al. (1987) Microsporidian hepatitis in the acquired immunodeficiency syndrome. Ann Intern Med 107:61–2.

Tirelli U, Vaccher E, Rezza G et al. (1989) Hodgkin's disease in association with acquired immunodeficiency syndrome (AIDS): a report on 36 patients. Gruppo Italiano Cooperativo AIDS and Tumori. Acta Oncol 28:637–9.

Tomita T, Chiga M, Lenahan M et al. (1990) Identification of cytomegalovirus infection in acquired immunodeficiency syndrome. Virchows Arch [B] 416:497–503.

Travis WD, Lack EE, Ognibene FP et al. (1989) Lung biopsy interpretation in the acquired immunodeficiency syndrome: experience of the National Institutes of Health with literature review. In: Rotterdam H, Sommers SC, Racy P et al, eds. Progress in AIDS pathology. Field and Wood: New York, 51–84.

Triana AF, Shapiro RS, Polk BF et al. (1987) Mucocutaneous findings in AIDS/ARC patients (letter). J Am Acad Dermatol 16:888–9.

Turner RR, Levine AM, Gill PS et al. (1987) Progressive histopathologic abnormalities in the persistent generalised lymphadenopathy syndrome. Am J Surg Pathol 11:625–32.

Ullrich R, Zeitz M, Heise W et al. (1989) Small intestinal structure and function in patients infected with human immunodeficiency virus (HIV): evidence for HIV-induced enteropathy. Ann Intern Med 111:15–21.

Valle SL. (1987) Dermatologic findings related to HIV infection in high risk individuals. J Am Acad Dermatol 17: 951–61.

Villar LA, Massanari RM, Mitros FA. (1984) Cytomegalovirus infection with acute erosive oesophagitis. Am J Med 76: 924–8.

Volpe F, Schwimmer A, Barr C. (1985) Oral manifestation of disseminated *Mycobacterium avium intracellulare* in a patient with AIDS. Oral Surg Oral Med Oral Pathol 60: 567–70.

Waisman J, Rotterdam H, Niedt GN et al. (1987) AIDS: an overview of the pathology. Pathol Res Pract 182:729–54.

Wang HH, Tollerud D, Danar D et al. (1986) Another Whipple-like disease in AIDS? N Engl J Med 314:1577–8.

Webber J, Philip S. (1983) Human cryptosporidiosis. N Engl J Med 309:1326.

Weber JR Jr, Dobbins WO. (1986) The intestinal and rectal epithelial lymphocyte in AIDS: an electron-microscopic study. Am J Surg Pathol 10:627–39.

Welch K, Finkbeiner W, Alpers CE et al. (1984) Autopsy findings in the acquired immunodeficiency syndrome. JAMA 252:1152–9.

Wells M, Robertson S, Lewis F et al. (1988) Squamous carcinoma arising in a giant peri-anal condyloma associated with human papillomavirus types 6 and 11. Histopathology 12:319–23.

Wexner SD, Smithy WB, Milsom JW et al. (1986) The surgical management of anorectal diseases in AIDS and pre-AIDS patients. Dis Colon Rectum 29:719–23.

Wheat LJ, Slama TG, Zeckel ML. (1985) Histoplasmosis in AIDS. Am J Med 78:203–10.

Wheeler AP, Gregg CR. (1986) *Campylobacter* bacteremia, cholecystitis, and the acquired immunodeficiency syndrome (letter). Ann Intern Med 105:804.

WHO. (1986) Acquired immunodeficiency syndrome (AIDS). WHO/CDC case definition for AIDS. Wkly Epidemiol 61:69–73.

Widy-Wirski R, Berkley S, Downing R et al. (1988) Evaluation of the WHO clinical case definition for AIDS in Uganda. JAMA 260:3286–9.

Winchester R, Bernstein DH, Fischer HD et al. (1987) The co-occurrence of Reiter's syndrome and acquired immuno-deficiency. Ann Intern Med 106:19–26.

Winkler JR, Murray PA, Greenspan D et al. (1986) Perio-dontal disease of male homosexuals as related to AIDS-virus infection. Int Conf on AIDS Paris Abstract 1009 June 1986:23–25.

Winkler JR, Murray PA, Greenspan D et al. (1987) Periodontal disease of male homosexuals as related to AIDS virus infection (abstract 1197). J Dent Res 66:256.

Wolfson JS, Richter JM, Waldron MA et al. (1985) Cryptosporidiosis in immunocompetent patients. N Engl J Med 312:1278–82.

Wong B, Edwards FF, Kiehn TE et al. (1985) Continuous high grade *Mycobacterium avium intracellulare* bacteraemia in patients with AIDS. Am J Med 78:35–40.

Zazzo J-F, Pichon F, Regnier B. (1987) HIV and the pancreas (letter). Lancet 2:1212–13.

Zender HO, Arrigoni E, Eckert J et al. (1989) A case of *Encephalitozoon cuniculi* peritonitis in a patient with AIDS. Am J Clin Pathol 92:352–6.

Ziegler JL, Beckstead JA, Volberding PA et al. (1984a) Non-Hodgkin's lymphoma in 90 homosexual men: relation to generalised lymphadenopathy and the acquired immuno-deficiency syndrome. N Engl J Med 311:565–70.

Ziegler JL, Templeton AC, Vogel CL. (1984b) Kaposi's sarcoma: a comparison of classical, endemic, and epidemic forms. Semin Oncol 11:47–52.

Zon LI, Groupman JE. (1988) Haematologic manifestations of the human immunodeficiency virus (HIV). Semin Haematol 25:208–18.

Zon LI, Arkin C, Groopman JE. (1987) Haematologic mani-festations of the human immune deficiency virus (HIV). Br J Haematol 66:251–6.

zur Hausen H. (1987) Papillomaviruses in human cancer. Cancer 59:1692–6.

6 Pathology of the Nervous System

F. Scaravilli, F. Gray, J. Mikol and E. Sinclair

Introduction

At the beginning of the study of the changes produced by the acquired immune deficiency syndrome (AIDS) in the nervous system, all complications were thought to be due exclusively to opportunistic infections. Indeed, the early paper by Snider et al (1983) conveyed this impression by describing known organisms as the cause of almost all abnormalities. Also the 'subacute encephalitis', characterized by 'subtle cognitive changes accompanied by malaise, lethargy . . .', was found to be an opportunistic infection as intranuclear inclusions of cytomegalovirus were found within some of the microglial nodules. A relationship between viral infection and lymphomas, already found in patients undergoing organ transplants (Hanto et al, 1981), was confirmed in AIDS patients, all of whom have positive serology for the Epstein–Barr virus (EBV) (Fauci et al, 1984). The role of EBV was subsequently confirmed by Shaerer et al (1985), who documented the evolution of an EBV-stimulated polyclonal to a monoclonal B-cell proliferation and eventually to monoclonal B-cell lymphoma.

The discovery of HIV (Barré-Sinoussi et al, 1983; Gallo et al, 1984) and the demonstration of its presence in the nervous system (Shaw et al, 1985), went together with the finding of multinucleated giant cells (MGC) in the brains of HIV-positive patients (Sharer et al, 1985). Because of the unusual features of the MGC (Gray et al, 1988a), these cells were considered unique to AIDS (Petito et al, 1986) and to be the hallmark of the immune deficiency syndrome (Budka, 1986). The identification of these cells and the localization of the HIV (Mirra et al, 1986; Sharer et al, 1986; Gray et al, 1987) within them and the macrophages from which they derive (Koenig et al, 1986) helped define a new entity, the HIV encephalopathy. This was recognized to be the morphological counterpart of the dementia, a disorder which develops in a proportion of HIV-positive patients (Navia et al, 1986a,b).

The identification of this new illness changed the way neuropathological lesions in AIDS are classified. Whether HIV encephalopathy is included within the broad group of viral infections (Petito et al, 1986; Vinters et al, 1989a) or considered as a separate group (Budka et al, 1987), it is obvious that neuropathological lesions have to be divided into two separate categories: one including those due to the direct action of HIV, the other secondary to opportunistic infections.

The demonstration that HIV can be isolated from the nervous system (Shaw et al, 1985), the clinical evidence that approximately 70 per cent of all AIDS patients may develop mental changes of variable severity (McArthur, 1987), the observation that macrophages and MGC within the brain can synthesize viral RNA and consequently that the brain is the reservoir of the virus all suggest that HIV is 'neurotropic', in spite of the fact that, to date, there is no firm evidence of the presence of this virus within cells of neuro-ectodermal origin (Gartner et al, 1986; Streicher and Joynt, 1986; Wiley et al, 1986a) (see also Chapters 7 and 12).

Unfortunately, the relationship between the virus and the nervous system cannot be reduced to the encephalopathy which develops in a number of HIV-positive patients. On the one hand, there are a number of symptoms and complications, at different stages of the illness (Price et al, 1988), that have no morphological counterpart; on the other hand, in cases in which the changes lack the characteristic features of both MGC encephalitis and of opportunistic infections, an HIV aetiology can, at present, be considered only speculative.

The use of monoclonal antibodies to HIV and the introduction of techniques such as in situ hybridization and polymerase chain reaction (PCR) have undoubtedly contributed to the understanding of the role of the virus in the brain. However, immuno-histochemistry has demonstrated disappointingly small amounts of virus in the tissues examined, and in situ hybridization has often given conflicting results. Moreover, the absence of tropism of HIV for glial cells (which is obvious in visna) makes it unlikely that myelin damage in HIV encephalopathy is the result of direct cytopathic effect on these cells. Therefore, other hypotheses have been put forward to explain brain lesions in HIV infection. Sharing of antigenic polypeptides between HIV and constituents of myelin could trigger an autoimmune process; however, there is no evidence of shared antigen(s). Another proposed mechanism, that of expression of viral proteins on the surface of oligodendrocytes, is not supported by evidence of tropism for oligodendrocytes. A 'bystander effect' has also been proposed, whereby myelin would be attacked by factors produced by macrophages, probably in response to antigens unrelated to myelin (a situation similar to the events leading to demyelination in Theiler's virus infection).

Although an alteration of the white matter as the cause of dementia is the mechanism gathering, at present, the largest consensus of opinion, other possibilities, such as competition of segments of the virus with trophic factors (Gurney et al, 1986a,b) or with neuropeptides (Pert et al, 1986) are also being considered (Dal Canto, 1989). These hypotheses are discussed in the context of dementia by Budka in Chapter 7 and Wiley in Chapter 11 of this book.

It has been reported that 40 to 50 per cent of AIDS patients, including adults and children, have neurological symptoms which, in 10 per cent of the cases, may represent the first mani-

festation of the disease (Bredesen and Messing, 1983), and that up to 80 per cent of HIV-positive patients who are examined pathologically show morphological abnormalities of the nervous system (Guarda et al, 1984; Kui et al, 1984; Moskowitz et al, 1984; Lang et al, 1989; Levy et al, 1985; Anders et al, 1986a; Petito et al, 1986; Budka et al, 1987).

The present chapter deals with the neuropathological changes observed in the brains of AIDS sufferers. They include those produced by HIV and those associated with opportunistic infections, as well as lymphomas. The description will be based on personal experience and on cases reported in the literature. Other chapters will be devoted to the AIDS dementia complex (Chapter 7), the pathology of the spinal cord (Chapter 8), the encephalopathy in children (Chapter 9), and the pathology of peripheral nerves and muscles (Chapter 10).

In the last few years it has emerged that in the brains of a group at risk of AIDS, the haemophiliacs, pathological lesions differ from those seen in other groups. The subject has been discussed recently by Esiri et al (1989) and will be addressed in this chapter. In a number of HIV-positive patients, the nervous system shows lesions due to different agents. The occurrence of multiple infections will also be described and its significance in relation to the brain biopsy discussed.

Infections of the Central Nervous System by HIV

The possibility of direct infection of the central nervous system (CNS) by the human immune deficiency virus (HIV) has been substantiated by the detection of HIV RNA and DNA within the brain using the following methods: Southern blot and in situ hybridization (Shaw et al, 1985; Koenig et al, 1986; Sharer et al, 1986; Wiley et al, 1986a; Ho et al, 1987; Vazeux et al, 1987); immunocytochemistry (Gabuzda et al, 1986; Wiley et al, 1986a; Budka et al, 1987; Hénin et al, 1987; Pumarola-Sune et al, 1987; Vazeux et al, 1987; Ward et al, 1987); ultrastructural demonstration of viral particles (Epstein et al, 1985; Koenig et al, 1986; Sharer et al, 1986; Budka et al, 1987; Gray et al, 1987; Gyorkey et al, 1987; Kato et al, 1987; Meyenhofer et al, 1987); isolation of HIV from cerebrospinal fluid (CSF) and brain tissue of patients with AIDS or AIDS-related complex (ARC) (Ho et al, 1985a; Levy et al, 1985; Chiodi et al, 1986); and evidence in these patients of intrathecal production of antibodies against HIV (Resnick et al, 1985; Ackermann et al, 1986).

There is now consensus of opinion that invasion of the brain by HIV may produce an encephalopathy. This has been referred to as 'AIDS encephalopathy' (Epstein et al, 1985; Sharer et al, 1985, 1986; Shaw et al, 1985) or 'AIDS dementia complex' (ADC) (Navia et al, 1986a,b). Subsequently, the term 'HIV-encephalopathy' was proposed by the CDC-WHO (1988) and, more recently, terminology (HIV-encephalitis or HIV-leukoencephalopathy) based on the morphological changes observed in the white matter has been adopted (Budka et al, 1991; see below). Its incidence varies from 16 to 65 per cent in American series (Snider et al, 1983; Levy et al, 1985; Navia et al, 1986a,b; McArthur, 1987). In Central Africa, severe dementia is observed in only 3.2 per cent of cases (Bélec et al, 1989), with the exception of Tanzania, where clinical features of dementia were found in 54 per cent of cases (Howlett et al, 1989). These discrepancies probably reflect geographical variations (Levy et al, 1988a; Lang et al, 1989), but are also partly due to differences in brain sampling and lack of strict diagnostic criteria (Janssen et al, 1989). However, they may simply be due to the fact that, in Africa, patients die at an early stage of the disease. Although HIV encephalitis/encephalopathy usually occurs in the late stages of the disease, in 16–25 per cent of cases dementia precedes the onset of opportunistic infections or tumours (Navia et al, 1986c; Levy and Bredesen, 1988a), and in 9 per cent it is the only manifestation of the disease (Navia and Price, 1987).

In spite of the different labels, the neuropathological changes described in HIV encephalitis/encephalopathy are relatively uniform and quite characteristic; indeed, they have not been observed previously in HIV-negative immunocompromised patients. However, the clinicopathological correlation of the lesions within the nervous system remains controversial, and their pathogenesis speculative (see Chapters 7 and 12).

Neuropathology

The most characteristic neuropathological feature in AIDS is the presence of distinctive multinucleated giant cells (MGC) which Budka (1986) considered to be the hallmark of HIV infection in the brain. These cells were initially described by Sharer et al (1985) in the brains of children in whom HIV had been demonstrated by in situ hybridization (Shaw et al, 1985). They are large, irregularly round, elongated or polyhedral, with eosinophilic cytoplasm more densely stained in the centre and vacuolated at the periphery. Multiple (up to 20) round or elongated strongly basophilic nuclei may form a circular or semicircular arrangement at the periphery of the cells, or may be scattered haphazardly. Immunohistochemical studies have demonstrated that these cells are of the monohistiocyte/macrophage lineage, which also includes rod cells or microglia (Budka, 1986, 1989; Dickson, 1986; Budka et al, 1987; Gray et al, 1987; Vazeux et al, 1987). MGC may have variable aspects, the extreme forms of which are 'macrophage-like' MGC (Figure 6.1a) with a large, well-demarcated cytoplasm containing peripheral lipid vacuoles, and the smaller 'microglia-like' multinucleated cells with prominent processes, scanty cytoplasm and elongated nuclei clustered at the centre of the cell (Figure 6.1b). Various techniques have demonstrated the presence of HIV genome or proteins in their cytoplasm (Koenig et al, 1986; Sharer et al, 1986; Wiley et al, 1986a; Budka et al, 1987; Pumarola-Sune et al, 1987; Vazeux et al, 1987), and ultrastructural studies have also shown retroviral particles either free in the cytoplasm or in cytoplasmic cisterns (Koenig et al, 1986; Budka et al, 1987; Gray et al, 1987; Meyenhofer et al, 1987). These particles are round or oval, have a diameter of approximately 100 nm, a double membrane and variable amounts of electrondense nucleoid. Mature forms with denser nucleoid, and a conical or tapering less dense shell in variable position are more frequently found (see Chapter 9, Figure 9.9), whereas immature forms with crescent-shaped densities and budding of HIV particles from membranes are less common. Koenig et al (1986), using co-culture, immunocytochemistry, in situ hybridization and electron microscopy, demonstrated that macrophages and macrophage-derived MGC are capable of synthesizing HIV RNA and may, therefore, be considered both the main reservoir and the vehicle of spread of the virus. The cytopathic effect of HIV on mononuclear macrophages, resulting in the formation of MGC, has been demonstrated in cultures infected with HIV (Lifson et al, 1986) and illustrated in human brain (Kato et al, 1987; Gray et al, 1988b,c). Nuclear bridges (Figure 6.2) have also been occasionally observed in MGC, suggesting that amitotic

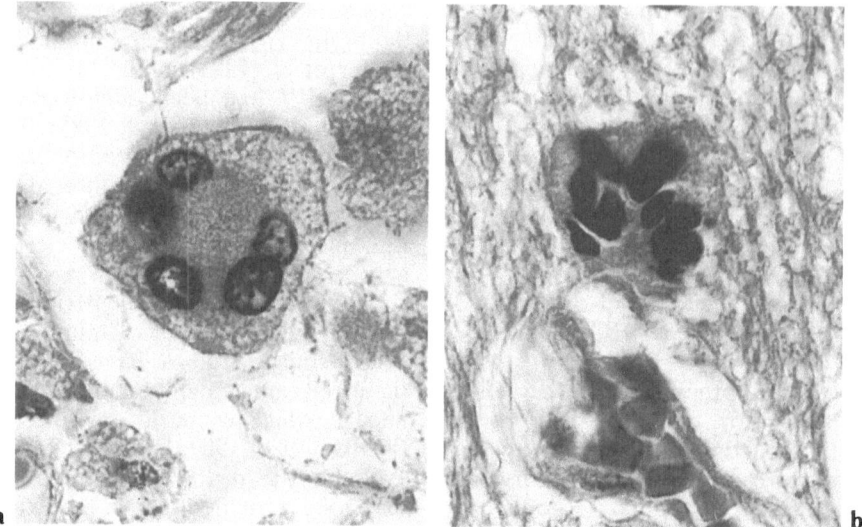

Figure 6.1a,b. Multinucleated giant cells. **a** 'Macrophage-like' multinucleated giant cell (MGC) of HIV encephalitis. Note the dense central area of cytoplasm, the ring of nuclei and the finely vacuolated peripheral cytoplasm. H&E, ×1200. **b** 'Microglia-like' MGC are smaller than the previous type and their clustered nuclei are surrounded by scanty cytoplasm. H&E, ×1200.

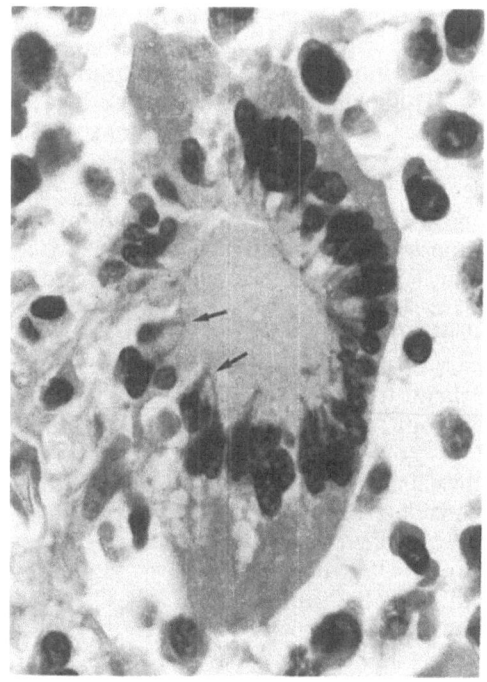

Figure 6.2. Multinucleated giant cells, some with nuclear bridges (*arrows*) which connect nuclei. H&E, ×1000.

nuclear division might account, in addition to cell fusion, for the formation of multinucleated cells (Gray et al, 1988b,c; Mizusawa et al, 1987, 1988). On the other hand, HIV proteins have been demonstrated by immunohistochemistry in brains of AIDS patients in the absence of MGC (Gabuzda et al, 1986; Hénin et al, 1987; Vazeux et al, 1987), suggesting that MGC represent a highly characteristic, but not essential, feature of HIV infection of the nervous system (see Chapter 7). However, as the number of MGC in AIDS brains varies considerably in different series, such observations may only reflect differences in brain sampling.

Cells with similar features have been observed in the liver, gastrointestinal tract, lymph nodes, spleen and brain of monkeys infected with the simian immunodeficiency virus (Sharer et al, 1987a). In contrast, in man, MGC have only exceptionally been observed outside the nervous system (Gray et al, 1990).

Other inclusions have been observed, at ultra-structural level, in the nervous system of HIV-positive patients. These have been named 'tubuloreticular structures' (TRS), and include tubuloreticular inclusions (TRI) and cylindrical confronting cisternae (CCC). TRI are composed of acidic glycoproteins and have been produced by α-interferon (Rich, 1981). They consist of aggregates of membranous tubules forming a compact honeycomb, whose single unit has a

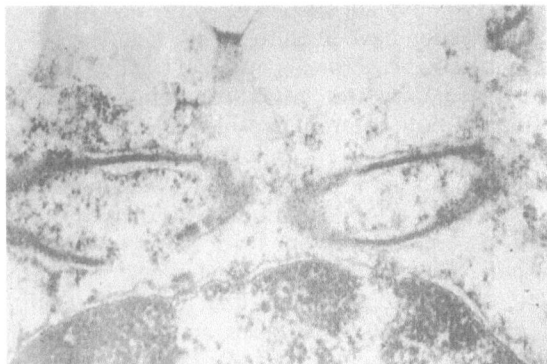

Figure 6.3. Cylindrical confronting cisternae in the cytoplasm of endothelial cells appear as relatively straight cylindrical structures, about 300 nm wide and of variable length. ×46 000.

uniform diameter of about 25 nm (see Figure 6.61 and Figure 10.16). TRI have been reported in systemic lupus erythematosus (Gyorkey et al, 1982), viral infections (Uzman et al, 1971) and autoimmune diseases (Helder and Feltkamp-Vroom, 1976), as well as in lymph node macrophages in one case of multiple sclerosis (Prineas and Wright, 1978). CCC appear as relatively straight cylindrical structures, about 300 nm wide, and of variable length (Figure 6.3); they are probably not directly due to viral proteins, but rather represent a cell response to viral infection (Jackson et al, 1979; Shamato et al, 1981; Kostianovsky et al, 1987). In patients suffering from AIDS, TRS have been demonstrated in extracerebral tissue (Sidhu et al, 1985; Kostianovsky et al, 1987), as well as in the brain (Lee et al, 1988; Scaravilli et al, 1989a) and in peripheral nerves (see Chapter 10). In nervous tissues, they are present in endothelial cells and there is no evidence of their presence in neuroectodermal cells.

Sidhu et al (1985) found TRI and CCC to be present in 85 and 41 per cent of AIDS patients respectively. The former were even more frequently found (92 per cent) in patients with the AIDS-related complex, whereas CCC were less common (8 per cent). On the basis of the high incidence of TRS in AIDS, it has been suggested that these inclusions could be considered a marker for HIV and of prognostic significance. However, it should be remembered that they are by no means specific for this illness.

In addition to the presence of MGC, another characteristic feature of HIV encephalitis/encephalopathy is the involvement of the white matter. Two main appearances, which may co-exist in one-third of cases (Figure 6.4), have been

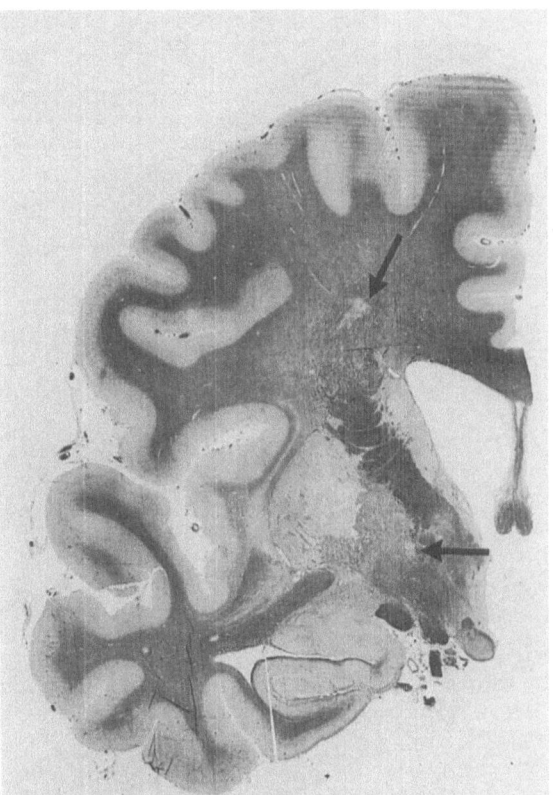

Figure 6.4. Coronal section of a cerebral hemisphere showing the co-existence of progressive diffuse leucoencephalopathy and HIV encephalitis. The former appears as myelin pallor of the centrum semiovale, the latter as small focal areas of necrosis (*arrows*).

described and are considered by some authors to be the extremes of a spectrum of HIV-induced pathological change (Budka et al, 1987; Lang et al, 1989). In a recently published consensus report (Budka et al, 1991) these two types of white matter lesions have been referred to as HIV-lep (leukoencephalopathy) and HIV-E (encephalitis) respectively. The former, which corresponds to the 'progressive diffuse leuko-encephalopathy' described by Kleihues et al (1985), is characterized by diffuse, ill-defined damage of the central white matter of both cerebral and cerebellar hemispheres, with sparing of the gyral white matter and compact myelin pathways (corpus callosum, internal capsule, optic pathways and cerebellar peduncles), and is occasionally associated with secondary degeneration of the pyramidal tracts (Horoupian et al, 1984; Dickson et al, 1989a). Microscopically, myelin loss of variable severity is associated with glial proliferation and diffuse and/or perivascular

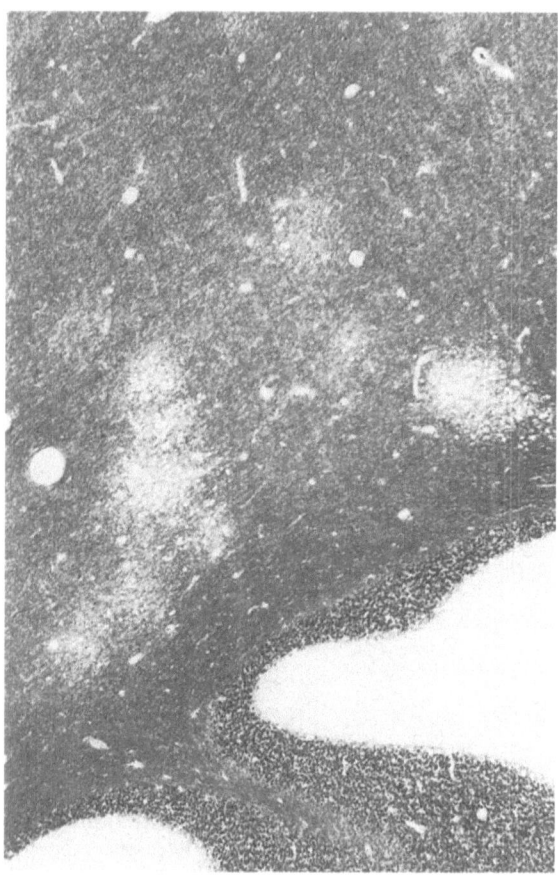

Figure 6.5. HIV encephalitis. Section of the cerebellum showing multiple areas of myelin necrosis. Bodian–Luxol, ×27.

infiltration by mono- and multinucleated cells of monohistiocyte/macrophage lineage (Budka et al, 1987, Gray et al, 1987; Lang et al, 1989), some of which may be topographically related to damaged myelin (Budka, 1989). The second type of lesion is characterized by more circumscribed, multifocal lesions (Figure 6.5) and has been previously termed 'subacute encephalitis' (Michaels et al, 1988) and 'multifocal giant cell encephalitis' (Budka 1986; Budka et al, 1987; Lang et al, 1989). In this form, clusters of MGC, rod cells, and/or macrophages, reactive gliosis, lymphocytic infiltrates and variable focal necrosis are disseminated in the CNS. Changes also involve basal ganglia and brainstem, and may show perivascular arrangement.

Clusters of MGC in the spinal cord have been mentioned in some instances (Grafe and Wiley, 1989; Rhodes et al, 1989; Rosenblum et al, 1989; Sharer et al, 1990). In one case, multifocal lesions with MGC restricted to the cord ('HIV myelitis') had caused a myelopathic syndrome (Geny et al,

1990). Other spinal lesions probably due to direct HIV infection have been reported. Rhodes (1987) described one case with pyramidal tract degeneration secondary to MGC-containing necrotic lesions of both internal capsules.

The incidence of encephalitis with MGC varies in different series: Hénin et al (1987) found one case in a series of 31 patients who died between June 1984 and November 1985; Petito et al (1986) described 42 cases (28 per cent) in a series of 149 patients collected between 1980 and 1985; Budka et al. (1987) found 38 cases among a series of 100 who died before August 1987; Lang et al (1989) observed 21 cases (15 per cent) in 135 brains between April 1981 and December 1987; and in Paris, we observed 40 cases in a series of 100 examined between August 1982 and December 1989. The possibility that these differences reflect also an actual yearly increase in incidence of MGC encephalitis, as pointed out by Petito et al (1986), is confirmed in our series: of the 22 patients who died before 1987, 6 (27 per cent) had MGC; of the 48 who died in 1987–8, the encephalitis was seen in 21 (41 per cent); and of the 30 cases who died in 1989, 13 (43 per cent) had MGC. This may also reflect changes in survival time related to improved diagnosis and treatment of opportunistic infections, such as *Pneumocystis carinii* pneumonia, cerebral toxoplasmosis or, to a lesser extent, cytomegalovirus infection.

In contrast to HIV-specific neuropathology, other changes of unknown etiology and pathogenesis have been inconstantly associated with the local presence of HIV:

Vacuolar Myelopathy

Vacuolar myelopathy (VM) (see also Chapter 8) was first described by Goldstick et al (1985), and Petito et al (1985). It is probably the most frequent spinal change found in AIDS patients, although its incidence varies considerably in the different series (Snider et al, 1983; Petito et al, 1985; Petito, 1988; Anders et al, 1986a,b; Berger and Resnick, 1987; de la Monte et al, 1987; Hénin et al, 1987; McArthur, 1987; Gray et al, 1988c; Dickson et al, 1989a; Grafe and Wiley, 1989; Lang et al, 1989; Sharer et al, 1990). These conflicting results are probably due to differences in the recruitment of patients: some series include exclusively adults, whilst others deal with children (Dickson et al, 1989a; Sharer et al, 1990); in some cases, studies have been made in neurological hospitals, in others investigations have been

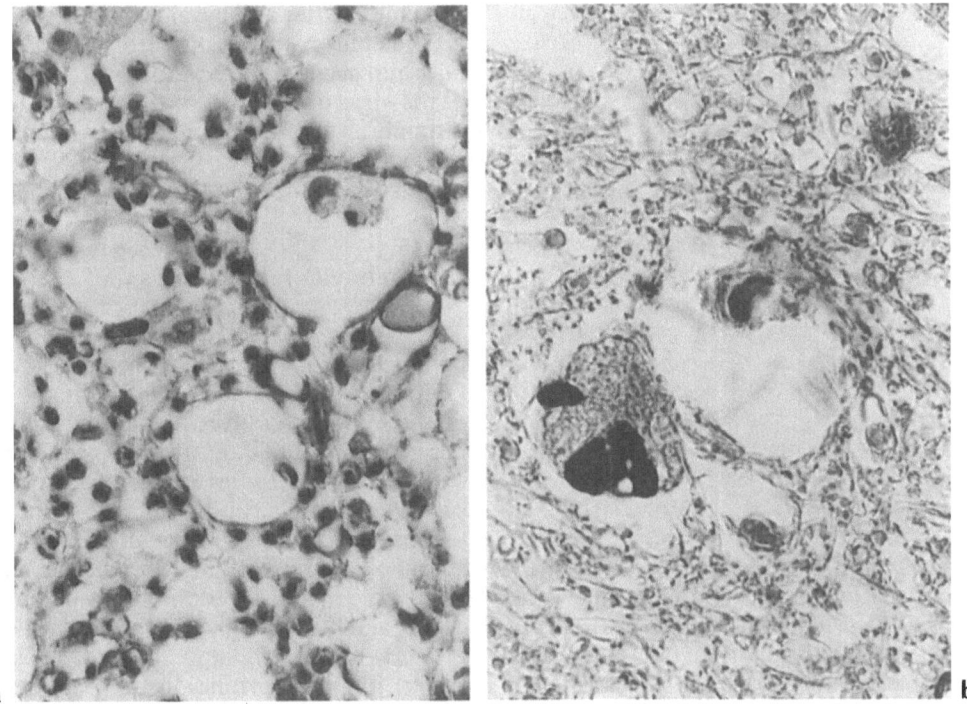

Figure 6.6a,b. Vacuolar myelopathy. **a** In the posterior columns of the thoracic cord many myelin sheaths are distended by vacuoles, one of which contains a macrophage. Bodian–Luxol, ×400. **b** An occasional vacuole may be lined by a MGC. H&E, ×1000.

carried out on patients admitted to general hospitals (Lang et al, 1989). In addition, they also probably reflect the variability of diagnostic criteria that the various investigators apply to the diagnosis of vacuolar myelopathy.

A detailed description of the morphological features of VM is given in Chapter 8. Suffice here to say that vacuoles usually contain lipid-laden macrophages (Figure 6.6a) and, ultrastructurally, they consist of intramyelin swelling with splitting of the lamellae and presence of lipid-laden macrophages. Axons are normal in mildly-affected areas, but may be swollen or disrupted in areas with severe vacuolation with consequent wallerian degeneration (Petito et al, 1985, 1986). The pathogenetic relationships of VM with HIV infection are debated. The frequent, although by no means constant, association of VM with HIV encephalitis (Price et al, 1986), the finding in some cases of MGC in close vicinity to the vacuoles (Budka et al, 1988; Gray et al, 1988c) (Figure 6.6b), and the observation of MGC, HIV antigens and HIV particles in the lesions (Maier et al, 1989) suggest that VM could be an HIV-induced spinal cord disease. On the other hand, the finding of VM in non-AIDS immuno-compromised patients (Kamin and Petito, 1988,

1991; Petito 1988), the rare occurrence of VM in children who usually have pure HIV encephalitis (Dickson et al, 1989a; Sharer et al, 1990), the presence of MGC in the spinal cord in the absence of vacuolation (Sharer et al, 1986, 1990; Geny et al, 1990), together with the dissociation between vacuolar changes and the presence of HIV antigens in some studies (Grafe and Wiley, 1989; Rosenblum et al, 1989) suggest that factors other than, or additional to, HIV are likely to take part in the production of VM.

Vacuolar lesions of the hemispheric white matter, with or without MGC, similar to those of VM, and usually associated with the latter, have been mentioned in a few cases (Petito et al, 1985; Budka et al, 1987; de la Monte et al, 1987; Rhodes 1987; Maier et al, 1989). In one case, 'multifocal vacuolar leukoencephalopathy' was the most prominent pathological finding, and its cortico-subcortical distribution appeared to mimic progressive multifocal leukoencephalopathy. However, immunohistochemistry and in situ DNA hybridization for papovaviruses were negative. In addition, immunostaining of mono- and multinucleated macrophages with HIV antibodies led some authors to propose that this entity could be a cerebral equivalent to VM

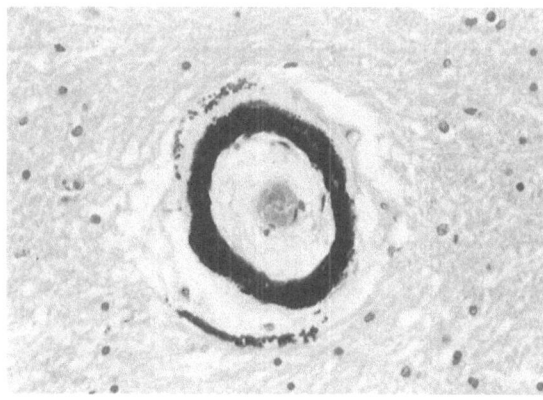

Figure 6.7. Mineralized vessel in the deep grey nuclei of a brain with HIV encephalitis. The lumen is considerably narrowed by proliferation of the intima. H&E, ×340.

(Schmidbauer et al, 1990). Another lesion of the spinal cord, strictly limited to the gracile tracts and appearing as symmetrical pallor of the white matter, was originally reported by Rance et al (1988). Its pathogenesis has been discussed by Scaravilli et al (1992), who compared the changes in the spinal cord with those in the sensory ganglia. They concluded that in spite of a correlation between sensory nerve cell loss and pallor of the tract, other mechanisms such as a 'dying back' process may contribute to the pathogenesis of the pallor. This subject will be further discussed in Chapter 8.

Meningitis

This type of aseptic meningitis, which may occur at the time of seroconversion and is usually attributed to HIV (Ho et al, 1985a; Levy et al, 1985; Gabudza and Hirsch, 1987), is described in Chapter 7. Probable morphological correlates include fibrous thickening of the leptomeninges, mononuclear inflammatory infiltrates and occasional MGC (Gray et al, 1988a).

Mineralization of Blood Vessel Walls

The mineralization of the blood vessel walls of the white matter and basal ganglia has been described in association with HIV encephalitis. In adults it is considered non-specific, but its high incidence in children deserves special consideration (Belman et al, 1986; see Chapter 9). A metabolic disturbance cannot be excluded. In two adult cases we observed massive mineralization (Figure 6.7) in

the white matter and basal ganglia associated with HIV nephropathy. Conversely, one must also take into account the possibility that the existence of the virus on the wall of the vessels may lead to intramural deposition of immune complex, with consequent vasculitis (Sotrel, 1989), a process similar to that seen in congenital rubella.

Involvement of the Grey Matter

This has received relatively little attention, and its pathogenesis is still unknown. de la Monte et al (1987) reported cortical gliosis in the majority of their cases. A diffuse poliodystrophy (DPD), described by Budka et al (1987) in about 50 per cent of their cases, was frequently associated with prominent gyral atrophy (Figure 6.8). It is characterized by diffuse astrocytic proliferation with variable amounts of glial fibrillary acidic protein in the grey matter, predominantly in the temporal cortex and basal ganglia, in the absence of hepatic or renal encephalopathy. Recently, Ciardi et al (1990) found a correlation between the severity of cortical gliosis and that of dementia and, in addition, noted that expression of GFAP was increased in asymptomatic cases. Cortical thinning and reduced number of cells per unit area, involving particularly the large neurons of the parietal lobe, was reported by Wiley et al (1990). Furthermore, in a morphological study, Ketzler et al (1990) showed an 18 per cent reduction of the neuronal density and a decrease of the perikaryon volume fraction by 31 per cent. These results were subsequently confirmed by Everall et al (1991) and Wiley et al (1991b); in addition the latter authors demonstrated a decrease in synaptic density in the cortex using immunohistochemical methods for synaptophysin as well as decreased dendritic branching using the Golgi method (Wiley et al, 1991a). Diffuse gliosis and nerve cell loss in the cortex of AIDS sufferers may result from a number of causes, including co-existing opportunistic infections, cerebrovascular disease or anoxia. However, in a number of cases with DPD studied by Sinclair and Scaravilli (1992), none of these causes were present. Moreover, detection of HIV by polymerase chain reaction in the cortex of these cases suggests that HIV may be involved in the pathogenesis of cortical pathology. Spongiform changes, of the type seen in Creutzfeldt–Jakob disease (CJD), have also been described. In one case, in which they were associated with glial proliferation (Schwenk et al, 1987), in situ hybridization revealed HIV DNA and RNA

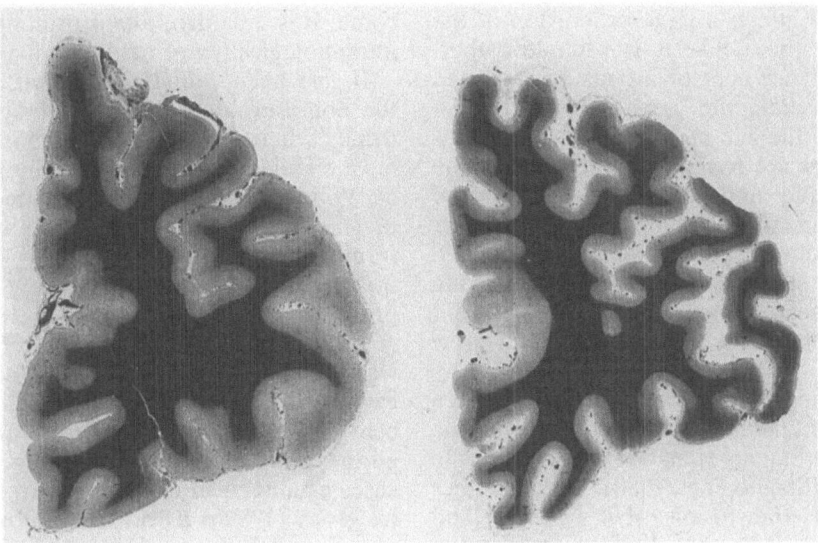

Figure 6.8. Diffuse poliodystrophy. The frontal lobe of a 28-year-old patient (*right*) showing considerable gyral atrophy, although myelin appears normal. A control section is seen on the left. Loyez.

(Gosztonyi et al, 1988). In the five additional cases reported by Artigas et al (1989), spongiform degeneration was present not only in the cortex, but also in the basal ganglia and thalamus. On the other hand, the white matter showed patchy areas of vacuolar leukoencephalopathy, and MGC could be found only in three cases. Diffuse necrotizing cortical encephalopathy with HIV antigens and HIV-like particles was also described in one case (Clague et al, 1988).

The most frequent type of pathology in HIV-positive patients is that due to opportunistic infections and reflects the severe defect in T-cell mediated immunity (Fauci et al, 1984, 1985; Lane et al, 1985). The frequency with which each opportunistic organism has been found by the various authors differs among areas and in different risk groups: toxoplasmosis, for instance, appears to occur more frequently in France than in Britain and, in America, it has been reported almost four times more frequently in Haitians with AIDS (Moskowitz et al, 1984) than in a homosexual male population (Nielsen et al, 1984; Lemann et al, 1985).

Opportunistic Infections: Viral

Viruses producing opportunistic infections in HIV-positive patients may either enter the body via the same route used by HIV or, being already present in the individual, be reactivated by the immunosuppression. Among the latter, cytomegalovirus (CMV) is the most commonly found (Hawley et al, 1983; Moskowitz et al, 1984; Nielsen et al, 1984; Morgello et al, 1987) and is followed in frequency by papovavirus (Hawley et al, 1983; Snider et al, 1983; Moskowitz et al, 1984; Levy et al, 1985; Morgello et al, 1987). Two other viruses of the herpes group, H. simplex and Varicella zoster virus (VZV), are also implicated in AIDS, but occur much less frequently than CMV or papovavirus. H. simplex encephalitis (Dix et al, 1985) and myelitis (Britton et al, 1985; Tucker et al, 1985) are usually found in association with CMV. Although VZV is rarely the cause of severe lesions, these may pose diagnostic problems since they may share clinical and radiological features with PML and cerebrovascular diseases.

Cytomegalovirus Encephalitis

Cytomegalovirus (CMV) belongs to the family of Herpesviridae and has been identified as a pathogen in several mammals, including mice, guinea pigs and primates. The human forms, several hundreds of which have been identified by restriction enzyme analysis (Huang et al, 1980; Yow et al, 1982), are enveloped viruses measuring approximately 200 nm in diameter and containing a 110 nm icosahedral capsid. CMV is ubiquitous, and infects a large proportion of the population. In the USA, it is estimated that, whereas only 1 per cent of infants are infected in utero and secrete the virus at birth (Stagno et al, 1982), by the age of 40 years complement-fixing antibodies are found in 80 per cent of the population (Gold and Nankervis, 1982). The infection predominates in people living in under-developed countries and in densely-populated areas (Stagno et al, 1982), as well as among sexually-promiscuous homosexuals, 90 per cent of whom have been found to be infected (Drew et al, 1981).

CMV infection may occur via the placenta (Alford and Britt, 1985), through contact with the cervix at birth (Stagno et al, 1975) and from the maternal milk (Stagno et al, 1980). Other sources include blood transfusion (Ho, 1982), and organ transplant (Fiala et al, 1975). However, the most important routes are the oral and respiratory, as demonstrated by outbreaks of infection in schools (Hutto et al, 1985). Sexual contacts have also been shown to be an important route, the diffusion taking place through cervical secretions or semen (Davis et al, 1975; Lang and Kummer, 1975; Persson et al, 1979; Willmott, 1975). CMV shares with HIV some biological characteristics, which are relevant for the pathogenesis of brain lesions among HIV-positive patients: it produces immunosuppression (Rinaldo et al, 1977; Carney et al, 1981; Schrier et al, 1986) by infecting T-cells (Schrier et al, 1985) and monocytes (Rice et al, 1984).

In healthy young adults, primary CMV infection may occur as an asymptomatic illness or may produce a mononucleosis-like infection (Klemola et al, 1969; Klemola, 1973), both of which are cleared by humoral as well as cell-mediated immunity (Alford and Britt, 1985); however, the virus becomes latent and, given appropriate conditions, can be reactivated.

Neuropathology

Before the appearance of AIDS, CMV encephalitis had been described in immunocompromised (Schneck, 1965; Dorfman, 1973; Schober and Herman, 1973; Yanagisawa et al, 1975; Hotson and Pedley, 1976; Linneman et al, 1978; Kauffman et al, 1979; Zaia and Lang, 1984) as well as in normal hosts (Perham et al, 1971; Chin et al, 1973; Phillips et al, 1977; Duchowny et al, 1979; Cohen and Corey, 1985). It must be emphasized, however, that, whereas in immunocompromised patients the illness has a fatal outcome, it is a benign, albeit prolonged, disease in immunologically normal individuals.

It has been found that about 7 per cent of the homosexual population excrete CMV in the urine, whereas almost all homosexual patients have circulating antibodies to the virus (Drew et al, 1981). Furthermore, CMV infection has been diagnosed in almost all HIV-positive patients (Quinnan et al, 1984), among whom it is a frequent cause of death. Indeed, CMV has become probably the most frequent neuropathological complication of AIDS (Anders et al, 1986a): it was found in 30 per cent of the cases in the series examined by Morgello et al (1987), and in 26 per cent of the brains studied by Petito et al (1986); an incidence of 25 per cent was found among the cases examined in London, whereas Vinters and his group (1989b) report CMV infection in 15–20 per cent of their cases. Conversely, the incidence in Switzerland (Lang et al, 1989) was only 10 per cent.

In most cases, the nervous system is one of several organs involved. Indeed, in all the cases examined by Morgello et al (1987) more than one organ appeared to be affected, and Vinters et al (1989b) found multiple lesions in 26 out of 31 cases. The other organs involved included, in order of frequency, the lungs (17 cases), the eye and adrenals (14 cases), the gastrointestinal tract (11 cases), the liver (4 cases), the pancreas, thyroid and spleen (3 cases), the lymph nodes and prostate (2 cases) and the kidney and pituitary (1 case). Recent investigations indicate that CMV in HIV-positive patients correlates with rapid progression of the illness (Webster et al, 1989).

Macroscopic findings are disappointing, and consist of small areas of softening in the centrum semiovale or in the periventricular regions. In one personal case, similar foci were present in the cerebellar folia and a larger area of necrosis occupied one side of the basis pontis. The latter lesion resembled the poorly-defined regions of necrosis shown in the midbrain and pons of one of the cases of Vinters et al (1989b). The same authors describe the unusual findings of haemorrhagic necrosis around the lateral ventricles and aqueduct, a finding similar to that reported by Morgello et al (1987). The latter, however, emphasize the point that in ten of their cases it was impossible to establish whether the lesions were due exclusively to CMV, because of the co-existence of other organisms.

Microscopic lesions were subdivided by Morgello et al (1987) into five groups: (1) microglial nodules (MGN); (2) isolated inclusion-bearing cells; (3) focal parenchymal necrosis; (4)

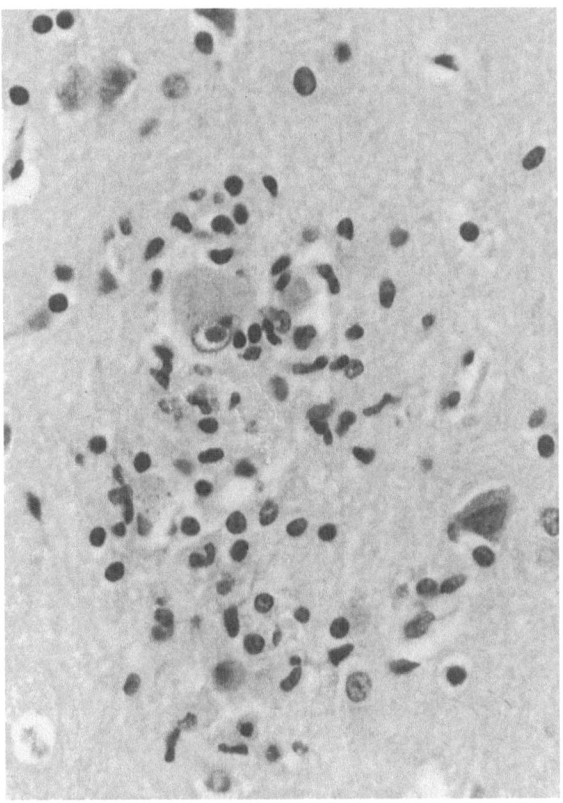

Figure 6.9. CMV encephalitis. A poorly-circumscribed microglial nodule in the cortical grey matter includes a large astrocyte bearing an intranuclear inclusion. H&E, ×300.

necrotic ventriculo-encephalitis; and (5) necrotizing radiculomyelitis. The present chapter will describe the first four lesions, radiculomyelitis being included among the discussions on spinal lesions (Chapter 8) and the peripheral nervous system (Chapter 10). In addition, it will deal with CMV lesions in the eye, an organ which appears to be affected in a considerable number of cases.

Microglial Nodules. The presence of MGN characterizes the 'subacute encephalitis' described by Snider et al (1983). They were found by Morgello et al (1987) in all their cases, and in 29 out of 31 patients studied by Vinters et al (1989b). The MGN are situated predominantly in the cortical grey, but also in the subjacent white matter and in the deep grey nuclei, and consist of rod cells and macrophages. They also include occasional astrocytes, some of which may contain intranuclear inclusions (Figure 6.9). Usually, the cells are densely packed, but on occasion they appear loosely arranged in a microcystic background. MGN vary in number from case to case,

and statistical analysis has revealed that they are most numerous in the basal ganglia, diencephalon and brainstem (Morgello et al, 1987). The same authors found that only as few as 6.5 per cent of the MGN contained viral inclusions, and that the frequency of inclusion-bearing nodules paralleled the frequency distribution of the MGN in the various regions of the brain.

Whether microglial nodule encephalitis is due to CMV or to a mixed CMV and HIV infection, is still a controversial subject for some authors. Nielsen and Davis (1988) describe subacute encephalitis as characterized by cortical nodules, some of which contain CMV, and by lesions of the white matter which include MGC. The problem was investigated and discussed by Vinters et al (1989b) who assessed the number of MGN, CMV inclusions and MGC in a number of brains. They found that MGN and inclusions co-localized in 12 per cent of the sections (chi-squared testing, however, failed to show a significant association), and evidence of both CMV and HIV infection in the same patient was found in 7 out of 31 brains, in which the two types of histological lesions were found in close association. Moreover, MGN were found with MGC more often than with CMV. The authors concluded that the MGN encephalitis in AIDS can be produced either by CMV or HIV, or by both. On the other hand, Schmidbauer et al (1989a) observed that in a number of nodules without obvious CMV inclusions or MGC, significant amounts of CMV DNA could be found by in situ hybridization. Furthermore, no CMV could be detected in cases with MGC. They conclude that MGC encephalitis is a separate entity from the 'subacute encephalitis'. It should be added that these nodules are similar to those described in cases of CMV encephalitis unrelated to AIDS (Schneck, 1965; Evans and Williams, 1968; Vortel and Plachy, 1968), and those produced in guinea pigs inoculated intracerebrally with the virus (Booss et al, 1988).

Isolated Inclusion-Bearing Cells. Morgello et al (1987) found isolated cells containing inclusions in an otherwise normal parenchyma in 50 per cent of their cases. In this respect, brains of HIV-positive patients differ from those described before the outbreak of AIDS, in which these findings were rarely seen. Previous studies (Myerson et al, 1984; Wiley et al, 1986b) had shown that the virus can be demonstrated in histologically normal cells; it becomes morphologically demonstrable only when full replication is possible, and this may take place only in

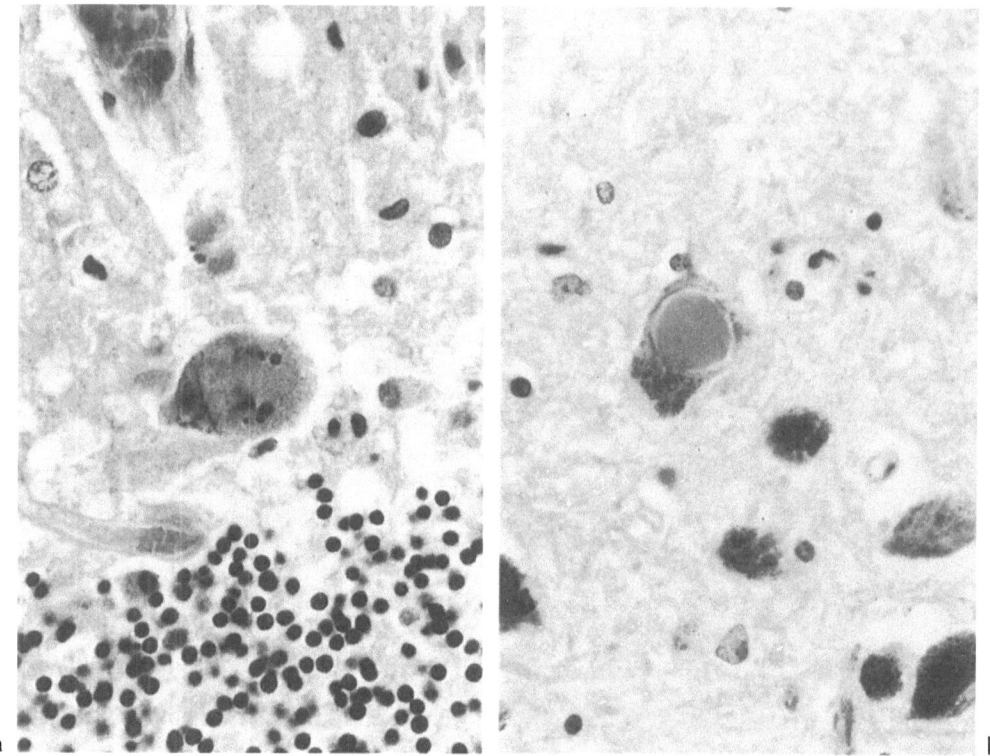

Figure 6.10a,b. CMV encephalitis. **a** Numerous small inclusions of CMV may also be seen in the cytoplasm of astrocytes. H&E, ×450. **b** The nucleus of a nerve cell in the substantia nigra is distended by an inclusion of CMV. H&E, ×470.

occasional cells in an otherwise non-permissive environment.

Inclusions are most frequent in astrocytes (Figure 6.9), where they are situated in the nucleus or, as multiple small formations, in the cytoplasm (Figure 6.10a). However, they may also be present in neurons (Figure 6.10b) and endothelial cells. In the latter, inclusions were seen by Morgello et al (1987) in one-third of the cases, but only 'rarely' by Vinters et al (1989b), even with in situ hybridization techniques. Association of endothelial inclusions and vasculitis is an even rarer event, which was described only in four cases (Vinters et al, 1989b). Similar findings had previously been reported by Goodman and Porter (1973) and Koeppen et al (1981) in HIV-negative patients.

Focal Parenchymal Necrosis. Areas of parenchymal necrosis remote from the ventricles are relatively uncommon: they were present in 2 out of 31 cases and 4 out of 30 cases in the series of Vinters et al (1989b) and Morgello et al (1987) respectively. In our cases four areas were found in three brains, and these were located in the

cerebellum (two) and pons (two). In the cerebellum, one area was situated in the hemispheric white matter and the other in the dentate nucleus. The two foci in the pons (Figure 6.11) involved the basis. Necrosis included neuronal and glial cells, which were replaced by foamy macrophages and reactive astrocytes, many of which contained intranuclear inclusions. Small inflammatory cells were scanty, except in one of the pontine lesions, which contained a massive number of polymorphs. It is believed that this presentation corresponds to a more invasive behaviour of the virus, and is considered by Nelson et al (1990) as unique to AIDS.

Ventriculo-encephalitis. This variety, sometimes associated with choroid plexitis, was found by Morgello et al (1987) in three brains, two of which showed association with HSV-1. In most cases, the necrotic area is pale and grey, and can often be overlooked macroscopically. Multiple foci may be present in the same brain. One of the cases examined by Vinters et al (1989b) was exceptional in showing a line of haemorrhagic necrosis in the lower part of the corpus callosum.

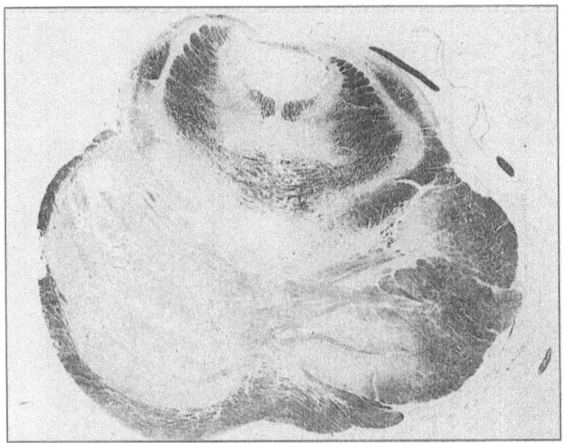

Figure 6.11. CMV encephalitis. Transverse section of the pons showing a large necrotic area resulting from CMV encephalitis. The lesion involves the whole left, and the medial part of the right half of the basis. Luxol fast blue/cresyl violet (LFB/N).

Using myelin staining, necrotic areas appear completely demyelinated and relatively well demarcated from the surrounding normal brain (Figure 6.12). The occurrence of pure demyelination in association with CMV has been reported by Moskowitz et al (1984), but the finding has not been confirmed by other laboratories. Additional histological features include patchy loss of the ependymal lining, oedema, and presence of large inclusion-bearing astrocytes (Figure 6.13a). An occasional ependymal cell may contain an eosinophilic intranuclear inclusion (Figure 6.13b). Inflammatory cells are usually scanty. Hawley et al (1983) and Post et al (1986) emphasize the

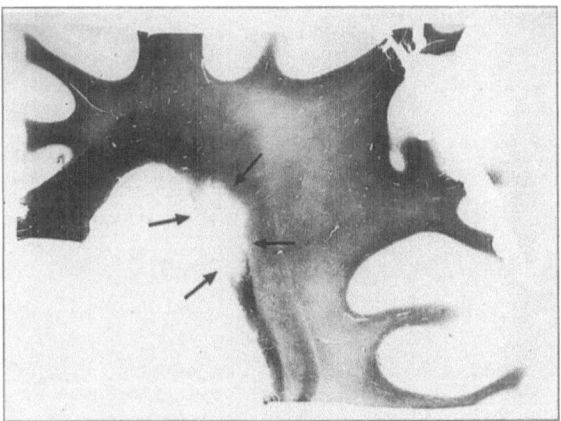

Figure 6.12. CMV ventriculitis. Coronal section of the cerebral hemisphere showing a well circumscribed area of CMV necrosis (*arrows*) adjacent to and impinging on the lateral ventricle. LFB/N.

affinity of CMV for the ependymal surfaces from which the virus moves in a ventriculofugal direction (Wiley et al, 1986b).

Cytomegalovirus Retinitis. This is the commonest cause of blindness in HIV-positive patients (Rosenberg et al, 1983; Khadem et al, 1984). CMV has a predilection for early involvement of the macula and the optic nerve head, and the lesion appears morphologically as foci of necrosis of the neurons of the retina, which are replaced by inflammatory small-cell exudate, with occasional aggregates of pigment and many spindle-shaped cells. At the edge of the necrosis, numerous retinal cells contain intranuclear (Figure 6.14) and intracytoplasmic CMV inclusions. The choroid contains only scanty inflammatory cells (Skolnik et al, 1989).

Although lesions of CMV can co-exist in brains with foci of toxoplasmosis, mycosis, PML and other infections they are not usually topographically related to one another. An exception to this rule is represented by the cases in which CMV is associated with HSV-1 (Pepose et al, 1984; Tucker et al, 1985; Laskin et al, 1987; Morgello et al, 1987). These resemble those described by Yanagisawa et al (1975) in HIV-negative patients.

The importance of CMV in AIDS, however, is not limited to its direct damage to the nervous system. Indeed, the presence of macrophages induced by this virus represents a crucial point in the progression of the HIV-produced illness (see Chapter 11). Macrophages derive from circulating monocytes and, if the latter are HIV-infected, they can introduce the virus into the nervous system. Transformation of monocytes into macrophages is an important step in activating HIV-replication in latently infected cells, and it has also been demonstrated in vitro that this change is necessary in order to obtain optimal transcriptional activity of HIV (Griffin et al, 1989). Furthermore, recent in vitro studies have shown a possible stimulation of HIV by co-infection with DNA viruses (Gendelman et al, 1986). In particular, it was found that CMV increases the expression of HIV (Gendelman et al, 1989), and that an immunoglobulin fraction induced by CMV allows immune complexes of HIV to infect fibroblasts, which are otherwise not permissive to HIV infection. A combined action of HIV and CMV has been suggested in HIV encephalopathy (Wiley and Nelson, 1988; Grafe and Wiley, 1989; Bélec et al, 1990a), as well as in demyelinating lesions of the peripheral nervous

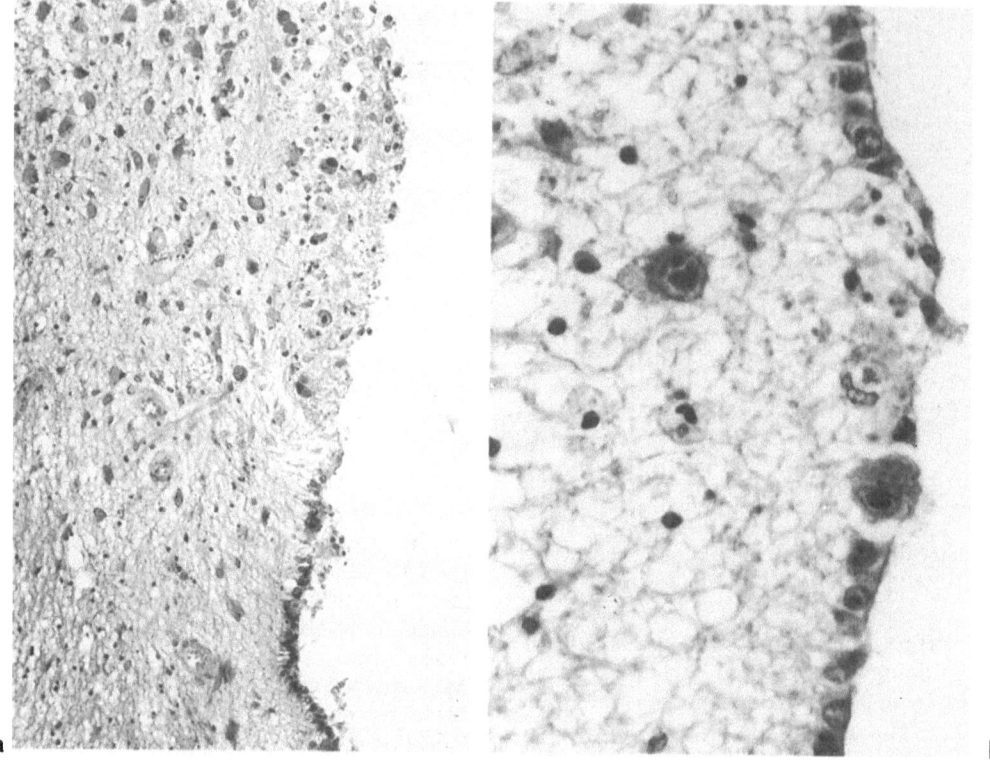

a   b

Figure 6.13a,b. CMV ventriculitis. **a** The ependymal lining is partly lost and the sub-ependymal region contains many large astrocytes, some of which show intranuclear inclusions. H&E, ×115. **b** Ependymal cell containing one intranuclear inclusion. H&E, ×470.

system (Dalakas and Pezeshkpour, 1988). The possibility of an interaction between the two viruses has been demonstrated in vitro (Skolnik et al, 1988), and is consistent with observations in vivo (Bélec et al, 1990a,b).

Clinico-pathological Correlations

In order to correlate the clinical presentation with the pathological findings produced by CMV in HIV-positive patients, lesions complicated by other organisms or tumours must be excluded.

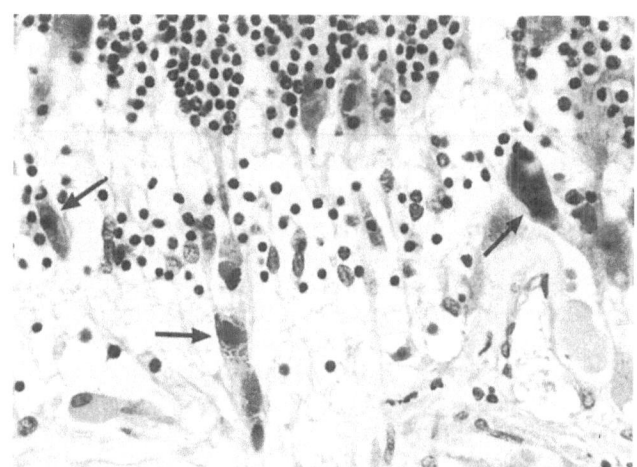

Figure 6.14. CMV retinitis. Numerous retinal cells show intranuclear inclusions (*arrows*). H&E, ×300.

Only half the cases studied by Vinters et al (1989a) were 'pure'. A review by Vinters et al (1989b) of the radiological findings (CT and MRI) showed no abnormalities in a large number, and cortical atrophy only in a minority. In a CT study of 200 patients with neurological symptoms, CMV encephalitis was confirmed in one case only, in which it co-existed with HSV-1 encephalitis, and CT showed only atrophy (Levy et al, 1986). Clinically, a number of cases were asymptomatic, and this was confirmed in the cases studied in London. Indeed, well-defined clinical signs, including hemiplegia and lower motoneuron facial palsy, were seen only in one patient with ventriculitis and pontine involvement.

Progressive Multifocal Leukoencephalopathy

Progressive multifocal leukoencephalopathy (PML), described originally by Åstrom et al (1958), is characterized morphologically by multiple confluent foci of demyelination, pleomorphic astrocytes, and enlarged oligodendrocytes containing intranuclear inclusions. In 1965, Silverman and Rubinstein, and ZuRhein and Chou observed that the inclusions contained icosahedral particles, 39 nm in diameter. These were interpreted by the latter authors as papovavirus, similar to the SV-40; in 1971 Padgett et al succeeded in cultivating the virus from a patient with PML onto primary cultures of human fetal glial cells. The viral isolate was called JC, after the initials of the original patient's name. Antibodies to JC are present in approximately 70 per cent of the population, although people living in remote areas of the world tend to be free from the infection (Brown et al, 1975).

After an initial, usually subclinical, infection in childhood, the virus persists in the kidneys and reaches the nervous system, probably via the blood stream. The virus exists as unintegrated forms of DNA sequences in oligodendrocytes, astrocytes, and possibly endothelial cells, in capillaries of the brain (Dorries et al, 1979), as well as in other organs of patients with PML (Grinnell et al, 1983a). Restriction enzyme cleavage analysis shows that JC viral genomes vary in different patients with PML (Grinnell et al, 1983b; Martin and Foster, 1984; Martin et al, 1985). The isolation of multiple such genomes from urine and brain of the same patient raises the question of whether the neurological illness is produced by a virus resident in the kidneys and reactivated by the immunosuppression, or by an exogenous virus (Dix and Bredesen, 1988).

Not all cases of PML are produced by JC virus. In two cases, Weiner et al (1972) isolated SV-40. This virus was initially thought to be a laboratory contaminant, but was subsequently found to be localized in the brain (Penny et al, 1972; Weiner et al, 1972), and it is estimated that 2–4 per cent of the human population have antibodies to it. However, to date, there are no cases in HIV-positive patients produced by this virus.

PML exists worldwide, and in the vast majority of cases is an opportunistic infection. It occurs in immunosuppressed individuals (Brook and Walker, 1984), and has also been described in immunosuppressed monkeys (Gribble et al, 1975). In a few cases, however, no known underlying illness could be detected (Silverman and Rubinstein, 1965; Fermaglich et al, 1969; Bolton and Rozdilsky, 1971). It begins insidiously and is accompanied by focal neurological signs which include sensory and motor abnormalities, impaired vision, dysarthria, mental deterioration and cerebellar signs. Headache and fever seldom occur. The disorder has a rapidly progressive course, which usually leads to death within one year. Exceptions to this rule have, however, been reported: Kepes et al (1975) and Hedley-White et al (1966) described patients who survived 10 and 5 years respectively. The patient reported by Price et al (1983) suffered from Hodgkin's lymphoma, and developed an encephalopathy whose nature was histologically confirmed to be PML. The unusual feature, in this case, was that after progressing for several months, the illness stabilized for almost one year before the patient further deteriorated and died. CT scans of the brain showed multiple focal areas of hypodensity localized to the white matter. Unlike the situation in tumours, there is neither mass effect nor contrast enhancement.

Neuropathology

PML was described in AIDS patients as early as 1982 by Miller et al. Further reports are those by Bedri et al (1983), Snider et al (1983), Bernick and Gregorios (1984), Katlama et al (1984), Ho et al (1984) and Blum et al (1985). In large series, the incidence varies from 2 per cent (Petito et al, 1986) and 5 per cent (Budka et al, 1987; Vinters et al, 1989a), to 7 per cent (Lang et al, 1989). In the brains examined in London, the incidence was 6 per cent. These figures compare with 8 per cent observed by Hooper et al

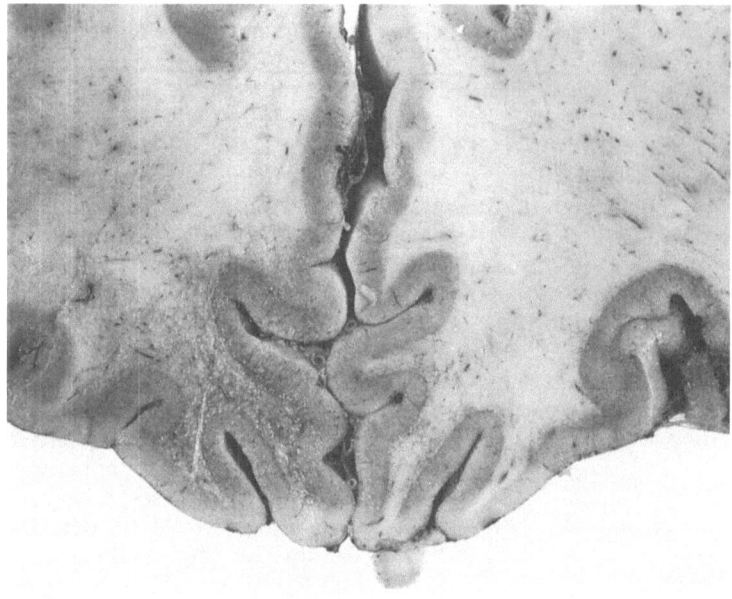

Figure 6.15. Progressive multifocal leukoencephalopathy. Coronal section of the frontal lobes showing grey discolouration and softening of the white matter of the orbital gyri.

(1982) in immunosuppressed HIV-negative patients.

Macroscopic lesions are similar to those occurring in HIV-negative patients, but are often more extensive. The white matter appears white–grey or yellow and granular (Figure 6.15), and its consistency is soft or even semiliquid.

Histological changes are almost exclusively confined to the white matter, and include con-

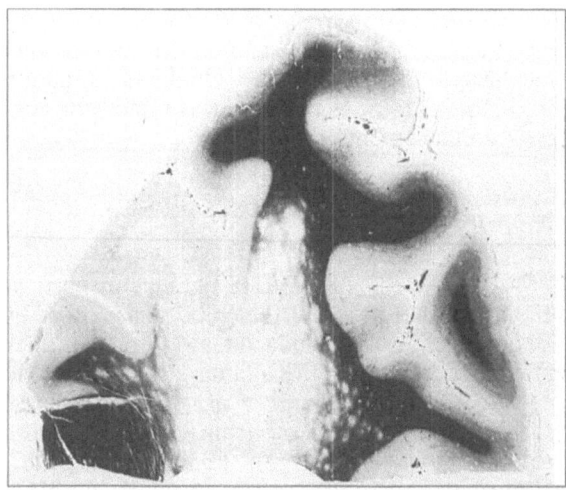

Figure 6.16. Coronal section of the brain showing multiple foci of myelin loss resulting in a large area of demyelination. LFB/N.

fluent foci of myelin loss (Figure 6.16) and presence of foamy macrophages and reactive astrocytes (Figure 6.17a). Many of the astrocytes may have bizarre hyperchromatic nuclei (Figure 6.17b) measuring up to 45 µm, and abundant cytoplasm. A number of oligodendrocytes show greatly enlarged nuclei containing homogeneous amphophilic inclusions, which displace the chromatin towards the nuclear membrane (Figure 6.17c). The inflammatory reaction is usually scanty and limited to small amounts of lympho-cytes around blood vessels (Figure 6.17d). The adjacent grey matter may show reactive gliosis and may contain oligodendrocytes with intra-nuclear inclusions.

Papovavirions can usually be demonstrated in brain tissue in cases of PML. In its typical ultra-structural appearance, the virus consists of round and rod-shaped particles (Figure 6.18a), 45 nm in diameter, within the nuclei of oligodendrocytes or, on occasion, also in processes of astrocytes (Figure 6.18b). Papovavirus usually develops in the following sequence within the host cells (Baker and Rayment, 1987): (1) viral precursor material assembles in the host cell cytoplasm; (2) this material then passes into the nucleus; (3) individual viral assembly occurs in the nucleus, both round and rod particles being formed; (4) with the increase in the number of mature par-ticles within the nucleus, the latter's membrane

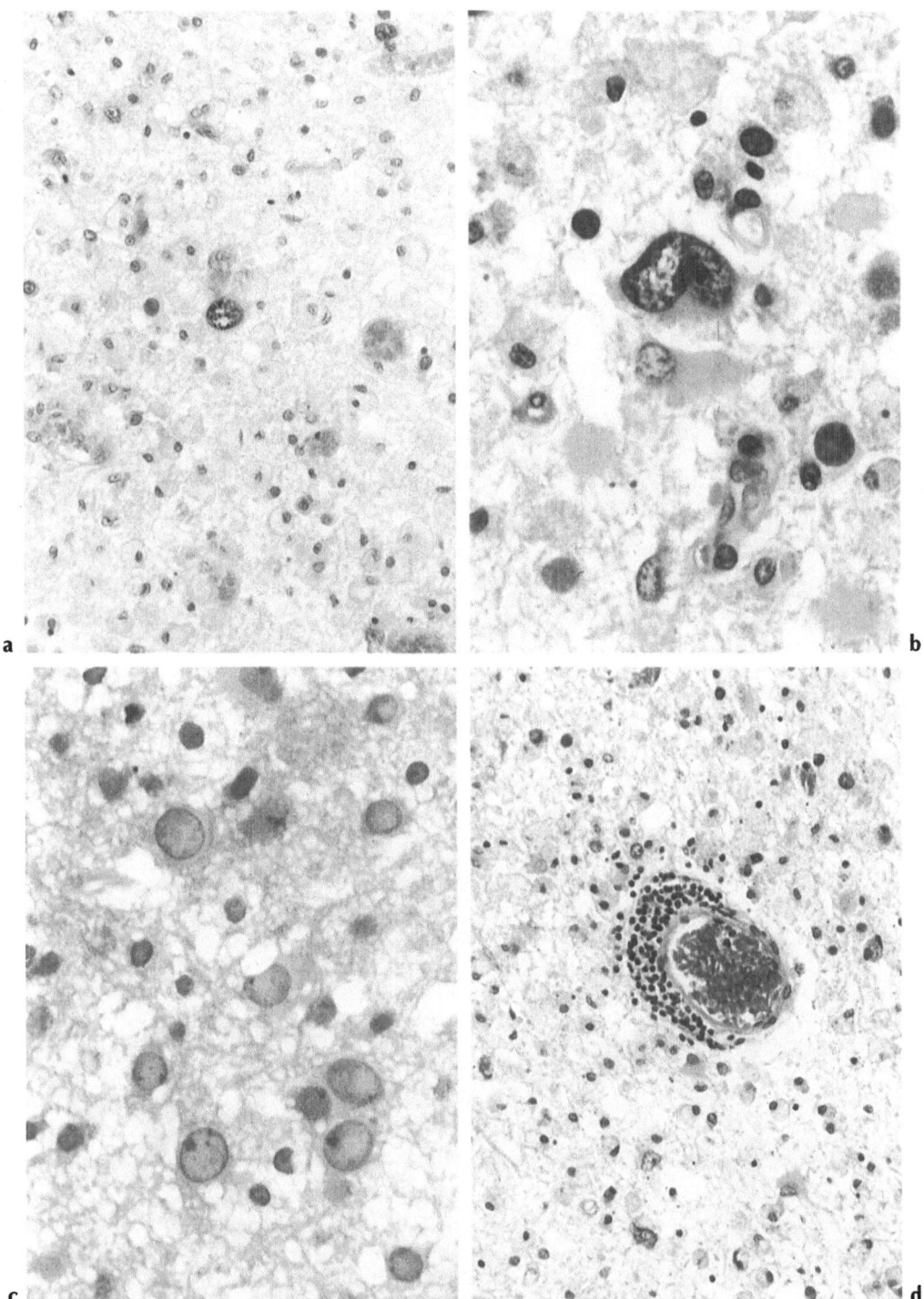

Figure 6.17a–d. Histological changes in progressive multifocal leukoencephalopathy (PML). **a** Severe myelin breakdown, oedema and presence of large foamy macrophages and reactive astrocytes. H&E, ×190. **b** Atypical astrocytes with bizarre hyperchromatic nuclei are a characteristic feature of PML. H&E, ×470. **c** Many oligodendrocytes have enlarged nuclei containing amphophilic inclusions. H&E, ×470. **d** In PML the inflammatory reaction is limited to moderate amounts of lymphocytes cuffing thin walled blood vessels. H&E, ×190.

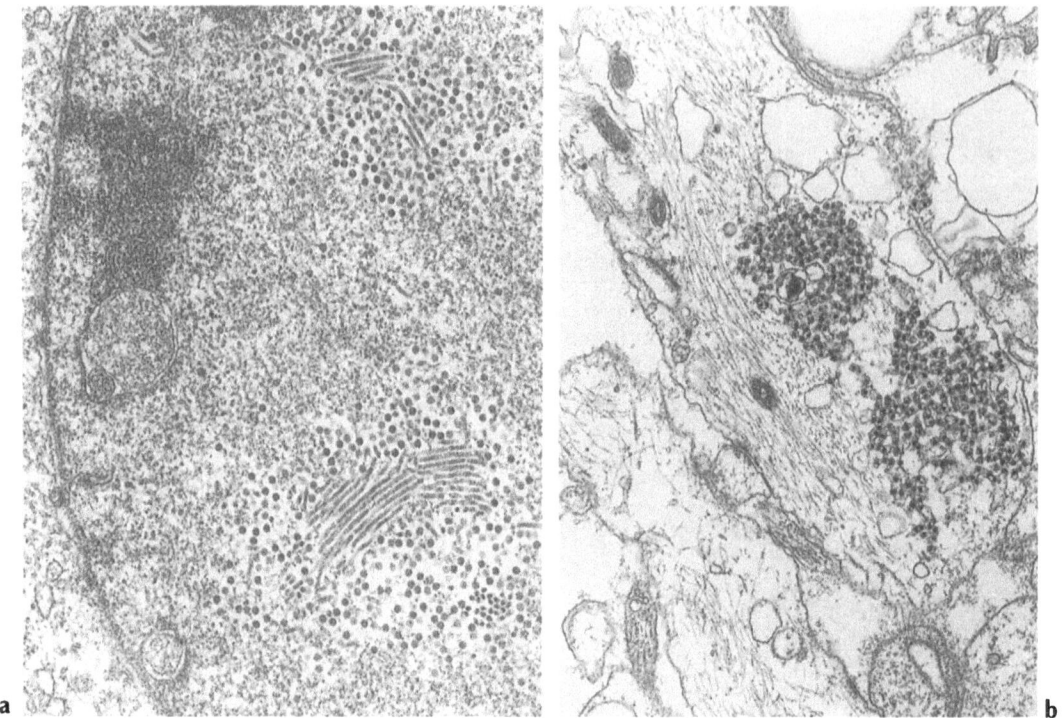

Figure 6.18a,b. Progressive multifocal leukoencephalopathy. **a** Electron micrograph showing round and elongated viral particles in the nucleus of an oligodendrocyte. ×22000. **b** Round viral particles are seen in an astrocytic process. ×15000.

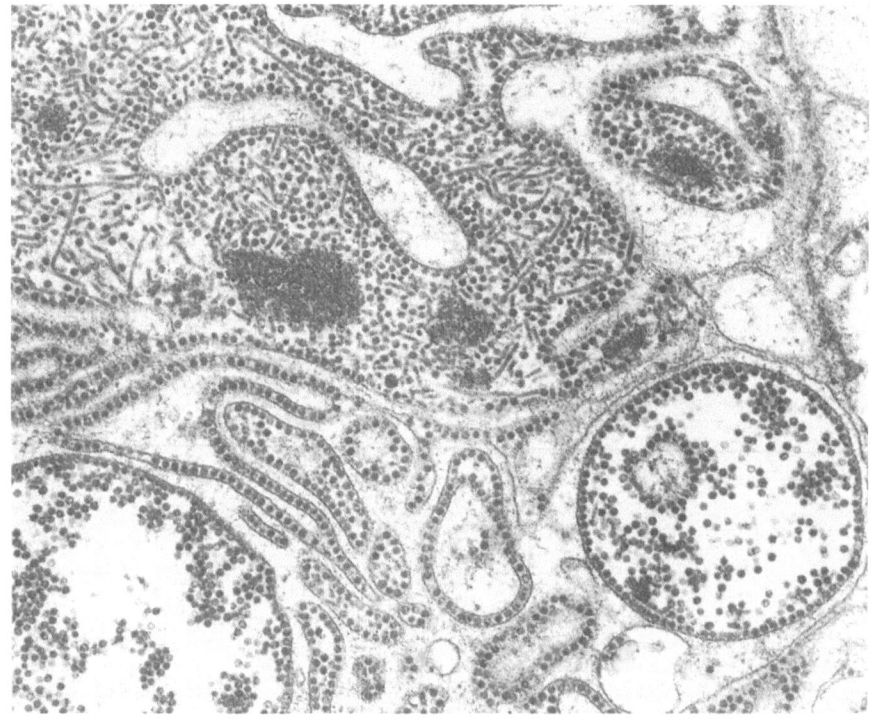

Figure 6.19. Electron micrograph showing the remains of a cell in which the normal architecture is replaced by masses of membrane systems and attached viral particles. ×24000.

starts to disintegrate; (5) virions, no longer enclosed by the nuclear membrane, become enveloped by a new cytoplasmic membrane system from which they are liberated on the final disintegration of their host cell. Steps 3–5 are usually seen in cases of PML, although unusual features were seen by Scaravilli et al (1989a) in brain biopsies of two HIV-positive patients. In one case, abnormally large aggregations of precursor material were seen lying within the cytoplasm; in the other, intranuclear viral assembly was achieved via large centralized precursor masses, and extensive new membrane systems developed within the host cell nuclei before the encompassing nuclear membranes had started to disintegrate. An additional unusual finding in the late stages of infected cells was the florid proliferation of intracytoplasmic membrane systems (Figure 6.19). The authors suggested that these unexpected features could be due to stimulation of JC by HIV as the latter was diagnosed at postmortem in one of the cases by the presence of MGC, and confirmed by immunohistochemistry.

Association between PML and HIV is not an uncommon finding and, indeed, the first cases in which HIV was detected from within the nervous system were cases of PML (Koenig et al, 1986). Orenstein and Jannotta (1988) reported a patient in whom HIV particles co-existed with papovavirus, and suggested a synergistic action of the two viruses. This possibility is supported by the observation that, in PML, a number of macrophages fuse into multinucleated giant cells, a finding that suggests that papovavirus may cause entry of HIV-infected macrophages into the brain (Nelson et al, 1990).

Herpes Simplex Virus

Herpes simplex virus (HSV) is an enveloped virus 120 nm in diameter, with a DNA genome of approximately 96×10^6 daltons (Roizman and Batterson, 1985). The virus is assembled in the nucleus of infected cells and acquires its envelope from a modification of the nuclear membrane.

Two types of HSV have been described (Dowdle et al, 1967; Plummer, 1964) which differ on biological, biochemical and immunological grounds. Type 1 is usually acquired during early childhood and is associated with ocular and oral lesions. Type 2 is acquired through sexual activity and produces genital disease. Antibodies to HSV are found with increasing frequency in people of older ages and of lower socio-economic levels. A

study of antibodies to HSV-2 has revealed that 80 per cent of female prostitutes, 60 per cent of adults of lower, and 10 per cent of higher, socio-economic groups, and 3 per cent of nuns were seropositive (Nahmias et al, 1970; Duenas et al, 1972). Between 44 and 85 per cent of sexually-active homosexual men were also found to be positive (Goodell et al, 1983; Mann et al, 1984).

Both types 1 and 2 are acquired via mucosal contacts, and from the primary site the virus reaches, through the peripheral nerves, the regional sensory ganglion (Rawls, 1985). The haematogenous route has rarely been documented in adults (Dix and Bredesen, 1988). HSV remains latent in the ganglia (Blyth and Hill, 1984): type 1 can be recovered from trigeminal, superior cervical and vagus ganglia (Bastian et al, 1972; Warren et al, 1978); type 2 is present in sacral ganglia (Baringer, 1974). Once reactivated by various external stimuli, the virus travels back to the corresponding dermatome using axonal transport, and produces a localized disease. In immunocompetent individuals this remains circumscribed and resolves within 1–2 weeks, but in immunocompromised and otherwise debilitated persons it becomes a more severe and generalized problem.

Two distinct clinico-pathological illnesses are produced by these viruses in man. HSV-1 is responsible for a relatively-common encephalitis in previously normal individuals. Pathological findings are characteristic and consist of haemorrhagic necrosis involving particularly the temporal lobes and the subfrontal regions. Histological examination during the acute stages reveals meningo-encephalitis with lipid-laden macrophages, lymphocytes and plasma cells. Vessels which may be involved in the necrotic process are cuffed by inflammatory cells. Nerve cells are conspicuously decreased in number and the remaining ones show shrinkage or appear to be undergoing neuronophagia. Inclusion bodies in the nuclei of nerve cells may be seen, but, according to Brownell and Tomlinson (1984) are neither common nor easy to see.

The localization of the pathological changes to the 'limbic system' has raised a number of speculations regarding the pathogenesis of the encephalitis. Johnson and Mims (1968) suggested the olfactory route through the cribriform plate, but spread along nerves has also been demonstrated (Wildy, 1967; Kristensson et al, 1978). A systematic study of the olfactory tract was undertaken by Dinn (1979, 1980) and by Esiri (1982a). Virus was found in all cases examined by the former, but only in 9 out of 15 by the latter.

Esiri (1982a,b) concludes that in some cases with negative findings in the olfactory tract, the encephalitis may originate from virus latent in the brain.

HSV-2 is responsible for aseptic meningitis associated with outbreaks of genital infections in immunocompetent people. The illness follows the cutaneous lesion by 3–12 days, is benign and self-limited, and is accompanied by fever, headache, vomiting, photophobia and nuchal rigidity. Investigations by Craig and Nahmias (1973) suggest that it results from viraemia. Aseptic meningitis has, however, also been associated with HSV-1 in patients suffering from leukaemia (Cappel and Klastersky, 1973), bronchiectasis (Harford et al, 1975) or ascending myelitis (Klastersky et al, 1972).

Recently, a new type of herpes virus (human herpesvirus-6, HHV6), originally designated human B-lymphotropic virus, has been isolated, among other subjects, from HIV-positive patients (Salahuddin et al, 1986). Like HIV, this virus has affinity for CD4-positive cells, and co-infection between the two viruses is probably increasing HIV expression in tissue culture (Lusso et al, 1989).

Neuropathology

Brain involvement associated with HSV in HIV-positive patients can present as subacute, acute, and subclinical encephalitis.

Subacute Encephalitis. Patients suffering from this type of encephalitis have been reported by Dix and Bredesen (1988), Pepose et al (1984), and Tucker et al (1985), and present a subacute illness lasting several weeks and associated, in all cases, with CMV. The case reported by Tucker et al (1985) was that of a 36-year-old homosexual man presenting with symptoms of progressive ascending myelitis. At postmortem, the predominant feature was CMV infection in the spinal cord, cerebral and cerebellar hemispheres, and brainstem. The authors emphasize the fact that the structures of the limbic system were not involved. Immunohistochemical studies, however, showed that, in addition to CMV, the cytoplasm of a number of cells stained with HSV-2 antibodies.

Of the other two cases in which HSV-1 was implicated, one presented with retinitis and encephalitis (Pepose et al, 1984), the other with encephalitis (Dix and Bredesen, 1988).

Acute Encephalitis. This was described by Dix et al (1985) in two homosexual men with persistent lymphadenopathy. Clinical signs included sudden changes in mental stage and seizures, and were associated with CSF pleocytosis. In both cases, brain biopsy showed inflammatory perivascular cuffing, proliferation of microglial cells and inclusions in the nuclei of nerve cells. In both patients cultures of biopsy material yielded HSV-2. Following treatment with acyclovir they recovered gradually and survived with major neurological sequelae. In a personal observation, the morphological changes seen in the brain of a 39-year-old homosexual man were the same as those described in immunocompetent patients. They included extensive necrosis involving particularly the limbic system. Cell infiltrate consisted predominantly of macrophages; intranuclear inclusion bodies were seen in most of the surviving nerve cells in the affected areas. The only other pathological condition accompanying herpes encephalitis was a focus of CMV infection in the conus and filum terminale.

Dix and Bredesen (1988) reviewed the literature and found seven cases of HSV-2 encephalitis: four of these occurred in immunocompetent subjects (Nahmias et al, 1982); two patients were immuno-compromised because of thymic dysplasia and immunosuppressive treatment and had disseminated viral infection (Sutton et al, 1974). In immunocompetent patients, brain and spinal cord showed perivascular infiltration by lymphocytes and plasma cells, but no inclusions. HSV, however, was isolated from specimens of brain. In one of the immunocompromised, a brain biopsy of the temporal lobe showed necrotizing lesions and perivascular cuffing by lymphocytes, mononuclear cells and polymorphs; there was considerable nerve cell loss, while other neurons contained intranuclear inclusions, seen also in endothelial cells. The last patient of the series, who was immunologically normal, developed a psychotic syndrome (Oommen et al, 1982). At postmortem, the brain was grossly normal. Histological examination showed abnormalities limited to the temporal lobe and consisting of focal lymphocytic infiltration of the subarachnoid spaces. There was microglial proliferation in the hippocampus, but no inclusion bodies. HSV was confirmed by immunohistochemistry.

Subclinical Encephalitis. HSV-1 was seen in the brains of two neurologically-normal homosexuals in a series of unselected AIDS cases using inoculation techniques (Dix and Bredesen, 1988). The same authors, however, do not draw any conclusion from this finding, and suggest that more work is needed to establish the incidence of a subclinical neurological syndrome secondary to HSV.

Tucker et al (1985) comment on the atypical pathological findings seen in their case, which, unlike those in immunocompetent patients, does not show the characteristic haemorrhagic necrosis, and has discrete inflammatory changes in spite of widespread involvement of the nervous system. Similar findings were seen in patients with cardiac transplant (Hatson and Pedley, 1976), as well as in a patient with Hodgkin's disease reported by Price et al (1973).

The hypothesis that the intense immune response required to eliminate the virus may be responsible for the devastating changes induced by HSV, and that a deficit of this type of immunity may limit the severity of the lesions, is supported by experimental studies by Townsend and Baringer (1979). In mice infected with HSV and pre-treated with immunosuppressive drugs, they found that the encephalitis was less severe and necrotic lesions were reduced.

Varicella Zoster Virus

Varicella zoster virus (VZV) has been classified amongst herpes viruses on the basis of morphological and biological characteristics. The virion measures 180–200 nm in diameter, and consists of a 100 nm icosahedral, symmetric nucleocapsid surrounded by a host-derived envelope (Gelb, 1985). VZV is extremely labile and difficult to retrieve from human specimens, but it has been successfully cultivated in human embryonic lung fibroblasts and in primary cultures of African green monkey kidney cells.

In man, VZV produces two different illnesses, varicella (chickenpox) and herpes zoster (shingles). Varicella is the primary, highly-contagious infection, occurring predominantly in children; herpes zoster is due to reactivation of the virus. The possibility that both syndromes are produced by the same virus is supported by the similarity between viral isolates obtained from the same patient during separate episodes of chickenpox and shingles (Iltis et al, 1977). In immunocompetent individuals, the virus produces an infection of the mucosa of the upper respiratory tract. This is followed by an asymptomatic viraemia, during which the virus spreads to the reticulo-endothelial system where it replicates; during the primary infection, it spreads to the regional sensory ganglia following the neural route (Gelb, 1985). In sensory neurons, it becomes latent and can be reactivated.

Neurological complications occur predominantly in elderly individuals (Jemsek et al, 1983), although cases have also been reported in children (Brunell et al, 1968). Immunosuppressed patients are particularly at risk (Rifkind 1966; Muller, 1967), and have more severe lesions (Jemsek et al, 1983). The spinal cord and posterior root ganglia are most frequently involved, but changes in the brain are also found.

Encephalitis. This is a rare complication in which morphological changes may be minimal or absent in the cortex, and are limited to the gyral white matter, deep grey nuclei and brainstem (Rose et al, 1964; McCormick et al, 1969). Histological examination shows oedema, haemorrhagic foci and perivascular cuffing by small cells; intranuclear inclusions have also been found. Several mechanisms have been proposed to explain the pathogenesis of the encephalitis, and include direct invasion by the virus and an immunological reaction of the brain to VZV.

The two cases reported by Horten et al (1981) are more interesting in relation to the findings in HIV-positive cases. Both patients suffered from neoplasms and developed progressive and fatal neurological illnesses. Pathological examination revealed confluent plaque-like lesions in the white matter, which resembled those of PML and contained intranuclear inclusions in oligodendrocytes, astrocytes and neurons. Presence of VZV was confirmed by electron microscopy and immunohistochemistry.

Of the three patients studied by Eidelberg et al (1986), two (n 1 and 3) suffered from lymphomas. They both developed cerebrovascular disease following spells of herpes zoster and, at postmortem, showed, respectively, organizing thrombosis of the middle cerebral artery, and narrowing and thrombosis of several arteries with subintimal lymphocytic infiltration. In patient 1, immunohistochemistry revealed the presence of VZV. Patient 2 was a 60-year-old homosexual, with no evidence of AIDS, who developed trigeminal zoster which resolved and was followed, 3 months later, by neurological symptoms. Brain biopsy revealed recent infarction and thrombosed meningeal arteries. The authors label these cases as 'thrombotic vasculopathy' following herpes zoster, and attribute the scarcity of previous pathological reports to the low acute mortality (2–4 per cent) and self-limited course.

Organizing infarcts were also present in the case reported by Linneman and Alvira (1980), but, in this case, the arteries showed granulomatous angiitis. VZV was confirmed by electron microscopy.

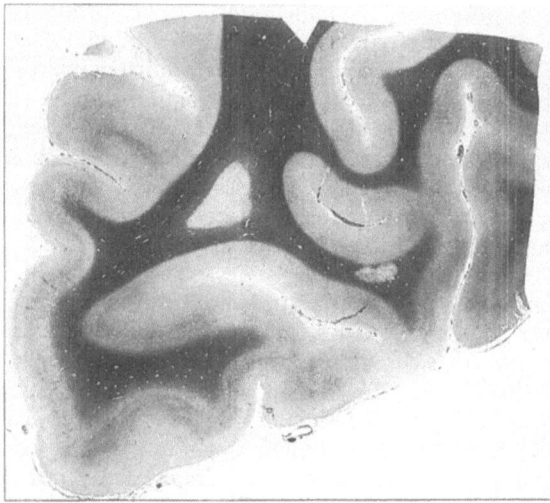

Figure 6.20. VZV encephalitis. Brain section showing two well-circumscribed plaque-like lesions in the white matter. LFB/N.

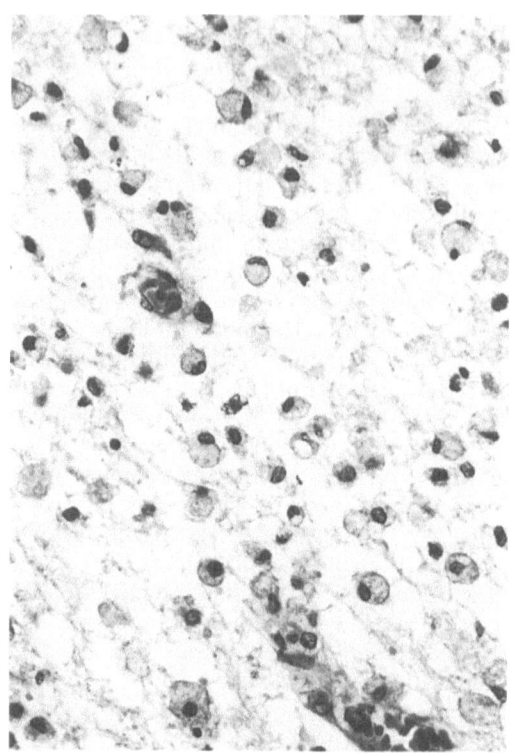

Figure 6.21. VZV encephalitis. Within a plaque-like lesion there is extensive destruction of the tissue and replacement with foamy macrophages. Blood vessels are also seen. H&E, ×300.

Ganglionitis. This is an acute illness which involves more commonly the upper cervical and thoracic levels than the enlargements. Usually one ganglion is predominantly affected, and appears macroscopically swollen and haemorrhagic. These changes may extend also to short segments of peripheral nerves and roots. Histological examination shows haemorrhagic infarction and thrombosis of some small arteries. Within this area, a number of ganglion cells have disappeared, and the remaining ones are shrunken, chromatolytic and vacuolated. In addition, an inflammatory exudate consisting of polymorphs and lymphocytes may be present. Involvement of the spinal cord can also follow (ganglioneuropathy) (Hogan and Krigman, 1973). Macroscopically, the region appears swollen over several spinal segments, and the entire transverse section, or only the grey and adjacent white matter, is soft and grey–yellow. Histological findings in still viable areas include mononuclear infiltrates of the subarachnoid spaces and perivascular cuffing; nerve cells show neuronophagia and chromatolysis. These changes have been likened to those seen in poliomyelitis, except that, with VZV, they tend to be unilateral and extend to the dorsal horn (Denny-Brown et al, 1944).

Late complications include post-herpetic neuralgia and a motor syndrome, both of which are secondary to abnormalities of the spinal cord. The pathological findings of the former were reported by Denny-Brown et al (1944) and Watson

et al (1988), whereas there are no postmortem descriptions of the latter.

Neuropathology

Encephalitis, as a complication of VZV in HIV-positive patients, is a rare event. The patient included by Levy et al (1985) in their series, had developed seizures while being treated for cryptococcal meningitis. Herpes virus was isolated from the CSF. Virus was also recovered by Dix et al (1985) from the CSF of a homosexual patient who had developed headache, aphasia, auditory hallucinations and seizures. In both cases, however, postmortem examination was not carried out.

The only convincing cases with pathological evidence of VZV encephalitis are those reported by Petito et al (1986), the single case described by Ryder et al (1986) and the four cases described by Gray et al (1992). No clinical details are given for the former. The case reported by Ryder et al was a 37-year-old homosexual man who developed the neurological illness 12 weeks after suffering

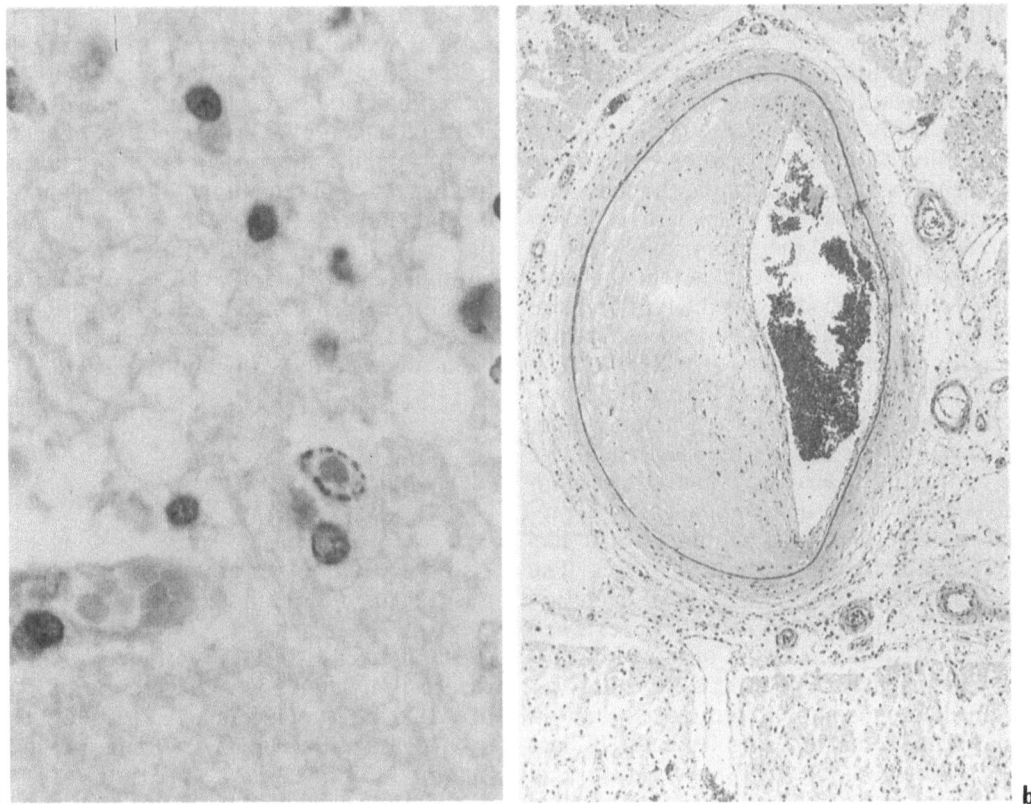

Figure 6.22a,b. VZV encephalitis. **a** Photomicrograph showing an intranuclear inclusion in a cortical nerve cell. H&E, ×800. **b** The lumen of this meningeal vessel is severely narrowed by a process of intimal fibrosis. Elastic/van Gieson, ×4.5.

from cutaneous zoster which had, meanwhile, disappeared. The patient presented with memory deficit, tetraparesis and mild sensory deficit. CT scan showed enlargement of the lateral and third ventricles, but no focal abnormalities. However, a repeat CT scan 1 month later showed a focal, low-density, non-enhancing lesion in the right cerebellar hemisphere.

All the cases contain multiple, well-circumscribed and 'plaque-like' areas of yellow–grey discoloration of the cerebral white matter (Figure 6.20), some of them appearing cavitated. Microscopic lesions are more extensive than those macroscopically identifiable, and consist of complete loss of myelinated fibres, which are replaced by foamy macrophages (Figure 6.21). In the case described by Ryder et al (1986), a lesion was also present in the thoracic cord. Usually, no inflammatory cells are seen within, or around, the areas which contain a number of eosinophilic intranuclear inclusions (Figure 6.22a) with the ultrastructural features of VZV. In addition, Petito et al (1986) described endarteritis in an occasional leptomeningeal artery in one patient;

furthermore, in one of the cases described by Gray et al (1992) a non-inflammatory vasculopathy involving medium-sized leptomeningeal vessels and producing severe reduction (Figure 6.22b) or total occlusion of their lumina was associated with circumscribed areas of infarction. We interpret the former as the result of a burnt-out arteritis.

It has already been mentioned that brain invasion may take place at any time during or after the onset of the cutaneous zoster. Although this complication is rare, it should nevertheless be considered in the differential diagnosis of space-occupying lesions in HIV-positive patients.

Opportunistic Infections: Fungal

Fungal infections of the brain are, in the vast majority of cases, secondary to a primary focus elsewhere, usually in the lungs or intestine. However, in some cases, the site of entry may remain unrecognized and indeed, cases have been reported in which the only evidence of infection was in the nervous system. An additional source of infection, which has appeared in the last few decades, is represented by prosthetic heart valves (Stulz et al, 1980; Kalish et al, 1982). In histological sections, organisms appear as yeasts of up to 20 μm in diameter, branching hyphae and pseudohyphae, the last of which are of intermediate size between the other two. Consequently, lesions in the nervous system take different appearances according to the type and size of organism involved: the smallest forms, such as *Blastomyces*, *Cryptococcus* and *Histoplasma*, can reach the capillaries and usually produce leptomeningitis; the large hyphae are trapped within, and occlude, vessels of intermediate and large calibre, giving rise to extensive infarcts; pseudohyphae, such as those of *Candida* obstruct preterminal vessels, producing multiple small foci which eventually evolve towards abscesses. The types and incidence of mycoses, their evolution and the changes they produce have been reviewed by Fraser et al (1979). Reviews of infectious diseases (including mycoses) affecting the nervous system in immunocompromised hosts, and in particular in patients suffering from AIDS, are those by Hooper et al (1982) and Gray et al (1988a), Levy et al (1988a) and Vinters et al (1989a), respectively.

Cryptococcosis

Cryptococcosis is caused by *Cryptococcus neoformans*, a fungus usually present in soil and the droppings of some birds. In routinely-stained sections (H&E and PAS), the organism is easily identified as a round structure, 4–7 μm in diameter, surrounded by a capsule 3–5 μm thick (Binford and Dooley, 1976). Small buds are attached to the main body through a short neck. The classic, simple way to demonstrate the capsule is by adding a few drops of diluted India ink to the fresh preparation. This provides a dark background against which the capsule is clearly seen. Cryptococcosis has a worldwide distribution, but most reported cases have occurred in the southern parts of the USA and in Australia. The disease affects individuals of all races, predominantly males, and sometimes develops spontaneously in previously-healthy people. However, in up to 85 per cent of cases, it is associated with debilitating illnesses (Collins et al, 1951; Hay et al, 1980). The portal of entry of the fungus is the respiratory tract by inhalation, and the pulmonary disease is often asymptomatic.

The commonest presentation of brain involvement is meningitis, affecting predominantly the base and carrying a severe prognosis with high mortality, although remissions have been described (Baker, 1972). In cases of longer survival, hydrocephalus may occur and organisms may produce cysts in the Virchow–Robin spaces. The inflammatory response consists of lymphocytes, plasma cells and eosinophils and may include multinucleated giant cells. Cryptococcomas (localized granulomas) have sometimes been described, and the literature on this subject has been reviewed by Selby and Lopes (1973).

Neuropathology

As observed in other groups of immunocompromised patients (Hooper et al, 1982), cryptococcosis appears to be the commonest type of cerebral mycosis in HIV-positive patients. Cryptococcal meningitis was the presenting symptom of AIDS in 26 per cent of 27 patients reported by Kovacs et al (1985). In neuropathological material, this mycosis was present in 2.6 per cent of the cases examined by Petito et al (1986), and in 4 per cent of those seen by Lang et al (1989) in Switzerland and by ourselves in London. A higher incidence (8 per cent) was found by Vinters et al (1989a) in Los Angeles. Although this organism appears to be sensitive to amphotericin B and flucytosine, CSF levels of these drugs fail to reach therapeutic levels (Poon et al, 1988).

Macroscopic changes are not always recognizable, and can be often absent or overlooked. The leptomeninges, predominantly at the base, may be distended by clear or opaque exudate which may extend to the ventricles. In the brain parenchyma, lesions appear as grey and gelatinous areas with a spongy or cribriform appearance, involving the deep grey nuclei and the adjacent white matter (Figure 6.23). Circumscribed lesions – cryptococcomas – may also be seen, but necrosis is not commonly seen.

Histological examination shows, in most cases, lack or minimal amounts of inflammatory cells, while granulomas are extremely rare. The commonest appearance is that of massive amounts of organisms, in the leptomeninges or surrounding parenchymal vessels, producing dilatation of the perivascular spaces (Figure 6.24) and behaving, on occasion, as a space-occupying lesion. Each organism consists of an intensely-stained central area surrounded by a thick mucinous capsule.

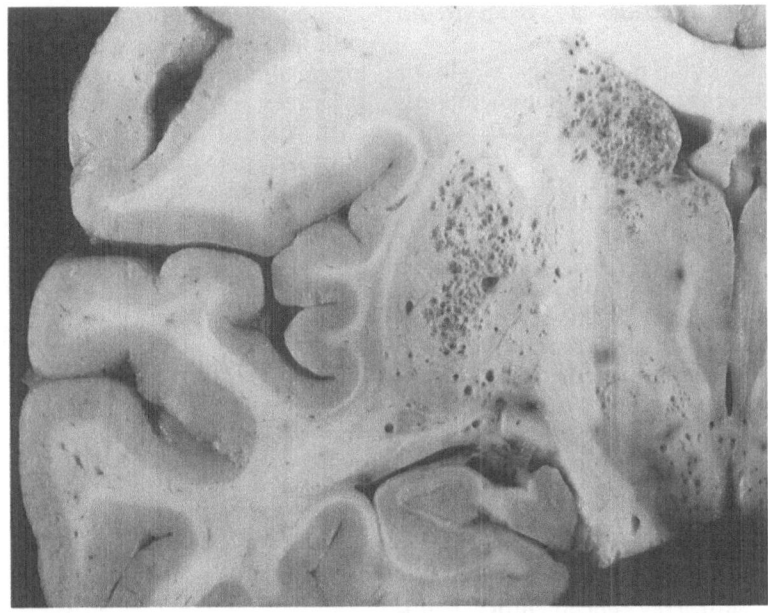

Figure 6.23. Cryptococcosis. Coronal section of a cerebral hemisphere in which the deep grey nuclei show multiple small cysts also involving the substantia nigra.

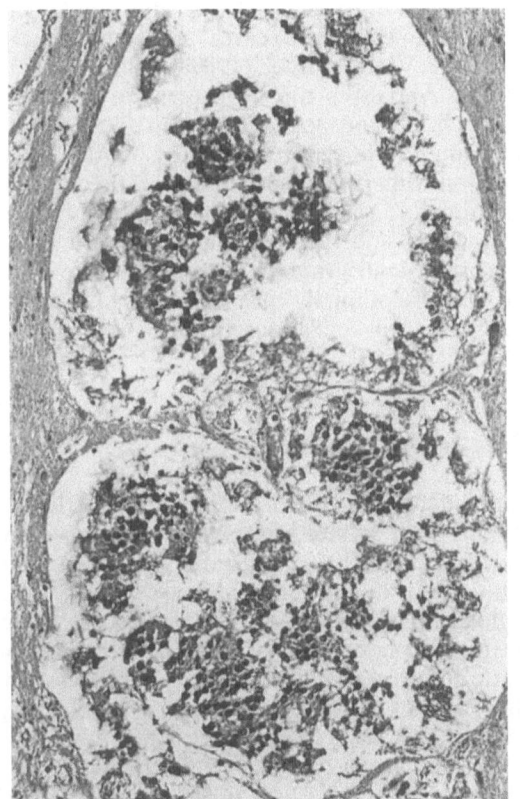

Figure 6.24. Cryptococcosis. In the brain parenchyma, perivascular spaces are distended by large amounts of organisms. Alcian blue/PAS, ×120.

The glial reaction of the surrounding brain is minimal.

Aspergillosis

Aspergillus is an ubiquitous fungus and includes over 350 species, although only a few are pathogenic to man (*A. fumigatus*, *A. niger*, *A. flavus* and *A. oryzae*). The organisms consist of branching, septate hyphae, 3–10 µm in diameter.

In man, the disease is becoming more common and is usually localized to one organ. A generalized form with involvement of the CNS is less frequently encountered (MacKenzie, 1979) and, as a rule, it occurs in patients debilitated by neoplastic or collagen disorders (Pillay et al, 1968; Aisner et al, 1979), drug addiction (Kaufman et al, 1976), alcoholism (Mukoyama et al, 1969) or immunosuppression (Ingwer et al, 1978). Although the true incidence of the cerebral involvement is difficult to ascertain, Grčević and Matthews (1959) and Carbone et al (1964) estimated that 60–70 per cent of cases with disseminated forms of aspergillosis have neurological lesions.

Infection may spread to the CNS either directly from the paranasal sinuses (Warder et al, 1975) or the orbits (Hedges and Leung, 1976), or is blood-borne from the lungs or gastrointestinal tract (Young et al, 1970). In some cases, however, no apparent source of infection has been found (Linares et al, 1971).

Infection in the brain presents as single or multiple abscesses, usually in immunocompromised patients. Granulomas also occur, sometimes as solitary lesions.

Neuropathology

Cases of cerebral infection by *Aspergillus* may be found in HIV-positive patients, but are exceedingly rare. Levy et al (1985) reported a single case among 320 patients examined. In a subsequent report, the same authors (Levy et al, 1988a) included four examples. Petito et al (1986) mentioned seven cases (4.5 per cent), whilst Vinters et al (1989a) and Lang et al (1989) included only one case each in groups of 158 and 135 brains, respectively. Only one case was seen in London; in this case several mycotic abscesses were associated with HIV encephalitis. Additional cases are those reported by Berger et al (1984), Gapen (1982), Henochowicz et al (1985) and Laulund et al (1986). Woods and Goldsmith (1990) described three HIV-positive patients with lesions of the nervous system due to *A. flavus* (two cases) and *A. fumigatus* (one case). In the latter case, the lesion was localized to the spinal cord and was considered to be an extension from a thoracic abscess. The two cases due to *A. flavus* had cerebral abscesses. The authors emphasize the point that patients with abscesses due to *Aspergillus* show multiple enhancing lesions on CT scanning, similar to those seen in cerebral lymphoma. However, patients with this type of neoplasm, as well as those with cerebral toxoplasmosis, tend not to develop headache and fever, which are commonly seen in cerebral aspergillosis.

This low incidence contrasts with that of almost 15 per cent reported by Hooper et al (1982) in non-HIV immunosuppressed patients.

Macroscopically, abscesses appear as areas of necrosis and haemorrhage, sometimes with a cavitated centre or, less frequently, as ill-defined areas of dusky discoloration of the white matter. Histologically, the most striking feature is the intensity of vascular invasion with thrombosis and, eventually, necrosis. Inflammation is usually scanty.

Candidosis

This is a worldwide infection due to *Candida spp.*, a normal constituent of human flora. It presents as pseudohyphae, chains of elongated cylindrical cells and oval budding cells 2–3 µm long. In most cases, the organism acts as an opportunist, and produces a disseminated illness in association with diabetes, leukaemia, lymphoma and drug abuse, or may present in the oral cavity of infants. The nervous system is only occasionally involved, usually by haematogenous dissemination from a primary focus in the respiratory or gastrointestinal tract. Cerebral lesions occur late and are usually complicated by those in the heart and kidney (Parker et al, 1981). However, Louria et al (1962) collected ten cases from the literature in which brain lesions had occurred in previously healthy persons.

Neuropathology

Cerebral involvement by *Candida* in HIV patients is an uncommon event, since these patients have unimpaired neutrophil function (Armstrong and Wong, 1982; Hooper et al, 1982). Most of the cases described in this group are abscesses, and have been reported by Snider et al (1983), Levy et al (1984), Petito et al (1986) and Budka et al (1987). However, no cases are present in the series by Vinters et al (1989a) and Lang et al (1989). Abscesses can be present in any region of the brain and appear as small foci of yellow softening, localized to the grey matter at the junction with the white matter. Histologically, inflammatory reaction consists of polymorphs and monocytes, and may surround a central area of necrosis. Small granulomas may also be found, but are exceptionally rare. An unusual case with meningitis has been described by Harris et al (1985).

Coccidioidomycosis

The fungus *Coccidioides immitis* is restricted to regions with a semi-arid climate. The disease is endemic in South America and in Mexico, and is still a major health problem in the south-western region of the United States. In tissues, fungi occur as round spherules, 20–35 µm in diameter, and mature forms which contain endospores. They also exist in nature and culture as hyphae with arthrospores. Budding is rarely observed.

The organism is both a pathogen and an opportunist, and initially produces a mild febrile illness followed, in a number of cases, by infection of the lungs. The nervous system may be involved in one-third to one-half of the cases, usually as a terminal event, and as the result of diffuse haematogenous spread.

Neuropathology

A number of cases of coccidiodomycosis in HIV-positive patients have been reported by Kovacs et al (1984), Abrams et al (1984), Roberts (1984), Post et al (1985), Bronnimann et al, (1987), Levy et al (1988b) and Jarvik et al (1988). In these cases, the illness appeared as meningitis with thickening and opacity of the meninges, particularly at the base, and formation of small nodules. Hydrocephalus or involvement of the vessels may follow. In a patient reported by Levy et al (1985), the illness presented as multiple cerebellar microabscesses, whereas Bronnimann et al (1987) and Jarvik et al (1988) described involvement of the brainstem. Only occasionally is the spinal cord involved.

Other Mycoses

Other fungi have been described in AIDS patients, but the incidence of these infections is very low.

Nocardia asteroides, associated with *Salmonella enteritidis*, was isolated and grown from brain abscesses in one of the cases reported by Holtz et al (1985). Case 8 of Sharer and Kapila (1985), a 34-year-old female drug abuser, presented with blurred vision, headache and dizziness. CT scan revealed a large contrast-enhancing mass with a low-density centre. A biopsy showed necrotic areas of various sizes containing polymorphs, newly-formed vessels and Gram-positive organisms consistent with *N. asteroides*.

In the single case of meningo-encephalitis due to *Histoplasma capsulatum* (Anders et al, 1986a), the brain showed ill-defined granulomatous formations containing the organisms.

Zygomycosis associated with AIDS has been described by Levy et al (1988a) in a child. In addition, Micozzi and Wetli (1985) reported five patients with cerebral mycosis, in whom the illness may have been related to AIDS.

Single cases of brain involvement by *Acremonium alabamensis* and *Cladosporidium bantianum* have been described by Wetli et al (1984) and Colon et al (1988) respectively. In the former, large foci of necrosis were present in the cerebral and cerebellar hemispheres. In the latter, necrotic areas in the cerebral hemisphere

and brainstem included granulomatous foci containing pigmented organisms.

Opportunistic Infections: Protozoal

Infections by this group of organisms are due, almost exclusively, to *Toxoplasma*.

Toxoplasmosis

The coccidian parasite *Toxoplasma gondii* is known to infect a wide variety of warm-blooded animals, including man. The incidence of human infection varies in different populations, but in most studies it has been of the order of 30–50 per cent (Feldman, 1968; Hay and Hutchinson, 1983). Remington and Klein (1976) found that the incidence of antibodies to *T. gondii* in women aged under 35 was 33 per cent in London and New York, whilst it reached 85 per cent in Paris.

The organism is an obligatory intracellular parasite whose definitive host is the cat family, where the entire life cycle is completed. Both sexual and asexual forms may be found in the intestinal mucosa. The organism exists as oocysts, found only in the cat, endozoites and cystozoites. Animals and man acquire the infection by eating raw meat from chronically-infected animals, or by ingesting oocysts deposited in soil, sand or litterpans of cats. However, only a minority of infected individuals develop the disease which, in man, does not include enteritis. Endozoites, which are typical of the acute stage of the infection, are oval or crescent-shaped and measure 2–4 by 4–8 µm. They usually appear clustered together in parasitophorous vacuoles, but a few may lie free within the host cell cytoplasm. The cavities that enclose the parasite (pseudocysts) are lined by a single plasma membrane and are the site of multiplication. It is generally agreed that the *Toxoplasma* divides by a process of internal budding, called 'endodyogeny', and not by binary fission as previously held (Ghatak et al, 1970). The cytoplasm of parasitized cells may become heavily packed with organisms and rupture, releasing them. Alternatively, organisms may persist within the cells for long periods without forming true cysts. Cysts are most often found in the CNS, and in the skeletal and cardiac muscle (Remington and Cavanaugh, 1965). They vary in size and contain a few, or up to 3000, cystozoites (Hay and Hutchinson, 1983). The cystozoites multiply slowly and may persist for months; indeed, they have been found in guinea pigs 5 years after inoculation (Lainson, 1959).

The development of toxoplasmosis in man depends very much on the host's susceptibility and immunological status. Of the two types of immunity, that due to circulating antibodies is the first to appear, and plays a role in limiting extracellular dissemination of the organism. Hypersensitivity to *Toxoplasma* is a further complication, and can lead to the death of non-infected cells. There are, therefore, different patterns and degrees of severity of the disease according to whether it takes place in normal individuals, in those who have become immunologically deficient, or in fetuses and children before they develop immunological competence.

Toxoplasmosis in the Adult

Although a large number of people are infected with the organism, most of the infections remain subclinical. Whenever present, symptoms include a mononucleosis-like illness (Skipper et al, 1954), an encephalitic syndrome (Sabin, 1941) or a form of fever reminiscent of typhus spotted fever, all of which resolve spontaneously.

Involvement of the CNS by *T. gondii* in previously healthy individuals is a rather uncommon event and may appear clinically as a non-specific encephalopathy, a diffuse meningo-encephalitis, or with features of a space-occupying lesion (Townsend et al, 1975). In a review of 21 cases from the literature, Hooper (1957) found that the brain and lungs were the organs most frequently involved (13 cases) after the heart (16 cases). However, Callahan et al (1946) state that, in acquired toxoplasmosis, lesions of extraneural organs are more severe than in congenital forms, in which the brain seems to be the primary target. Neuropathological lesions (Kass et al, 1952) include areas of necrosis, variously associated with infiltration by polymorphs and mononuclear cells, and proliferation of astrocytes and microglial cells. Vessels may sometimes undergo fibrinoid necrosis (Mashaly et al, 1983), and perivascular infiltration is usually minimal. Cysts, 20–100 μm in diameter, are found within or near the inflammatory nodules, at the periphery of, or at some distance from, the necrotic areas. Leptomeninges also may be involved in the process and show mild infiltration by small cells.

An atypical case has been reported by Bobowski and Reed (1958) in a patient with moderately-elevated dye-test and absence of fever. In this case a space-occupying lesion included numerous cysts, areas of necrosis and multinucleated giant cells; the latter are an uncommon finding in cerebral toxoplasmosis. Foci of asymptomatic toxoplasmosis may occasionally occur in brains of patients dying from unrelated causes (Plant, 1946). The presence of *Toxoplasma* cysts in otherwise normal brains is of special interest because it allows relapse to occur if immunity fails.

Congenital Toxoplasmosis

Cerebral toxoplasmosis in the fetus and newborn babies shows completely different features from these described above. Of the two factors acting in the adult form – proliferation of the organism with rupture of the cysts and spreading of the antigen in an immunocompetent background – only the first is present in immature brains. Additional factors to be considered in the pathogenesis of the encephalitis are the limited effect of maternal immunity, the extreme vulnerability and lack of regeneration of neurons and, possibly, the delayed development of the immune competence in the nervous system.

Congenital toxoplasmosis occurs only if maternal infection is acquired during pregnancy, and is transmitted through the placenta. Clinical cases are found with a frequency of about 1:4000 births in the USA, and 2–3:1000 births in France (Larroche, 1977). Transmission of the organism to the fetus may occur at any time during pregnancy; however, it does not take place during the first 2 months, probably because the placenta has not yet been formed (Robertson, 1962). During the following 3 months, the risk of infection is very high (1 in 3), and the damage severe. During the 6th and 7th months the risk is even higher, but the damage is less severe and associated with a high rate of prematurity. Towards the end of the pregnancy the fetus is seldom affected.

In the nervous system, *Toxoplasma* proliferates in the ependymal and sub-ependymal tissues, and spreads widely because of the absence of circulating antibodies. Contact between maternal antibodies and *Toxoplasma* antigen in the vessel walls may account for the perivascular inflammation and thrombosis. The relative sparing of the fourth ventricle by the necrotizing process is thought to be related to the low protein content of the CSF at that level, due to drainage through the foramina.

Affected children present Sabin's tetrad of signs: convulsion, chorioretinitis, cerebral calcification and internal hydrocephalus (the hydrocephalus probably being initiated by destruction of the aqueduct). Morphological features include: necrosis of single cells or of large areas of tissue, following invasion by the parasite; presence of foci of complete destruction of grey and white matter secondary to vascular thrombosis; periventricular and periaqueductal ulceration and necrosis with calcification and obstruction; presence of moderate amounts of lymphocyte and polymorphs in the subarachnoidal spaces. Microcephaly is found in cases in which destruction of the brain has been severe.

Toxoplasmosis in Immunocompromised and HIV-Positive Patients

Toxoplasmosis, a relatively-benign disease in normal adults, is often fatal in patients treated with antimitotic or immunosuppressive drugs. Severe involvement of the nervous system is the major complication, and has been found in about 5 per cent of such patients (Krik and Remington, 1978). Cerebral toxoplasmosis has, however, remained a relatively uncommon disease (Hooper et al, 1982) until, with the appearance of AIDS, it has become the commonest cause of intracerebral mass lesions (Snider et al, 1983; Luft et al, 1984; Petito et al, 1986; Levy et al, 1988a). Its incidence ranges between 2 and 13 per cent in American series, and varies according to the risk group of patients and their geographical location. The highest frequency is found in Florida and among Haitians (Levy et al, 1988b). Cerebral toxoplasmosis represents the most frequent neurological complication of AIDS in some European series also (Hénin et al, 1987; Gray et al, 1988c; Lang et al, 1989).

Neuropathology. The most characteristic lesion in HIV-positive, as in HIV-negative, patients (Mashaly et al, 1983; Navia et al, 1986c) is the presence of multiple foci of infection. These involve predominantly the cerebral hemispheres, particularly the frontoparietal areas, the basal ganglia and the cortex–white matter junction, but cerebellar lesions are not uncommon. On the other hand, involvement of the brainstem (Kure et al, 1989) and spinal cord (Mehren et al, 1988; Herskovitz et al, 1989) have only rarely been described, usually in the setting of multiple cerebral lesions (Navia et al, 1986c; Lang et al, 1989).

Toxoplasmosis of the brain has been subdivided by Navia et al (1986c) into three morphological types, according to the stage of infection and degree of tissue reaction. However, these different presentations may occur together depending on the modalities of treatment, degree of immunosuppression and recurrence of the disease.

1. Necrotizing abscesses are predominantly found in patients who die in the acute stage, whether or not any treatment has been initiated. Macroscopically, the lesions appear as poorly-circumscribed areas of softening and brown discoloration with variable amount of haemorrhage and perifocal oedema (Figure 6.25). Diffuse forms with ill-defined, confluent areas of necrosis were referred to as 'necrotizing encephalitis' (Lang et al, 1989). Gross meningitis and/or ventriculitis (Figure 6.26) may be found, mainly in fulminating forms (Wongmongkolrit et al, 1983). Microscopic examination shows coagulation necrosis with variable numbers of petechiae, usually a scanty inflammatory response and, invariably, numerous tachyzoites and encysted bradyzoites at the periphery of the necrosis. Small arteries in and around the necrotic lesions may show hypertrophy of the intima, and thrombosis and frequent fibrinoid necrosis of the wall (Figure 6.27), accompanied by polymorph infiltration. Immunohistochemistry and electron microscopy have established that tachyzoites are present in large numbers in the hypertrophic arterial wall. They support the hypothesis that the organisms may invade the arteries, causing arteritis, and that ischaemic infarction, and not abscess, is the main pathogenetic process in toxoplasmosis of the nervous system (Huang et al, 1988).

2. Organizing abscesses are found in patients who have been treated for 2 weeks or more, and have more chronic infections. Macroscopically, lesions consist of well-demarcated, usually large, areas with central necrosis and a halo of congestion (Figure 6.28). Microscopically, a central acellular, eosinophilic area of coagulative necrosis is surrounded by a granulomatous reaction, which includes a rim of tightly-packed lipid- and, occasionally, haemosiderin-laden macrophages, vascular proliferation and variable inflammatory infiltrates, and proliferation of astrocytes and microglial cells. In a minority of cases, organisms may be present. In this variety, occlusive hypertrophic arteritis may also be seen within the necrosis, as well as in the peripheral reaction, where the density of lymphocytes may mimic a lymphoma (Figure 6.29) (Sharer and Kapila, 1985).

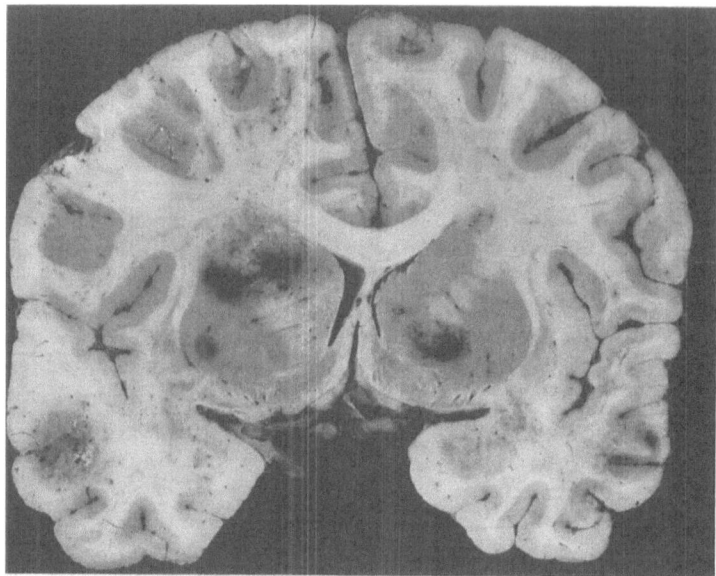

Figure 6.25. Toxoplasmosis. Acute lesion consisting of ill-defined areas of necrosis and haemorrhage in the deep grey nuclei, and at the border between cortical grey and white matter.

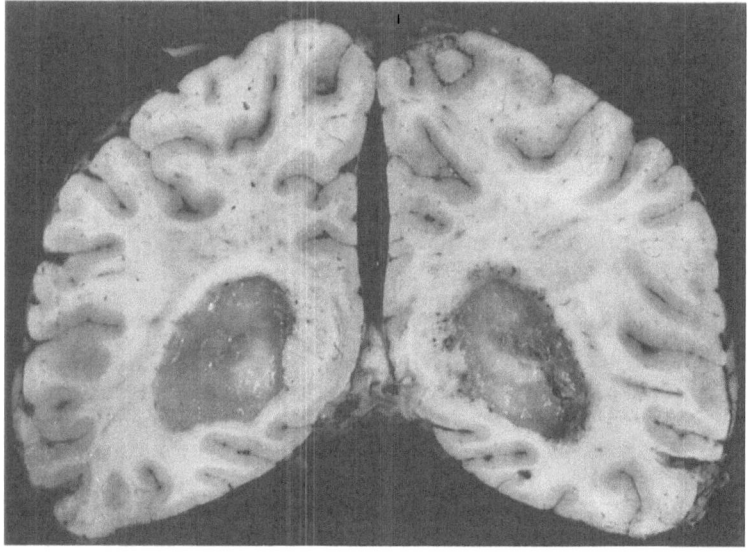

Figure 6.26. Toxoplasmosis. Coronal section of the occipital lobes. The lateral ventricles are the site of an inflammatory reaction appearing as gelatinous material and resulting in hydrocephalus.

3. Chronic abscesses are, at present, observed with increasing frequency in patients who have been treated for one month or more. Macroscopically, they appear as well-demarcated cystic cavities (Figure 6.30) or as small (<5 mm in diameter), orange–yellow, linear scars. Histologically, the cystic spaces contain variable amounts of lipid-laden, and occasionally haemosiderin-containing, macrophages and are surrounded by a marked astrocytic reaction. Mineralization is frequently seen, predominantly in vessel walls. Organisms are only occasionally found.

While most of the reported cases fulfil these clinico-pathological criteria, a few do not comply with the previous description.

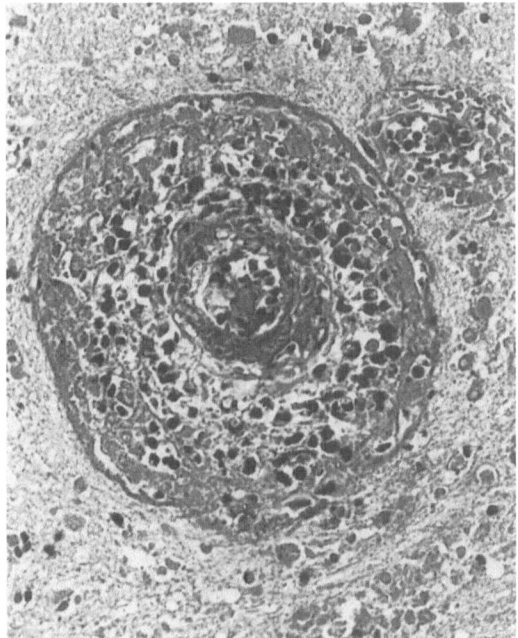

Figure 6.27. Toxoplasmosis. Fibrinoid necrosis of a vessel. Note the thickening of the wall. H&E, ×150.

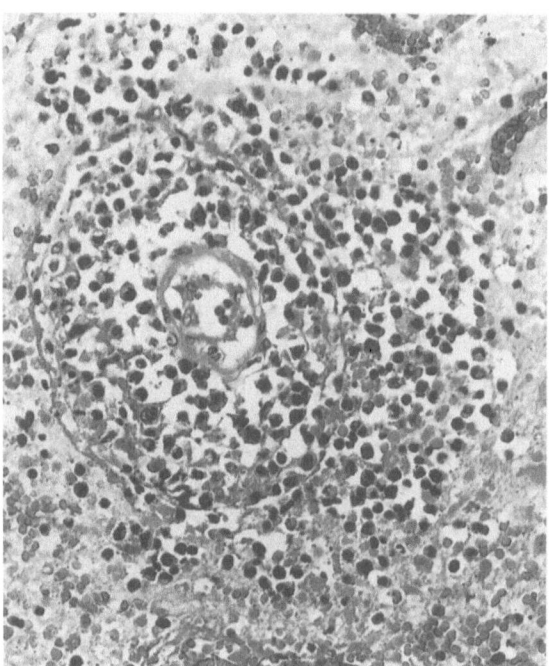

Figure 6.29. Toxoplasmosis. The layers of the arterial wall are infiltrated and dissociated by a florid lymphoid proliferation, sometimes mimicking the appearance of a lymphoma. H&E, ×140.

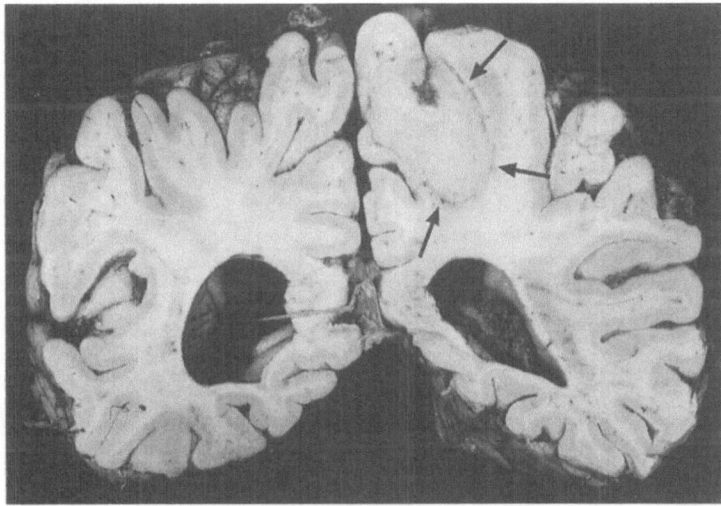

Figure 6.28. Toxoplasmosis. Organizing abscess consisting of a necrotic core surrounded by a ring of congestion (*arrows*).

1. *Toxoplasma* cysts, widely disseminated in the CNS without parenchymal reaction (Figure 6.31a) are not uncommon in AIDS (Anders et al, 1986a; Navia et al, 1986c; Hénin et al, 1987; Gray et al, 1988c). In some cases, they may be the only manifestation of cerebral involvement by the organism. These lesions are always asymptomatic and represent incidental findings at postmortem.

2. Diffuse microglial nodules with scattered encysted bradyzoites and a few tachyzoites within

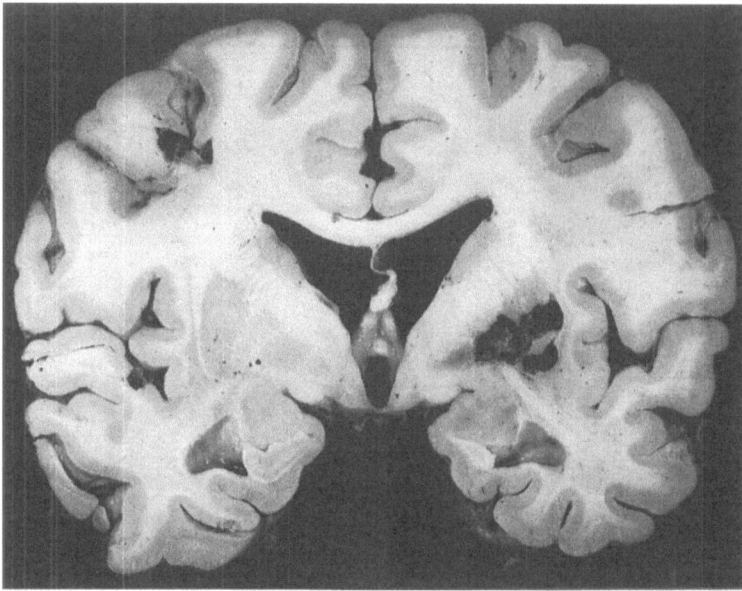

Figure 6.30. Toxoplasmosis. In patients dying during the chronic stage, cerebral lesions may appear as multiple cystic cavities with yellow–orange walls.

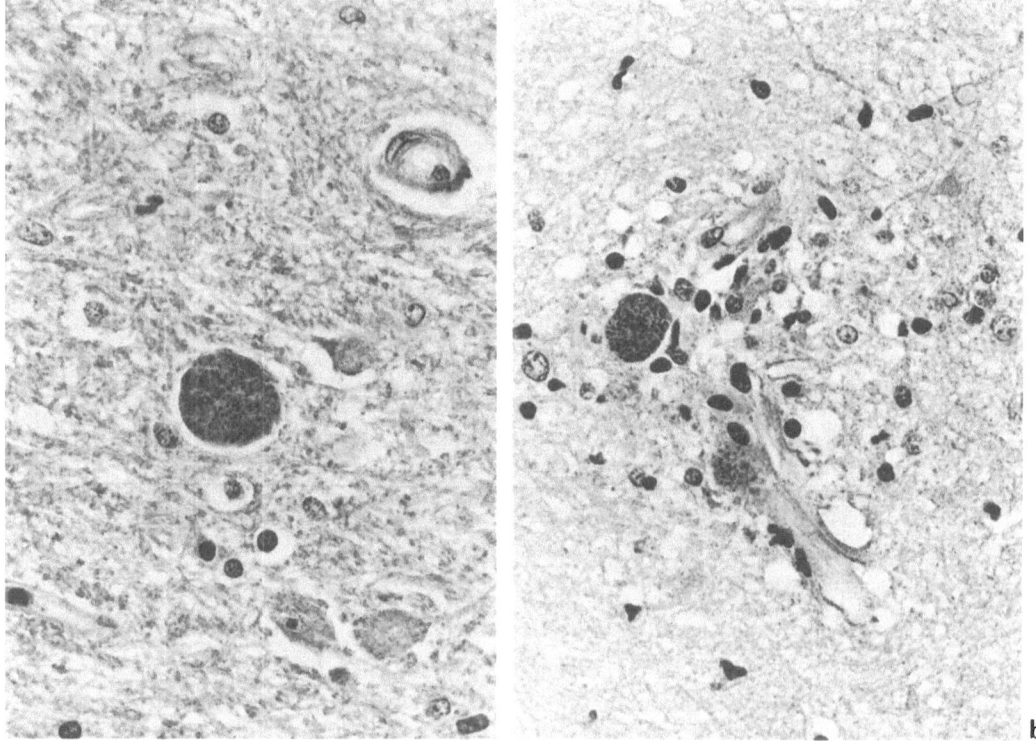

Figure 6.31a,b. Toxoplasmosis. **a** A cyst of *Toxoplasma* may be occasionally seen in an otherwise normal brain tissue. H&E, ×250. **b** In the diffuse encephalitic form, microglial nodules are seen in close proximity to blood vessels. Organisms are present within cysts. H&E, ×250.

(Figure 6.31b), or in the immediate vicinity of, the nodules may be associated with focal lesions (Navia et al, 1986c) or be the only morphological manifestation of cerebral toxoplasmosis (Gray et al, 1989; Lang et al, 1989). Such diffuse, non-necrotic 'encephalitic' illness appears unique to AIDS, and may be the pathological counterpart of a rapidly progressive and diffuse encephalopathy in these patients (Gray et al, 1989). On the other hand, diffuse microglial nodules in brain parenchyma, the so-called 'nodular encephalitis', are a common finding in viral encephalitis (Poirier et al, 1990) and, in AIDS patients, are associated with CMV infection (Schmidbauer et al, 1989b), although they have also been found in a variety of pathological processes, including herpes simplex, HIV encephalitis, candidosis or lymphomas (Matthiessen et al, 1988). In addition to the presence of the specific organism within, or in proximity to, the nodules, other features are suggestive of 'Toxoplasma encephalitis'. These include the occurrence of central areas of necrosis or the finding of a vessel within some nodules (Figure 6.31b); the association, in most cases, with haematogenous visceral dissemination of organisms; and the presence of occasional parasitic emboli (Gray et al, 1989).

Cerebral toxoplasmosis is sometimes associated with one or more other neurological complication, including PML (Gray et al, 1987), HIV encephalitis (Gray et al, 1987, 1988b, 1989; Schmidbauer et al, 1990), and tuberculosis (Fischl et al, 1985).

Cerebral toxoplasmosis does not represent a diagnostic problem when organisms can be identified, or when the presence of the characteristic coagulative necrosis associated with hypertrophic arteritis strongly suggests this diagnosis (Gaston et al, 1985). In atypical cases, however, and when the parasite is not apparent, particularly in brain biopsies, immunohistochemical techniques (Figure 6.32a) (Conley et al, 1981) and electron microscopy (Horowitz et al, 1983; Sauron et al, 1983) may be particularly helpful. With the latter technique, a typical endozoite (Figure 6.32b) shows a 35–40 nm thick pellicle, which consists of a thick outer membrane separated by a 15 nm gap from two inner membranes. At the anterior pole of the endozoite, a nipple-shaped elevation of the pellicle covers the conoid, a structure composed of spirally-wound microtubules and two preconoidal rings. The conoid is the part of the organism structurally equipped to facilitate entry of the parasite into the cell (Powell et al, 1978). Sub-pellicular microtubules extend from the conoid through the equatorial region and into the posterior half of the body of the parasite. Each endozoite contains a single mitochondrion, a prominent Golgi apparatus and scanty rough endoplasmic reticulum and shows, in front of the nucleus, numerous micronemes and rhoptries. The cystozoite is the parasitic form of the chronic stage of the disease, and is contained within cysts. The latter are formations lined by a wall derived from the organism (Matsubayashi and Akao, 1963), and consist of a thin outer and a thicker, but poorly-defined, inner layer measuring together 200–400 nm (Ghatak et al, 1970).

Other Protozoal Infections

Another member of this phylum known to produce lesions in an HIV-positive patient is *Acanthamoeba*. This is one of four genera of amoeba (*Entamoeba*, *Naegleria*, *Acanthamoeba* and *Vahlkampfia*), which produce lesions in man. The term 'primary encephalitis' was introduced by Butt (1966) and Carter (1968), and now includes primary amoebic meningo-encephalitis due to *Naegleria fowleri* and granulomatous encephalitis due to the *Hartmannella–Acanthamoeba* group.

Naegleria produces an acute necrotizing and haemorrhagic meningo-encephalitis in previously-healthy young individuals, with scanty inflammatory cells and presence of large numbers of organisms. The infection occurs in summer and autumn, as isolated cases or in small outbreaks and is acquired by swimming in warm lakes or pools contaminated with the organism. Lesions may be found anywhere in the CNS, but are particularly severe over the orbital surface of the frontal lobes around the olfactory area, thus supporting the early hypothesis, further confirmed by experimental work (Martinez et al, 1973), that protozoa enter the nervous system via the nasal mucosa.

Amoebae of the *Hartmannella–Acanthamoeba* group produce a subacute or chronic illness, predominantly in patients suffering from liver disease, diabetes mellitus, Hodgkin's disease or receiving treatments which include antibiotics, radiotherapy or steroids. The route of entry into the brain is not completely understood: the organism probably exists in man as part of normal flora (Visvesvara and Balamuth, 1975) or in the environment (Kingston and Warhurst, 1969; Lawande et al, 1979). Wang and Feldman (1967) isolated the organism from human pharynx; invasion from an ocular focus was emphasized by Jones et al (1975), whilst a pulmonary origin with haematogenous spread was considered a poss-

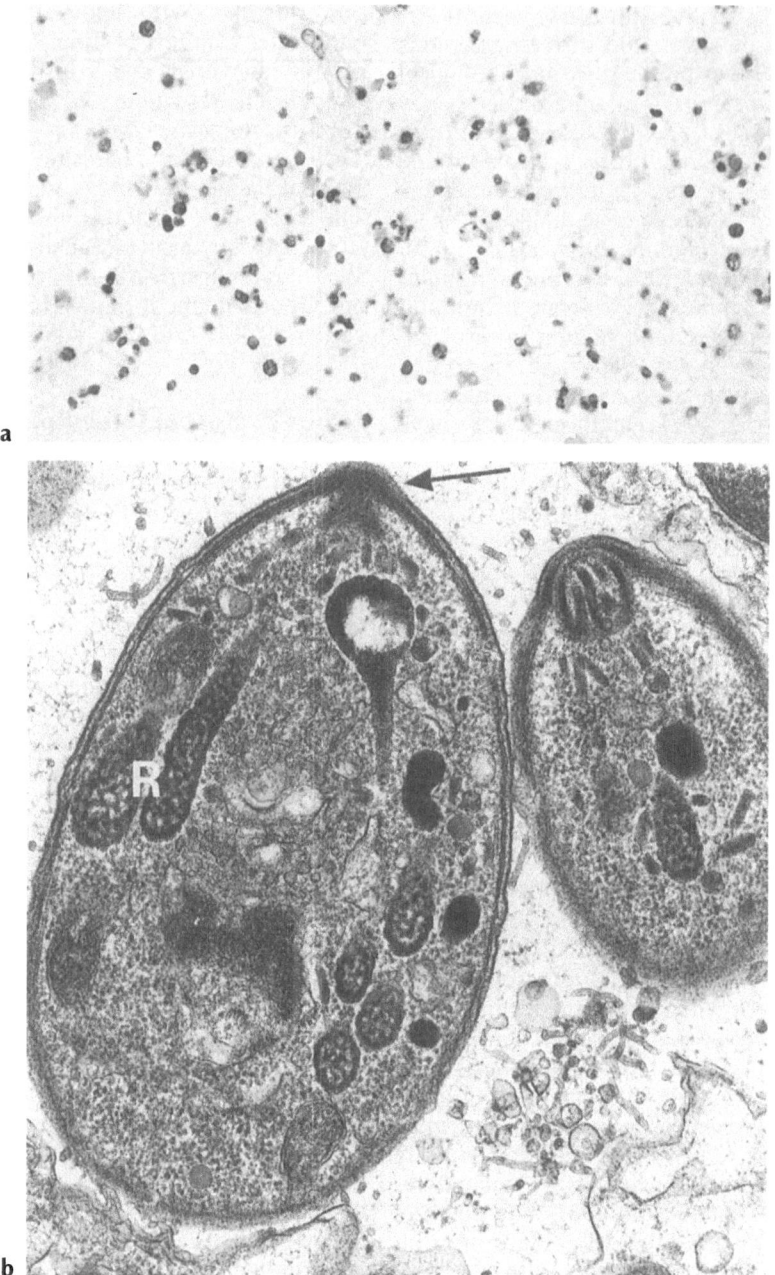

Figure 6.32a,b. Toxoplasmosis. **a** Anti-*Toxoplasma* antibodies reveal the presence of antigen in cases in which morphological evidence of the organism is lacking. ×450. **b** Endozoites of *Toxoplasma gondii*. The larger shows the conoid (*arrow*), numerous rhoptries (R) and a distinct double layered pellicle. ×26 600.

ibility by Martinez et al (1980). The tropism of these organisms for the brain of immunocompromised hosts is confirmed by experimental work (Martinez et al, 1975; Markowitz et al, 1978).

Neuropathology

In the case described by Wiley et al (1987), the patient was a 34-year-old homosexual man who died after a 6-week-long illness. Symptoms, which included hemiparesis and focal clonic movements, were followed by rapid decline of the mental status. CT scanning showed three well-circumscribed cortical lesions.

At postmortem, the gyri of both cerebral hemispheres were flattened, and there were foci of haemorrhagic softening in both cerebral and cerebellar hemispheres, accompanied by purulent meningitis. Histologically, the main feature was necrosis of the walls of vessels which, in addition, contained trophozoites. These usually measured 15–45 µm in diameter, and had a vesicular nucleus with dense central nucleolus and cytoplasm containing granules and vacuoles. Cystic forms were not described.

In cases of *Acanthamoeba* encephalitis, the brain may also contain focal granulomas. However, such foci were absent from this case, as well as from a number of non-HIV-positive immuno-suppressed patients. Wiley et al (1987) speculate that this atypical response could have been due to the severe immunosuppression. *Acanthamoeba culbertsoni* was the organism responsible for this encephalitis. Another case of 'acute amoebic necrotizing encephalitis', described by Robinson et al (1987) in a patient suffering from AIDS was caused by an unspecified organism of the *Hart-mannella–Acanthamoeba* group. Furthermore, a fatal case of meningoencephalitis due to a leptomyxoid amoeba of another order of free-living amoebae, leptomyxida, was reported by Anzil et al (1991).

An association between American trypanoso-miasis (Chagas' disease) and AIDS has been reported by Gluckstein et al (1988) and Del Castillo et al (1990). The latter group described a 19-year-old haemophiliac presenting with a space-occupying lesion, suspected to be a glioma. In both cases, diagnosis was made at craniotomy and histological examination showed aggregates of histiocytes containing leishmania forms of *Trypanosoma cruzi*.

The relationship between HIV and cerebral malaria has been investigated by Leaver et al (1990) in a prospective study in Lusaka. Results suggest that the outcome of adult cerebral malaria is not affected by HIV.

Pneumocystis carinii predominantly involves the lungs, although in HIV-positive patients the organism has also been found in the retina, bone marrow, small intestine, mastoid and skin, as well as producing generalized infections (Grimes et al, 1987; Unger et al, 1988; Matsuda et al, 1989).

To date, a single case with cerebral infection has been reported (Mayayo et al, 1990) in a 30-year-old drug abuser, in whom CT appear-ances were suggestive of multiple *Toxoplasma* abscesses. At postmortem, the brain showed four abscesses containing, in addition to occasional cysts of *T. gondii*, *P. carinii* organisms. The authors emphasize the fact that *Pneumocystis carinii* could not be detected in any other organ.

Sharer et al (1987b) have reported a case of meningo-encephalitis in an HIV-positive patient, which was produced by the achloric alga *Proto-theca*.

Opportunistic Infections: Bacterial

The bacteria which most commonly cause lesions of the nervous system in HIV-positive patients are mycobacteria. These are anaerobic bacilli which do not form spores, and are non-motile. Their cell wall, made of mycolic acids, gives them a characteristic staining property known as 'acid resistance'. Whilst tuberculosis, in general, was decreasing in the last decades, it has become a hazard in HIV-positive patients, in whom both the typical *M. tuberculosis* and atypical bacteria, predominantly *M. avium–intracellulare*, play an important role.

M. tuberculosis

In developing countries, infections due to this bacillus are decreasing in children, but are more common in adults. In Britain, Swart et al (1981) observed a fall from 400 cases per year in England and Wales in the mid-1960s to about 100, most of the cases occurring among immigrants of Asian origin.

Tuberculosis of the nervous system usually presents as an acute or subacute meningitis, and can affect previously healthy individuals or immunocompromised patients. *M. tuberculosis* reaches the nervous system through the respiratory or alimentary tracts, where the primary lesion takes place. Spread of the infection depends on the immune response and the number of bacteria present, and can occur immediately after the primary infection or take several years. The components of the nervous system which are involved include the dura, the leptomeninges and the brain tissue.

Epidural tuberculosis is rare in the cranial cavity, but is relatively more frequent in the spinal canal, where it arises from a focus in the vertebrae. The subdural form of tuberculosis,

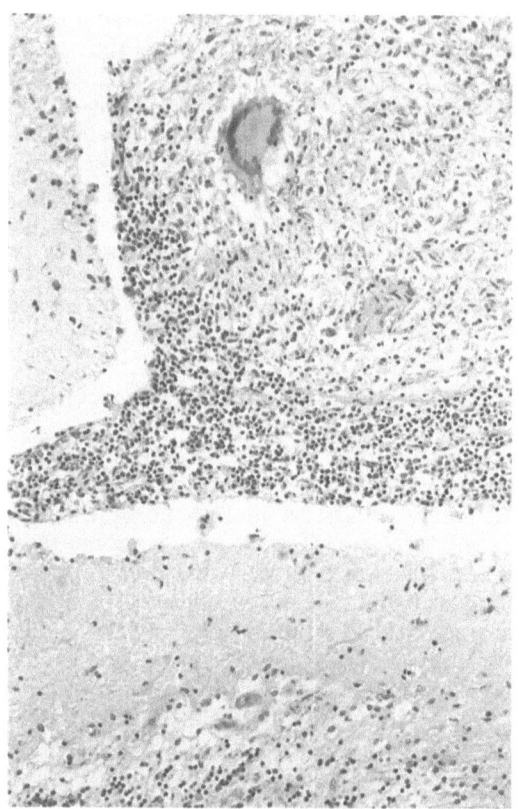

Figure 6.33. Tuberculous meningo-encephalitis. The granuloma involving the leptomeninges over the cerebellum consists of epithelioid cells, scattered among which are multinucleated giant cells. The nodule is surrounded by lymphocytes. H&E, ×120.

event. Among the 15 cases reported in the literature (Kelly and Brant-Zawadzki, 1983; Levy et al, 1984; Fischl et al, 1985; Jordan et al, 1985; Bishburg et al, 1986), 6 have occurred in Haitians and the remaining 9 in drug abusers. Of these cases, 10 showed single or multiple lesions, whilst meningitis was present in 4 and no description was given for the last patient.

CT scan appearances (see Chapter 4) include both single and multiple hypodense, sometimes ring-enhancing hemispheric lesions, which, as in two cases described by Bishburg et al (1986), may be mistaken for tumours.

Macro- and microscopic appearances are similar to those seen in cases not associated with AIDS. The leptomeninges may be occupied by a thick layer of yellow–grey pus. Lesions in the meninges (Figure 6.33), as well as in the parenchyma, consist of nodules, often with a caseous centre and surrounded by a florid granulomatous reaction, which includes small inflammatory, epithelioid and multinucleated giant cells.

Since Bishburg et al (1986) described improvement in 3 in 10 of their patients after antituberculous therapy, they consider brain biopsy to be mandatory in suspected cases.

Atypical Mycobacteria

Mycobacterium avium and *Mycobacterium intracellulare* are non-photochromogen bacteria, so closely similar to each other in common laboratory tests that they are referred to as the *M. avium–intracellulare* complex (MAI) (Runyon, 1971). The organism is ubiquitous in the environment, but the actual source of infection caused by MAI is usually not apparent. In immunocompromised patients, the organism produces disseminated infections involving the lungs, liver, spleen and bone marrow, and indeed, up to 50 per cent of HIV-positive patients have evidence of disseminated MAI infection at postmortem (Armstrong et al, 1985); yet, infection of the nervous system remains rare.

A small number of cases have been reported to date (Masur et al, 1982; Zakowski et al, 1982; Berger et al, 1984; Pitlik et al, 1984), most of them in Haitians. They present as disseminated illness and diffuse encephalitis. In some cases, MAI was diagnosed by brain biopsy or aspiration of abscesses. In the large majority of cases, however, diagnosis was made at postmortem.

Treatment of MAI infections is usually unsuccessful (Hawkins et al, 1986), even when new agents (ansamycin and clofazimine) are used.

known as 'tuberculoma en plaque', often involves the cortex. Isolated tuberculomas are occasionally found in the leptomeninges and in the brain, and can be large enough to become a space-occupying lesion. Multiple small tuberculomas of the meninges occur frequently in patients who die from tuberculosis without clinical evidence of intracranial involvement. Tuberculous meningitis is the commonest presentation within the nervous system and, whilst it is now a rare event, it is also becoming more difficult to diagnose (Parsons, 1982). It may follow miliary tuberculosis or the discharge of a small meningeal tuberculoma into the subarachnoid spaces.

Neuropathology

Like in other immunosuppressed patients (see Hooper et al, 1982), tuberculosis of the nervous system in HIV-positive patients is an uncommon

Neuropathology

Involvement of the nervous system by MAI is usually part of a general infection. Zakowski et al (1982) describe diffuse, non-suppurative granulomatous lesions. Bacteria are contained within macrophages which are sometimes localized around blood vessels, and are intensely PAS-positive.

Other Bacterial Infections

Except for mycobacteria, no other bacterial infections were reported by Snider et al (1983) and Levy et al (1985), and have rarely been described ever since.

Listeria monocytogenes, the commonest cause of bacterial meningitis in immunocompromised hosts, only rarely produces infections in HIV-positive patients (Jacobs and Murray, 1986; Mullin and Sheppell, 1987). It acts as an intracellular parasite. Three HIV-positive patients with *Listeria* were reported by Levy et al (1988a). The first presented with altered mental status, fever and focal neurological signs. CT scanning showed a single enhancing lesion, which was considered an abscess. The other two patients, as well as the patient reported by Koziol et al (1986), showed signs of meningitis.

Two cases of *Escherichia coli* involving the nervous system, and producing meningitis and meningo-encephalitis, have been reported by Post et al (1985) and Berger et al (1984). In a case examined in London, the brain of a patient who had died with generalized strongyloidiasis (Harcourt-Webster et al, 1991) (see Chapter 5) was found to have a diffuse purulent meningitis. Although no attempts at cultivating bacteria were made, it is known that infestation by this worm can be associated with *E. coli* infection (Brown and Voge, 1982).

A few cases of neurosyphilis have been reported among HIV-positive patients: one of meningo-encephalitis by Berger et al (1984), and four (two of meningovascular syphilis, one of acute meningitis and one asymptomatic) by Johns et al (1987). More recently, Joyce et al (1989) described the case of a 24-year-old homosexual man who developed soreness of the right eye, photophobia and diminished visual acuity. Fundoscopy revealed retinitis and evidence of localized vasculitis. Secondary syphilis was diagnosed and treatment was started with prompt improvement.

Johns et al (1987) emphasize the fact that four cases of neurosyphilis during a period of 18 months represent a considerable increase in incidence, and suggest that syphilis may show a particularly aggressive behaviour in HIV-positive patients. This is further supported by the fact that meningovascular syphilis, which usually occurs 5–12 years after primary infection, presented, in one of these patients, only 4 months after primary infection. Johns et al (1987) believe that, in cases of HIV and syphilitic infection, the immunosuppression produced by the former organism is worsened by the treponeme (Pavia et al, 1978). Alternatively, the meningitis produced by either organism might facilitate the penetration of the other into the brain.

An HIV-positive patient suffering from Lyme disease was reported by García-Monco et al (1989). He presented with neurological symptoms which were successfully treated.

Jankovic (1986) reported the only known case of Whipple's disease of the nervous system associated with AIDS.

Cerebrovascular Disorders

Cerebrovascular accidents among HIV-positive patients are well documented in the literature. Six (12 per cent of the total number of cases) were reported by Snider et al (1983), and Levy et al (1986) found that 18 among 94 unselected patients (19 per cent) suffered from this complication. Cho et al (1987) described non-inflammatory intimal proliferation of leptomeningeal arteries associated with brain infarcts in four patients with HIV encephalitis, and Budka et al (1987) reported 11 cases with infarcts and haemorrhages among 100 patients. Lang et al (1989) included 'vascular' cases under non-specific lesions: the overall figure was 21 and represents 16 per cent of the total. Vinters et al (1988) reported one case with widespread necrotizing vasculitis of the nervous system, causing multiple foci of haemorrhagic necrosis in the spinal cord and cauda equina in a patient with AIDS-related complex, and proposed that the angiitis was a response of the blood vessel wall to HIV infection. This observation may correlate with the demonstration of HIV replication in necrotizing vasculitis of peripheral nerves (Gherardi et al, 1989). In a subsequent publication, Vinters et al (1989a) included 33 patients (21 per cent) suffering from 'vascular complications'. However, if the cases

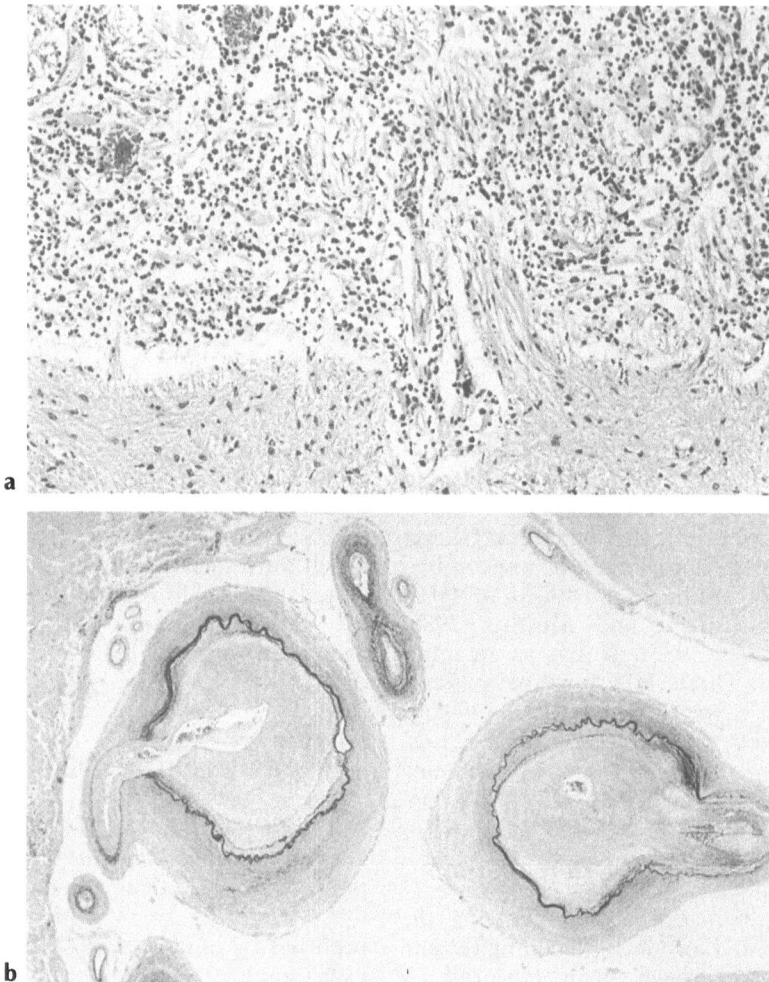

Figure 6.34a,b. Cerebrovascular disease. **a** Leptomeningeal exudate around the optic nerve consists of lymphocytes and plasma cells and includes some epithelioid cells. H&E, ×120. **b** Section of the base of a brain in coronal plane. Both anterior cerebral arteries are almost completely occluded by intimal proliferation. The elastic lamina is intact. Elastic/van Gieson, ×12.

with 'anoxic–ischaemic changes' are added, the total amounts to 42 (27 per cent). On the other hand, Petito et al (1986) mention that 30 (19 per cent) of their cases had suffered from anoxic–ischaemic encephalopathy, but do not speculate further on the type or extent of the lesions. In the series of 100 cases examined in London, only three patients were found to have suffered major strokes; a fourth patient, a haemophiliac, showed two small asymptomatic foci of infarction in the caudate nuclei.

The subject has been reviewed by Engstrom et al (1989) and Berger et al (1990). The latter examined 181 HIV-positive patients (154 adult and 27 children). Cerebrovascular lesions were described in 13 (8 per cent) of the adults, and included those due to ischaemia and to haemorrhage. Ischaemic changes were secondary to emboli, the heart being the main source of the latter. Clinical studies of AIDS patients have revealed a number of functional abnormalities of the heart (Reichert et al, 1989) and, in a pathological study of 54 patients, cardiac changes were found in 55 per cent (Roldan et al, 1987). The conclusion by Berger et al (1990), that an 8 per cent incidence in cerebrovascular diseases among HIV-positive subjects represents a considerable increase when compared with the general population, is shared by Engstrom et al (1989). However, when Berger et al (1990) compared their results with a 'control' group of HIV-negative, terminally-ill patients, they showed that the spectrum of cerebrovascular diseases appeared similar in the two groups. The only difference was that cases of vasculitis could be found only in AIDS patients. In addition to these two cases, cerebral vasculitis without identifiable cause was observed by Rhodes (1987) in eight brains (8 per cent); the case reported by Vinters et al (1988) showed diffuse necrotizing vasculitis, affecting predominantly the spinal cord and cauda equina, and Yankner et al (1986) described multiple non-enhancing lesions in a 42-year-old homosexual. Morphological examination of the brain showed narrowing of large and medium-sized arteries. The case reported by Scaravilli et al (1989b) was that of a 55-year-old HIV-positive man who suffered multiple strokes. CT scanning showed bilateral ventricular dilatation and two non-enhancing lesions in the frontal and temporal lobes. At postmortem there were multiple foci of infarction of different ages, basal meningitis consisting of lymphocytes, plasma cells and epithelioid cells (Figure 6.34a), and vasculitis. The vasculitis included multinucleated giant cells and produced occlusion of large (Figure 6.34b) and medium-sized vessels.

Vasculopathy, in association with HIV infection, has also been described in children, among whom Joshi et al (1987) reported coronary arteriopathy resembling Kawasaki's disease. More recently, Park et al (1990) found pathological evidence of cerebrovascular disease in 6 children, representing 24 per cent of a group of 25 HIV-positive patients. Of these, 4 had intracerebral haemorrhages, associated with immune thrombocytopenia, and all 6 showed non-haemorrhagic infarcts. Microscopic changes of the vessels included intimal fibroplasia with medial thickening, and destruction or reduplication of the internal elastic lamina. In one patient, the necrotic tissue contained multinucleated giant cells that were likened by the authors to those characteristic of xanthomas (Touton cells). In addition, in another of the cases, immunohistochemical techniques revealed mononuclear cells reacting with an antibody to gp41, a major glycoprotein of HIV.

Perhaps the most typical change of the vessels in HIV-positive children is calcification of medium-sized and small vessels. This has been mentioned earlier in this chapter and will be described in more detail in Chapter 9.

Although HIV has been considered the cause of most cases of vasculitis among AIDS sufferers, other causes may be operative and have to be excluded in each case. The roles of herpes zoster and syphilis have been discussed above. Another possible cause of brain thrombosis is the presence of lupus anticoagulant. This is a circulating IgG or IgM immunoglobulin, which produces thrombosis (Shapiro and Thiagarrajan, 1982) with subsequent infarction (Kelley et al, 1984; Fisher and McGhee, 1986). Although this immunoglobulin has been found in HIV-positive patients (Bloom et al, 1986; Cohen et al, 1986), it has not been reported in association with stroke (Berger et al, 1990).

Brain haemorrhage can follow thombocytopenia, which is a well-recognized complication in HIV-positive patients (Walsh et al, 1985) and may aggravate the coagulation deficit in haemophiliacs (see below). Thrombocytopenia may be an autoimmune phenomenon, be iatrogenic, or be secondary to intravascular coagulation.

Neoplasms of the Central Nervous System

Of the two main types of neoplasms occurring in HIV-positive patients – lymphomas and Kaposi's sarcoma (KS) – the former are by far the most frequently observed within the CNS, the latter being only rarely found and representing metastases from a focus elsewhere. Lymphomas in this group of patients are non-Hodgkin's (NHL) in type; leukaemias and Hodgkin's lymphomas (HL) have also been described, but association of these neoplasms with AIDS is considered coincidental. However, the recent publication of single case reports of HL in HIV-positive drug abusers in Italy and Spain deserves attention, and will be discussed below.

Non-Hodgkin's Lymphomas

Lymphomas of the CNS have been known since the beginning of the century (Bailey, 1929), whereas their morphological relationship to lymphomas arising elsewhere in the body was not recognized until 1972 (Schaumburg et al, 1972). They have previously been rare neoplasms, accounting for 0.3–1.5 per cent of all intracranial tumours, and 7.6 per cent of all NHL. Before the AIDS epidemic, the majority of cases were seen among patients with immunological deficits, either congenital or acquired, and recent epidemiological studies have even suggested an

Table 6.1. Incidence of primary malignant non-Hodgkin's lymphomas in adult AIDS patients

Authors	Year	No. of patients	No. of biopsy cases	No. of autopsy cases	PNHL No.	PNHL % patients	PNHL % autopsy	Country	Remarks
Snider et al	1983	50	–	–	3	6	–	USA	
Ziegler et al	1984	90	11	13	21	23.3	–	USA	
Koppel et al	1985	121	2	3	3	2.5	100	Sweden	
Levy et al	1985	352	–	–	9	2.5	–	USA	
Anders et al	1986b	–	–	86	6	–	7	USA	
Petito et al	1986	–	–	153	6	–	4	USA	
So et al	1986	1631	11	9	20	1.23	–	USA	
Budka et al	1987	–	–	100	6	–	6	Austria/Italy	
Hénin et al	1987	–	–	31	1	–	3.2	France	
Kato et al	1987	–	–	53	1	–	1.9	USA	
Rhodes	1987	–	–	100	8	–	8	USA	
Anzil et al	1988	–	–	144	4	–	2.8	USA	
Gonzales and Davis	1988	–	21	–	10	47.6	–	USA/Australia	Include part of So et al, 1986
Gray	1988c	–	–	38	1	–	2.6	France	
Knowles et al	1988	105	–	–	8	7.6	–	USA	
Loureiro et al	1988	–	2	20	8	–	40	USA	Series of lymphoma
Lowenthal et al	1988	63	2	7	9	14.3	–	USA	Include Swider et al, 1983
Stavrou et al	1989	–	–	61	0	–	0	RFA	
Esiri et al	1989	–	–	42	0	–	0	UK	
Lang et al	1989	–	–	135	3	–	2.2	Switzerland	Haemophiliacs
Gildenberg	1990a	–	61	–	17	–	–	USA	
Mikol	1990a	–	30	–	9	30	–	France	
Totals		2412	140	995	79 / 76 / 64	3.27 / 45.7	7.6		

It was difficult to ascertain that results concerning the same patients were not included in different series.
a Unpublished material.

increasing incidence among this group (Eby et al, 1988). The first single case reports in HIV-positive patients have been followed by larger series (Ziegler et al, 1984; Ioachim et al, 1985), in which the high incidence of primary and metastatic NHL of the CNS has been emphasized. This has produced a revision of the definition of AIDS by the Centers for Disease Control (CDC), which now also includes NHL (CDC, 1985).

A variety of primary and metastatic NHL of the CNS have been observed in patients with AIDS, both derived from B-cells.

Primary NHL of the CNS

Incidence. The incidence of primary NHL of the CNS has been reported to be up to 3.27 per cent in patients with ARC or AIDS (Table 6.1), 4.7 per cent in those with neurological involvement (Levy et al, 1985), and between 2.5 and 10 per cent in radiological series (Rosenberg S et al, 1986). A higher incidence is found in the group of patients undergoing cerebral biopsy (45.7 per cent) (Gonzales and Davis, 1988; unpublished observations, Mikol J), whereas in autopsy series, the mean incidence is as high as 7.6 per cent (Table 6.1). It has been estimated that primary NHL associated with AIDS will become more frequent than low-grade astrocytomas, and will have an incidence equal to that of meningiomas (Schoenberg, 1983; Rosenblum et al, 1988). NHL are more frequent in men than in women, and this is in keeping with the distribution in the general population. The age distribution appears to be similar to that of the AIDS population (between 25 and 40 years), whereas the average age at onset in non-AIDS patients is 57–59 years (Eby et al, 1988). Unlike Kaposi's sarcoma, which predominates among homosexual men, NHL are observed with the same frequency in homosexuals and intravenous drug abusers, and occur in the late stages of the illness (Levy and Bredesen, 1988b). Furthermore, they have not been described in haemophiliacs (Esiri et al, 1989). Two cases have been reported in homosexual partners (Capanna et al, 1987). In children, their incidence is between 3 and 6 per cent (Epstein et al, 1988; Kozlowski et al, 1990).

Clinical Presentation. Symptoms, either focal or generalized, are described only in a proportion of patients, in 0.6 per cent of whom they may be the presenting manifestation of AIDS (Rosenblum et al, 1988). They consist of alteration of cognitive functions, memory loss, confusion, lethargy or even dementia, and are more noticeable in

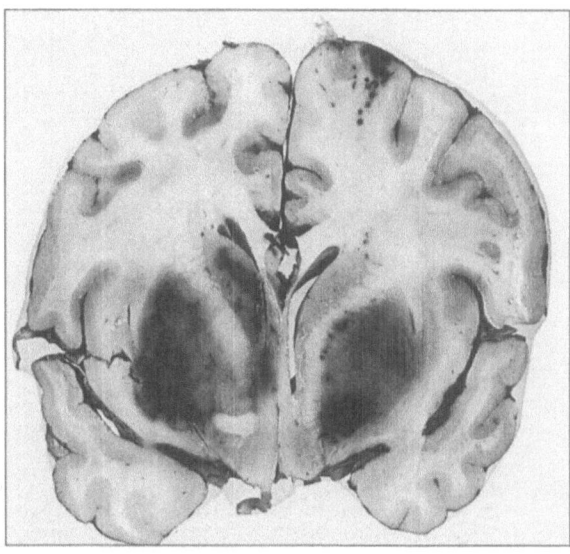

Figure 6.35. Lymphoma. Coronal section of a brain showing the deep grey nuclei of both sides almost completely replaced by a necrotic and haemorrhagic lymphoma, which produces compression of the left lateral ventricle and displacement of the right internal capsule.

younger subjects in whom they are associated with seizures. However, symptoms may be lacking, and the tumour is often found incidentally at postmortem. CT brain scanning demonstrates single or multiple (So et al, 1988; Poon et al, 1989) iso- or hypodense areas, with contrast enhancement. Magnetic resonance imaging (MRI) has proved a more effective method to detect intracranial abnormalities (Trotot et al, 1988; Levy et al, 1990), but cannot differentiate unequivocally lymphomas from other space-occupying lesions, such as *Toxoplasma* abscesses (see also Chapter 4) and, on occasion, may be normal. Lymphoma deposits, however, are said to be larger than abscesses (Gill et al, 1987). They usually respond to therapy, especially irradiation (2000 to 5000 cGy), and this may be confirmed by CT scanning. However, clinical improvement is usually of short duration, and patients experience recurrences. The mean survival time is extremely short, that is, 5 (Rosenblum et al, 1988) and 5.5 months (Formenti et al, 1989).

Pathology. Macroscopically, primary NHL of the CNS in AIDS patients are usually multicentric (So et al, 1986; O'Neill and Illig, 1990), and appear as fleshy, grey–pink nodular masses with variable amounts of necrosis and haemorrhage. They are localized predominantly in the deep grey nuclei of the cerebral hemispheres (Figure 6.35) and may extend to the optic nerves, but any

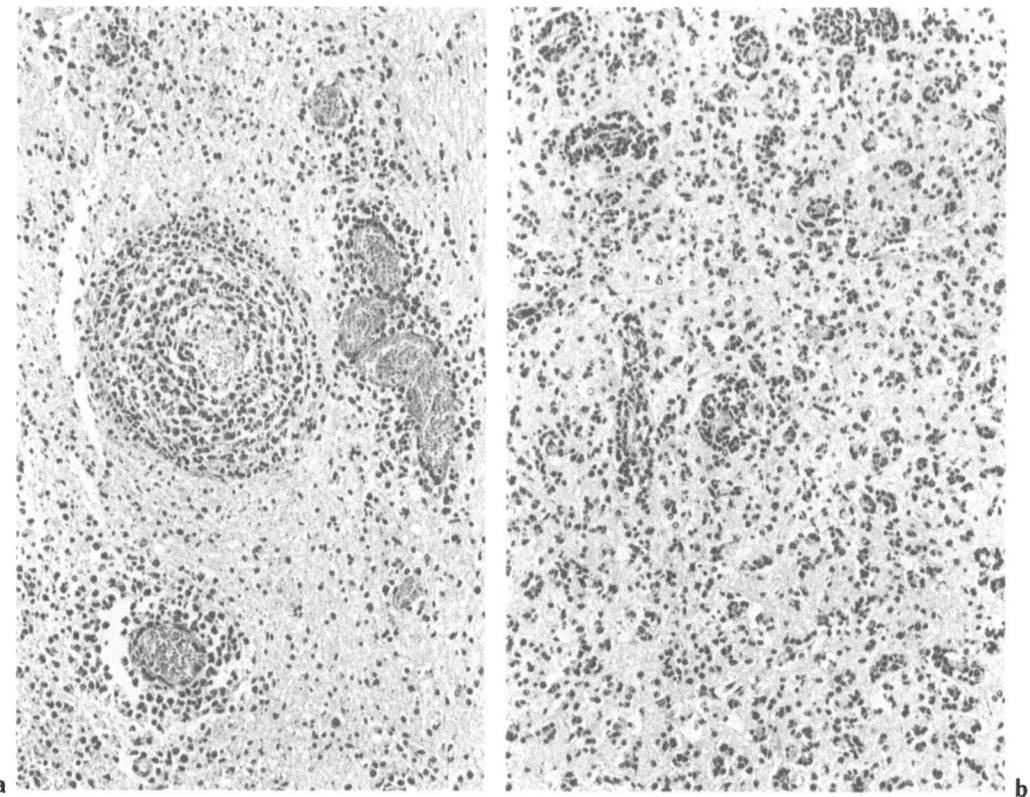

Figure 6.36a,b. Lymphoma. **a** Lymphoma cells showing the characteristic perivascular arrangement. H&E, ×120. **b** From the perivascular region, tumour cells infiltrate the adjacent neuropil. H&E, ×120.

region of the CNS may be involved. Cranial and peripheral nerves are only occasionally infiltrated (Ziegler et al, 1984). Involvement of the spinal cord has been reported by Iglesias-Rozas et al (1991), who described two cases. The lesions, which are often poorly circumscribed, are localized mainly in the parenchyma, but on occasion may extend to the leptomeninges. Sometimes gross examination is unremarkable and the diagnosis of NHL is made only by microscopic examination. Primary NHL do not generally metastasize to extracranial sites.

On histological examination the neoplastic lymphoid cells have a characteristic perivascular arrangement (Figure 6.36a) and tend to infiltrate (Figure 6.36b) or replace the surrounding parenchyma, which reacts showing variable degree of glial cell proliferation and contains, in addition, macrophages laden with cellular debris. Tumour cells may also infiltrate the vessel wall (Figure 6.37a), with consequent narrowing of the lumina. However, this does not usually result in necrosis of the vessel or of the surrounding tissue. Continuous growth of the tumour progressively

replaces the brain tissue, and the masses may eventually undergo necrosis. Indeed, AIDS-related lymphomas generally show more necrosis than non-AIDS tumours (Rennick et al, 1990; Goldstein et al, 1991).

The perivascular arrangement of tumour cells is emphasized by reticulin staining (Figure 6.37b), which shows an increase of fibrils forming concentric layers separated by tumour cells. Work by Kalimo et al (1985) and Kocki et al (1986) suggests that the perivascular network in lymphomas is produced not by the neoplastic cells, but by resident cells of mononuclear origin.

Tumour cells are arranged in irregular aggregates and lack the follicular or nodular pattern present in the extracerebral forms. In most cases they are of high-grade malignancy, immunoblastic, centroblastic mono- and polymorphic subtypes, diffuse with large cells and diffuse with small non-cleaved (Burkitt and Burkitt-like) cells (Figure 6.38). Using the classification of NHL by the National Cancer Institute USA (1982), So et al (1986) classified their 20 cases into large cell, immunoblastic (30 per cent), small cell, non-

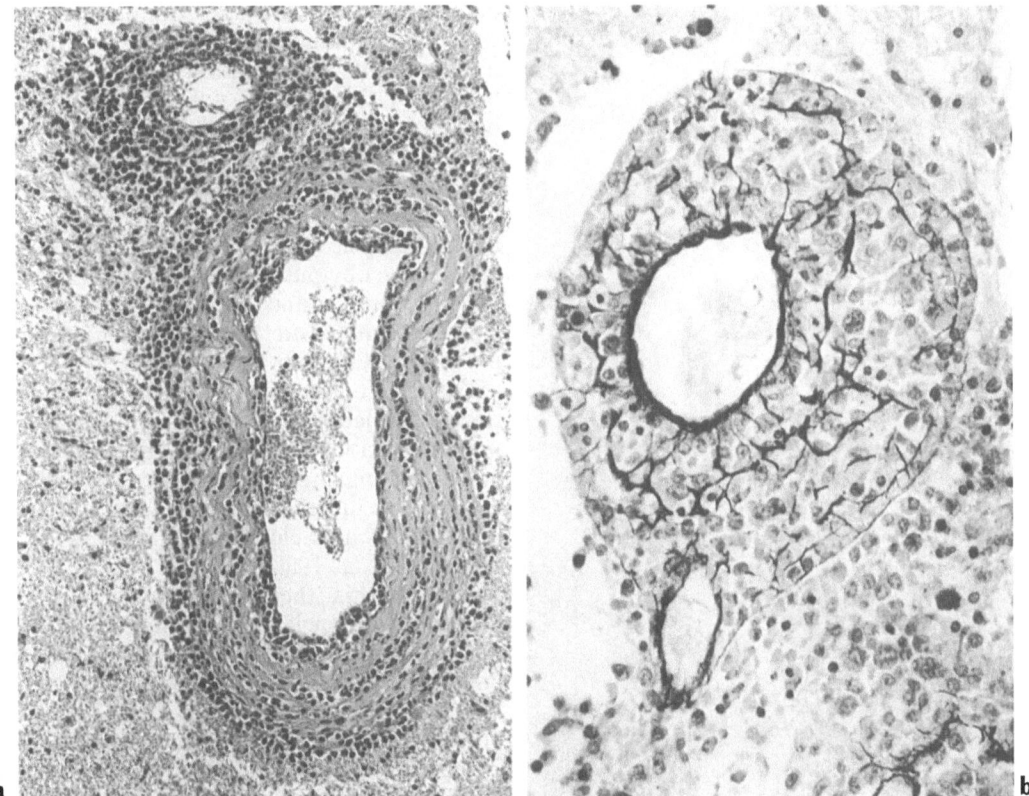

Figure 6.37a,b. Lymphoma. **a** In lymphomas, brain vessels may also show their walls infiltrated by tumour cells. In this microphotograph, neoplastic cells have reached the lumen. H&E, ×120. **b** Reticulin proliferates around vessels cuffed with lymphoma cells and forms concentric layers. It can also be seen surrounding single cells. Gordon and Sweet silver impregnation, ×300.

cleaved (Burkitt and non-Burkitt) (60 per cent), and unclassified (10 per cent). However, they also observed that a precise sub-classification of cerebral lymphomas cannot be as easily achieved as in extracerebral forms, due to the co-existence among tumour cells of a number of non-tumour lymphoid and other inflammatory cells.

An unusual case is that described by Dozic et al (1990) in a 12-year-old haemophiliac. The tumour presented as a form of angio-endotheliomatosis of the CNS, associated with HIV encephalitis, and meningocerebral cryptococcosis.

A small number of cases among the 30 biopsies examined in Paris (see Table 6.1) were also studied with the electron microscope. Tumour cells, with the features of centroblasts or immunoblasts, contained a variety of organelles (Golgi apparatus, rough endoplasmic reticulum, scattered mitochondria and polyribosomes). No HIV particles could be identified within neoplastic cells; however, tubulo-reticular inclusions were seen in endothelial cells and macrophages, as well as in association with the endoplasmic reticulum

of tumour cells in a patient who had previously been treated with interferon (unpublished observation, Mikol J) (Figure 6.39).

The lymphoma cells are shown by immunohistochemistry (Table 6.2) to be B-cells (Figure 6.40) (Ziegler et al, 1984; Gill et al, 1985; Ioachim et al, 1985; Di Carlo et al, 1986; Egerter and Beckstead, 1988) and of monoclonal lineage, as

Table 6.2. Antibodies reactive with B- and T-cells in paraffin-embedded tissues

Primary antibody	Polyclonal (P)/ monoclonal (M)	Cell type recognized
LCA (CD45)	M	Leucocytes
UCHL 1	M	T-cell
CD3	P	T-cell
MB2	M	B-cell
LN1	M	B-cell
LN2	M	B-cell
L26	M	B-cell

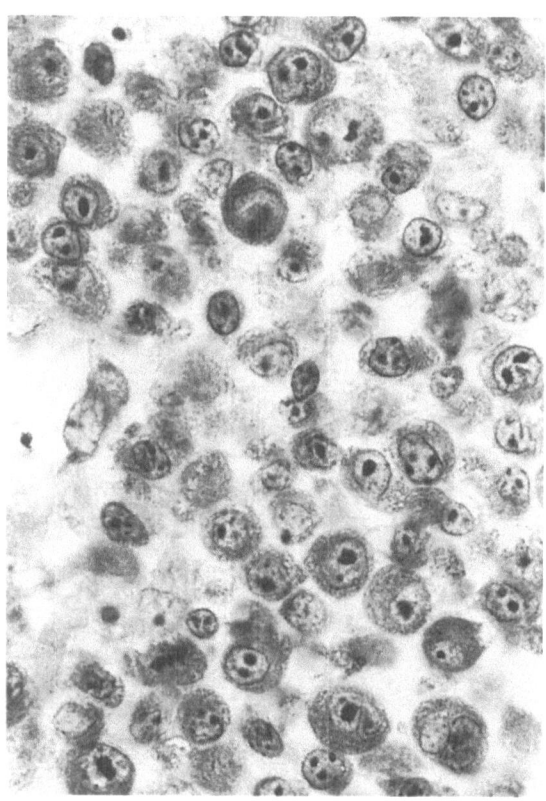

Figure 6.38. Lymphoma showing a mixed population of immunoblasts and centroblasts. H&E, ×900.

demonstrated by the use of antibodies to heavy and light immunoglobulin chains. The B-cell nature of the proliferation can be further confirmed by the positive staining of plasma membranes for adenosine triphosphatase, a B-associated marker (Egeter and Beckstead, 1988). The use of immunohistochemical techniques may be particularly rewarding in the diagnosis of small-needle biopsies, when the histological pattern may not be obvious and the pathologist has to rely upon the appearance of small numbers of cells. Two immunoblastic T-cell NHL have been reported by Rao et al (1989).

Cytogenetic studies of systemic lymphomas have revealed chromosomal abnormalities in a number of clones in NHL (Kalter et al, 1985), and gene rearrangement investigations have proved to be valuable in defining lymphomas and establishing lineage and clonality (Arnold et al, 1983; Cossman et al, 1988).

NHL within the nervous system have been reported in association with opportunistic infections, usually toxoplasmosis, CMV encephalitis and PML (Levy et al, 1985; Lang et al, 1989; Rodesch et al, 1989). Cases have also been reported in which lymphoma occurred with HIV encephalitis, either in separate regions of the brain (Budka, 1986) or in close vicinity (Misuzawa et al, 1987; Gray et al, 1990). Large multinucleated

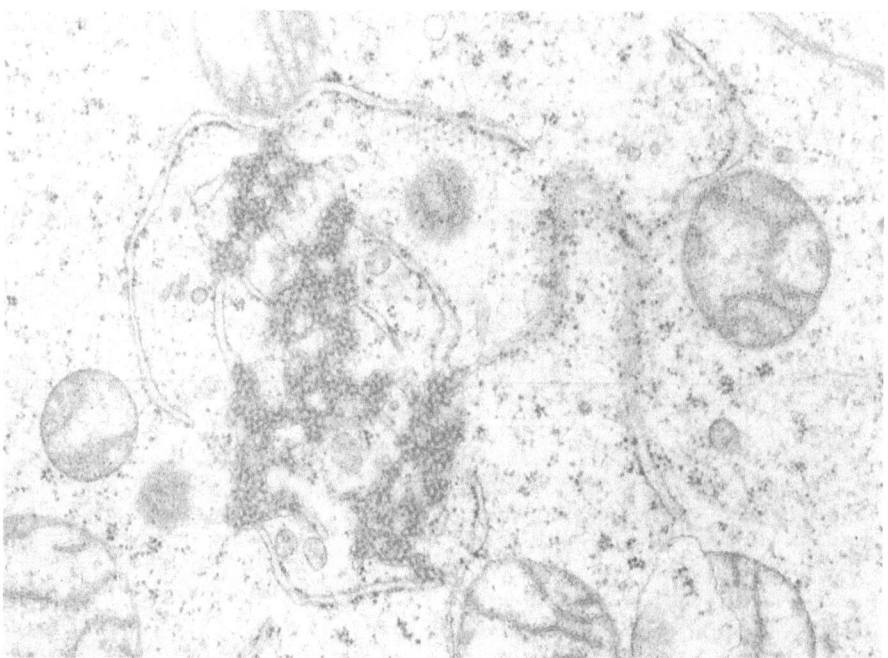

Figure 6.39. Lymphoma. Electron micrograph showing tubulo-reticular inclusions in continuity with the rough endoplasmic reticulum of a lymphoma cell. ×21 900.

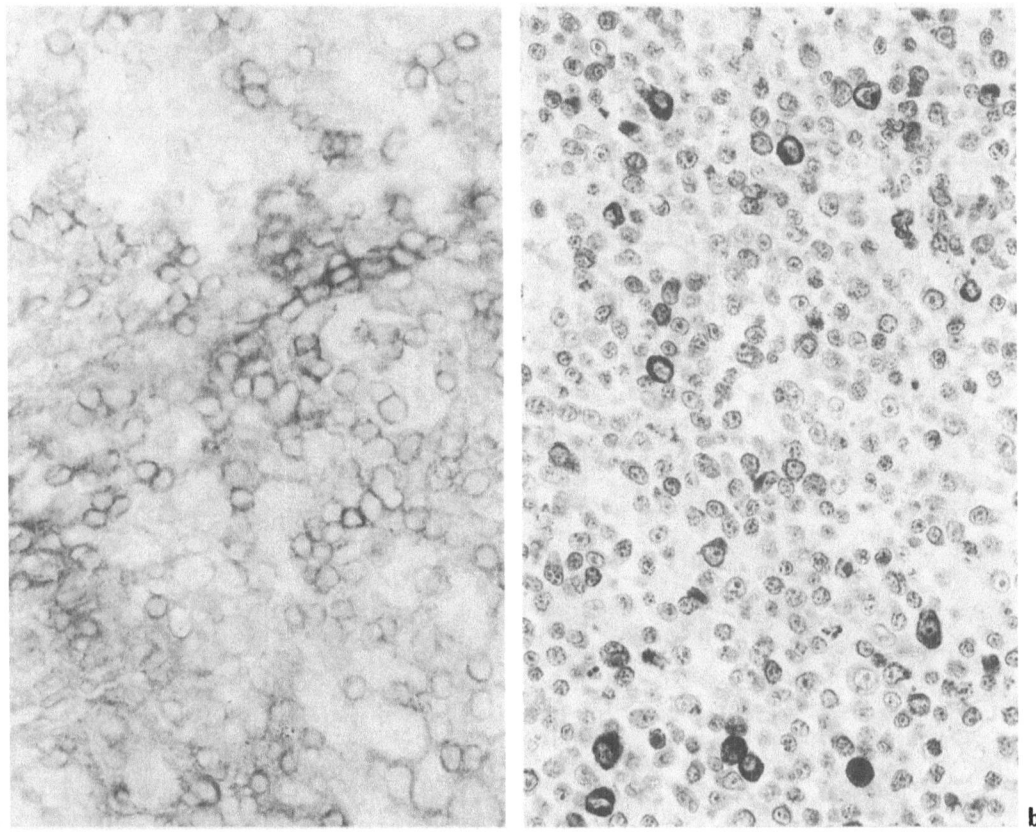

a   b

Figure 6.40a,b. Lymphoma cells immunostained with antibody to leukocyte common antigen in frozen section (**a**), and to LN2 for B cells (**b**) in paraffin section. ABC method, ×390.

cells in the tumour can be either tumour cells (Figure 6.41a), or MGC typical of HIV infection (Figure 6.41b).

The association between immunodeficiency, as it occurs in AIDS, and the development of tumours is well recognized (Penn, 1979), and it takes place both in primary and acquired syndromes. As already mentioned, the majority of tumours are NHL (Louie and Schwartz, 1978; Penn, 1983; Filipovich et al, 1984).

Viruses have been suggested as possible aetiological agents of lymphomas, and many observations suggest EBV as the most probable culprit in these neoplasms. EBV is an ubiquitous transforming herpesvirus, that replicates in stratified oropharyngeal epithelium (Sixbey et al, 1984) and in the epithelial cells of the ducts of the salivary glands (Wolf et al, 1984). Infection of these epithelial cells is productive, and the release of virions ensures that EBV-receptor positive B-cells are infected. The virus can probably transform B-cells, producing any of the major immunoglobulin isotypes: IgM, and to a lesser

degree, IgG-I clones are the most commonly found. Moreover, EBV glycoproteins gp350 and gp220 favour attachment to the complement receptor-type 2 (CR2) of B-cells, and gp85 probably plays a role in the fusion of the viral envelopes with cell membranes with consequent endocytosis. Sero-epidemiological investigations have shown that up to 90 per cent of the world's population has contracted this infection, which persists in lymphocytes (Essex, 1984).

Shaerer et al (1985) studied a 12-year-old boy who had undergone bone marrow transplant. At postmortem, multiple tumour-like B-cell nodules were found throughout the body, and EBV was isolated from the patient's pharyngeal secretions. Ziegler et al (1982) found serological evidence of EBV infection in four patients with AIDS and Burkitt-like lymphoma, and Rosenberg NL et al (1986) found evidence of EBV genome in the brain tissue of an HIV-positive patient suffering from cerebral lymphoma.

The steps leading to the appearance of lymphoma in EBV-positive patients have been

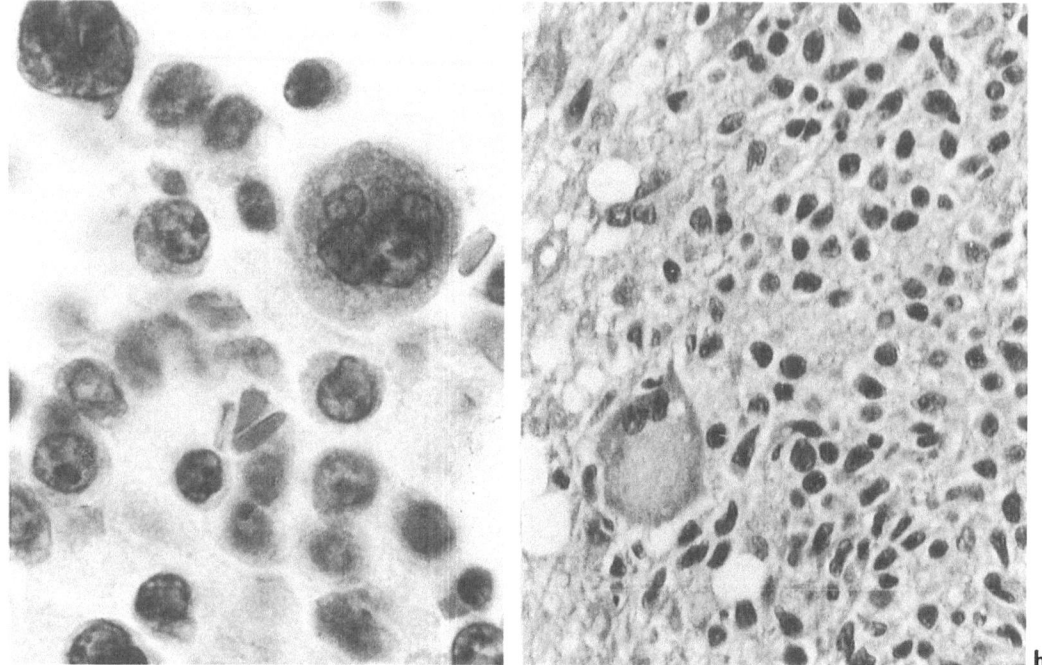

Figure 6.41a,b. Multinucleated lymphoma cells in a centroblastic lymphoma (**a**), and MGC of HIV encephalitis at the periphery of a focus of centroblastic lymphoma (**b**). H&E, ×1385 (**a**), and ×940 (**b**).

followed both in HIV-negative and positive patients. In individuals undergoing transplant operations, the appearance of these tumours is preceded by a benign B-cell proliferation. This has been observed by Hanto et al (1981, 1983), who found that, in spite of their aggressive behaviour, lymphomas consisted of polyclonal cells. Furthermore, of 15 malignant lymphomas described by Starzl et al (1984), four were polyclonal.

Similar changes were seen in lymph nodes of homosexual men with persistent generalized lymphadenopathy, which consisted predominantly of hyperplasia of B-cells (Egerter and Beckstead, 1988). Since the activation of B-cells is not checked by the T-cells, which are deficient in AIDS, a cytogenetic error may occur, resulting in a transformation of the polyclonal B-cell proliferation into a monoclonal malignant form (Purtilo, 1980; Egerter and Beckstead, 1988). These changes may happen more easily in the brain in view of its privileged position as a 'sanctuary', in which the normal mechanisms of immune surveillance might not be as efficient as elsewhere (Ciobanu and Wiernik, 1986).

In addition, it has been suggested that HIV could also directly infect B-lymphocytes (Pahwa et al, 1986; Witt et al, 1987), thus exhibiting synergism with EBV.

CNS Involvement in Systemic NHL

Incidence. Secondary CNS lesions have been observed in 40 to 47 per cent of systemic lymphomas in HIV-positive patients (Safai et al, 1987; Serke et al, 1988), and have been described also in children (Dickson et al, 1989b).

Clinical Presentation. Neurological symptoms are produced by leptomeningeal spread of the neoplasm and include headache and infiltration of cranial nerves and spinal roots, which may precede the discovery of the systemic lymphoma. After a systemic lymphoma has been diagnosed, prophylactic treatment to prevent spreading to the CNS has been proposed, either by lumbar puncture (So et al, 1988) or by intraventricular reservoir (Haaxma-Reiche and van Imhoff, 1989).

Pathology. The infiltrates tend to occur in the leptomeninges with subsequent extension along the Virchow–Robin spaces, the cranial and spinal nerves, and/or the CSF as in non-AIDS NHL (Mead et al, 1986). Although tumour cells are localized to sites near the CNS (roots, proximal trunks, spinal ganglia), direct infiltration of peripheral nerves has been shown in two cases.

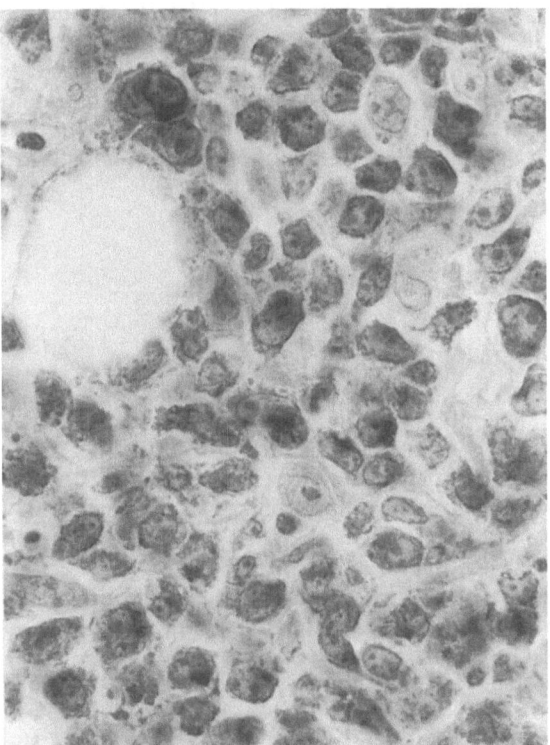

Figure 6.42. Immunostaining with MB2 antibody (for B-cells) of a metastatic epidural lymphoma appearing as cytoplasmic granules. ABC method, ×820.

Indeed, on occasion, the neuropathy may precede the diagnosis of lymphoma (Gold et al, 1988), an event that has been shown to occur also in non-AIDS patients (Gherardi et al, 1986; Zuber et al, 1987) (see Chapter 10). In the absence of nerve biopsy, it would be difficult to distinguish direct invasion of the peripheral nerves by lymphoma from a paraneoplastic syndrome. According to the new classification of peripheral neuropathies and lymphoma without monoclonal gammopathy, these neoplasms should be included in group 5 (Vital et al, 1990). A paraspinal plasma cell tumour, producing spinal cord compression, has also been reported by Israel et al (1983). A similar personal case of compression appeared to be produced by a B-cell tumour (Figure 6.42).

Hodgkin's Disease

Although single case reports or small series (Knowles et al, 1988; Lowenthal et al, 1988) have described Hodgkin's disease (HD) in HIV-positive patients, its incidence was not thought to be greater than that in the general population (Egerter and Beckstead, 1988). However, recent reports of a number of cases among intravenous drug abusers in Italy (Tirelli et al, 1988) and Spain (Serrano et al, 1990), have suggested an increased incidence amongst this risk group.

In most cases, patients were at an advanced stage of HD, with a severe decrease of helper T-lymphocytes and a predominance of suppressor cells (Unger and Strauchen, 1986). Only in one case was the CNS the initial site of involvement (Tirelli et al, 1988). The overall survival rate was poor in all reported cases, in which the most frequent localization was epidural. Review of histological data in 44 cases of systemic HD in AIDS has shown the predominance of mixed cellularity (66 per cent) and nodular sclerosis (30 per cent) in systemic disease (Diebold and Audouin, 1990). No such data are available for the CNS. Furthermore, it has been shown that a number of the HD cases of the CNS published before immunolabelling came into use (Henry et al, 1974), were in fact primary T-cell lymphomas (Taylor et al, 1976). The differential diagnosis between some forms of HD and T-cell lymphomas may be very difficult, and has to include extensive studies with markers (Gledhill et al, 1990).

Primary lymphomas of the CNS in immunocompetent and immunosuppressed (including AIDS) patients have recently been reviewed by Grant and Isaacson (1992).

Lymphomatoid Granulomatosis

Lymphomatoid granulomatosis (LG), first described by Liebow et al in 1972, predominantly affects middle-aged patients with a male to female ratio of 2:1, and is sometimes seen in association with immunosuppression (Jauregui, 1978; Cohen et al, 1979; Michaud et al, 1983). The disease usually runs a rapid course, with only few patients surviving more than 2 years (Katzenstein et al, 1979). The lungs are the organs primarily involved, and consequently patients die of respiratory failure.

Involvement of the nervous system in LG is part of a multisystem process, and has been reported in 20–30 per cent of cases, whilst in 20 per cent neurological symptoms may represent the onset of the disease (Katzenstein et al, 1979; Hogan et al, 1981). Moreover, in some cases the illness has remained confined to the CNS

(Schmidt et al, 1984; Kokmen et al, 1977; Sackett et al, 1979).

Symptoms vary considerably according to the site of the lesion in the brain, spinal cord or peripheral nerves. Macroscopically, changes are seen in the cortex and white matter; they do not differ from those seen in lymphomas and include single or multiple foci of softening and necrosis.

Histological examination shows a pleomorphic inflammatory infiltrate, angiocentric in type. Cells include lymphocytes, histiocytes, plasma cells, cells of lympho-plasmacytoid appearance and a number of atypical monocytes with scattered mitotic figures (Liebow et al, 1972; Anders et al, 1989). According to Katzenstein et al (1979), the presence of immature histiocytes and numerous mitoses is associated with a bad prognosis.

Whilst the angiocentric pattern and the infiltration of the surrounding brain tissue are features similar to those seen in NHL, LG may also show infiltration of the small- and medium-sized arteries and veins by inflammatory infiltrate, with disruption of the walls, narrowing or occlusion of the lumen by thrombosis, as well as presence of fibrinoid necrosis. The impairment of the blood flow is followed by infarction. However, infiltration of the vessel walls is not pathognomonic of LG and has been described also in some NHL (see Figure 6.37a).

Immunohistochemical staining methods show the majority of lymphocytes, both in HIV-negative (Nichols et al, 1982; Minase et al, 1985; Montilla et al, 1987) and HIV-positive (Anders et al, 1989) patients, to be T-cells. The inflammatory infiltrate contains polyclonal light (kappa and lambda) chains.

The nosological classification of LG is still uncertain, and further complicated by the fact that in about 12 per cent of cases the disease evolves into (or is associated with) a lymphoma (Liebow et al, 1972; Katzenstein et al, 1979; Michaud et al, 1983), usually monomorphic large-cell 'histiocytic' or high-grade immunoblastic type (Reddick et al, 1978; Sordillo et al, 1982; Kapanci and Toccanier, 1983), most of them of B-cell lineage with a monoclonal light chain pattern (Bender and Jaffe, 1980; Ironside et al, 1984; Petras et al, 1985). Nichols et al (1982), Nonomura et al (1983), and Kadin and Said (1988) consider LG a T-cell lymphoma, an angiocentric T-cell lymphoma according to Kadin and Said (1988). Veltri et al (1982) have a different opinion and consider LG a reactive process, possibly related to EBV. Hood et al (1982) suggest that LG begin as a polyclonal infiltrate, the T-cell abnormality leading to hyperproliferation of B-cells, and eventually to B-cell lymphoma. The immunohistochemical pattern exhibited by the cells, and the observation that some patients recover from this illness without treatment (Liebow et al, 1972; Ilowite et al, 1986), support the hypothesis of LG being, at least at its onset, a non-neoplastic process.

Kaposi's Sarcoma

Before 1980 KS was predominantly seen among male Jews and Italians, and in Bantus in Congo. Although the disease has already been known to affect immunocompromised patients, it is the spreading of AIDS that has made it a common complication affecting all risk groups (Centers for Disease Task Force on Kaposi's sarcoma, 1982; McNutt et al, 1983; Pitchenick et al, 1983), except haemophiliacs (see below). Indeed, the tumours appear to behave extremely aggressively in young homosexuals (McNutt et al, 1983).

In spite of this, however, localization of the neoplasm within the nervous system has been reported in only a limited number of cases, and in all it appears as part of a metastatic spread from a primary site outside the nervous system (see Chapter 5). No proven cases with cerebral localization were reported before 1983 (Snider et al, 1983), since in the case described by Hymes et al (1981) the nature of the frontal mass could not be confirmed by postmortem examination. Welch et al (1984) reported two patients, in a group of 36 with AIDS: one of them was found to have multiple cerebral deposits of the tumour; in the other, the only cerebral mass was in the right frontal lobe. Two patients with cerebral metastases of KS are also described by Gorin et al (1985). The first was a 47-year-old homosexual man with seven well-circumscribed haemorrhagic lesions, six in the cerebral and one in the cerebellar hemispheres. In the second case, that of a 60-year-old man with no apparent risk factor, the brain contained two haemorrhagic lesions in the cerebral and cerebellar hemispheres. Ariza and Kim (1988) reported a case of sarcoma of the dura mater in a 35-year-old HIV-positive man suffering from immunoblastic lymphoma and KS of the skin. He presented with subdural haematoma, which was evacuated. Histological examination of a small nodule attached to the dura showed proliferation of spindle-shaped cells and slit-like blood vessels, the features of KS.

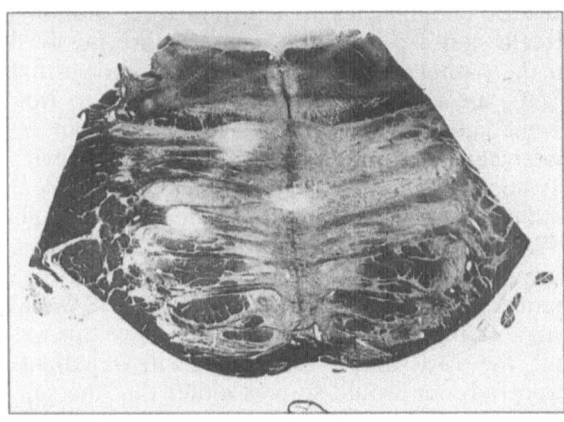

Figure 6.43. Transverse section of the pons in a case of focal pontine leukoencephalopathy in which 3 small foci of myelin loss appear in the parenchyma; 2 of them are at some distance from the midline. LFB/N.

Other Lesions

Pontine lesions have been described either in the form of central pontine myelinolysis (CPM) (Anders et al, 1986a) secondary to prolonged hyponatraemia, or as focal pontine leukoencephalopathy (FPL) (Vinters et al, 1987). The latter group of authors reported three cases in HIV-positive patients, and emphasized the point that FPL had been previously described in patients who had received chemotherapy or radiotherapy to the nervous system (Breuer et al, 1978). Subsequently Anders et al (1990) described ten more patients with the same pathological findings, five with AIDS, two with lymphoma and status post-bone marrow transplant, and the remaining immunosuppressed as a result of illness or treatment. Pathological features are limited to the pons, and consist of several small and well-demarcated areas (Figure 6.43) of vacuolation involving the ponto-cerebellar tracts, with loss of myelin and calcification of axons. These lesions do not seem to have any predilection for the central areas of the pons.

Vinters et al (1987) differentiate FPL from CPM by the presence of axonal calcification and necrosis in the former, and central localization in the pons in the latter. They admit, however, that in their third case, the lesion was centrally located. Unfortunately, they also omitted to mention the condition of the nerve cells within or near the lesions; the neurons are described as being relatively normal in CPM. In two such

cases, observed in London, lesions were asymmetrically seen in close relation to the midline, and included loss of myelin and relative preservation of the axons, a few of which showed spheroids. Conversely, neurons were well preserved.

Cases of Wernicke's encephalopathy and hepatic encephalopathy showing Alzheimer's type II astrocytes have been described by Rosenberg S et al (1986) and Petito et al (1986), respectively. In addition, a single case report of the former condition has been described by Davtyan and Vinters (1987) following treatment with zidovudine.

Although a number of data suggest the possibility of early HIV infection of the CNS (Chiodi et al, 1986), the occurrence of an illness such as aseptic meningitis, encephalopathy or myelopathy at the time of seroconversion or in the early stages of HIV infection is a rare and usually reversible event (Carne et al, 1985; Ho et al, 1985b; Denning et al, 1987), which makes it difficult to establish their pathogenesis. A case of acute fatal leukoencephalopathy in a seronegative patient in whom HIV was cultured from CSF, probably corresponding to a very early stage of HIV infection, was described by Jones et al (1988). Changes occurred predominantly in the white matter, and consisted of well-defined areas of demyelination with better preservation of axons and a possible perivascular pattern associated with acute neuronal damage and petechial haemorrhages. The authors considered this illness a direct consequence of acute HIV infection. However, MGC were not found and immunostaining for HIV was equivocal.

The occurrence of lesions indistinguishable from plaques of multiple sclerosis (MS) in HIV-positive patients has been reported by Berger et al (1989) and Gray et al (1991). The former group of authors described seven patients, in four of whom the neurological disease had preceded by several months the demonstration of HIV, and suggested a coincidental association of the two illnesses. In the remaining three positive serology was shown concomitantly or within 3 months from the onset of the neurological disorder. The two cases described by the latter group were most unusual since in one of them (Figure 6.44) the occurrence of the first attack of acute MS was at the age of 66 years, while the other patient was from Brazil, a country where the incidence of MS is particularly low (Kurtzke, 1980). Berger et al (1989) stressed the point that, in their cases (two brain biopsies and one autopsy case), demyelination was perivascular and ac-

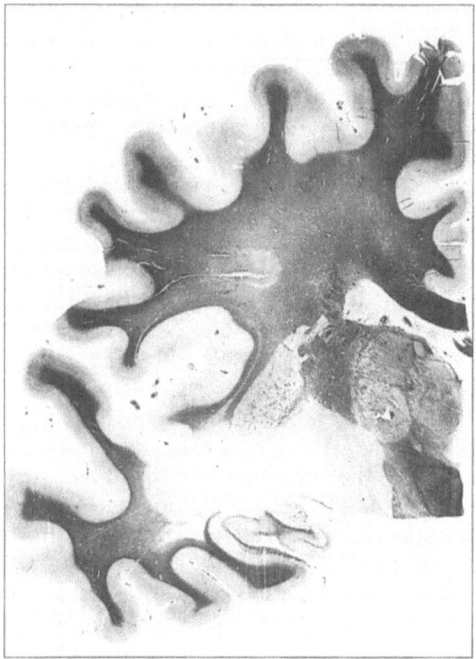

Figure 6.44. Coronal section of the left cerebral hemisphere showing a large plaque of demyelination involving the internal capsule, globus pallidus, optic tract, and the white matter of the temporal lobe above the temporal horn of the lateral ventricle. Loyez.

companied by macrophage infiltration and preservation of axons. In the two cases of Gray et al (1991), the demyelinating lesions were found around the lateral ventricle. Neither multinucleated giant cells nor MGN were found, and immunohistochemical staining for HIV by Gray et al (1991) was negative.

The significance of plaques in brains of HIV-positive patients remains, to date, obscure. A fortuitous association is mentioned by Berger et al (1989), and could probably be the case in four of their patients. The age of one and the country of origin of the other patient reported by Gray et al (1991) make this eventuality less likely. Moreover, the association, however coincidental, as well as the acute course of the cases described by Berger et al (1989) suggest the possibility that early HIV infection of the CNS may induce an immunopathological process similar to that which has been proposed in the Guillain–Barré type of polyradiculoneuropathies appearing at the time of seroconversion (Piette et al, 1986; Vendrell et al, 1987; Persuy et al, 1988). These events raise some intriguing questions about the

role of lymphocytes in demyelinating diseases. T-cells and T-cell-related immunity are involved in the pathogenesis of MS, and it is known that EAE, an experimental disorder related to MS, is mediated by cell-related immunity, can be passively transferred by sensitized lymphocytes (Paterson, 1960; Åstrom and Waksman, 1962) and cannot be induced in T-cell-depleted animals (Waksman et al, 1961). In view of these findings, one would expect that, since HIV-positive patients have a decreased T-cell count, MS would run a more chronic course in them. Since this was not the case in all the HIV-positive patients reported, one would suspect either that the suppression of one subset of lymphocytes by HIV could induce the acute progression of the disease, or that HIV itself is responsible for the appearance of the plaques.

Lesions of the Pituitary Gland

Reports of postmortem examination of the pituitary gland of AIDS sufferers are not as numerous as those of brain. Indeed, apart from personal communications, very little is known in the literature.

Ferreiro and Vinters (1988) examined the pituitary of 88 HIV-positive patients, in three of whom they described nuclear and cytoplasmic inclusions of CMV in the epithelial cells of the adenohypophysis without inflammatory changes. Other abnormalities seen in these cases were: one focus of cryptococcosis in the anterior lobe of one patient; microglial nodules in five patients; and one MGC in the pars nervosa one patient. The authors emphasize the fact that all three patients with CMV lesions showed signs of diffuse CMV infection. Furthermore, necrosis of the adenohypophysis was observed in ten cases, seven and two of which were accompanied by severe and mild brain changes respectively.

In a smaller group of pituitaries (eight cases) examined in London, one showed inflammatory cells forming irregular aggregates at the anterior edge of the pars nervosa, extending into the anterior part (Figure 6.45). The aggregates consisted mainly of lymphocytes, plasma cells and some epithelioid cells. However, granulomas, MGN and MGC were not found, and no lesions were described in the corresponding brain. Of the remaining seven specimens, one pituitary contained a microadenoma and another showed two

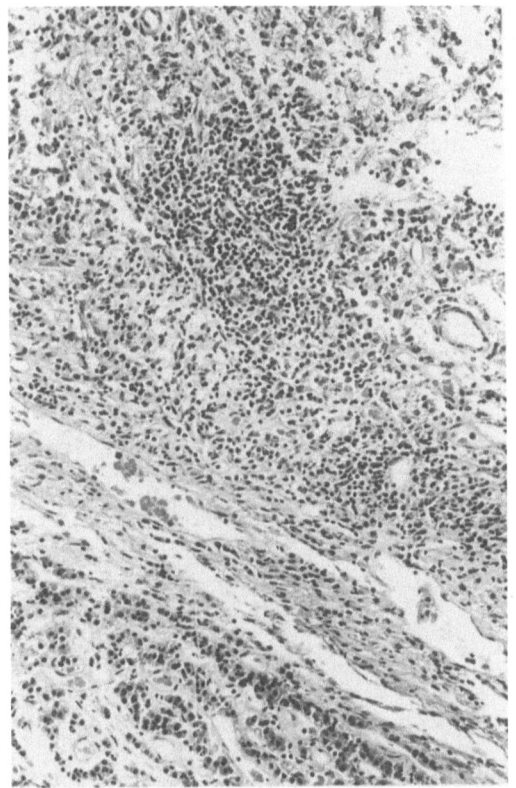

Figure 6.45. Microphotograph of a pituitary gland of an HIV-positive patient showing infiltration of the posterior lobe by small cells. H&E, ×120.

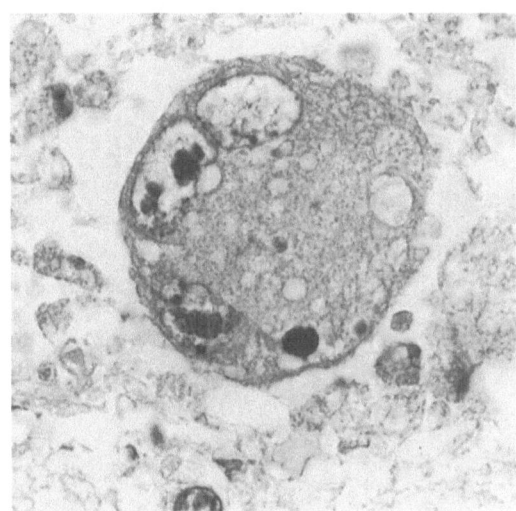

Figure 6.46. MGC containing intranuclear inclusions of CMV. H&E, ×1200.

well-circumscribed foci of infarction, the latter possibly related to the poor respiratory condition of the patient, leading to a hypoxic state.

To date, there are no studies correlating pathological changes in the pituitary with clinical syndromes.

Multiple Intracranial Lesions

A characteristic, though by no means exclusive, neuropathological feature in AIDS is the presence of lesions due to multiple agents. They are known to occur in other immunocompromised patients, and Hooper et al (1982) found an incidence of 12 per cent in a group of 49 HIV-negative individuals. Levy et al (1983) reported two cases, one with toxoplasmosis and CMV encephalitis, the other with cryptococcosis and CMV, and, in another report (Levy et al, 1985),

toxoplasmosis was found associated with lymphoma. Other cases have been described in which the associations were between HIV on the one hand and, on the other, CMV (Kleihues et al, 1985), the two viruses having been seen exceptionally in the same cell (Figure 6.46) (Bélec et al, 1990b), papovavirus (Figure 6.47a) (Gray et al, 1987), *Cryptococcus* (Figure 6.47b) (Gray et al, 1988b,c), *Toxoplasma* (Figure 6.48a) (Gray et al, 1988b) or with lymphoma (Figure 6.48b) (Mizusawa et al, 1987). The association does not seem to follow any rule, and Pitlick et al (1984) described a case in which brain abscesses contained *Toxoplasma*, *Candida* and *Staphylcoccus epidermidis*. An early incidence of 13.5 per cent of cases with lesions due to multiple organisms was considered an underestimate by Levy and Bredesen (1988a). Subsequent investigations produced a more likely figure of 29 per cent (Levy and Bredesen, 1988b) which also includes a number of patients with three or more unrelated processes. Multiple infections may appear simultaneously or at different stages of the illness, in the same or in different organs, or even in the same area.

The possible enhancement of one organism by another has been discussed above under the different headings (HIV, CMV, papovavirus). The presence of multiple lesions has important implications for the diagnosis and subsequent management of the patients. This subject is further discussed below in the section on brain biopsy.

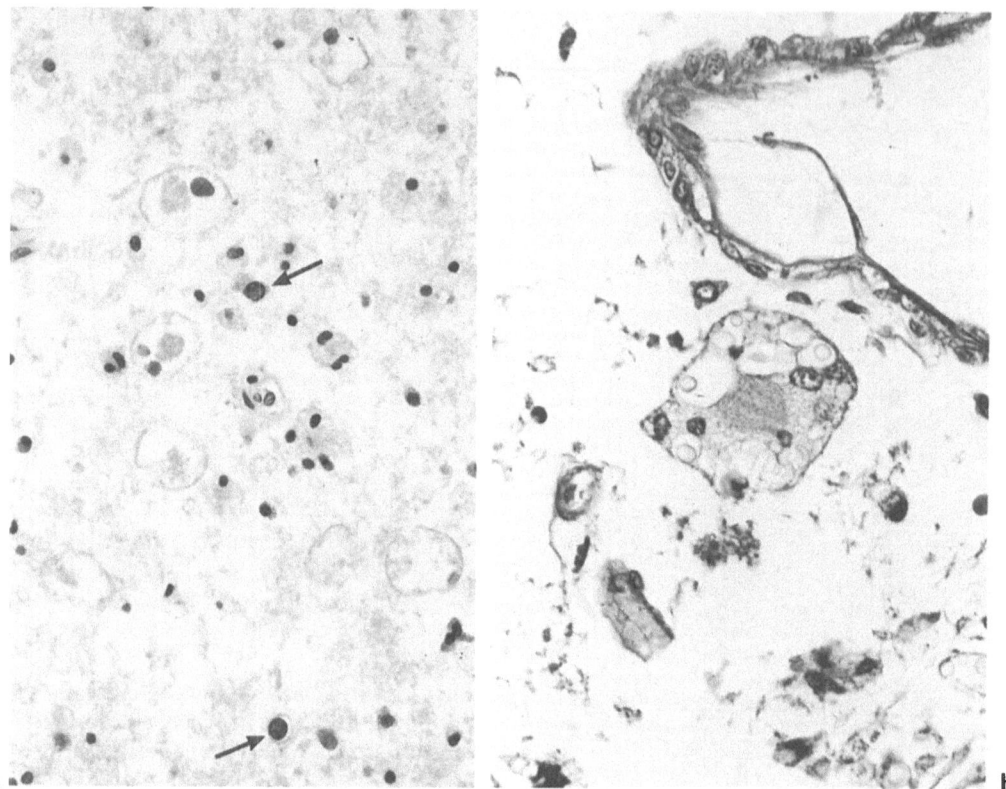

Figure 6.47. **a** Co-existence of HIV encephalitis and PML. Note the presence of MGC and of two swollen oligodendrocytes with intranuclear inclusions of papovavirus (*arrows*). H&E, ×250. **b** This MGC contains several cryptococci. Masson trichrome. ×1000.

Haemophilia and AIDS

In 1983, Davis et al described the case of a 59-year-old haemophilic man, who developed the immune deficiency syndrome. A small number of similar cases have subsequently been reported in all the large series. However, it has never been clear whether both the neurological illness and neuropathological changes in this group of patients were the same as those described in the other risk groups. Epstein et al (1985) and Sharer et al (1986) reported two cases with CMV encephalitis, and Rosenberg S et al (1986), Ries (1986) and Jones et al (1985) described a single case each with cryptococcosis, PML and toxoplasmosis, respectively. Cases with clinical and radiological signs attributable to HIV encephalitis were reported by Rahemtulla et al (1986) and Jones et al (1985), but they had no neuropathological examination.

More recently, the subject has been investigated by Esiri et al (1989) who studied a group

of HIV-positive patients, 11 of whom were haemophiliacs, with 29 patients belonging to other risk groups. They found that, whereas opportunistic infections (CMV and *Toxoplasma* encephalitis, and PML) were present in half of the patients of the latter group, they were absent in the former. Moreover, they observed that 45 and 36 per cent of the brains of the haemophiliacs contained acute and chronic intracranial haemorrhages, respectively, which were not a feature in the brains of the other group. HIV encephalopathy was found in only one haemophiliac, and in 8 out of 31 (26 per cent) of the other subjects. Outside the nervous system, cirrhosis of the liver was present only among the haemophiliacs, in whom no cases of KS were described.

The low incidence of opportunistic infections and HIV encephalopathy in the brains of haemophiliacs is explained by Esiri et al (1989) by the fact that these patients die at an earlier (pre-AIDS) stage of the immunodeficiency syndrome than other patients. There are a number of possible reasons for an early death. Haemophiliacs

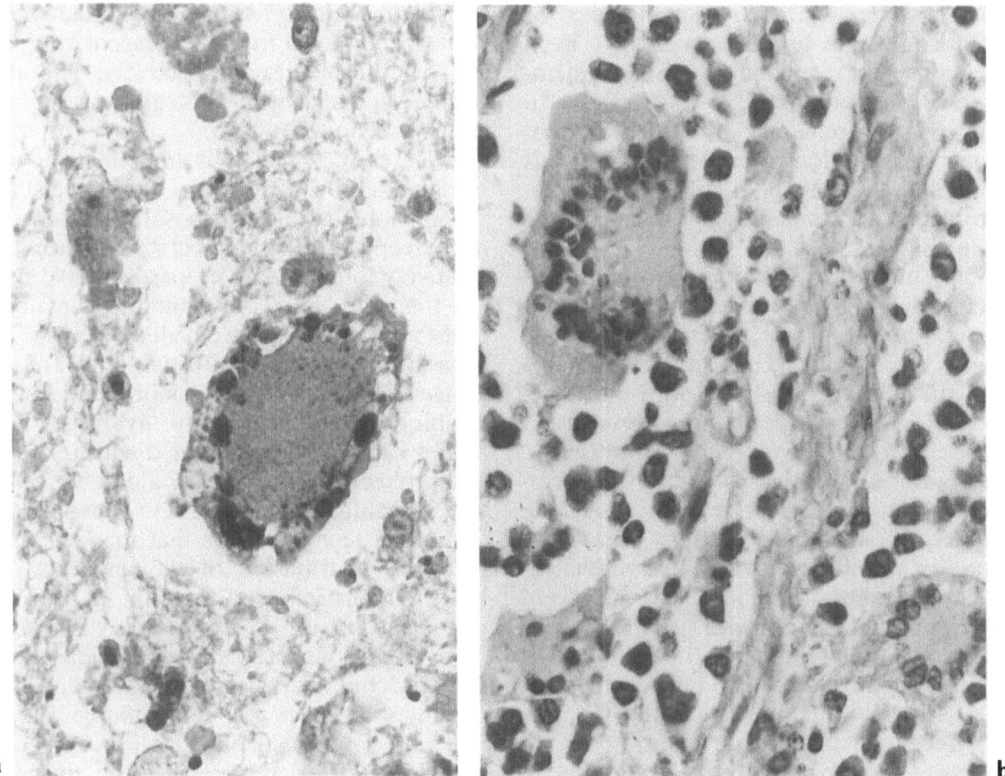

Figure 6.48. a *Toxoplasma* organisms within the cytoplasm of a MGC. H&E, ×1000. **b** MGC in a cerebral lymphoma. H&E, ×1000.

have a higher risk of developing intracranial haemorrhages as a consequence of their deficiency in clotting factors. There is no clear evidence that HIV infection enhances this risk, although in some patients it does produce thrombocytopenia (Morris et al, 1982; Abrams et al, 1986; Desforges and Mark, 1987). It is known that thrombocytopenia in HIV-positive haemophiliacs has a particularly poor prognosis when combined with HIV antigenaemia and low titre of p24 antibody. Another reason is the high risk, shared with HIV-negative haemophiliacs, of developing viral hepatitis and cirrhosis.

Brain Biopsy

The advances made in recent years in the field of neuro-imaging have produced a considerable improvement in the diagnosis of the neurological complications of HIV. It is possible to follow the evolution of the changes in the brains of patients with the AIDS-dementia complex, and to com-

pare the severity of the atrophy with the clinical signs, although it may not be always possible to confirm, at postmortem, the diagnosis of brain atrophy made on radiological grounds. There are, however, a not-negligible number of cases in which neuro-imaging does not show any pathology, or, when abnormalities are present, fail to provide the necessary information for the differential diagnosis. CMV encephalitis is the typical example of an opportunistic infection which can produce damage in both grey and white matter that may not be seen either with CT or MRI. On the other hand, multiple ring-enhancing lesions suggest toxoplasmosis, mainly when present in Haitians (Pitchenick et al, 1985). However, these appearances are similar to those seen in cases of lymphoma. It is estimated that between 5 and 12 per cent of scans fail to show mass lesions or other specific abnormalities (Post et al, 1983; Whelan et al, 1983; Wanke et al, 1984). The most reliable results in cases of toxoplasmosis are obtained with MRI, which can reveal lesions undetectable by CT scanning (Levy et al, 1985; Zee et al, 1985) (see Chapter 4). Unfortunately, even MRI is unable to give an answer in all cases, and brain biopsy has become a necessary complement for the diagnosis. Indeed, brain biopsies have been performed since the beginning of the investigations into AIDS (Rodan et al, 1984; Levy et al, 1985). It has to be remembered, however, that the number of conditions for which a biopsy is required is limited, that there still remain diagnostic problems, and that surgery is not exempt from some risks.

The main indication for brain biopsy is the differential diagnosis between toxoplasmosis and lymphoma. There are two main schools of thought on this subject. Levy et al (1985) recommend that all AIDS patients with intracerebral mass lesions should have a biopsy. However, Pitchenick et al (1983) have a more conservative approach, and suggest that an empirical treatment should precede surgery in cases where the location of the lesions makes biopsy hazardous. This cautious attitude seems to have prevailed, and patients in whom *Toxoplasma* is suspected are now submitted to a trial with specific therapy for 2 weeks (see Chapter 3). Only if, at the end of this period, they have shown no improvement, is a biopsy carried out. However, if the cerebral lesions produce mass effect requiring the use of steriods, an early biopsy may be preferable to treatment with antibiotics, since the morphological features resulting from the combined treatment might be difficult to interpret. In a series of 92 HIV-positive patients with mass lesions reported by Levy and Bredesen (1988b), only 53 per cent were found to have toxoplasmosis. The same authors extrapolate that, assuming 29 per cent of the patients with toxoplasmosis have multiple pathologies (Levy and Bredesen, 1988b), then 62 per cent of the total number with mass lesions have illnesses other than toxoplasmosis. In large series, combined treatment with pyrimethamine and sulfonamides is successful in approximately 75 per cent of cases, whilst the remaining 25 per cent require a brain biopsy. However, even in the most successful cases, the overall mortality after 1 year is almost 100 per cent and results from other opportunistic infections.

Unfortunately, a brain biopsy cannot guarantee the diagnosis in every case, as it is dependent on the fact that the specimen may have been taken from an area of complete necrosis or, alternatively, from the area of brain surrounding the lesion, in which only non-specific change may be present. In both cases, it is unlikely that any reliable diagnosis will be made. We find it very useful to examine rapid smear preparations (see Appendix for details); the finding of non-specific gliosis or necrotic tissue suggests that the specimen was removed from an adjacent area, or from the necrotic centre, respectively, and that the needle should be re-directed towards a more suitable region.

In cases in which it is impossible to reach a morphological diagnosis, pathologists have several tools at their disposal. We find electron microscopy of little practical value, since it takes a relatively long time to prepare suitable material. On the other hand, immunohistochemical techniques are much more valuable and, on one occasion, it was possible to visualize *Toxoplasma* antigen in the absence of morphologically identifiable organisms. In situ hybridization techniques can now be performed with relative ease, and probes are available for a number of viruses. The advantage they offer in comparison with immunohistochemistry is that they show a more extensive distribution of the antigen. This has been demonstrated by Aksamit et al (1987) for papovavirus and by Wiley et al (1986a) for CMV. Two limitations of the needle biopsy are represented by the associated morbidity and the possible co-existence of multiple infections. Uncontrollable bleeding following a needle biopsy could be triggered by the thrombocytopenia described in some of these subjects (Morris et al, 1982; Abrams et al, 1986; Desforges and Mark, 1987). In cases in which several infectious agents are present, it is unlikely that they will all be revealed by the needle biopsy, and it is theoretically possible that the successful treatment of one of them may be counterbalanced by the unmasking and worsening of the other.

Appendix. Morphological Demonstration of Infectious Organisms Within the CNS in AIDS

Tissue Preparation

Organisms producing lesions in the CNS can be demonstrated in sections obtained both from frozen and formalin-fixed, paraffin-embedded material. Unfortunately, all tissue specimens from HIV-positive patients are potentially infectious, and preliminary fixation with 10 per cent formalin has been recommended (MRC) as this fixative has been found to inactivate HIV effectively. Nevertheless, the potential danger of contamination should not deter the pathologist from using fresh material for rapid diagnosis (see below). Whenever frozen samples have to be used for other reasons, for example to visualize HIV by in situ hybridization or by polymerase chain reaction (PCR), they should be handled within adequate containment facilities and each laboratory in which investigations on AIDS are carried out should ideally have access to such facilities. Fortunately, the development of antibodies and nucleic acid probes that work well on paraffin sections has helped considerably to alleviate this problem.

All the methods described below, which are in routine use in the Department of Neuropathology at The National Hospital, Queen Square in London, can be performed on formalin-fixed, paraffin-embedded material.

Rapid Preparations

Of the two possible methods of processing unfixed material for rapid diagnosis (smear preparations and frozen sections), we prefer the former for two reasons. Firstly, it is quicker and does not require the use of a cryostat, which subsequently must be decontaminated. Secondly, the specimens are collected and immediately dealt with in a room adjacent to the operating theatre. They usually consist of a small amount of soft or semiliquid material, which has been removed by needle biopsy. The specimen is transferred to a Petri dish and divided into 3 parts for smear preparation, formalin and glutaraldehyde fixation respectively. Smears are prepared immediately and slides immersed in a water-tight Coplin jar containing methylated spirit. The jar, as well as the other containers used to store material, are wiped with glutaraldehyde (to neutralize the virus); used blades, the Petri dish and other contaminated objects are put in appropriate bags and sent for incineration. The fixed smear preparations can then be taken safely to the laboratory where they are stained with toluidine blue, following the standard technique.

Correct diagnosis was obtained with this method in 5 out of 8 cases: 3 PML (Figure 6.49a), 1 lymphoma (Figure 6.49b) and 1 case of toxoplasmosis in which cysts could be identified (Figure 6.49c). Of the remaining 3 cases, PML was suspected in one, but the degree of pleomorphism of the large astrocytes did not appear striking enough to warrant such diagnosis beyond doubt, although the diagnosis was subsequently confirmed by in situ hybridization. In a second case, the co-existence of necrotic material and inflammatory cells, including macrophages and epithelioid cells (Figure 6.49d) in the appropriate clinical and radiological setting, suggested the diagnosis of toxoplasmosis, which was confirmed by immunohistochemistry, using anti-toxoplasma antibodies. In the last case no diagnosis could be made.

Bacteria and Fungi

The bacteria most commonly found in the nervous system in AIDS are mycobacteria. *M. tuberculosis* can be easily demonstrated with Ziehl–Neelsen or an auramine-rhodamine stain. *M. avium–intracellulare* (MAI), on the other hand, is only weakly positive with these methods, but is intensely PAS-positive.

Other organisms, including *E. coli* and *Listeria monocytogenes*, have been occasionally reported in HIV-positive patients (see text). The Gram stain is still the most reliable method for visualizing these opportunists.

The most common fungus associated with AIDS brains is *Cryptococcus neoformans*. Like other fungi, this organism is positive with the Grocott method; however, it also stains strongly with alcian blue.

Protozoa

The only organism of this family which is commonly found in brains of HIV-positive patients is *Toxoplasma gondii*. Cysts of various sizes or isolated organisms are usually easily identified in routine preparations. Identification can be helped by PAS or Giemsa methods. Whenever organisms cannot be seen by routine methods, immunohistochemistry with polyclonal anti-*Toxoplasma* antibody (Bioquote, UK) highlights non-viable organisms. Electron microscopy may also show the parasite, but, as already mentioned, this technique requires longer times than immunohistochemistry and should not be used when prompt therapy depends on rapid diagnosis.

Viruses

Immunohistochemistry and in situ hybridization techniques are now widely used for the identification of viruses. Antibodies are available for

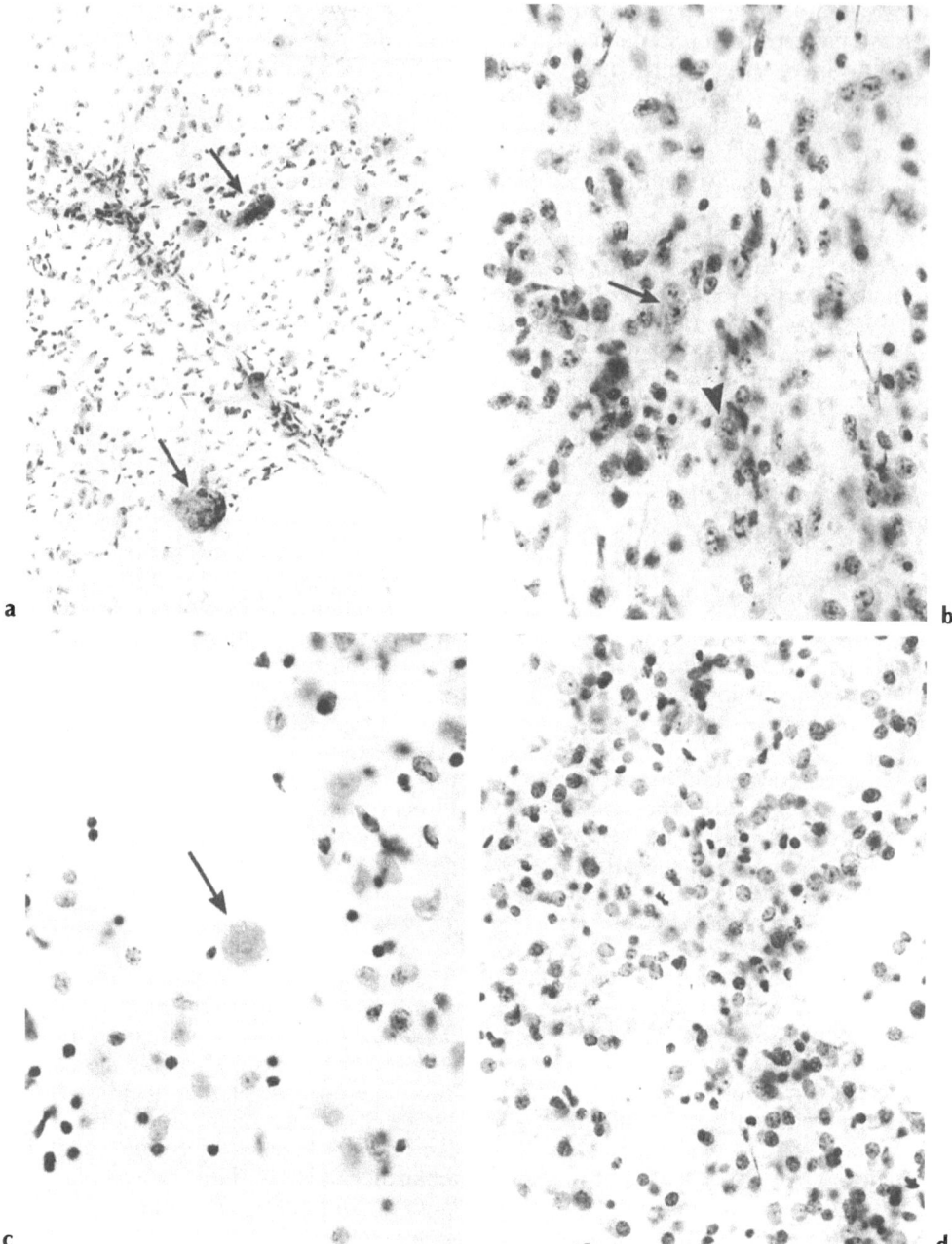

Figure 6.49a–d. Smear preparations of different pathologies of the nervous system in HIV-positive patients. **a** PML is strongly suggested by the presence of scattered pleomorphic astrocytes (*arrows*) in a background which includes reactive astrocytes and macrophages. **b** Lymphoma showing large cells with one or two (arrow) nuclei, prominent nucleoli and ill-defined cytoplasm. Some nuclei are indented (*arrowhead*). **c** The presence of a large cyst (*arrow*), containing oval structures and surrounded by small inflammatory cells and macrophages, is diagnostic of *Toxoplasma gondii*. **d** Non-specific inflammatory aggregates consisting of lymphocytes and macrophages and including small amounts of necrotic tissue. These appearances helped exclude some of the disorders commonly found in AIDS (lymphoma, PML). *Toxoplasmosis* could only be suspected and was eventually confirmed (see text). Toluidine blue, **a** ×120; **b** ×300; **c** ×480; **d** ×300.

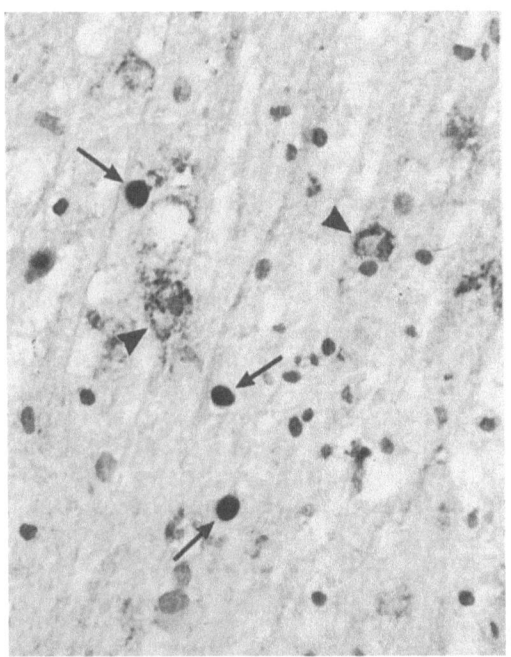

Figure 6.50. Combined in situ hybridization (*arrows*) for JC virus and immunohistochemical method (*arrowheads*) for HIV. ×300.

Detection of Viral DNA by In Situ Hybridization with Biotinylated DNA Probes (after Schmidbauer et al, 1989a)

Hybridization

1. Dewax in fresh xylene and take to absolute ethanol
2. Inhibit endogenous peroxidase in methanol containing 1 per cent H_2O_2 (30 min)
3. Wash in ice-cold PBS
4. Digest in proteinase K (Sigma protease type XXVIII) at 37°C for 15 to 60 min
5. Stop digestion in ice-cold PBS
6. Air dry sections 1 hr at room-temperature
7. Apply hybridization buffer 12 µl per slide and cover with salinated coverslips
8. Denature at 95°C for 5 min
9. Hybridize at room temperature for 30 min
10. Wash in 2 × SSC to remove coverslips
11. Rinse in 3 changes 2 × SSC
12. Wash in 0.1 × SSC with 20 per cent formamide at 37°C for 10 min
13. Rinse in 3 changes 2 × SSC.

Visualization

14. Rinse in PBS, 3 changes
15. Incubate in avidin-peroxidase conjugate (Dako, UK), 1/500 with 10 per cent normal swine serum
16. Wash in PBS, 3 changes
17. Incubate in 0.05 per cent 3,3-diaminobenzidine-tetrahydrochloride (DAB) containing 32 µl 30 per cent hydrogen peroxide, 5 min
18. Counterstain nuclei in Meyer's Haematoxylin

Solutions

Hybridization buffer: 50 per cent de-ionized formamide 2 × SSC
                10 per cent Dextran sulphate
20 × SSC: 3.0 M sodium chloride, 0.3 M sodium citrate.
PBS: 0.01 M phosphate-buffered normal saline, pH 7.0–7.2.

CMV (Dako, UK) and HSV-1 and 2 (Dako, UK). Immunocytochemical studies have also been performed on JC virus (Budka and Shah, 1983; Wiley et al, 1988) and Herpes zoster virus (Eidelberg et al, 1986) with antibodies which, unfortunately, are not commercially available.

Probes for in situ hybridization are now becoming easily obtainable, and methods are quicker and more straightforward. For many applications it is no longer necessary to use radioactive labels, colorimetric methods being well established and making the detection of nucleic acids more accessible to routine histology laboratories. Probes can be labelled with biotin or other haptens such as deoxygenin and kits for labelling are available from several suppliers (Enzo Biochem, New York; Universal Biological Ltd, UK; Boehringer, Mannheim, FRG). Enzo Biochem also supplies 'Bioprobes', ready biotinylated probes for a range of viruses including CMV, HSV and JC (Figure 6.50). Using an Enzo bioprobe, Schmidbauer et al (1989a) detected CMV DNA in nodular encephalitis without visible inclusion bodies or otherwise demonstrable antigen. The method given below, which works well with Enzo Biochem probes on paraffin sections, is a modification of the protocol recommended by the makers.

HIV

Antibodies for the demonstration of HIV are now available. However, most of them do not work reliably on paraffin sections and, of those that do, some give high background staining. Dupont market two monoclonal antibodies against HIV core proteins, p17 and p24, which work on paraffin-embedded sections. In our experience, p24 has given the most reliable results (Figure 6.50). The amount of staining is usually very low, except in cases showing severe changes of HIV encephalitis. Only scattered microglia, macrophages and multinucleated giant cells appear positively stained, with little background staining. The p17 antibody stains fewer cells than p24, and, very rarely, has given positive results where p24 had been negative; in addition, it gives some background staining. Genetic System's envelope glycoprotein, gp41, has been used in a number of studies (Scaravilli et al, 1989b; Budka, 1990). This antibody stains more cells than Dupont's p24. However, when we used it, it also stained, on some occasions, macrophages in HIV-negative controls, and had to be carefully controlled. The protocol below, which is recommended for use with these antibodies on paraffin sections of brain material, is the result of testing a number of methods.

In situ hybridization for HIV has been performed in several laboratories on postmortem brains (see Chapter 12); however, it is extremely precarious, probably because of the rapid breaking down of RNA, and is not routinely used.

Modified Avidin–Biotin Immunohistochemical Method, for HIV

1. Dewax in fresh xylene and take to absolute ethanol
2. Block endogenous peroxidase with 1 per cent H_2O_2 in methanol
3. Wash in ice-cold distilled water
4. Pre-incubate in tris buffer saline (TBS) with 10 per cent fetal bovine serum (Sigma)
5. Incubate with primary antibody, 1 hr at room temperature
6. Wash in TBS, 3 changes
7. Incubate with biotinylated rabbit anti-mouse antibody (Dako) 1/200, 1 hr at room temperature
8. Wash in TBS, 3 changes

9. Incubate with peroxidase conjugated avidin (Dako, UK), 1/500, 30 min at room temperature
10. Wash in TBS, 3 changes
11. Incubate in 0.05 per cent DAB containing 32 µl 30 per cent H_2O_2, 5 min.

All antibodies diluted in TBS with 10 per cent fetal bovine serum.

Acknowledgements. We would like to thank for their assistance Dr Meena Battacharjee, Dr F. Morinet, Miss J. Simpson, Mrs M. Favolini, Mrs I. Coquart, the Technical Staff of the Department of Neuropathology at The National Hospital, Queen Square, the Medical Research Council (MRC), The Wellcome Trust, the PNRS, the Scientific Council of Lariboisière (Paris VII) and the Ligue contre le Cancer (Comité de Paris).

References

Abrams DA, Robia M, Blumenfeld W et al. (1984) Disseminated coccidioidomycosis in AIDS. N Engl J Med 310:986.

Abrams DA, Kiprov DD, Goedert JJ et al. (1986) Antibodies to human T-lymphotropic virus type III and development of acquired immune deficiency syndrome in homosexual men presenting with immune thrombocytopenia. Ann Intern Med 104:47–59.

Ackermann R, Nekic M, Jürgens R. (1986) Locally synthesized antibodies in cerebrospinal fluid of patients with acquired immune deficiency syndrome. J Neurol 233:140–1.

Aisner J, Murilla J, Schimpff SC et al. (1979) Invasive aspergillosis in acute leukaemia: correlation with nose cultures and antibiotic use. Ann Intern Med 90:4–9.

Aksamit AJ, Major EO, Ghatak NR et al. (1987) Diagnosis of progressive multifocal leucoencephalopathy by brain biopsy with biotin labelled DNA: DNA *in situ* hybridization. J Neuropathol Exp Neurol 46:556–66.

Alford CA Jr, Britt WJ. (1985) Cytomegalovirus. In: Fields BN, Knipe DM, Chanock RM, et al, ed. Virology. Raven Press: New York, 629–60.

Anders KH, Guerra WF, Tomiyasu U et al. (1986a) The neuropathology of AIDS. UCLA experience and review. Am J Pathol 124:537–58.

Anders KH, Steinsapir KD, Iverson DJ et al. (1986b) Neuropathologic findings in the acquired immunodeficiency syndrome (AIDS). Clin Neuropathol 5:1–20.

Anders KH, Latta H, Chang BS et al. (1989) Lymphomatoid granulomatosis and malignant lymphoma of the central nervous system in the acquired immunodeficiency syndrome. Hum Pathol 20:326–34.

Anders KH, Holden JL, Becker PS et al. (1990) Focal pontine leukoencephalopathy: a review of 10 cases. J Neuropathol Exp Neurol 49:315 (abstr).

Anzil AP, Kozlowski PB, Wrzolek M et al. (1988) The central nervous system in AIDS: sex differences in an autopsy study of 144 adult cases (abstr). J Neuropathol Exp Neurol 47:387.

Anzil AP, Rao C, Wrzolek MA et al. (1991) Amebic meningo-encephalitis in a patient with AIDS caused by a newly recognized opportunistic pathogen, *Leptomyxid ameba*. Arch Pathol Lab Med 115:21–5.

Ariza A, Kim JH. (1988) Kaposi's sarcoma of the dura mater. Hum Pathol 19:1461–3.

Armstrong D, Wong B. (1982) Central nervous system infections in immunocompromised hosts. Annu Rev Med 33:293–308.

Armstrong D, Gold JW, Dryjanski J et al. (1985) Treatment of infections in patients with AIDS. Ann Intern Med 103:738–43.

Arnold A, Cossman J, Bakhshi A et al. (1983) Immunoglobulin-gene rearrangements as unique clonal markers in human lymphoid neoplasms. N Engl J Med 309:1593–9.

Artigas J, Niedobitek F, Grosse G et al. (1989) Spongiform encephalopathy in AIDS dementia complex: report of five cases. J AIDS 2:374–81.

Åstrom KE, Waksman BH. (1962) The passive transfer of experimental allergic encephalomyelitis and neuritis with living lymphoid cells. J Pathol Bacteriol 83:89–106.

Åstrom KE, Mancall EL, Richardson EP. (1958) Progressive multifocal leukoencephalopathy. A hitherto unrecognized complication of chronic lymphatic leukemia and Hodgkin's disease. Brain 81:93–111.

Bailey PR. (1929) Intracranial sarcomatous tumors of leptomeningial origin. Arch Surg 18:1359–402.

Baker RD. (1972) Fungus infection of the central nervous systems. In: Minckler J, ed. Pathology of the nervous system, vol 3. McGraw-Hill: New York, 2476–503.

Baker TS, Rayment I. (1987) Animal virus structure. In: Nermut MV, Steven AC, ed. Perspectives in medical virology. Elsevier: Amsterdam, 335–48.

Baringer JR. (1974) Recovery of herpes simplex virus from human sacral ganglions. N Engl J Med 291:828–30.

Barré-Sinoussi F, Nugeyre M, Dauguet C et al. (1983) Isolation of a T-lymphotropic retrovirus from a patient at risk for acquired immune deficiency syndrome. Science 220:868–71.

Bastian FO, Rabson AS, Yee CL et al. (1972) Herpesvirus hominis: isolation from human trigeminal ganglion. Science 178:306–97.

Bedri J, Weistein W, DeGregorio P et al. (1983) Progressive multifocal leukoencephalopathy in acquired immuno-deficiency syndrome. N Engl J Med 309:492–3.

Bélec L, Martin PMV, Vohito MD et al. (1989) Low prevalence of neuro-psychiatric clinical manifestations in central African patients with acquired immune deficiency syndrome. Trans R Soc Trop Med Hyg 83:844–6.

Bélec L, Tayot J, Mikol J et al. (1990a) Cytomegalovirus encephalopathy in an infant with congenital acquired immuno-deficiency syndrome. Neuropediatrics 21:124–9.

Bélec L, Gray F, Tron P et al. (1990b) Cytomegalovirus (CMV) myeloradiculitis and human immunodeficiency virus (HIV) encephalitis. Presence of HIV and CMV co-infected multinucleated giant cells. Acta Neuropathol 81:99–104.

Belman AL, Lantos G, Horoupian D et al. (1986) AIDS: calcification of the basal ganglia in infants and children. Neurology 36:1192–9.

Bender BL, Jaffe R. (1980) Immunoglobulin production in lymphomatoid granulomatosis and relation to other 'benign' lymphoproliferative disorders. Am J Clin Pathol 73:41–7.

Berger JR, Resnick L. (1987) HTLV-III/LAV related neuro-logical diseases. In: Broder S, ed. AIDS: modern concept and therapeutic challenges. Marcel Dekker Inc: New York, 263–83.

Berger JR, Moskowitz L, Fischl M et al. (1984) The neuro-logic complications of AIDS: frequently the initial mani-festation. Neurology 34 (suppl 1):134–5.

Berger JR, Sheremata WA, Resnik L et al. (1989) Multiple sclerosis-like illness occurring with human immunode-ficiency virus infection. Neurology 39:324–9.

Berger JR, Harris JO, Gregorios J et al. (1990) Cerebro-vascular disease in AIDS: a case-control study. AIDS 4:239–44.

Bernick C, Gregorios JB. (1984) Progressive multifocal leukoencephalopathy in a patient with acquired immune deficiency syndrome. Arch Neurol 41:780–792.

Binford CH, Dooley JR. (1976) Diseases caused by fungi and actinomycetes. Deep mycoses. In: Binford CH, Connors DH, ed. Pathology of tropical and extraordinary diseases, vol 2. AFIP: Washington DC, 551–609.

Bishburg E, Sunderam G, Reichman LB et al. (1986) Central nervous system tuberculosis with the acquired immunodeficiency syndrome and its related complex. Ann Intern Med 105:210–13.

Bloom EJ, Abrams DI, Rodgers G. (1986) Lupus anticoagulant in the acquired immunodeficiency syndrome. JAMA 256:491–3.

Blum LW, Chambers RA, Schwartzman RJ et al. (1985) Progressive multifocal leukoencephalopathy in acquired immune deficiency syndrome. Arch Neurol 42:137–9.

Blyth WA, Hill TJ. (1984) Establishment, maintenance, and control of herpes simplex virus (HSV) latency. In: Rouse BT, Lopez C, ed. Immunobiology of herpes simplex virus infection. CRC Press: Boca Raton, Florida, 10–32.

Bobowski SJ, Reed WG. (1958) Toxoplasmosis in the adult, presenting as a space-occupying cerebral lesion. Arch Pathol 65:460–4.

Bolton CF, Rozdilsky B (1971) Primary progressive multifocal leukoencephalitis. Neurology 21:72–7.

Booss J, Dann PR, Griffith BP et al. (1988) Glial nodule encephalitis in the guinea pig: serial observations following cytomegalovirus infection. Acta Neuropathol 75:465–473, 1988.

Bredesen DE, Messing R. (1983) Neurological syndromes heralding the acquired immune deficiency syndrome. Ann Neurol 14:141 (abstr).

Breuer AC, Blank NK, Schoene WC. (1978) Multifocal pontine lesions in cancer patients treated with chemotherapy and CNS radiotherapy. Cancer 41:2112–20.

Britton CB, Mesa-Tejada R, Fenoglio CM et al. (1985) A new complication of AIDS: thoracic myelitis caused by herpes simplex virus. Neurology 35:1071–4.

Bronnimann DA, Adam RD, Galgiani JN et al. (1987) Coccidioidomycosis in the acquired immunodeficiency syndrome. Ann Intern Med 106:372–9.

Brook BR, Walker DL. (1984) Progressive multifocal leukoencephalopathy. Neurol Clin 2:299–313.

Brown P, Tsai T, Gajdusek DC. (1975) Seroepidemiology of human papoviruses. Discovery of virgin populations and some unusual patterns of antibody prevalence among remote peoples of the world. Am J Epidemiol 102:331–40.

Brown WJ, Voge M. (1982) Neuropathology of parasitic infections. Oxford University Press: Oxford.

Brownwell B, Tomlinson AH. (1984) Virus diseases of the central nervous system. In: Adams JH, Crosellis JAN, Duchen LW, ed. Greenfield's neuropathology. Edward Arnold: London, 260–303.

Brunell PA, Miller LH, Lovejoy F. (1968) Zoster in children. Am J Dis Child 115:432–7.

Budka H. (1986) Multinucleated giant cells in brain: a hallmark of the acquired immune deficiency syndrome (AIDS). Acta Neuropathol 69:253–8.

Budka H. (1989) Human immunodeficiency virus (HIV)-induced disease of the central nervous system: pathology and implications for pathogenesis. Acta Neuropathol 77:225–36.

Budka H. (1990) Human immunodeficiency virus (HIV) envelope and core proteins in CNS tissues of patients with the acquired immune deficiency syndrome (AIDS). Acta Neuropathol 79:611–19.

Budka H, Shah KV. (1983) Papovavirus antigens in paraffin sections of PML brains. In: Sever JL, Madden DL, ed. Polyomaviruses and human neurological diseases. Alan R Liss Inc: New York, 299–309.

Budka H, Costanzi G, Cristina S et al. (1987) Brain pathology induced by infection with the human immunodeficiency virus (HIV). Acta Neuropathol 75:186–98.

Budka H, Maier H, Pohl P. (1988) Human immunodeficiency virus in vacuolar myelopathy of the acquired immunodeficiency syndrome. N Engl J Med 319:1667–8.

Budka H, Wiley CA, Kleihues P et al. (1991) HIV associated disease of the nervous system: review and nomenclature and proposal for neuropathology-based terminology. Brain Pathol 1:143–52.

Butt CG. (1966) Primary amebic meningoencephalitis. N Engl J Med 274:1473–6.

Callahan WP, Russell WD, Smith MG. (1946) Human toxoplasmosis. A clinicopathological study with presentation of five cases and review of the literature. Medicine (Baltimore) 251:343–97.

Capanna AH, La Mancusa J Jr, Erculei F et al. (1987) Primary cerebral lymphoma in two consortial partners afflicted with acquired immune deficiency syndrome. Neurosurgery 21:920–3.

Cappel R, Klastersky J. (1973) Herpetic meningitis (type 1) in a case of acute leukemia. Arch Neurol 28:415–16.

Carbone PP, Sabesin SM, Sidranski H et al. (1964) Secondary aspergillosis. Ann Intern Med 60:556–67.

Carne CA, Tedder RS, Smith A et al. (1985) Acute encephalopahy coincident with seroconversion for anti-HTLV-III. Lancet ii:1206–8.

Carney WP, Rubin RH, Hoffman RA et al. (1981) Analysis of T-lymphocyte subsets in cytomegalovirus mononucleosis. J Immunol 126:2114–6.

Carter RF. (1968) Primary amoebic meningo-encephalitis: clinical, pathological and epidemiological features of six fatal cases. J Pathol Bacteriol 96:1–25.

Centers for Disease Control. (1985) Revision of the case definition of acquired immunodeficiency syndrome for national reporting in the United States. Ann Intern Med 103:402–3.

Centers for Disease Control. (1988) Acquired immunodeficiency syndrome (AIDS), 1987 revision of CDC/WHO case definition for AIDS. Wkly Epidem Rec 63:1–8.

Centers for Disease Task Force on Kaposi's Sarcoma and Opportunistic Infections. (1982) Epidemiologic aspects of the current outbreak of Kaposi's sarcoma and opportunistic infections. N Engl J Med 306:248–52.

Chin W, Magoffin R, Frierson G et al. (1973) Cytomegalovirus infection. A case of meningoencephalitis. JAMA 225:740–1.

Chiodi F, Asjö B, Fenyö E-M et al. (1986) Isolation of human immunodeficiency virus from cerebrospinal fluid of antibody-positive virus carrier without neurological symptoms. Lancet ii:1276–7.

Cho E-S, Sharer LR, Peress NS et al. (1987) Intimal proliferation of leptomeningeal arteries and brain infarcts in subjects with AIDS. J Neuropathol Exp Neurol 46:385 (abstr).

Ciardi A, Sinclair E, Scaravilli F et al. (1990) Involvement of the cerebral cortex in HIV-encephalopathy. A morphological and immuno-histochemical study. Acta Neuropathol 81:51–9.

Ciobanu N, Wiernik PH. (1986) Malignant lymphomas, AIDS, and the pathogenic mechanism of Epstein–Barr virus. Mt Sinai J Med (NY) 53:627.

Clague CPT, Ostrowski MA, Deck JHN et al. (1988) Severe diffuse necrotizing cortical encephalopathy in acquired immune deficiency syndrome (AIDS): an immunocytochemical and ultrastructural study. J Neuropathol Exp Neurol 47:346 (abstr).

Cohen AJ, Phillips TM, Kessler CM. (1986) Circulating coagulation inhibitors in the acquired immunodeficiency syndrome. Ann Intern Med 104:175–80.

Cohen JI, Corey GR. (1985) Cytomegalovirus infection in the normal host. Medicine 64:100–14.

Cohen ML, Dawkins RL, Henderson DW et al. (1979) Pulmonary lymphomatoid granulomatosis with immunodeficiency terminating as malignant lymphoma. Pathology 11:537–50.

Collins VP, Gellhorn A, Trimble JR. (1951) The coincidence of cryptococcosis and disease of the reticuloendothelial and lymphatic systems. Cancer 4:883–9.

Colon L, Lasala G, Kanzer MD et al. (1988) Cerebral cladosporiosis in AIDS. J Neuropathol Exp Neurol 47:387 (abstr).

Conley FG, Jenkins KA, Remington JS. (1981) *Toxoplasma gondii* infection of the central nervous system. Use of peroxidase-antiperoxidase method to demonstrate *Toxplasma* in formalin fixed, paraffin embedded tissue sections. Human Pathol 12:690–8.

Cossman J, Uppenkamp M, Sundeen J et al. (1988) Molecular genetics and the diagnosis of lymphoma. Arch Pathol Lab Med 112:117–27.

Craig CP, Nahmias AJ. (1973) Different patterns of neurological involvement with herpes simplex virus type 1 and 2: isolation of herpes simplex virus type 2 from the buffy coat of two adults with meningitis. J Infect Dis 127:365–372.

Dalakas MC, Pezeshkpour GH. (1988) Neuromuscular diseases associated with human immunodeficiency virus infection. Ann Neurol 23 (suppl):S38–48.

Dal Canto MC. (1989) AIDS and the nervous system. Current status and future perspectives. Hum Pathol 20:410–18.

Davis KC, Horsburgh CR Jr, Hasiba U et al. (1983) Acquired immunodeficiency syndrome in a patient with haemophilia. Ann Intern Med 98:284–6.

Davis LE, Stewart JA, Garvin S. (1975) Cytomegalovirus infection: a seroepidemiologic comparison of nuns and women from a venereal disease clinic. Am J Epidemiol 102:327–30.

Davtyan DG, Vinters HV. (1987) Wernicke's encephalopathy in AIDS patients treated with zidovudine. Lancet i:919–20 (Lett.).

Del Castillo M, Mendoza G, Oviedo J et al. (1990) AIDS and Chagas' disease with central nervous system tumour-like lesion. Am J Med 88:693–4.

Denning DW, Anderson J, Rudge P et al. (1987) Acute myelopathy associated with primary infection with human immunodeficiency virus. Br Med J 294:143–4.

Denny-Brown D, Adams RD, Fitzgerald PJ. (1944) Pathologic features of herpes zoster. A note on 'geniculate herpes'. Arch Neurol Psychiat 51:216–31.

Desforges J, Mark EJ. (1987) Case record 41-1987. N Engl J Med 317:946–53.

Di Carlo E, Amberson JB, Metroka CE et al. (1986) Malignant lymphomas and the acquired immunodeficiency syndrome. Evaluation of 30 cases using a working formulation. Arch Pathol Lab Med 110:1012–16.

Dickson DW. (1986) Multinucleated giant cells in acquired

immunodeficiency syndrome encephalopathy: origin from endogenous microglia? Arch Pathol Lab Med 110:967–8.

Dickson DW, Belman AL, Kim TS et al. (1989a) Spinal cord pathology in pediatric acquired immunodeficiency syndrome. Neurology 39:227–35.

Dickson DW, Belman AL, Parky D et al. (1989b) Central nervous system pathology in pediatric AIDS: an autopsy study. APMIS 8 (suppl):40–57.

Diebold J, Audouin J. (1990) Lymphomes malins et autres proliférations malignes des organes hématopoïétiques chez les sujets HIV positifs. Sem Hôp (Paris) 9:467–73.

Dinn JJ. (1979) Distribution of herpes simplex virus in acute necrotizing encephalitis. J Pathol 129:135–8.

Dinn JJ. (1980) Transolfactory spread of virus in herpes simplex encephalitis. Br Med J 281:1392.

Dix RD, Bredesen DE. (1988) Opportunistic viral infections in acquired immunodeficiency syndrome. In: Rosenblum ML, Levy RM, Bredesen DE, ed. AIDS and the nervous system. Rowen Press: New York, 221–61.

Dix RD, Bredesen DE, Erlich KS et al. (1985) Recovery of herpes viruses from cerebrospinal fluid of immunodeficient homosexual men. Ann Neurol 18:611–14.

Dorfman LJ. (1973) Cytomegalovirus encephalitis in adults. Neurology 23:136–43.

Dorries K, Johnson RT, ter Meulen V. (1979) Detection of polyoma virus DNA in PML-brain tissue by in situ hybridization. J Gen Virol 42:49–57.

Dowdle WR, Nahmias AJ, Harwell RW et al. (1967) Association of antigenic type of herpesvirus hominis with site of viral recovery. J Immunol 99:974–80.

Dozic S, Suvakovic V, Cvetlovic D et al. (1990) Neoplastic angioendotheliomatosis (NAE) of the CNS in a patient with AIDS subacute encephalitis, diffuse leucoencephalopathy and meningocerebral cryptococcosis. J Neuropathol Exp Neurol 49:340 (abstr).

Drew WL, Mintz L, Miner RC et al. (1981) Prevalence of cytomegalovirus infection in homosexual men. J Infect Dis 143:188–92.

Duchowny M, Caplan L, Siber G. (1979) Cytomegalovirus infection of the adult nervous system. Ann Neurol 5:458–61.

Duenas A, Adam E, Melnick JL et al. (1972) Herpesvirus type 2 in a prostitute population. Am J Epidemiol 95:483–9.

Eby NL, Grufferman S, Flannelly CM et al. (1988) Increasing incidence of primary brain lymphoma in the US. Cancer 62:2461–5.

Egerter DA, Beckstead JH. (1988) Malignant lymphomas in the acquired immunodeficiency syndrome. Arch Pathol Lab Med 112:602–6.

Eidelberg D, Sotrel A, Horoupian S et al. (1986) Thrombotic cerebral vasculopathy associated with herpes zoster. Ann Neurol 19:7–14.

Engstrom JW, Lowenstein DH, Bredesen DE. (1989) Cerebral infarctions and transient neurologic deficit associated with acquired immunodeficiency syndrome. Am J Med 86:528–32.

Epstein LG, Sharer LR, Joshi VV et al. (1985) Progressive encephalopathy in children with acquired immune deficiency syndrome. Ann Neurol 17:488–96.

Epstein LG, DiCarlo FJ, Joshi VV. (1988) Primary lymphoma of the central nervous system in children with acquired immunodeficiency syndrome. Pediatrics 82:355–63.

Esiri MM. (1982a) Herpes simplex encephalitis: an immunohistological study of the distribution of viral antigen within the brain. J Neurol Sci 54:209–26.

Esiri MM. (1982b) Viruses and Alzheimer's disease. J Neurol Neurosurg Psychiatry 45:759.

Esiri MM, Scaravilli F, Millard PR et al. (1989) Neuropathology of HIV infection in haemophiliacs: comparative necropsy study. Br Med J 299:1312–15.

Essex M. (1984) Viral etiology for naturally occurring leukemias and lymphomas. In: Magrath I, ed. Pathogenesis of leukemias and lymphomas: environmental influences. Raven Press: New York, 315–29.

Evans DJ, Williams ED. (1968) Cytomegalic inclusion disease in the adult. J Clin Pathol 21:311–16.

Everall I, Luthert PJ, Lantos PL. (1991) Neuronal loss in the frontal cortex in HIV infection. Lancet ii:1119–21.

Fauci AS, Macher AM, Longo DL et al. (1984) Acquired immunodeficiency syndrome: epidemiologic, clinical, immunologic and therapeutic considerations. Ann Intern Med 100:92–106.

Fauci AS, Masur H, Gelmann EP (1985) The acquired immunodeficiency syndrome: an update, Ann Intern Med 102:800–13.

Feldman HA. (1968) Toxoplasmosis. N Engl J Med 279:1370–5, 1431–7.

Fermaglich J, Hardman JM, Earle KM. (1969) Progressive multifocal leukoencephalopathy. Neurology 17:287 (abstr).

Ferreiro J, Vinters HV. (1988) Pathology of the pituitary gland in patients with acquired immune deficiency syndrome (AIDS). Pathology 20:211–15.

Fiala M, Payne JE, Berne TV et al. (1975) Epidemiology of cytomegalovirus infection after transplantation and immunosuppression. J Infect Dis 132:421–33.

Filipovich AH, Zerbe D, Spector BD et al. (1984) Lymphomas in persons with naturally occurring immunodeficiency disorders. In: Magrath I, ed. Pathogenesis of leukemias and lymphomas: environmental influences. Raven Press: New York, 225–34.

Fischl MA, Pitchenik AE, Spira TJ. (1985) Tuberculous brain abscess and Toxoplasma encephalitis in a patient with the acquired immunodeficiency syndrome. JAMA 253:3428–30.

Fisher M, McGhee W. (1986) Cerebral infarct, TIA and lupus inhibitor. Neurology 36:1234–7.

Formenti SC, Gill PS, Lean E et al. (1989) Primary central nervous system lymphoma in AIDS. Cancer 63:1101–7.

Fraser DW, Ward JI, Ajello L et al. (1979) Aspergillosis and other systemic micosis. The growing problem. JAMA 242:1631–5.

Gabudza DH, Hirsch MS. (1987) Neurologic manifestations of infection with human immunodeficiency virus. Ann Intern Med 107:383–91.

Gabudza DH, Ho DD, de la Monte SM et al. (1986) Immunohistochemical identification of HTLV-III antigen in brains of patients with AIDS. Ann Neurol 20:289–95.

Gallo RC, Salahuddin SZ, Popovic M et al. (1984) Frequent detection and isolation of cytopathic retrovirus (HTLV-III) from patients with AIDS and at risk for AIDS. Science 224:500–3.

Gapen P. (1982) Neurological complications now characterizing many AIDS victims. JAMA 248:2941–2.

Garcia-Monco JC, Frey HM, Fernandez Villar B et al. (1989) Lyme disease concurrent with human immunodeficiency virus infection. Am J Med 87:325–8.

Gartner S, Markovits P, Markovitz DM et al. (1986) The role of mononuclear phagocytes in HTLV-III/LAV infection. Science 233:215–19.

Gaston A, Gherardi R, N'Guyen JP et al. (1985) Cerebral toxoplasmosis in acquired immuno-deficiency syndrome. Neuroradiology 27:83–6.

Gelb LD. (1985) Varicella/zoster virus. In: Fields BN, Knipe DM, Chanock RM, et al, ed. Virology. Raven Press: New York, 591–627.

Gendelman HE, Phelps W, Feigenbaum L et al. (1986) Transactivation of the human immunodeficiency virus long terminal repeat sequence by DNA viruses. Proc Natl Acad Sci USA 83:9759–63.

Gendelman HE, Orenstein JM, Baca LM et al. (1989) The macrophage in the persistence and pathogenesis of HIV infection. AIDS 3:475–95.

Geny C, Boudes P, Lionnet F et al. (1990) Multifocal giant cell myelitis in an AIDS patient (Abstract). J Neurol 237 (suppl 1):S67–8.

Ghatak NR, Poon TP, Zimmerman HM. (1970) Toxoplasmosis of the central nervous system in the adult. A light and electron microscopic study of three cases. Arch Pathol 89:337–48.

Gherardi R, Gaulard P, Prost C. (1986) T-cell lymphoma revealed by a peripheral neuropathy. A report of two cases with an immunohistologic study on lymph node and nerve biopsies. Cancer 58:2710–16.

Gherardi R, Lebargy F, Gaulard P et al. (1989) Necrotizing vasculitis and HIV replication in peripheral nerve. N Engl J Med 321:103–6.

Gill PS, Levine AM, Meyer PR et al. (1985) Primary central nervous system lymphoma in homosexual men. Clinical, immunologic and pathologic findings. Am J Med 78:742–8.

Gill PS, Levine AM, Krailo M et al. (1987) AIDS related malignant lymphoma: result of prospective treatment trials. J Clin Oncol 5:1322–8.

Gledhill S, Krajewski AS, Dewar E. (1990) Analysis of T-cell receptor and immunoglobulin gene rearrangements in the diagnosis of Hodgkin's and non-Hodgkin's lymphoma. J Pathol 161:345–54.

Gluckstein D, Ruskin J, Cifferi F et al. (1988) Chagas' disease: a new cause of cerebral mass in AIDS. Presented at ICAAC, Los Angeles.

Gold E, Nankervis GA. (1982) Cytomegalovirus. In: Evans AS, ed. Viral infections of humans: epidemiology and control. Plenum: New York, 167–86.

Gold JE, Jimenez E, Zalusky R. (1988) Human immunodeficiency virus-related lymphoreticular malignancies and peripheral neurologic disease. A report of four cases. Cancer 61:2318–24.

Goldstein JD, Dickson DW, Moser FG et al. (1991) Primary central nervous system lymphoma in acquired immune deficiency syndrome. A clinical and pathologic study with results of treatment with radiation. Cancer 67:2756–65.

Goldstick L, Mandybur TI, Bode R. (1985) Spinal cord degeneration in AIDS. Neurology 35:103–6.

Gonzales MF, Davis RL. (1988) Neuropathology of acquired immunodeficiency syndrome. Neuropathol Appl Neurobiol 14:345–63.

Goodell SE, Quinn TC, Mkritichian PAC et al. (1983) Herpes simplex virus proctitis in homosexual man. N Engl J Med 308:868–71.

Goodman MD, Porter DD. (1973) Cytomegalovirus vasculitis with fatal colonic hemorrhage. Arch Pathol 96:281–4.

Gorin FA, Bale JF, Halks-Miller M et al. (1985) Kaposi's sarcoma metastatic to the CNS. Arch Neurol 42:162–5.

Gosztonyi G, Lamperth L, Webster HF. (1988) In situ hybridization study of HIV DNA and RNA distribution in AIDS encephalopathy lesions. J Neuropathol Exp Neurol 47:346 (abstr).

Grafe MF, Wiley CA. (1989) Spinal cord and peripheral nerve pathology in AIDS: the roles of cytomegalovirus and human immunodeficiency virus. Ann Neurol 25:561–6.

Grant JW, Isaacson PJ (1992) Primary central nervous system lymphoma. Brain Pathol 2:97–109.

Gray F, Gherardi R, Baudrimont M et al. (1987) Leucoencephalopathy with multinucleated giant cells containing human immune deficiency virus-like particles and multiple opportunistic cerebral infections in one patient with AIDS. Acta Neuropathol 73:99–104.

Gray F, Gherardi R, Scaravilli F. (1988a) The neuropathology of the acquired immune deficiency syndrome (AIDS). Brain 111:245–66.

Gray F, Fenelon G, Gherardi R et al. (1988b) Etude neuropathologique de 15 cas d'encéphalite à cellules géantes multinucléées au cours du syndrome d'immunodéficience acquise (SIDA). Ann Pathol 8:281–9.

Gray F, Gherardi R, Keohane C et al. (1988c) Pathology of the central nervous system in 40 cases of acquired immune deficiency syndrome (AIDS). Neuropathol Appl Neurobiol 14:365–80.

Gray F, Gherardi R, Wingate E et al. (1989) Diffuse 'encephalitic' cerebral toxoplasmosis in AIDS. Report of four cases. J Neurol 236:273–7.

Gray F, Gaulard P, Le Bezu M et al. (1990) HIV encephalitis-like multinucleated giant cells in a nodal lymphoma in AIDS. Histopathology 16:402–5.

Gray F, Mohr M, Chimelli L et al. (1991) Fulminating multiple sclerosis-like leukoencephalopathy revealing human immunodeficiency virus infection. Report of two cases. Neurology 41:105–9.

Gray F, Mohr M, Rozenberg F et al. (1992) Varicella-zoster virus encephalitis in acquired immunodeficiency syndrome: a report of 4 cases. Neuropathol Appl Neurobiol (in press).

Grĉević N, Matthews WF. (1959) Pathologic changes in acute disseminated aspergillosis, particularly involvement of the central nervous system. Am J Clin Pathol 32:536–51.

Gribble DH, Haden CC, Schwartz LW et al. (1975) Spontaneous progressive multifocal leukoencephalopathy (PML) in macaques. Nature 254:602–4.

Griffin GE, Leung K, Folks TM et al. (1989) Activation of HIV gene expression during monocyte differentiation by induction of NF-KB. Nature 339:70–3.

Grimes MM, La Pook JD, Bar MM et al. (1987) Disseminated *Pneumocystis carinii* infection in a patient with acquired immunodeficiency syndrome. Hum Pathol 18:307–8.

Grinnell BW, Padgett BL, Walker DL. (1983a) Distribution of nonintegrated DNA from JC papovirus in organs of patients with progressive multifocal leukoencephalopathy. J Infect Dis 147:669–75.

Grinnell BW, Padgett BL, Walker DL. (1983b) Comparison of infectious JC virus DNAs cloned from human brain. J Virol 45:299–308.

Guarda LA, Luna MA, Smith JL et al. (1984) Acquired immune deficiency syndrome: postmortem findings. Am J Clin Pathol 81:549–57.

Gurney ME, Apatoff BR, Spear GT et al. (1986a) Neuroleukin: a lymphokine product of lectin-stimulated T cells. Science 234:574–81.

Gurney ME, Heinrich SP, Lee MR et al. (1986b) Molecular cloning and expression of neuroleukin, a neurotrophic factor for spinal and sensory neurons. Science 235:566–74.

Györkey F, Sinkovics JG, Györkey P. (1982) Tubuloreticular structures in Kaposi's sarcoma. Lancet ii:984–5.

Györkey F, Melnick JL, Györkey P. (1987) Human immunodeficiency virus in brain biopsies of patients with AIDS and progressive encephalopathy. J Infect Dis 155:870–6.

Haaxma-Reiche H, van Imhoff GW. (1989) Results of intraventricular central nervous system prophylaxis and treatment in non-Hodgkin's lymphoma. Reg Cancer Treat 2:32–6.

Hanto DW, Frizzera G, Purtillo DT et al. (1981) Clinical spectrum of lymphoproliferative disorders in renal transplant recipients and evidence for the role of Epstein–Barr virus. Cancer Res 41:4253–61.

Hanto DW, Gayl-Peczalska KJ et al. (1983) Epstein–Barr virus induced polyclonal and monoclonal B-cell lymphoproliferative diseases occurring after renal transplantation. Am Surg 198:356–69.

Harcourt-Webster JN, Scaravilli F, Darwish AH (1991) *Strongyloides stercoralis* hyperinfection in an HIV-positive patient. J Clin Pathol 44:346–8.

Harford CG, Wellinghoff W, Weinstein RA. (1975) Isolation of Herpes simplex virus from the cerebrospinal fluid in viral meningitis. Neurology 25:198–200.

Harris AA, Segreti J, Levin S. (1985) Central nervous system infections in patients with the acquired immune deficiency syndrome (AIDS). Clin Neuropharmacol 8:201–10.

Hatson JR, Pedley TA. (1976) The neurological complications of cardiac transplantation. Brain 99:673–94.

Hawkins C, Gold JM, Whimbey E et al. (1986) Mycobacterium avium complex infections in patients with AIDS. Ann Intern Med 105:184–8.

Hawley DA, Schafer JF, Schulz DM et al. (1983) Cytomegalovirus encephalitis in acquired immunodeficiency syndrome. Am J Clin Pathol 80:874–7.

Hay J, Hutchinson WM. (1983) *Toxoplasma gondii*: an environmental contaminant. Ecol Dis 2:33–43.

Hay RJ, Mackenzie DWR, Campbell CK et al. (1980) Cryptococcosis in the United Kingdom and the Irish Republic, an analysis of 69 cases. J Infect 2:13–22.

Hedges TR, Leung LE. (1976) Parasellar and orbital apex syndrome caused by aspergillosis. Neurology 26:117–20.

Hedley-White ET, Smith BP, Tyler HR et al. (1966) Multifocal leukoencephalopathy with remission and 5-year survival. J Neuropathol Exp Neurol 25:107–15.

Helder AW, Feltkamp-Vroom TM. (1976) Tubuloreticular structures and antinuclear antibodies in autoimmune and non-autoimmune disorders. J Pathol 119:49–56.

Hénin D, Duyckaerts C, Chaunu MP et al. (1987) Etude neuropathologique de 31 cas de syndrome d'immunodépression acquise. Rev Neurol (Paris) 143:631–42.

Henochowicz S, Mustafa M, Lawrinson WE et al. (1985) Cardiac aspergillosis in acquired immune deficiency syndrome. Am J Cardiol 55:1239–90.

Henry JM, Heffner RR Jr, Dillard SH et al. (1974) Primary malignant lymphomas of the central nervous system. Cancer 34:1293–1302.

Herskovitz S, Siegel SE, Schneider AT et al. (1989) Spinal cord toxoplasmosis in AIDS. Neurology 39:1552–3.

Ho DD, Sarngadharan MG, Resnick L et al. (1985a) Primary human T-lymphotropic virus type III infection. Ann Intern Med 103:880–3.

Ho DD, Rota TR, Schooley RT et al. (1985b) Isolation of HTLV-III from cerebrospinal fluid and neural tissues of patients with neurological syndromes related to the acquired immunodeficiency syndrome. N Engl J Med 313:1493–7.

Ho DD, Pomerantz RJ, Kaplan JC. (1987) Pathogenesis of infection with human immunodeficiency virus. N Engl J Med 317:278–86.

Ho JL, Poldre PA, McEniry D et al. (1984) Acquired immunodeficiency syndrome with progressive multifocal leukoencephalopathy and monoclonal B-cell proliferation. Ann Intern Med 100:693–6.

Ho M. (1982) Epidemiology of cytomegalovirus infection in man. In: Greenough WB III, Merigan TC, ed. Cytomegalovirus, biology and infection: current topics in infectious diseases. Plenum: New York, 79–104.

Hogan EL, Krigman MR. (1973) Herpes zoster myelitis: evidence for viral invasion of spinal cord. Arch Neurol 29:309–13.

Hogan PJ, Greenberg MK, McCarty GE. (1981) Neurologic complications of lymphomatoid granulomatosis. Neurology 31:619.

Holtz HA, Lavery DP, Kapila R. (1985) Actinomycetales infection in the acquired immunodeficiency syndrome. Ann Intern Med 102:203–5.

Hood J, Wilson ER Jr, Alexander et al. (1982) Lymphomatoid granulomatosis manifested as a mass in the cerebello-pontine angle. Arch Neurol 39:319–20.

Hooper AD. (1957) Acquired toxoplasmosis. Report of a case with autopsy findings, including a review of previously reported cases. Arch Pathol 64:1–10.

Hooper DC, Pruitt AA, Rubin RH. (1982) Central nervous system infection in the chronically immunosuppressed. Medicine 61:166–88.

Horoupian DS, Pick P, Spigland I et al. (1984) Acquired immune deficiency syndrome and multiple tract degeneration in a homosexual man. Ann Neurol 15:502–5.

Horowitz SL, Bentson JR, Benson F et al. (1983) CNS toxoplasmosis in acquired immunodeficiency syndrome. Arch Neurol 40:649–52.

Horten B, Price RW, Jimenez D. (1981) Multifocal varicella-zoster virus leukoencephalitis temporally remote from herpes zoster. Ann Neurol 9:251–66.

Hotson JR, Pedley TA. (1976) The neurological complications of cardiac transplantation. Brain 99:673–94.

Howlett WP, Nkya WM, Mmuni KA et al. (1989) Neurological disorders in AIDS and HIV disease in the northern zone of Tanzania. AIDS 3:289–96.

Huang ES, Alford CA, Reynolds DW et al. (1980) Molecular epidemiology of cytomegalovirus infections in women and their infants. N Engl J Med 303:958–62.

Huang TE, Chou SM. (1988). Occlusive hypertrophic arteritis as the cause of discrete necrosis in CNS toxoplasmosis in the acquired immunodeficiency syndrome. Hum Pathol 19:1210–14.

Hutto C, Ricks R, Garvie M et al. (1985) Epidemiology of cytomegalovirus infections in young children: day care vs home care. Pediatr Infect Dis J 4:149–52.

Hymes KB, Cheung T, Green JB, et al. (1981) Kaposi's sarcoma in homosexual men: a report of eight cases. Lancet 2:598–600.

Iglesias-Rozas JR, Bantz B, Adler T et al. (1991) Cerebral lymphoma in AIDS. Clinical, radiological, neuropathological and immunopathological study. Clin Neuropathol 10:65–72.

Ilowite NT, Fligner CL, Ochs HD et al. (1986) Pulmonary angiitis with atypical lymphoreticular infiltrates in Wiskott–Aldrich syndrome: possible relationship of lymphomatoid granulomatosis and EBV infection. Clin Immunol Immunopathol 41:479–84.

Iltis JP, Oakes JE, Hyman RW et al. (1977) Comparison of the DNAs of varicella-zoster viruses isolated from clinical cases of varicella and herpes zoster. Virology 82:345–52.

Ingwer I, McLeish KR, Tight RR et al. (1978) Aspergillus fumigatus epidural abscess in a renal transplant recipient. Arch Intern Med 138:153–4.

Ioachim HL, Cooper MC, Hellman GC. (1985) Lymphomas in men at high risk for acquired immune deficiency syndrome (AIDS). A study of 21 cases. Cancer 56:2831–2.

Ironside JW, Martin JF, Richmond J et al. (1984) Lymphomatoid granulomatosis with cerebral involvement. Neuropathol Appl Neurobiol 10:397.

Israel AM, Koziner B, Strauss DJ. (1983) Plasmacytoma and the acquired immunodeficiency syndrome. Ann Intern Med 99:635–6.

Jackson D, Tabor E, Garety RJ. (1979) Acute non-A, non-B hepatitis: specific ultrastructural alterations in endoplasmic reticulum of infected hepatocytes. Lancet i:1249–50.

Jacobs JL, Murray HW. (1986) Why is *Listeria monocytogenes* not a pathogen in acquired immunodeficiency symdrome? Arch Intern Med 146:1299–300.

Jankovic J. (1986) Whipple's disease of the central nervous system in AIDS. N Engl J Med 315:1029–30.

Janssen RS, Cornblath DR, Epstein LG et al. (1989) Human immunodeficiency virus (HIV) infection and the nervous system: report from the American Academy of Neurology AIDS Task Force. Neurology 83:119–22.

Jarvik JG, Hesselink JR, Wiley C et al. (1988) Coccidioidomycotic brain abscess in an HIV-infected man. West J Med 149:83–6.

Jauregui HO. (1978) Lymphomatoid granulomatosis after immunosuppression for pemphigus. Arch Dermatol 114:1052–5.

Jemsek J, Greenberg SB, Taber L et al. (1983) Herpes zoster-associated encephalitis: clinicopathologic report of 12 cases and review of the literature. Medicine (Baltimore) 62:81–97.

Johns DR, Tierney M, Felsenstein D. (1987) Alteration in the natural history of neurosyphilis by concurrent infection with the human immunodeficiency virus. N Engl J Med 316:1569–72.

Johnson RT, Mims CA. (1968) Pathogenesis of viral infections of the nervous system. N Engl J Med 278:23–30.

Jones DB, Visvesvara GS, Robinson NM. (1975) *Acanthamoeba polyphaga* keratitis and *Acanthamoeba* uveitis associated with fatal meningoencephalitis. Trans Ophthalmol Soc UK 95:221–32.

Jones. HR Jr, Ho DD, Forgacs P et al. (1988) Acute fulminating fatal leukoencephalopathy as the only manifestation of human immunodeficiency virus infection. Ann Neurol 23:519–22.

Jones P, Hamilton PJ, Bird G et al. (1985) AIDS and haemophilia: morbidity and mortality in a well defined population. Br Med J 291:695–9.

Jordan BD, Navia BA, Petito C et al. (1985) Neurological syndromes complicating AIDS. Front Radiat Ther Oncol 19:82–7.

Joshi VV, Pawel B, Conner E et al. (1987) Arteriopathy in children with acquired immune deficiency syndrome. Pediatr Pathol 7:261–75.

Joyce PW, Haye KR, Ellis ME. (1989) Syphilitic retinitis in a homosexual man with concurrent HIV infection: case report. J Neurol Neurosurg Psychiatry 65:244–7.

Kadin ME, Said J. (1988) T-cell lymphomas and leukemias of post-thymic differentiation. Clin Lab Med 8:135–50.

Kalimo H, Lehto M, Näntö-Salonen K et al. (1985) Characterization of the perivascular reticulin network in a case of primary brain lymphoma. Immunohistochemical demonstration of collagen types I, III, IV and V; laminin and fibronectin. Acta Neuropathol 66:299–305.

Kalish SB, Goldschmidt R, Curtis L. (1982) Infective endocarditis caused by Paecilomyces. Am J Clin Pathol 78:249–52.

Kalter SP, Riggs SA, Cabanillas F et al. (1985) Aggressive non-Hodgkin's lymphomas in immunocompromised homosexual males. Blood 66:655–9.

Kamin SS, Petito CK. (1988) Vacuolar myelopathy in immunocompromised, non-AIDS patients. J Neuropathol Exp Neurol 47:385 (abstr).

Kamin SS, Petito CK (1991) Idiopathic myelopathies in non-acquired immunodeficiency patients. Hum Pathol 22:816–24.

Kapanci Y, Toccanier MF. (1983) Lymphomatoid granulomatosis of the lung. An immunohistochemical study. Appl Pathol 1:97–114.

Kass EH, Andrus SD, Adams RA et al. (1952) Toxoplasmosis in the human adult. Arch Intern Med 89:759–82.

Katlama C, Matheron S, Gaultier T et al. (1984) Manifestations neurologiques du SIDA. In: Zittoun R, ed. Syndrome Immuno Déficitaire Acquis. Boin: Paris, 81–98.

Kato T, Hirano A, Llena JF et al. (1987) Neuropathology of acquired immune deficiency syndrome (AIDS) in 53 autopsy cases with particular emphasis on microglial nodules and multinucleated giant cells. Acta Neuropathol 73:287–94.

Katzenstein AA, Carrington CB, Liebow AA. (1979) Lymphomatoid granulomatosis. A clinicopathologic study of 152 cases. Cancer 43:360–73.

Kauffman CA, Linnemann CC, Alvira MM. (1979) Cytomegalovirus encephalitis associated with thymoma and immunoglobulin deficiency. Am J Med 67:724–8.

Kaufman DM, Thal LJ, Farmer PM. (1976) Central nervous system aspergillosis in two young adults. Neurology 26:484–8.

Kelley RE, Gilman PB, Kovacs A. (1984) Cerebral ischaemia in the presence of lupus anticoagulant. Arch Neurol 41:521–3.

Kelly WN, Brant-Zawadzki M. (1983) Acquired immunodeficiency syndrome: neurologic findings. Radiology 149:485–91.

Kepes JJ, Chon SM, Price LW. (1975) Progressive multifocal leukoencephalopathy with 10-year survival in a patient with non-tropical sprue. Neurology 25:1006–12.

Ketzler S, Weis S, Haug H et al. (1990) Loss of neurons in the frontal cortex in AIDS brains. Acta Neuropathol 80:92–4.

Khadem M, Kalish SB, Goldsmith J et al. (1984) Ophthalmologic findings in acquired immune deficiency syndrome (AIDS). Arch Ophthalmol 102:201–6.

Kingston D, Warhurst DC. (1969) Isolation of amoebae from the air. J Med Microbiol 2:27–36.

Klastersky J, Cappel R, Snoeck JM et al. (1972) Ascending myelitis in association with herpes-simplex virus. N Engl J Med 287:182–4.

Kleihues P, Lang W, Burger PC et al. (1985) Progressive diffuse leukoencephalopathy in patients with acquired immune deficiency syndrome (AIDS). Acta Neuropathol 68:333–9.

Klemola E. (1973) Cytomegalovirus infection in previously healthy adults. Ann Intern Med 79:267–8.

Klemola E, von Essen R, Wager O et al. (1969) Cytomegalovirus mononucleosis in previously healthy individuals. Five new cases and follow-up of 13 previously published cases. Ann Intern Med 71:11–9.

Knowles DM, Chamulak GA, Subar M et al. (1988) Lymphoid neoplasia associated with the acquired immunodeficiency syndrome (AIDS), The New York University Medical Center Experience with 105 patients (1981–1986). Ann Intern Med 108:744–53.

Kocki N, Budka H, Radaszkiewicz T. (1986) Development of stroma in malignant lymphomas of the brain compared with epidural lymphomas. An immunohistochemical study. Acta Neuropathol 71:125–9.

Koenig S, Gendelman HE, Orenstein JM et al. (1986) Detection of AIDS virus in macrophages in brain tissue from AIDS patients with encephalopathy. Science 233:1089–93.

Koeppen AH, Lansing LS, Peng S-K et al. (1981) Central nervous system vasculitis in cytomegalovirus infection. J Neurol Sci 51:395–410.

Kokmen E, Billman JK Jr, Abell MR. (1977) Lymphomatoid granulomatosis clinically confined to the CNS. A case report. Arch Neurol 34:782–4.

Koppel BS, Wormser GP, Tuchman AJ et al. (1985) Central

nervous system involvement in patients with acquired immune deficiency syndrome (AIDS). Acta Neurol Scand 71:337–53.

Kostianovsky M, Orenstein JM, Schaff Z et al. (1987) Cytomembranous inclusions observed in acquired immunodeficiency syndrome. Clinical and experimental review. Arch Pathol Lab Med 111:218–23.

Kovacs A, Forthal DN, Kovacs JA et al. (1984) Disseminated coccidioidmycosis in a patient with acquired immune deficiency syndrome. West J Med 140:447–9.

Kovacs JA, Kovacs AA, Polis M et al. (1985) Cryptococcosis in the acquired immunodeficiency syndrome. Ann Intern Med 103:533–8.

Koziol K, Rielly KS, Bonin RA et al. (1986) *Listeria monocytogenes* meningitis in AIDS. Can Med Assoc J 135:43–4.

Kozlowski PB, Sher JH, Dickson DW et al. (1990) Central nervous system in children with AIDS. A multicenter study. J Neuropathol Exp Neurol 49:350 (abstr).

Krick JA, Remington JS. (1978) Current concepts in parasitology. Toxoplasmosis in the adult. An overview. N Engl J Med 298:550–553.

Kristensson K, Vahlne A, Persson LA et al. (1978) Neural spread of herpes simplex virus types 1 and 2 in mice after corneal or subcutaneous (foot-pad) inoculation. J Neurol Sci 35:331–40.

Kui AN, Koss MN, Meyer PR. (1984) Necropsy findings in acquired immunodeficiency syndrome. Hum Pathol 15:670–6.

Kure K, Harris C, Morin LS et al. (1989) Solitary midbrain toxoplasmosis and olivary hypertrophy in a patient with acquired immunodeficiency syndrome. Clin Neuropathol 8:35–40.

Kurtzke JF. (1980) The geographic distribution of multiple sclerosis – an update with special reference to Europe and the Mediterranean regions. Acta Neurol Scand 62:65–80.

Lainson R. (1959) A note on the duration of toxoplasma infection in the guinea pig. Ann Trop Med Parasitol 53:120–1.

Lane HC, Depper JM, Greene WC et al. (1985) Qualitative analysis of immune function in patients with the acquired immunodeficiency syndrome: evidence for a selective defect in soluble antigen recognition. N Engl J Med 313:79–84.

Lang DJ, Kummer JF. (1975) Cytomegalovirus in semen: observations in selected populations. J Infect Dis 132:472–3.

Lang W, Miklossy J, Deruaz JP et al. (1989) Neuropathology of the acquired immune deficiency syndrome (AIDS): a report of 135 consecutive autopsy cases from Switzerland. Acta Neuropathol 77:379–90.

Larroche JC. (1977) Developmental pathology of the neonate. Excerpta Medica, Amsterdam, 1977.

Laskin OL, Stahl-Bayliss CM, Morgello S. (1987) Concomitant herpes simplex virus type I and cytomegalovirus ventriculoencephalitis in acquired immunodeficiency syndrome. Arch Neurol 44:843–7.

Laulund S, Visfeldt J, Klinken L. (1986) Patho-anatomical studies in patients dying of AIDS. APMIS 94:201–21.

Lawande RV, Duggan MB, Constantinidou M et al. (1979) Primary amoebic meningoencephalitis in Nigeria (report of two cases in children). J Trop Med Hyg 82:84–8.

Leaver RJ, Haile Z, Watters DAK. (1990) HIV and cerebral malaria. Trans R Soc Trop Med Hyg 84:201.

Lee S, Harris C, Hirshfield A et al. (1988) Cytomembranous inclusions in the brain of a patient with the acquired immunodeficiency syndrome. Acta Neuropathol 76:101–6.

Lemann W, Cho E-S, Nielsen S et al. (1985) Neuropathologic findings in 104 cases of acquired immunodeficiency syn-

drome. J Neuropathol Exp Neurol 44:349 (abstr).

Levy RM, Bredesen DE. (1988a) Central nervous system dysfunction in acquired immunodeficiency syndrome. J AIDS 1:41–64.

Levy RM, Bredesen DE. (1988b) Central nervous system dysfunction in acquired immunodeficiency syndrome. In: Rosenblum JL, Levy RM, Bredesen DE, ed. AIDS and the nervous system. Raven Press: New York, 29–63.

Levy RM, Pons VG, Rosenblum ML. (1983) Intracerebral-mass lesions in the acquired immunodeficiency syndrome (AIDS). N Engl J Med 309:1459–955.

Levy RM, Pons VG, Rosenblum ML. (1984) Central nervous system mass lesions in the acquired immunodeficiency syndrome (AIDS). J Neurosurg 61:9–16.

Levy RM, Bredesen DE, Rosenblum ML. (1985) Neurological manifestations of the acquired immunodeficiency syndrome (AIDS): experience at UCSF and review of the literature. J Neurosurg 62:475–95.

Levy RM, Rosenblum S, Perrett LV. (1986) Neuroradiologic findings in AIDS: a review of 200 cases. AJNR 7:833–9.

Levy RM, Bredesen DE, Rosenblum ML. (1988a) Opportunistic central nervous system pathology in patients with AIDS. Ann Neurol 23 (suppl):s7–12.

Levy RM, Janssen RS, Bush TJ et al. (1988b) Neuroepidemiology of acquired immunodeficiency syndrome. J AIDS 1:31–40.

Levy RM, Mills CM, Posin JP et al. (1990) The efficacy and clinical impact of brain imaging in neurologically symptomatic AIDS patients: a prospective CT/MRI study. J AIDS 3:461–71.

Liebow AA, Carrington CRB, Friedman PJ. (1972) Lymphomatoid granulomatosis. Hum Pathol 3:457.

Lifson JD, Reyes GR, McGrath MS et al. (1986) AIDS retrovirus induced cytopathology: giant cell formation and involvement of CD4 antigen. Science 232:1123–7.

Linares G, McGarry PA, Baker RD. (1971) Solid solitary aspergillotic granuloma of the brain. Report of a case due to *Aspergillus candidus* and review of the literature. Neurology 21:177–84.

Linneman CC Jr, Alvira MM. (1980) Pathogenesis of varicella-zoster angiitis in the CNS. Arch Neurol 37:239–40.

Linneman CC Jr, Dunn CR, First MR et al. (1978) Late onset of fatal cytomegalovirus infection after renal transplantation. Primary or reactivation infection? Arch Intern Med 138:1247–50.

Louie S, Schwartz RS. (1978) Immunodeficiency and the pathogenesis of lymphoma and leukemia. Semin Hematol 15:117–38.

Louria DB, Stiff DP, Bennett B. (1962) Disseminated moniliasis in the adult. Medicine (Baltimore), 41:307–27.

Louriero C, Gill PS, Meyer PR et al. (1988) Autopsy findings in AIDS-related lymphoma. Cancer 62:735–9.

Lowenthal DA, Straus DJ, Campbell SW et al. (1988) AIDS related lymphoid neoplasia. The Memorial Hospital Experience. Cancer 61:2325–37.

Luft BJ, Brooks RG, Conley FK et al. (1984) Toxoplasmic encephalitis in patients with acquired immune deficiency syndrome. JAMA 252:913–17.

Lusso P, Ensoli B, Markham PD et al. (1989) Productive dual infection of human DC4+T lymphocytes by HIV-1 and HHV-6. Nature 337:370–3.

Mackenzie DWR. (1979) Imported fungal infections. Postgrad Med J 55:595–7.

Maier H, Budka H, Lassmann H et al. (1989) Vacuolar myelopathy with multinucleated giant cells in the acquired immune deficiency syndrome (AIDS). Light and electron microscopic distribution of human immunodeficiency virus (HIV) antigens. Acta Neuropathol 78:497–503.

Mann SL, Meyers JD, Holmes KL et al. (1984) Prevalence and incidence of herpes virus infections among homosexually active men. J Infect Dis 149:1026–7.

Markowitz SH, Sobieski T, Martinez AJ et al. (1978) Experimental *Acanthamoeba* infections in mice pretreated with methyl prednisolone or tetracycline. Am J Pathol 92:733–44.

Martin JD, Foster GC. (1984) Multiple JC virus genomes from one patient. J Gen Virol 65:1405–11.

Martin JD, King DM, Slauch JM et al. (1985) Differences in regulatory sequence of naturally occurring JC virus variants. J Virol 53:306–11.

Martinez AJ, Duma RJ, Nelson EC et al. (1973) Experimental *Naegleria* meningoencephalitis in mice. Penetration of the olfactory mucosal epithelium by *Naegleria* and pathologic changes produced: a light and electron microscope study. Lab Invest 29:121–33.

Martinez AJ, Markowitz SM, Duma RJ. (1975) Experimental pneumonitis and encephalitis caused by *Acanthamoeba* in mice: pathogenesis and ultrastructural features. J Infect Dis 131:692–9.

Martinez AJ, Garcia CA, Halks-Miller M et al. (1980) Granulomatous amebic encephalitis presenting as a cerebral mass lesion. Acta Neuropathol 51:85–91.

Mashaly R, Gray F, Poisson M et al. (1983) Toxoplasmose acquise de l'adulte. Etude clinique et neuropathologique. Rev Neurol (Paris) 139:561–8.

Masur H, Michelis MA, Wormser GP et al. (1982) Opportunistic infection in previously healthy women. Ann Intern Med 97:533–9.

Matsubayashi H, Akao S. (1963). Morphological studies on the development of toxoplasma cysts. Am J Trop Med Hyg 12: 321–33.

Matsuda S, Urata Y, Yamada M et al. (1989) Disseminated infection of *Pneumocystis carinii* in a patient with the acquired immunodeficiency syndrome. Virchows Arch B 414:523–7.

Matthiessen L, Labrousse F, Marche C et al. (1988) Morphology and etiology of microglial nodules in 27 AIDS autopsy cases. Clin Neuropathol 7:187 (abstr).

Mayayo E, Vidal F, Alvira R et al. (1990) Cerebral *Pneumocystis carinii* infection in AIDS. Lancer ii:1592.

McArthur JC. (1987) Neurologic manifestations of AIDS. Medicine (Baltimore) 66:407–37.

McCormick WF, Rodnitzky RL, Schoechet SS et al. (1969) Varicella-zoster encephalomyelitis: a morphologic and virologic study. Arch Neurol 21:559–70.

McNutt NS, Fletcher V, Conant M. (1983) Early lesions of Kaposi's sarcoma in homosexual men. Am J Pathol 114:62–77.

Mead GM, Kennedy P, Smith JL et al. (1986) Involvement of the central nervous system by non-Hodgkin's lymphoma in adults. A review of 36 cases. Q J Med 231:699–714.

Mehren M, Burns PJ, Mamani F et al. (1988) Toxoplasmic myelitis mimicking intramedullary spinal cord tumor. Neurology 38:1648–50.

Meyenhofer MF, Epstein LG, Cho E-S et al. (1987) Ultrastructural morphology and intracellular production of human immunodeficiency virus (HIV) in brain. J Neuropathol Exp Neurol 46:474–84.

Michaels J, Price RW, Rosenblum MK. (1988) Microglia in the giant cell encephalitis of acquired immune deficiency syndrome: proliferation, infection and fusion. Acta Neuropathol 76:373–9.

Michaud J, Banerjee D, Kaufmann JCE. (1983) Lymphomatoid granulomatosis involving the central nervous system: complication of a renal transplant with terminal monoclonal B-cell proliferation. Acta Neuropathol 61:141–7.

Micozzi MS, Wetli CV. (1985) Intravenous amphetamine abuse, primary cerebral mucormycosis and acquired immunodeficiency. J Forensic Sci 30:504–10.

Miller JR, Barrett RE, Britton CB et al. (1982) Progressive multifocal leukoencephalopathy in a male homosexual with T-cell immune deficiency. N Engl J Med 307:1436–8.

Minase T, Ogasawara M, Kikuchi T et al. (1985) Lymphomatoid granulomatosis. Light microscopic electron microscopic and immunohistochemical study. Acta Pathol Jpn 35:711–21.

Mirra SS, Spira TJ, Anand R. (1986) HTLV-III/LAV infection presenting as giant cell encephalopathy. J Neuropathol Exper Neurol 45:331 (abstr).

Mizusawa H, Hirano A, Llena JF et al. (1987) Nuclear bridges in multinucleated giant cells associated with primary lymphoma of the brain in acquired immune deficiency syndrome (AIDS). Acta Neuropathol 75:23–6.

Mizusawa H, Hirano A, Llena F. (1988) Nuclear bridges within multinucleated giant cells in subacute encephalitis of acquired immune deficiency syndrome (AIDS). Acta Neuropathol 76:166–9.

Monte SM de la, Ho DD, Schooley RT et al. (1987) Subacute encephalomyelitis of AIDS and its relation to HTLV-III infection. Neurology 37:562–9.

Montilla P, Dronda F, Moreno S et al. (1987) Lymphomatoid granulomatosis and the acquired immunodeficiency syndrome. Ann Intern Med 106:166.

Morgello S, Cho E-S, Nielsen S et al. (1987) Cytomegalovirus encephalitis in patients with acquired immunodeficiency syndrome: an autopsy study of 30 cases and a review of the literature. Hum Pathol 18:289–97.

Morris S, Distenfeld A, Amorosi E et al. (1982) Autoimmune thrombocytopenic purpura in homosexual men. Ann Intern Med 96:714–17.

Moskowitz LB, Gregorios JB, Hensley GT et al. (1984) Cytomegalovirus-induced demyelination associated with acquired immune deficiency syndrome. Arch Pathol Lab Med 108:873–7.

Mukoyama M, Gimple K, Poser CM. (1969) Aspergillosis of the central nervous system. Report of a brain abscess due to *A. fumigatis* and review of the literature. Neurology 19:967–74.

Muller SA. (1967) Association of zoster and malignant disorders in children. Arch Dermatol 96:657–64.

Mullin GE, Sheppell AL. (1987) *Listeria monocytogenes* and the acquired immunodeficiency syndrome. Arch Intern Med 147:176.

Myerson D, Hackman RC, Nelson JA et al. (1984) Widespread presence of histologically occult cytomegalovirus. Hum Pathol 15:430–9.

Nahmias AJ, Josey WE, Naib ZM et al. (1970) Antibodies to herpes virus hominis types 1 and 2 in humans. Am J Epidemiol 91:539–46.

Nahmias AJ, Whitley RJ, Visintine AJ et al. (1982) Herpes simplex virus encephalitis: laboratory evaluations and their diagnostic significance. J Infect Dis 145:829–36.

National Cancer Institute. (1982) The non-Hodgkin's lymphoma pathologic classification project. National Cancer Institute sponsored study of classification of non-Hodgkin's lymphomas: summary and description of a working formulation for clinical usage. Cancer 49:2112–35.

Navia BA, Price RW. (1987) The acquired immunodeficiency syndrome dementia complex as the presenting or sole manifestation of human immunodeficiency virus infection. Arch Neurol 44:65–9.

Navia BA, Cho E-S, Petito CK et al. (1986a) The AIDS dementia complex: II. Neuropathology. Ann Neurol 19:525–35.

Navia BA, Jordan BD, Price RW. (1986b) The AIDS dementia complex: I. Clinical features. Ann Neurol 19:517–24.

Navia BA, Petito CK, Gold JWM et al. (1986c) Cerebral toxoplasmosis complicating the acquired immune deficiency syndrome: clinical and neuropathological findings in 27 patients. Ann Neurol 19:224–38.

Nelson JA, Ghazal P, Wiley CA. (1990) Role of opportunistic viral infections in AIDS. AIDS 4:1–10.

Nichols PW, Koss M, Levine AM et al. (1982) Lymphomatoid granulomatosis: a T-cell disorder? Am J Med 72:467–71.

Nielsen SL, Davis RL. (1988) Neuropathology of acquired immunodeficiency syndrome. In: Rosenblum ML, Levy RM, Bredesen DE, ed. AIDS and the nervous system. Raven Press: New York, 155–81.

Nielsen SL, Petito CK, Urmacher CD et al. (1984) Subacute encephalitis in acquired immune deficiency syndrome: a postmortem study. Am J Clin Pathol 82:678–82.

Nonomura A, Matsubara F, Nakamura Y et al. (1983) T cell lymphoma presenting clinical and morphological features resembling polymorphic reticulosis and lymphomatoid granulomatosis. Acta Pathol Jpn 33:1289–301.

O'Neill BP, Illig JJ. (1990) Primary central nervous system lymphoma. Mayo Clin Proc 64:1005–20.

Oommen KJ, Johnson PC, Ray CG. (1982) Herpes simplex type 2 virus encephalitis presenting as psychosis. Am J Med 73:445–8.

Orenstein JM, Jannotta F. (1988) Human immunodeficiency virus and papovavirus infections in acquired immunodeficiency syndrome. Hum Pathol 19:350–61.

Padgett BL, Walker DL, ZuRhein GM et al. (1971) Cultivation of papova-like virus from human brain with progressive multifocal leukoencephalopathy. Lancet i:1257–60.

Pahwa S, Fikrig S, Menez R et al. (1986) Pediatric acquired immunodeficiency syndrome: demonstration of B-lymphocyte defect in vitro. Diagn Immunol 4:24–30.

Park YD, Belman AL, Kim T-S et al. (1990) Stroke in pediatric acquired immunodeficiency syndrome. Ann Neurol 28:303–11.

Parker JC Jr, McCloskey JJ, Lee RS. (1981) Human cerebral candidosis. A postmortem evaluation of 19 patients. Hum Pathol 12:23–8.

Parsons M. (1982) Diagnoses not to be missed. Tuberculous meningitis. Br J Hosp Med 27:682–4.

Paterson PY. (1960) Transfer of experimental allergic encephalo-myelitis in rats by means of lymph node cells. J Exp Med 111:119–36.

Pavia CS, Folds JD, Baseman JB. (1978) Cell-mediated immunity during syphilis: a review. Br J Vener Dis 54:144–50.

Penn I. (1979) Tumour incidence in human allograft recipients. Transplant Proc 11:1047–51.

Penn I. (1983) Lymphomas complicating organ transplantation. Transplant Proc 15 (suppl):2790–7.

Penny JB Jr, Weiner LP, Herndon RM et al. (1972) Virions from progressive multifocal leukoencephalopathy: rapid serological identification by electron microscopy. Science 178:60–2.

Pepose JS, Hiborne LH, Cancilla PA et al. (1984) Concurrent herpes simplex and cytomegalovirus retinitis and encephalitis in the acquired immune deficiency syndrome (AIDS). Opthalmology 91:1669–77.

Perham TGM, Caul EO, Clarke SKR et al. (1971) Cytomegalovirus meningoencephalitis. Br Med J 2:50.

Persson K, Hansson H, Herre B et al. (1979) Prevalence of nine different microorganisms in the female genital tract: a comparison between women from a venereal disease clinic and from a health control department. Br J Vener Dis 55:429–33.

Persuy Ph, Arnott G, Fortier B et al. (1988) Syndrome de Guillain et Barré d'évolution favorable dans un cas d'infection récente par le virus de l'immunodéficience humaine. Rev Neurol (Paris) 144:32–5.

Pert CB, Hill JM, Ruff MR et al. (1986) Octapeptides deduced from the neuropeptide receptor-like pattern of antigen T4 in brain potently inhibits human immunodeficiency virus receptor binding and T-cell infectivity. Proc Natl Acad Sci USA 83:9254–8.

Petito CK. (1988) Review of central nervous system pathology in human immunodeficiency virus infection. Ann Neurol 23 (suppl):s54–7.

Petito CK, Navia BA, Cho E-S et al. (1985) Vacuolar myelopathy pathologically resembling subacute combined degeneration in patients with the acquired immuno-deficiency syndrome. N Engl J Med 312:874–9.

Petito CK, Cho E-S, Lemann W et al. (1986) Neuropathology of acquired immunodeficiency syndrome (AIDS): an autopsy review. J Neuropathol Exp Neurol 45:635–46.

Petras RE, Tubbs RR, Gephardt GW et al. (1985) T lymphocyte proliferation in lymphomatoid granulomatosis. Clev Clin Q 52:137–46.

Phillips CA, Fanning WL, Gump DW et al. (1977) Cyto-megalovirus encephalitis in immunologically normal adults. Successful treatment with vidarabine. JAMA 21:2299–300.

Piette AM, Tusseau F, Vignon D et al. (1986) Acute neuropathy coincident with seroconversion for anti-LAV/HTLV-III. Lancet i:852 (lett).

Pillay VKG, Wilson DM, Ing TS et al. (1968) Fungus infection in steroid-treated systemic lupus erythematosus. JAMA 205:261–65.

Pitchenik AE, Fischl MA, Dickinson GM et al. (1983) Opportunistic infections and Kaposi's sarcoma among Haitians: evidence of a new acquired immunodeficiency state. Ann Intern Med 98:277–84.

Pitchenik AE, Fischl MA, Walls KW. (1985) Evaluation of cerebral-mass lesions in acquired immunodeficiency syndrome. N Engl J Med 308:1099.

Pitlik SD, Rios A, Hersh EM et al. (1984) Polymicrobial brain abscess in a homosexual man with Kaposi's sarcoma. South Med J 77:271–2.

Plant A. (1946) The problem of human toxoplasma carriers. Am J Path 22:427–31.

Plummer G. (1964) Serological comparison of the herpes viruses. Br J Exp Pathol 45:135–41.

Poirier J, Gray F, Escourolle R. (1990) Manual of basic neuropathology 3rd edn. WB Saunders Company: Philadelphia.

Poon M, Cronin DC II, Wormser GP et al. (1988) In vitro susceptibility of Cryptococcus neoformans isolates from patients with acquired immunodeficiency syndrome. Arch Pathol Lab Med 112:161–2.

Poon T, Matoso I, Tschertkoff V et al. (1989) CT features of primary cerebral lymphoma in AIDS and non-AIDS patients. J Comput Assist Tomogr 13:6–9.

Post MJ, Chan JC, Hensley GT et al. (1983) Toxoplasma encephalitis in Haitian adults with acquired immuno-deficiency syndrome: a clinical-pathological-CT correlation. Am J Radiol 140:861–8.

Post MJ, Kursunoglu SJ, Hensley GT et al. (1985) Cranial CT in acquired immunodeficiency syndrome: spectrum of diseases and optimal contrast enhancement technique. Am J Radiol 145:929–40.

Post MJ, Hensley GT, Moskowitz LB et al. (1986) Cytomegalic inclusion virus encephalitis in patients with AIDS: CT, clinical and pathologic correlation. Am J Radiol 146:1229–34.

Powell HC, Gibbs CJ Jr, Lorenzo AM et al. (1978) Toxoplasmosis of the central nervous system in the adult. Electron microscopic observations. Acta Neuropathol 41:211–16.

Price R, Chernik NL, Horta-Barbosa L et al. (1973) Herpes simplex encephalitis in an anergic patient. Am J Med 54:222–7.

Price RW, Nielsen S, Horten B et al. (1983) Progressive multifocal leukoencephalopathy: a burnt-out case. Ann Neurol 13:485–90.

Price RW, Navia BA, Cho E-S. (1986) AIDS encephalopathy. Neurol Clin 4:285–301.

Price RW, Brew B, Sidtis J et al. (1988) The brain in AIDS: central nervous system HIV-1 infection and AIDS dementia complex. Science 239:586–92.

Prineas JW, Wright RG. (1978) Macrophages, lymphocytes and plasma cells in the perivascular compartment in chronic multiple sclerosis. Lab Invest 38:409–21.

Pumarola-Sune T, Navia BA, Cordon-Cardo C et al. (1987) HIV antigen in the brains of patients with the AIDS dementia complex. Ann Neurol 21:490–6.

Purtilo DT. (1980) Epstein–Barr virus-induced oncogenesis in immune deficient individuals. Lancet i:300–3.

Quinnan GU, Masur H, Rook AH et al. (1984) Herpes virus infections in the acquired immune deficiency syndrome. JAMA 252:72–7.

Rahemtulla A, Durrant STS, Hows JM. (1986) Subacute encephalopathy associated with human immunodeficiency virus in haemophilia. Br Med J 293:993.

Rance NE, McArthur JC, Cornblath DR et al. (1988) Gracile tract degeneration in patients with sensory neuropathy and AIDS. Neurology 38:265–71.

Rao C, Wrzolek M, Kozlowski PB et al. (1989) Primary T cell central nervous system lymphoma in AIDS. J Neuropathol Exp Neurol 48:302 (abstr).

Rawls WE. (1985) Herpes simplex virus. In: Fields BN, Knipe DM, Chanock RM et al, ed. Virology. Raven Press: New York, 527–62.

Reddick RL, Fauci AS, Valsamis MP et al. (1978) Immunoblastic sarcoma of the central nervous system in a patient with lymphomatoid granulomatosis. Cancer 42:652.

Reichert S, Visser C, Danner S et al. (1989) Echo-cardiographic abnormalities in AIDS. 5th Internal Conf AIDS, Montreal (abstr).

Remington JS, Cavanaugh EM. (1965) Isolation of the encysted form of *Toxoplasma gondii* from human skeletal muscle and brain. N Engl J Med 273:1308–10.

Remington JS, Klein JD, ed. (1976) Infectious diseases of the fetus and newborn infant. WB Saunders: Philadelphia.

Rennick SC, Diamond C, Migliozzi JA et al. (1990) Primary central nervous system lymphoma in patients with and without the acquired immune deficiency syndrome. Medicine (Baltimore) 69:345–60.

Resnick I, Di Marzo-Veronese F, Schüpback J et al. (1985) Intra-blood-brain-barrier synthesis of HTLV-III-specific IgG in patients with neurologic symptoms associated with AIDS or AIDS-related complex. N Engl J Med 313:1498–504.

Rhodes RH. (1987) Histopathology of the central nervous system in acquired immunodeficiency syndrome. Hum Pathol 16:636–43.

Rhodes RH, Ward JM, Cowan RP et al. (1989) Immunohistochemical localization of human immuno-

deficiency viral antigens in formalin-fixed spinal cords with AIDS myelopathy. Clin Neuropathol 8:22–7.

Rice GPA, Schrier RD, Oldstone MBA. (1984) Cyto-megalovirus infects human lymphocytes and monocytes: virus expression is restricted to immediate-early gene products. Proc Natl Acad Sci USA 81:6134–8.

Rich SA. (1981) Human lupus inclusions and interferon. Science 213:772–4.

Ries F. (1986) Progressive multifokale Leukoenzephalopathie bei Haemophilia A. Besteht und Zusammenhang mit AIDS. Nervenarzt 56:442–8.

Rifkind D. (1966) The activation of varicella-zoster virus infections by immuno-suppressive therapy. J Lab Clin Med 68:463–74.

Rinaldo CR Jr, Black PH, Hirsch MS. (1977) Virus-leukocyte interactions in cytomegalovirus mononucleosis. J Infect Dis 136:677–8.

Roberts CJ. (1984) Coccidioidomycosis in acquired immune deficiency syndrome: depressed humoral as well as cellular immunity. Am J Med 76:734–6.

Robertson JS. (1962) Toxoplasmosis. Develop Med Child Neurol 4:507–18.

Robinson G, Wilson SE, Williams RA (1987) Surgery in patients with acquired immunodeficiency syndrome. Arch Surg 122:170–5.

Rodan BA, Cohen FL, Bean WJ. (1984) CT biopsy of cerebral toxoplasmosis in AIDS. J Fla Med Assoc 71:158–60.

Rodesch G, Parizel PM, Farber CM et al. (1989) Nervous system manifestations and neuroradiologic findings in acquired immunodeficiency syndrome (AIDS). Neuroradiology 31:33–9.

Roizman B, Batterson W. (1985) Herpesviruses and their replication. In: Fields BN, Knipe DM, Chanock RM et al, ed. Virology. Raven Press: New York, 497–526.

Roldan EO, Moskowitz L, Hensley GT. (1987) Pathology of the heart in AIDS. Arch Pathol Lab Med 111:943–6.

Rose FC, Brett EM, Burston J. (1964) Zoster encephalomyelitis. Arch Neurol 11:155–72.

Rosenberg NL, Hockberg FH, Miller G et al. (1986) Primary central nervous system lymphoma related to Epstein–Barr virus in a patient with acquired immune deficiency syndrome. Ann Neurol 20:98–102.

Rosenberg PR, Uliss AE, Friedland GH et al. (1983) Acquired immune deficiency syndrome. Ophthalmic manifestions in ambulatory patients. Ophthalmology 90:874–8.

Rosenberg S, Lopes MBS, Tsanaclis AM. (1986) Neuropathology of acquired immunodeficiency syndrome (AIDS): analysis of 22 Brazilian cases. J Neurol Sci 76:187–98.

Rosenblum M, Scheck AC, Cronin K et al. (1989) Dissociation of AIDS-related vacuolar myelopathy and productive HIV-1 infection of the spinal cord. Neurology 39:892–6.

Rosenblum ML, Levy RM, Bredesen DE et al. (1988) Primary central nervous system lymphomas in patients with AIDS. Ann Neurol 23 (suppl):13–16.

Runyon EH. (1971) Whence mycobacteria and mycobacteriosis? (Editorial). Ann Intern Med 75:467–8.

Ryder JW, Croen D, Kleinschmidt-DeMasters BK et al. (1986) Progressive encephalitis 3 months after resolution of cutaneous zoster in a patient with AIDS. Ann Neurol 19:182–8.

Sabin AB. (1941) Toxoplasmic encephalitis in children. JAMA 116:801–7.

Sackett JF, ZuRhein GM, Bhimani SM. (1979) Lymphomatoid granulomatosis involving the central nervous system: radiologic-pathologic correlation. Am J Radiol 132:823–6.

Safai B, Lynfield R, Lowenthal DA et al. (1987) Cancers associated with HIV infection. Anticancer Res 7:1055–68.

Salahuddin SZ, Ablashi DV, Markham PD et al. (1986) Isolation of a new virus, HDLV, in patients with lymphoproliferative disorders. Science 234:596–601.

Sauron B, Rancurel G, Gray F. (1983) Encéphalopathie subaiguë fébrile au cours d'une maladie de Hodgkin. Confrontation de la Salpêtrière du 4 Novembre 1982. Rev Neurol (Paris) 139:597–604.

Scaravilli F, Ellis DS, Tovey G et al. (1989a) Unusual development of polyoma virus in the brains of two patients with the acquired immune deficiency syndrome (AIDS). Neuropathol Appl Neurobiol 15:407–18.

Scaravilli F, Daniel SE, Harcourt-Webster JN et al. (1989b) Chronic basal meningitis and vasculitis in acquired immunodeficiency syndrome. Arch Pathol Lab Med 113:192–5.

Scaravilli F, Sinclair E, Arango J-C et al. (1992) The pathology of the posterior root ganglia in AIDS and its relationship to the pallor of the gracile tract. Acta Neuropathol (in press).

Schaumburg HH, Planck CR, Adams RD. (1972) The reticulum cell sarcoma-microglioma group of brain tumors. A consideration of their clinical features and therapy. Brain 95:199–212.

Schmidbauer M, Budka H, Ulrich W et al. (1989a) Cytomegalovirus (CMV) disease of the brain in AIDS and connatal infection: a comparative study by histology, immunocytochemistry and in situ DNA hybridization. Acta Neuropathol 79:286–93.

Schmidbauer M, Budka H, Ambros P. (1989b) Herpes simplex virus (HSV) DNA in microglial nodular brainstem encephalitis. J Neuropathol Exp Neurol 48:645–52.

Schmidbauer M, Budka H, Okeda R et al. (1990) Multifocal vacuolar leukoencephalopathy: a distinct HIV-associated lesion of the brain. Neuropathol Appl Neurobiol 16:437–43.

Schmidt BJ, Meagher-Villemure K, Del Carpio J. (1984) Lymphomatoid granulomatosis with isolated involvement of the brain. Ann Neurol 15:478–81.

Schneck SA. (1965) Neuropathological features of human organ transplantation. J Neuropathol Exp Neurol 24:415–29.

Schober R, Herman MM. (1973) Neuropathology of cardiac transplantation. Lancet i:962–7.

Schoenberg BS. (1983) The epidemiology of CNS tumors. In: Walker MD ed. Oncology of the nervous system. Martinus-Nijhoff: Boston, 1–29.

Schrier RD, Nelson JA, Oldstone MBA. (1985) Detection of human cytomegalovirus in peripheral blood lymphocytes in a natural infection. Science 230:1048–51.

Schrier RD, Rice G, Oldstone MBA. (1986) Suppression of NK activity and T-cell proliferation induced by fast isolates of cytomegalovirus. J Infect Dis 153:1084–91.

Schwenk J, Cruz-Sanchez F, Gosztonyi G et al. (1987) Spongiform encephalopathy in a patient with acquired immune deficiency syndrome (AIDS). Acta Neuropathol 74:389–92.

Selby RC, Lopés NM. (1973) Torulomas (cryptococcal granulomata) of the central nervous system. J Neurol 38:40–6.

Serke M, Huhn D, Dienermann D et al. (1988) HIV-related malignant lymphomas. A clinical and pathological study of 13 cases. Klin Wochenschr 66:682–5.

Serrano M, Bellas C, Campo E et al. (1990) Hodgkin's disease in patients with antibodies to human immunodeficiency virus. A study of 22 patients. Cancer 15:2248–54.

Shaerer WT, Ritz J, Finegold MJ et al. (1985) Epstein–Barr virus-associated B-cell proliferations of diverse clonal origins after bone marrow transplantation in a 12-year-old patient with severe combined immunodeficiency. N Engl J Med 312:1151–9.

Shamato M, Murakami S, Zenke T. (1981) Adult T-cell leukemia in Japan: an ultrastructural study. Cancer 47:1804–11.

Shapiro SS, Thiagarrajan P. (1982) Lupus anticoagulants. Prog Hemost Thromb 6:263–85.

Sharer LR, Kapila R. (1985) Neuropathologic observations in acquired immunodeficiency syndrome (AIDS). Acta Neuropathol 66:188–98.

Sharer LR, Cho E-S, Epstein LG. (1985) Multinucleated giant cells and HTLV-III in AIDS encephalopathy. Hum Pathol 16:760.

Sharer LR, Epstein LG, Cho E-S et al. (1986) Pathologic features of AIDS encephalopathy in children: evidence for LAV/HTLV-III infection of brain. Hum Pathol 17:271–284.

Sharer LR, Epstein LG, Cho E-S et al. (1987a) Comparative pathology of HIV and SIV encephalitides. J Neuropathol Exp Neurol 46:349 (abstr).

Sharer LR, Kaminski Z, Cho E-S et al. (1987b) Case 1, 28th Annual Diagnostic Slide Session, American Association of Neuropathologists, Seattle, Washington, June 1987.

Sharer LR, Dowling PC, Michaels J et al. (1990) Spinal cord in disease in children with HIV-1 infection: a combined molecular and neuropathological study. Neuropathol Appl Neurobiol 16:317–31.

Shaw GM, Harper ME, Hahn BH et al. (1985) HTLV-III infection in brains of children and adults with AIDS encephalopathy. Science 227:177–82.

Sidhu GS, Stahl RE, El-Sadr W et al. (1985) The acquired immunodeficiency syndrome: an ultrastructural study. Hum Pathol 16:377–86.

Silverman L, Rubinstein LJ. (1965) Electron microscopic observations on a case of progressive multifocal leukoencephalopathy. Acta Neuropathol (Berl) 5:215–24.

Sinclair E, Scaravilli F. (1992) Detection of HIV proviral DNA in cortex and white matter of AIDS brains by non-isotopic polymerase chain reaction: correlation with disuse poliodystrophy. AIDS 6 (in press).

Sixbey JW, Nedrud JG, Raab-Tralb N. (1984) Epstein–Barr virus replication in oropharyngeal epithelial cells. N Engl J Med 310:1225–30.

Skipper E, Beverley JKA, Beattie CP. (1954) Acquired toxoplasmosis with a report of two cases simulating glandular fever and one possible case resembling typhus. Lancet i:287–90.

Skolnik PR, Kosloff BR, Hirsh MS. (1988) Bidirectional interactions between human immunodeficiency virus type I and cytomegalovirus. J Infect Dis 157:508–14.

Skolnik PR, Pomerantz RJ, de la Monte SM et al. (1989) Dual infection of retina with human immunodeficiency virus type 1 and cytomegalovirus. Am J Ophthalmol 107:361–72.

Snider WD, Simpson DM, Nielsen S et al. (1983) Neurological complications of acquired immune deficiency syndrome: analysis of 50 patients. Ann Neurol 14:403–18.

So YT, Beckstead JH, Davis RL. (1986) Primary central nervous system lymphoma in acquired immune deficiency syndrome: a clinical and pathologic study. Ann Neurol 20:566–72.

So YT, Choucair A, Davis RL. (1988) Neoplasms of the central nervous system in acquired immunodeficiency syndrome. In: Rosenblum ML, Levy RM, Bredesen DE, ed. AIDS and the nervous system. Raven Press: New York, 285–300.

Sordillo PP, Epremian B, Koziner B et al. (1982) Lymphomatoid granulomatosis. An analysis of clinical and immunologic characteristics. Cancer 49:2070–6.

Sotrel A. (1989) The nervous system. In: Harawi SJ, O'Hara CJ, ed. Pathology and pathophysiology of AIDS and HIV related diseases. Chapman and Hall: New York, 201–66.

Stagno S, Reynolds DW, Tsiantos A et al. (1975) Cervical cytomegalovirus excretion in pregnant and non pregnant women: suppression in early gestation. J Infect Dis 131:522–7.

Stagno S, Reynolds DW, Pass RF et al. (1980) Breast milk and the risk of cytomegalovirus infection. N Engl J Med 302:1073–6.

Stagno S, Pass RF, Dworsky ME et al. (1982) Maternal cytomegalovirus infection and perinatal transmission. In: Knox GE, ed. Clinical obstetrics and gynecology, JB Lippincott: Philadelphia, 563–76.

Starzl TE, Porter KA, Iwatsuki S et al. (1984) Reversibility of lymphomas and lymphoproliferative lesions developing under cyclosporine-steroid therapy. Lancet i:583–7.

Stavrou D, Mehraein P, Mellert W et al. (1989) Evaluation of intracerebral lesions in patients with acquired immunodeficiency syndrome. Neuropathological and experimental data. Neuropathol Appl Neurobiol 15:207–22.

Streicher HZ, Joynt RJ. (1986) HTLV-III/LAV and the monocyte/macrophage. JAMA 256:2390–1.

Stulz P, Hasse J, Mihatsch J et al. (1980) Candida endocarditis after heart valve replacement. J Cardiovasc Surg (Torino) 21:255–60.

Sutton AL, Smithwick EM, Seligman SJ et al. (1974) Fatal disseminated herpes virus hominis type 2 infection in an adult with associated thymic dysplasia. Am J Med 56:545–53.

Swart S, Briggs RS, Millac PA. (1981) Tuberculous meningitis in Asian patients. Lancet ii:15–16.

Taylor CR, Russell R, Lukes RJ et al. (1976) An immunohistological study of immunoglobulin content of primary central nervous system lymphomas. Cancer 41:2197–205.

Tirelli U, Vaccher E, Rezza G et al. (1988) Hodgkin disease and infection with the human immunodeficiency virus (HIV) in Italy. Ann Intern Med 108:309–10.

Townsend JJ, Baringer JR. (1979) Morphology of central nervous system disease in immunosuppressed mice after peripheral herpes simplex virus inoculation. Lab Invest 40:178–82.

Townsend JJ, Wolinsky JS, Baringer JR et al. (1975) Acquired toxoplasmosis: a neglected cause of treatable nervous system disease. Arch Neurol 32:335–43.

Trotot PM, Cabanis EA, Cabee AE. (1988) Apport de la résonance magnétique à l'étude des lésions cérébrales au cours du SIDA. In: Trotot PM ed. Imagerie médicale du SIDA et des rétrovirus. Vigot: Paris, 61–73.

Tucker T, Dix RD, Katzen C et al. (1985) Cytomegalovirus and herpes simplex virus ascending myelitis in a patient with acquired immune deficiency syndrome. Ann Neurol 18:74–9.

Unger PD, Strauchen JA. (1986) Hodgkin's disease in AIDS complex patients. Report of four cases and tissue immunologic marker studies. Cancer 58:821–5.

Unger PD, Rosenblum M, Krown SE. (1988) Disseminated Pneumocystis carinii infection in a patient with acquired immunodeficiency syndrome. Hum Pathol 19:113–16.

Uzman BG, Saito H, Kasak M. (1971) Tubular arrays in the endoplasmic reticulum in human tumour cells. Lab Invest 24:492–8.

Vazeux R, Brousse N, Jarry A et al. (1987) AIDS subacute encephalitis. Identification of HIV-infected cells. Am J Pathol 126:403–10.

Veltri RW, Raich PC, McClung et al. (1982) Lymphomatoid granulomatosis and Epstein-Barr virus. Cancer 50:1513–17.

Vendrell J, Heredia C, Pujol M et al. (1987) Guillain–Barré syndrome associated with seroconversion for anti-HTLV-III (letter). Neurology 37:544.

Vinters HV, Anders KH, Barach P. (1987) Focal pontine leukoencephalopathy in immunosuppressed patients. Arch Pathol Lab Med 111:192–6.

Vinters HV, Guerra WF, Eppolito L et al. (1988) Necrotizing vasculitis of the nervous system in a patient with AIDS-related complex. Neuropathol Appl Neurobiol 14:417–22.

Vinters HV, Tomiyasu U, Anders KH. (1989a) Neuropathologic complications of infection with the human immunodeficiency virus (HIV). Adv AIDS Pathol 1:101–30.

Vinters HV, Kwok MK, Ho HW et al. (1989b) Cytomegalovirus in the nervous system of patients with the acquired immune deficiency syndrome. Brain 112:245–68.

Visvesvara GS, Balamuth W. (1975) Comparative studies on related free-living and pathogenic amebae with special reference to Acanthamoeba. J Protozool 22:245–56.

Vital C, Vital A, Julien J et al. (1990) Peripheral neuropathies and lymphoma without monoclonal gammopathy: a new classification. J Neurol 237:177–85.

Vortel V, Plachy V. (1968) Glial-nodule encephalitis associated with generalized cytomegalic inclusion body disease. Am J Clin Pathol 49:319–24.

Waksman BM, Arbouys S, Arnason BG. (1961) The use of specific 'lymphocyte' antiserum to inhibit hypersensitive reactions of the 'delayed' type. J Exp Med 114:997–1022.

Walsh C, Krigel R, Lennette E et al. (1985) Thrombocytopenia in homosexual patients: prognosis, response to therapy and prevalence of antibody to a retrovirus associated with acquired immunodeficiency syndrome. Ann Intern Med 103:542–5.

Wang SS, Feldman HA. (1967) Isolation of Hartmannella species from human throats. N Engl J Med 277:1174–9.

Wanke CA, Tuazon CV, Levy C et al. (1984) Clinical course of tissue documented cerebral toxoplasmosis in patients with acquired immune deficiency syndrome (AIDS). Proc Intersci Conf on Antimicrobiol Agents and Chemotherapy (abstr), 230.

Ward JM, O'Leary TJ, Baskin GB et al. (1987) Immunohistochemical localization of human and simian immunodeficiency viral antigens in fixed tissue sections. Am J Pathol 127:199–205.

Warder FR, Chikes PG, Hudson WR. (1975) Aspergillosis of the paranasal sinuses. Acta Otolaryngol (Stockh) 101:683–5.

Warren KG, Brown SM, Wroblewska Z et al. (1978) Isolation of latent herpes simplex virus from superior cervical and vagus ganglions of human beings. N Engl J Med 298:1068–9.

Watson CP, Evans RJ, Watt VR. (1988) Post-herpetic neuralgia and topical capsaicin. Pain 33:333–40.

Webster A, Cook DG, Emery VEC et al. (1989) Cytomegalovirus infection and progression towards AIDS in haemophiliacs with human immunodeficiency virus infection. Lancet ii:63–6.

Weiner LP, Herndon RM, Narayan O et al. (1972) Further studies of a simian virus 40-like virus isolated from human brain. J Virol 10:147–9.

Welch K, Finkbeiner W, Alpers CE et al. (1984) Autopsy findings in the acquired immune deficiency syndrome. JAMA 252:1152–9.

Wetli CV, Weiss SD, Cleary TJ et al. (1984) Fungal cerebritis from intravenous drug abuse. J Forensic Sci 29:260–8.

Whelan MA, Kricheff II, Handler M et al. (1983) Acquired immunodeficiency syndrome: cerebral computed tomographic manifestations. Radiology 149:477–84.

Wildy P. (1967) The progression of herpes simplex virus to the central nervous system of the mouse. J Hyg 65:173–92.

Wiley CA, Nelson JA. (1988) Role of human immunodeficiency virus and cytomegalovirus in AIDS encephalitis. Am J Pathol 133:73–81.

Wiley CA, Schrier RD, Nelson JA et al. (1986a) Cellular localization of human immunodeficiency virus infection within the brain of acquired immune deficiency syndrome patients. Proc Natl Acad Sci USA 83:7089–93.

Wiley CA, Schrier RD, Denara FJ et al. (1986b) Localization of cytomegalovirus proteins and genome during fulminant central nervous system infection in an AIDS patient. J Neuropathol Exp Neurol 45:127–39.

Wiley CA, Safrin RE, Davis CE et al. (1987) Acanthamoeba meningoencephalitis in a patient with AIDS. J Infect Dis 155:130–3.

Wiley CA, Grafe M, Kennedy C et al. (1988) Human immunodeficiency virus (HIV) and JC virus in acquired immune deficiency syndrome (AIDS) patients with progressive multifocal leukoencephalopathy. Acta Neuropathol 76:338–46.

Wiley CA, Masliah E, Hansen L et al. (1990) Cerebral cortical abnormalities in AIDS. J Neuropathol Exp Neurol 49:350 (abstr).

Wiley CA, Ge N, Morey M et al. (1991a) Golgi impregnation studies of dendritic pathology in HIV encephalitis. J Neuropathol Exp Neurol 50:324 (abstr).

Wiley CA, Masliah E, Morey M et al. (1991b) Neocortical damage during HIV infection. Ann Neurol 29:651–7.

Willmott FE. (1975) Cytomegalovirus in female patients attending a venereal disease clinic. Br J Vener Dis 51:278–80.

Witt DJ, Craven DE, McCabe WR. (1987) Bacterial infections in adult patients with the acquired immune deficiency syndrome (AIDS) and AIDS-related complex. Am J Med 82:900–6.

Wolf H, Hans M, Wilmes E. (1984) Persistence of Epstein–Barr virus in the parotid gland. J Virol 51:795–8.

Wongmongkolrit T, McPherson S-L, El-Naggar A et al. (1983) Acute fulminating toxoplasma meningoencephalitis in a homosexual man. Acta Neuropathol (Berl) 60:305–8.

Woods GL, Goldsmith JC. (1990) Aspergillus infection of the central nervous system in patients with acquired immunodeficiency syndrome. Arch Neurol 47:181–4.

Yanagisawa ZN, Toyokura Y, Shiraki B. (1975) Double encephalitis with herpes simplex virus and cytomegalovirus in an adult. Acta Neuropathol 33:153–64.

Yankner BA, Skolnik PR, Shoukimas GM et al. (1986) Cerebral granulomatous angiitis associated with isolation of human T-lymphotropic virus type III from the central nervous system. Ann Neurol 20:362–4.

Young RC, Bennett JE, Vogel CL et al. (1970) Aspergillosis. The spectrum of the disease in 98 patients. Medicine (Baltimore) 49:147–73.

Yow MD, Lakeman AD, Stagno S et al. (1982) Use of restriction enzymes to investigate the source of primary cytomegalovirus infection in a pediatric nurse. Pediatrics 70:713–16.

Zaia JA, Lang DJ. (1984) Cytomegalovirus infection of the fetus and neonate. Neurol Clin 2:387–410.

Zakowski P, Fligiel S, Berlin GW et al. (1982) Disseminated mycobacterium avium–intracellulare in homosexual men dying of acquired immunodeficiency syndrome. JAMA 248:2980–2.

Zee C, Segall HD, Rogers C et al. (1985) MR imaging of cerebral toxoplasmosis: correlation of computed tomography and pathology. J Comput Assist Tomogr 9:797–9.

Ziegler JL, Miner RC, Rosenbaum E et al. (1982) Outbreak of Burkitt's-like lymphoma in homosexual men. Lancet ii:631–3.

Ziegler JL, Beckstead JH, Volberding PA et al. (1984) Non-Hodgkin's lymphoma in 90 homosexual men. Relation to generalized lymphadenopathy and the acquired immunodeficiency syndrome. N Engl J Med 311:565–70.

Zuber M, Bherardi R, Imbert M. (1987) Peripheral neuropathy with distal nerve infiltration revealing a diffuse pleomorphic malignant lymphoma. J Neurol 235:61–2.

ZuRhein GM, Chou SM. (1965) Particles resembling papovaviruses in human cerebral demyelinating disease. Science 148:1477–9.

7 HIV-Related Dementia: Pathology and Possible Pathogenesis

H. Budka

Introduction

Soon after HIV-1 was identified as the cause of AIDS, the notion that some neurological disorders described in these patients (Johnson et al, 1988; Michaels et al, 1988a; Price et al, 1988a) could be due to the neurotropism of the virus (Shaw et al, 1985; Price et al, 1988a) became widely accepted. However, viral neurotropism is a concept encompassing the ability to invade and cause disease of the nervous system, as well as the capacity to replicate in neuroectodermal cells (Johnson et al, 1988). With regard to the in vivo situation of HIV infection, the third criterion still awaits confirmation as discussed in detail below.

The aim of this chapter is to review the morphological features of the damage to the brain by HIV, which may give rise to dementing symptoms; it is based on personal experience and includes discussion of the possible pathogenesis of the lesions.

Incidence and Definition of HIV-Related Dementia

In spite of the increasing experience acquired in recent years in the study of patients with AIDS, proposals for a generally accepted nomenclature for HIV-associated neurological disease were made only very recently for clinical (Janssen et al, 1991) and neuropathological (Budka et al, 1991) conditions including case definitions. The major clinical syndrome associated with the virus has been given various names such as AIDS dementia complex (ADC) (Navia et al, 1986a; Navia and Price, 1987; Pumarola-Sune et al, 1987; Price et al, 1988a) or HIV-1-associated cognitive/motor complex (Janssen et al, 1991), which encompass other terms such as AIDS dementia (Johnson et al, 1988), AIDS encephalopathy (Sharer et al, 1985, 1986; Koenig et al, 1986), HIV encephalopathy (Michaels et al, 1988a), or vacuolar myelopathy (Navia and Price, 1987; Price et al, 1988a).

Earlier data on the frequency and severity of HIV-related dementia seem to have been somewhat biased by experience with a selected clinical population (Price et al, 1988b), as well as by lack of generally accepted criteria for such a diagnosis; the term 'dementia' seems to be used in a broader sense in the USA than in Europe. Consequently, initial reports on ADC indicated a rather high incidence in almost two-thirds of AIDS patients (Navia et al, 1986b), whereas more recent observations indicate a lower frequency than previously assumed.

In the largest published series of 1286 AIDS patients from the University of California at

San Francisco (UCSF), the overall incidence of neurological disease was given as 37 per cent (Levy and Bredesen, 1988). Among the many types of neurological complications, HIV encephalopathy was described as occurring in 21 per cent of subjects, that is, 7.8 per cent of all AIDS patients. In 90 per cent of these patients (7 per cent of all AIDS patients), dementia, defined as alteration in consciousness or cognition ranging from subtle alterations in memory to confusion, hallucinations, psychotic behaviour, stupor and coma (Levy and Bredesen, 1988), was the most common symptom of HIV encephalopathy. In the early studies by Navia et al (1986a,b), ADC was initially described as a variable, yet characteristic, constellation of abnormalities of cognitive, motor and behavioural functions; diagnosis of dementia was made according to DSM-III criteria (American Psychiatric Association, 1980). ADC was estimated to be the presenting or only sign of overt AIDS in 25 per cent of patients (Navia et al, 1986b). On the other hand, HIV-related dementia tends to occur more frequently during the later stages of the infection (Johnson et al, 1988; Michaels et al, 1988a; Price et al, 1988a). Indeed, the possibility cannot be excluded that most or all AIDS patients would be afflicted with CNS disease, were they to survive for a long enough period.

HIV-Associated Neuropathology

Most investigators agree that there is more than one set of neuropathological changes in HIV-related dementia. Thus, it is felt that to refer to the clinical syndrome as HIV-related is a little premature (Sidtis and Price, 1990).

Almost all our knowledge of HIV-related neuropathology is derived from autopsy studies which, by usually examining a late stage of the infection, introduce some kind of selection. Table 7.1 summarizes types and frequencies of pathological central nervous system (CNS) changes in 132 patients studied by the author of this chapter.

Differences in frequency of various types of neuropathology in AIDS among several large series (Lang et al, 1989) are likely to be the result of differences in the amount of tissue studied in each case, and in the methods employed. The latter must include, in addition to standard histopathological methods and electron microscopy, modern morphological techniques such as immunohistochemistry (see Figures 7.1–7.3 (facing

Table 7.1. Neuropathological diagnoses in 132 adult HIV (128 AIDS) patients (many patients had more than one type of CNS pathology)

Type of lesion/infectious agent	Patients	
	No.	%
HIV-specific CNS lesions	49[a]	37
HIV encephalitis (HIVE)	16	12
HIV leukoencephalopathy (HIV-lep)	17	13
Combined HIVE and HIV-lep	16	12
HIV-associated (possibly HIV-induced) CNS lesions	66	50
Diffuse poliodystrophy	62	47
Lymphocytic meningitis	7	5
Multifocal vacuolar leukoencephalopathy	4	3
Vacuolar myelopathy	4	3
Opportunistic CNS infections (agent found in lesion)	66	50
Toxoplasmosis	25	19
CMV	22	17
Papovavirus/PML	14	11
Cryptococcosis	9	7
Candida albicans	2	1.5
Tuberculosis	2	1.5
HSV	1	0.8
Nodular encephalitis (to be distinguished from HIVE)	44	33
Agent found in lesions (CMV, *Toxoplasma*, HSV)	21	16
Unspecific, no agent found in lesion	23[b]	17
Other lesions	37	28
Infarcts, haemorrhages	14	11
Malignant lymphoma	8	6
Hepatic encephalopathy	5	4
Necroses of undefined character	5	4
Spongy leukoencephalopathy	1	0.8
Acute epidural haematoma	1	0.8
Central pontine myelinolysis	1	0.8
Acute Wernicke encephalopathy	1	0.8
No lesion (except oedema, acute ischaemic changes)	5	4

[a] Isolated (no other inflammatory or infectious pathology) in 19 patients (39 per cent of HIV-specific CNS lesions).
[b] In 11 of these patients, circumstantial evidence for appropriate opportunistic infection (other infectious lesions in CNS elsewhere, or disseminated infection in other organs, usually CMV).
HIVE – HIV encephalitis (Budka 1990a, b), is synonymous with multifocal giant cell encephalitis (MGCE: Budka 1986, 1989; Budka et al, 1987; Lang et al, 1989), subacute encephalitis with multinucleated cells (Petito et al, 1986), multinucleated cell encephalitis (Price et al, 1988a), and giant cell encephalitis (Michaels et al, 1988a).
HIV-Lep – HIV leukoencephalopathy (Budka 1990a,b; Smith et al, 1990), is synonymous with progressive diffuse leukoencephalopathy (PDL: Kleihues et al, 1985; Budka et al, 1987; Budka, 1989; Lang et al, 1989) and seems to overlap at least in part with diffuse white matter/myelin 'pallor' (Petito et al, 1986; Johnson et al, 1988; Michaels et al, 1988b, Price et al, 1988a).

p. 174), 7.5, 7.6) and in situ nucleic acid hybridization, to demonstrate or exclude HIV and/or opportunistic agents and to identify cell types (Budka 1989, 1990a,b). As multiple pathologies frequently co-exist in AIDS patients, the amount of tissue available for examination is crucial for the diagnosis. Although small stereotactic brain biopsies have become indispensable in the clinical management of HIV patients, the small amount of tissue made available for this technique can offer limited information. Autopsies, on the other hand, produce an abundant supply of tissue. Even then, the amount of tissue sampled for examination may be of importance for the final diagnosis, especially when discrete or early changes are under consideration. It is our experience that the study of blocks of tissue of only a few cm^2 in size, as is routine in most pathology laboratories, may occasionally show only non-specific pathology, even when many blocks are examined. The use of large uni- or bi-hemispheric sections (see Figure 7.4), on the other hand, has, in several of our cases, emphasized HIV- or cytomegalovirus (CMV)-specific changes (Budka et al, 1987; Budka, 1990a), which were inconspicuous in smaller blocks.

A final factor is the diligence with which abnormalities are sought; careful histopathological examination is a time-consuming task – the search for small lesions such as glial nodules in a large bi-hemispheric section may well need an hour or more.

HIV-Specific Central Nervous System Pathology

In 37 per cent of brains from our material, histological abnormalities include changes, each of which, taken in isolation, may be non-specific. Considered together, however, this complex can claim to be specific for HIV (Budka 1990a,b). Evidence for the specificity of HIV-induced neuropathology includes:

1. Local productive HIV infection consistently revealed by immunohistochemistry (see Figures 7.1–7.2, 7.5b,c, 7.6c) (Gabuzda et al, 1986; Wiley et al, 1986; Budka et al, 1987; Pumarola-Sune et al, 1987; Vazeux et al, 1987; Ward et al, 1987; Michaels et al, 1988a,b; Price et al, 1988a), by in situ nucleic acid hybridization (Koenig et al, 1986; Sharer et al, 1986; Wiley et al, 1986; Vazeux et al, 1987) and electron microscopy (Koenig et al, 1986; Sharer et al, 1986; Budka

et al, 1987; Meyenhofer et al, 1987), and histologically characterized by the presence of multi-nucleated giant cells (MGC) (Kleihues et al, 1985; Sharer et al, 1985; Budka, 1986, 1987, 1989; Petito et al, 1986; Rhodes, 1987; Budka et al, 1987; Gray et al, 1988; Lang et al, 1989). These demonstrate the capacity of HIV-infected cells for fusion in the nervous system, analogous to the formation of syncytia in infected permissive cell cultures (Popovic et al, 1984).

2. The occurrence of these changes in isolation within the CNS, in the absence of other infectious agents (Budka et al, 1987; Budka, 1989; Schmidbauer et al, 1989a) and without any other pathology (Budka et al, 1987). Indeed, the neurological disease may represent in some cases the initial or sole manifestation of HIV infection (Navia et al, 1986b).

3. The fact that these characteristic changes were never described in the pre-AIDS era (Budka 1987, 1989).

HIV-specific neuropathology includes two different patterns of lesions, which represent end points of a morphological spectrum of mixed and transitional changes (Budka et al, 1987; Budka, 1989; Lang et al, 1989): the multifocal lesion of HIV encephalitis (HIVE) (Figure 7.1a), and the diffuse changes of HIV leukoencephalopathy (HIV-lep) (Figure 7.1b). The inflammatory character of the neuropathological changes, well seen in HIVE, is usually lacking in the diffuse damage of the white matter. Thus, the term 'leukoencephalopathy' appears appropriate for the latter. To designate both lesions simply as 'HIVE' or 'subacute encephalitis' (de la Monte et al, 1987; Vazeux et al, 1987; Gray et al, 1988; Hénin and Hauw, 1989) is not correct on strictly neuropathological grounds (Budka et al, 1987; Budka, 1989; Vinters and Anders, 1990). Moreover, the term 'subacute encephalitis' has caused confusion with nodular encephalitis due to CMV (Snider et al, 1983; Schmidbauer et al, 1989a), and with non-specific nodular encephalitis other than HIVE, when the agent cannot be demonstrated, although this is most likely to be CMV (Budka et al, 1987; Budka, 1989) (see discussion on CMV in Chapter 6). This type of nodular encephalitis was clearly shown not to be correlated with dementia (Navia et al, 1986a). 'Subacute encephalitis' of AIDS was also used indiscriminately to include changes such as gliosis of grey matter and atypical oligodendrocytes (de la Monte et al, 1987), which are not generally recognized features of HIVE. The many synonyms used in the literature on HIV neuropathology are listed in the legends of Table 7.1.

The neuropathological results summarized in Table 7.1 reflect the situation in adult HIV patients only. In children with AIDS (see Chapter 9) brain damage attributed to HIV is observed with higher frequency and similar histopathology. However, the development of cerebral calcifications is said to be characteristic, whereas opportunistic infections develop less frequently in children than in adults (Navia et al, 1986a; Johnson et al, 1988).

HIV Encephalitis

HIVE is characterized by multiple microgranulomatous foci irregularly disseminated in both grey and white matter (Figure 7.4), usually in a perivascular position. These foci consist of loosely aggregated microglia, and mono- and multinucleated macrophages, all of which show intense production of HIV (Figure 7.5). Lymphocytes are usually sparse and some reac'e astrogliosis is usually seen (Budka et al, 1987; Gray et al, 1988; Johnson et al, 1988; Michaels et al, 1988a,b; Lang et al, 1989; Budka, 1989, 1990a,b). Damage to the local nervous parenchyma is either absent or involves myelin sheaths (Budka, 1988), whereas neurons remain morphologically intact even when contacted by HIV-producing cellular processes (Figure 7.2). In rare cases, a massive diffuse destruction of the grey and white matter of the brain is observed (Giangaspero et al, 1989).

HIV Leukoencephalopathy

A characteristic diffuse histopathological triad of myelin reduction, reactive astrogliosis, and infiltration of mono- and multinucleated microglia/macrophages with HIV production (Figure 7.1b) (Budka, 1990b) involves the deep white matter of the cerebral hemispheres, usually in a symmetrical distribution (Budka et al, 1987; Budka, 1989, 1990a; Smith et al, 1990; Kleinhues et al, 1985; Lang et al, 1989). Microvascular changes with thickening of the walls, increased cellularity, and enlargement and pleomorphism of endothelial cells have been most recently described in three demented patients with HIV-lep (Smith et al, 1990). We have also found in our material occasional microvascular changes associated with HIV-specific neuropathology, usually some fibro-

sis and, rarely, extravasation of fibrinoid material (Budka et al, 1987). Such vascular changes may reflect important pathogenetic events at the blood–brain barrier level (see page 181).

The question arises whether HIV-specific neuropathology can be diagnosed in the absence of MGC, which are its histopathological hallmark (Sharer et al, 1985; Budka, 1986, 1989; Budka et al, 1987; de la Monte et al, 1987; Lang et al, 1989; Rhodes, 1987). It must be conceded that MGC are rare and inconspicuous in some brains which otherwise suggest HIV-specific neuropathology. Indeed, MGC were found by the author in one brain only after large amounts of HIV proteins in microglia/macrophages, as seen by immunohistochemistry, had suggested HIV-specific neuropathology; at the initial evaluation, MGC had escaped attention. If one accepts that development of MGC and of HIV-specific neuropathology is the result of increased HIV production, as discussed in detail below, then immunohistochemistry for HIV antigens should be a valuable diagnostic aid in difficult cases. Indeed, the presence of large amounts of HIV proteins is restricted, with two exceptions in our series, to HIV-specific neuropathology with or without diffuse poliodystrophy, to a single case of vacuolar myelopathy or vacuolar leukoencephalopathy, and to rare cases of opportunistic infections with massive HIV co-infection (Budka, 1990b; Schmidbauer et al, 1990a). The two exceptions with many HIV-producing cells in our series were focal microglial accumulations without MGC in one case, and

---

Figure 7.1a,b. Double-label immunohistochemistry reveals the focal character of HIV production (brown label) in HIV encephalitis (a), and diffuse HIV production in microglia in HIV-lep (b), whereas astroglia (blue label) does not appear to be affected. Avidin-biotin (AB) technique with monoclonal anti-HIV gp41 antibody, visualized with DAB as brown chromogen, followed by AAAP technique with anti-GFAP antiserum, visualized with blue chromogen. Original magnifications ×250 (a) and ×160 (b).

Figure 7.2. HIV-producing microglia/macrophages are scattered among intact neurons. Some labelled cellular processes are directly attached to the neuronal surfaces. AB technique with anti-gp41, slight haematoxylin counterstain. Original magnification ×400.

Figure 7.3. Cryptococcal meningo-encephalitis. Some of the many HIV-producing macrophages in perivascular space contain fungi, which are recognizable by their blue–green staining. Technique as for Figure 7.2 with additional Alcian blue stain. Original magnification ×630.

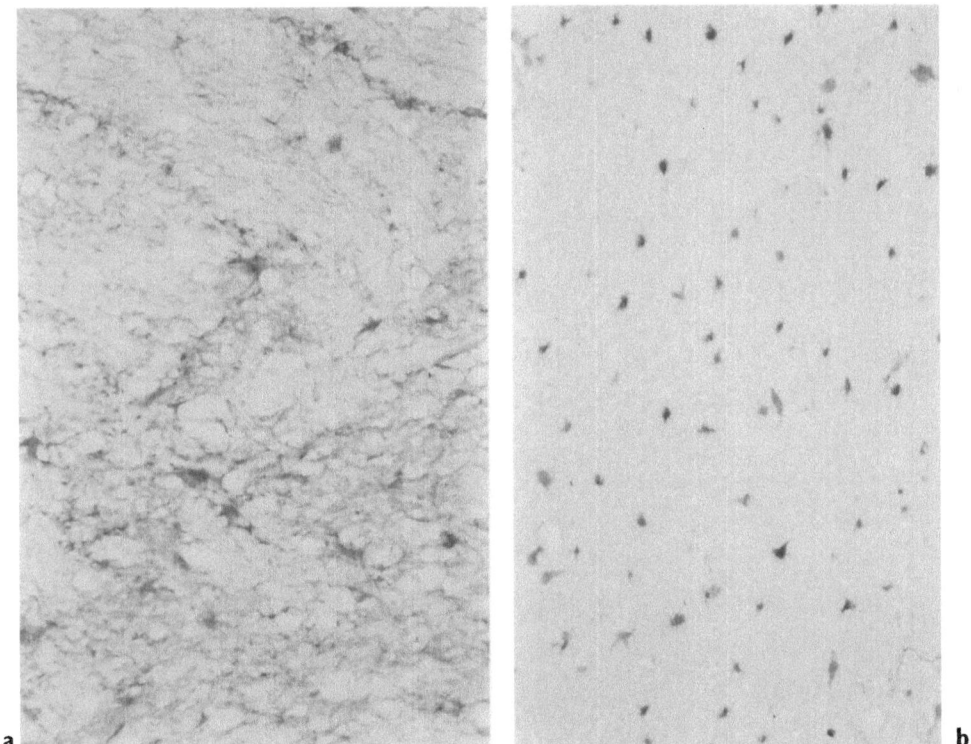

a

b

Fig. 7.1 a,b

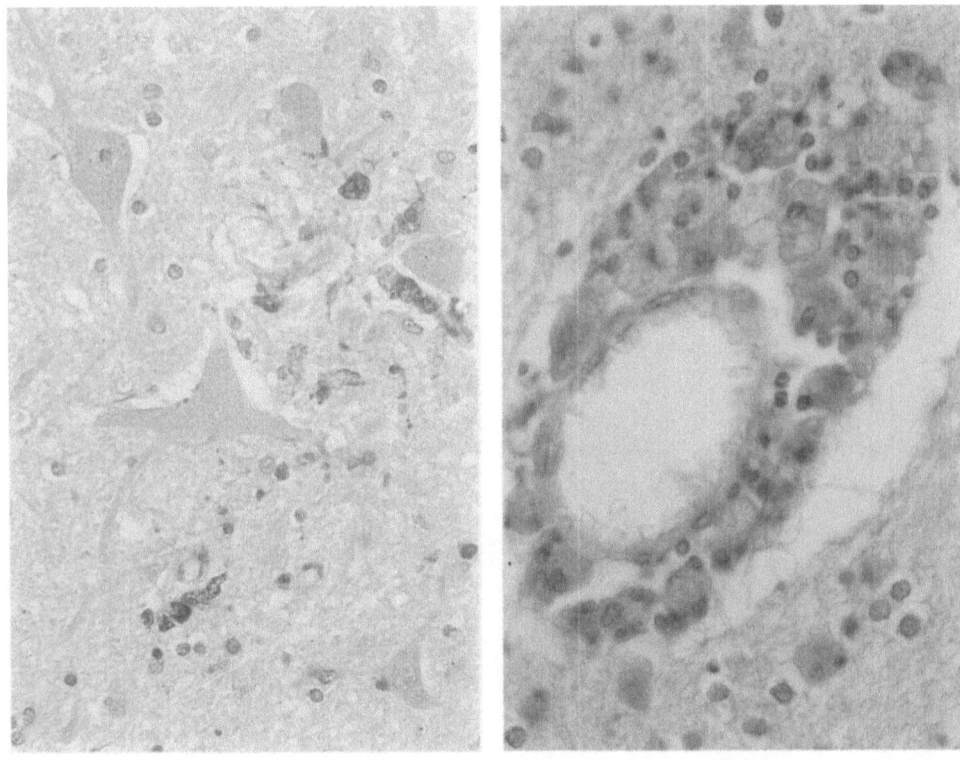

Fig. 7.2

Fig. 7.3

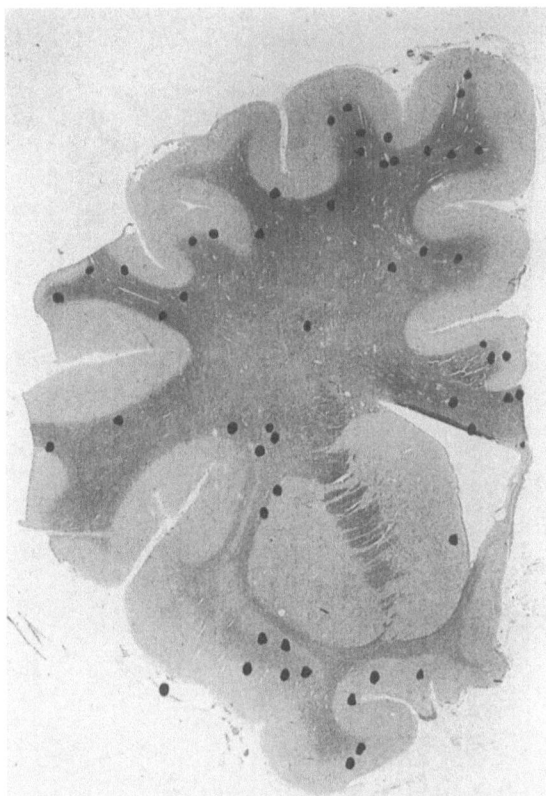

Figure 7.4. Many foci of HIV encephalitis (each marked with a dot) are irregularly disseminated, predominantly in the subcortical white matter. Cerebral hemispheric section, Luxol fast blue/neutral red.

non-specific necrosis without MGC in another (Budka, 1990b); the nature of these changes remains obscure. Thus, we believe that only in exceptional cases of HIV-induced neuropathology are MGC not revealed in the course of a thorough search. Immunocytochemistry should then show large numbers of HIV-producing cells in typical distribution, confirming diagnosis. In routine stains, however, HIV-specific neuropathology should be diagnosed only in the presence of MGC.

HIV-Associated (Possibly HIV-Induced) Central Nervous System Changes

The histopathology of this group of lesions is non-specific, and its pathogenesis obscure. Since HIV is not a constant finding in these lesions (Budka, 1990b), it is probable that these neuropathological syndromes do not represent uniform entities.

Lymphocytic Meningitis

An acute or chronic 'aseptic' meningitis may occur in the pre-AIDS stages of HIV infection, but is rare in AIDS (Johnson et al, 1988) and has found only limited attention among neuropathologists. Its aetiology is obscure. However, it has been found by us in a few brains in association with prominent collections of HIV-producing cells in the leptomeninges (Figure 7.6). The association, in these cases, of leptomeningitis with prominent HIV-specific neuropathology, suggests that, at least in some cases, HIV pathology may spread to the leptomeninges.

Vacuolar Myelopathy

This characteristic clinico-neuropathological entity is described in Chapters 6 and 8.

Multifocal Vacuolar Leukoencephalopathy

Multifocal vacuolation of the cerebral white matter with demonstrable HIV antigens was recently proposed as the cerebral equivalent of VM (Schmidbauer et al, 1990b). These distinct lesions have a distribution similar to extrapontine myelinolysis, and the authors speculate that lesions may initially have a pathogenesis unrelated to HIV, such as those operating in myelinolysis, but are later aggravated by the local presence of HIV (Schmidbauer et al, 1990b).

Diffuse Poliodystrophy

Most neuropathological studies emphasize damage to the white matter in AIDS brains (Navia et al, 1986a; Petito et al, 1986; Pumarola-Sune et al, 1987; Michaels et al, 1988a; Price et al, 1988a). However, it cannot be denied that the grey matter, including the cerebral cortex, may also be damaged, showing astrogliosis (de la Monte et al, 1987), proliferation of microglia (Figure 7.6b), and occasional reduction of nerve cells (Budka et al, 1987; Budka, 1989). These changes, which sometimes overlap, may be present in variable proportions and reach different degrees of severity. Such uncharacteristic lesions resemble metabolic glio-neuronal dystrophies (Seitelberger, 1975). It must be admitted that the subjective evaluation of such minor

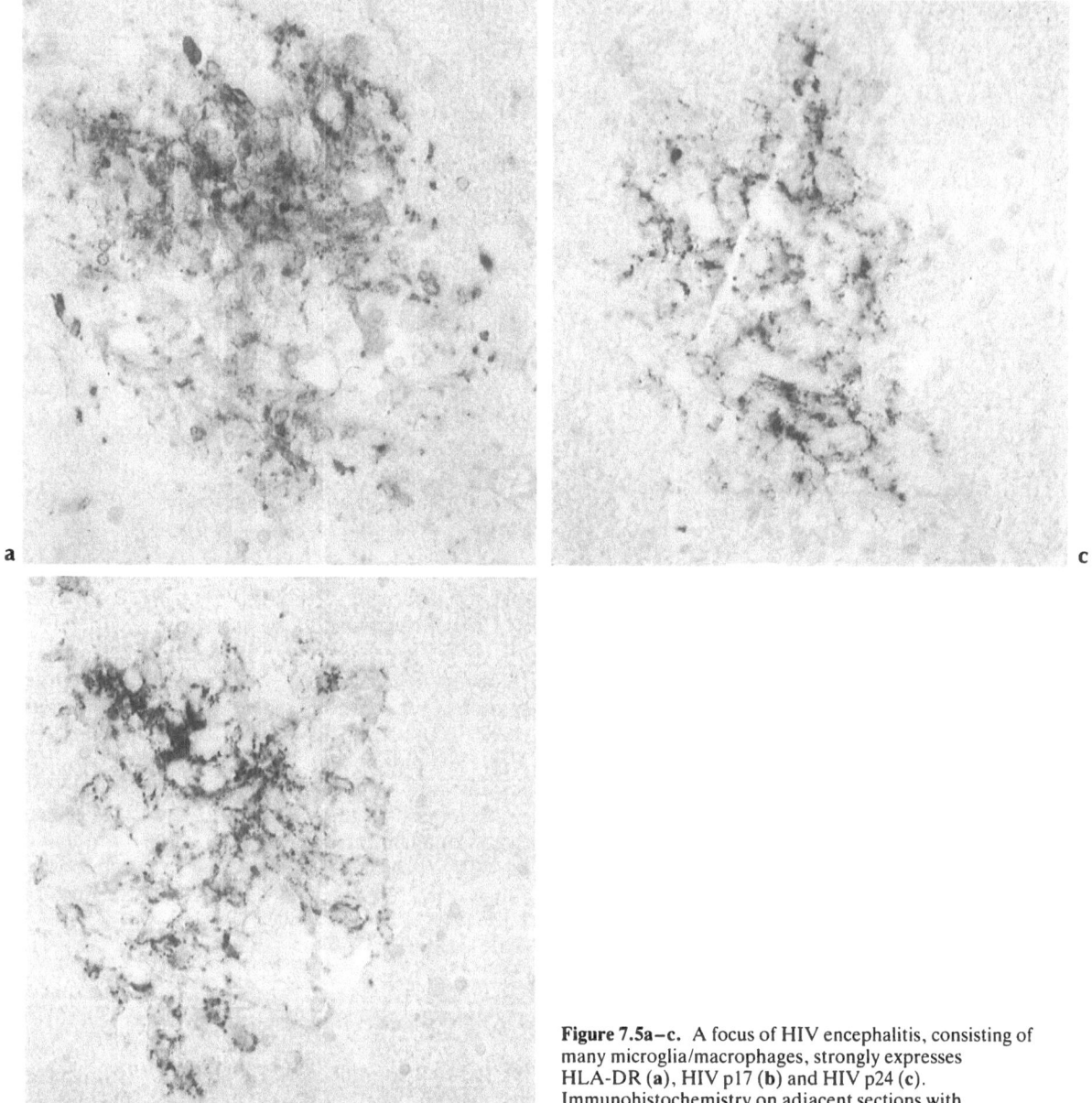

Figure 7.5a–c. A focus of HIV encephalitis, consisting of many microglia/macrophages, strongly expresses HLA-DR (**a**), HIV p17 (**b**) and HIV p24 (**c**). Immunohistochemistry on adjacent sections with appropriate monoclonal antibodies (Budka, 1990b). Original magnification ×380.

qualitative changes is extremely difficult in routine histological preparations. However, a recent morphometric study has confirmed that a significant loss of nerve cells may occur in the frontal cortex of AIDS patients (Ketzler et al, 1990). It is tempting to speculate that such a neuronal loss is a correlate of the diffuse atrophy of AIDS brains (Budka et al, 1987). Moreover, another recent immunohistochemical and morphometric study has shown increased numbers of GFAP-positive astroglia and of RCA-labelled microglia in the cerebral cortex of HIV-positive patients with and without clinical and/or pathological evidence of encephalopathy (Ciardi et al, 1990). Spongiform changes of the grey matter have also been described in five patients as another possible morphological correlate of ADC and compared to changes seen in Creutzfeldt-Jakob disease (Artigas et al, 1989a). In our material, however, we did not encounter such changes.

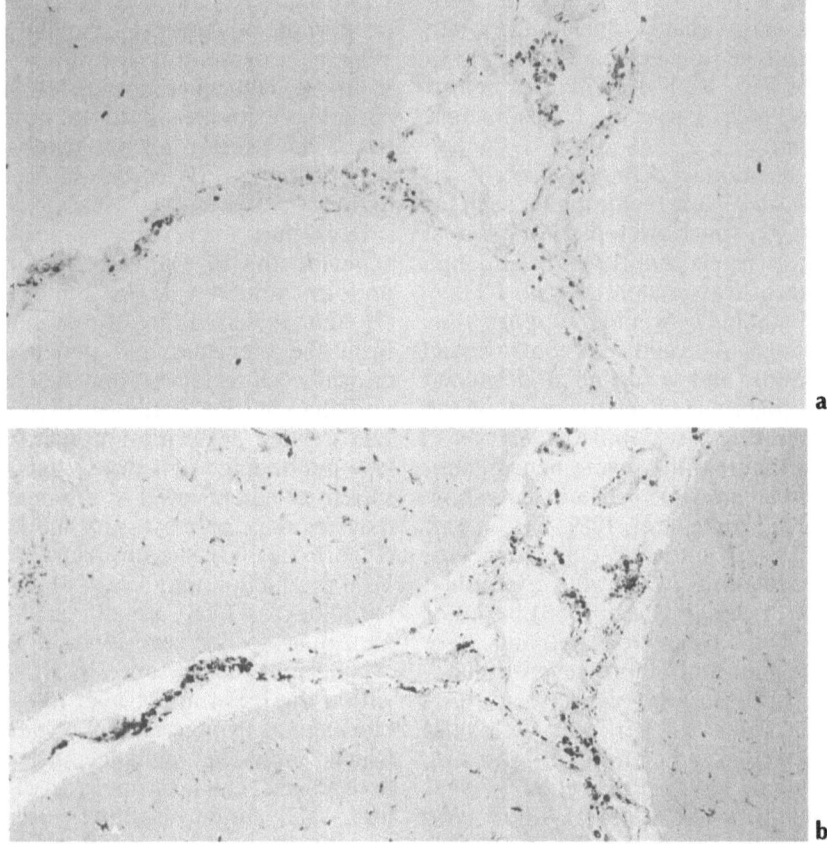

Figure 7.6a,b. Many macrophages in cerebellar leptomeninges labelled with a monoclonal anti-macrophage antibody (**a**, MAC 387) also contain large amounts of HIV gp41, as do cerebellar microglial cells (**b**). Original magnification ×63.

Clinico-neuropathological Correlation

Correlation of HIV-specific neuropathology with clinical disease related to cerebral HIV infection is far from established (Price et al, 1988a; Budka, 1989). In fact, this is the major problem in our understanding of HIV-related neurological disease. The lack of uniformity in the diagnosis of HIV-related dementia and of HIV-related neuropathology is one major obstacle for their correlation; available data must be interpreted with caution.

Longitudinal studies suggest that obvious dementia in the absence of HIV-related systemic symptoms or immunodeficiency is rare (Janssen et al, 1989). Whereas neuropathological correlation of fully-established HIV-related dementia in AIDS has been reported by several groups (see below), data on the neuropathology of subtle

cognitive dysfunction are scarce or non-existent. Such discrete involvement may be revealed only by neuropsychological testing; and it is a matter of controversy as to how frequently these subtle deficits may already be found in the early stages of the HIV infection (McArthur et al, 1989; Miller et al, 1990; Selnes et al, 1990; Wilkie et al, 1990). Astrogliosis in cerebral white matter was described (Lenhardt et al, 1988) in an HIV-infected individual with ARC but without dementia, who died accidentally. In our material, similar changes were found in an asymptomatic HIV-infected haemophiliac, who died from acute epidural haematoma; only a few non-specific glial nodules were seen in the brain at autopsy (Budka et al, 1987). Recently, diffuse pallor of the white matter and gliosis without multinucleated giant cells were described in a patient with early ARC, who was classified as suffering from mild HIV dementia (McArthur et al, 1989). Since these changes on their own are non-specific and HIV was not locally demonstrated, the significance of this observation remains to be established. Similar non-specific changes including white matter pallor, acute neuronal changes, and scanty cellular response have been described as neuropathological correlates of an acute HIV infection (Jones et al, 1988).

HIV-related dementia, defined clinically as 'subcortical', usually lacks symptoms, such as aphasia and apraxia, which are typical of classical 'cortical' dementia (Navia et al, 1986b). Although the frequent subcortical predominance of HIV-specific neuropathology would support this concept, its pathogenetic validity is controversial (Whitehouse, 1986), and is further undermined by increasing recognition of involvement of the cerebral cortex in AIDS (see above).

Most groups agree that the severe neuropathological changes found in AIDS brains usually show up clinically (de la Monte et al, 1987; Price et al, 1988a; Budka, 1990a), although detailed neuropathological correlation of newly established grades of ADC (Price and Brew, 1988) have not yet been published. Demented patients were shown to have significantly more severe inflammatory lesions of 'subacute encephalitis' than non-demented patients: 82 per cent with mild changes had no recognized neurological disorder (de la Monte et al, 1987). Price et al (1988a) found moderate to severe 'diffuse pallor' with multinucleated cells (obviously representing HIV-lep) in patients with moderate to severe ADC, but only mild to moderate 'pallor' without multinucleated cells in mild to moderate dementia, and absent to mild 'pallor' when ADC was absent. They estimate the percentages of these three groups among all AIDS autopsies as 25 per cent, 50 per cent, and 25 per cent, respectively. In contrast with this good, but crude, clinico-neuropathological correlation, we could observe a multitude of HIVE foci in a patient who had no clinical report of neurological abnormality (Figure 7.4 shows 52 foci in one hemispheric section; if foci were distributed in uniform density throughout the brain, their total number could be estimated as some 250 000). Moreover, one-third of patients with ADC were reported to have a bland brain histology, just as many non-demented patients showed pathology similar to that found in those who were clinically disabled (Navia et al, 1986a). Finally, productive HIV infection of the brain was detected by Price et al (1988a) only in a subset of patients with ADC. This is somewhat at variance with the neuropathological experience of the author in AIDS (Budka, 1990a), including immunostaining of brain tissue for HIV proteins (Budka, 1990b).

Unfortunately, clinical data are available only in a proportion of patients from the author's series. Whenever dementing symptoms were mentioned in these clinical reports, HIV-specific neuropathology with constant HIV production in brain tissue was seen in all but two cases. These two patients had DPD as the only neuropathological finding.

It is thus likely that clinically manifest encephalopathy in HIV-infected people is not a uniform entity with a uniform pathogenesis (Budka, 1990a). When the available information from the literature and personal experience is critically assessed as done below, a broad and unifying concept of the possible pathogenesis of HIV-related dementia emerges which comprises two major lines of cause–effect relations. In a system as complicated as a human patient, these may overlap, co-exist, and influence each other (Figure 7.7). On the one hand, locally enhanced HIV production may lead to HIV-specific neuropathology, which usually manifests clinically according to its severity and extent. On the other hand, the presence of HIV or its products within the brain might lead to clinically-relevant blockage of trophic factors or neurotransmitters, which could be signalled by morphologically non-specific changes such as DPD, nerve cell loss, and brain atrophy. Recent extension of our morphometric study to the cerebral cortex includes correlation with clinical and neuropathological data, and we have shown clearly that AIDS patients with dementia have a significantly larger loss of neurons in the frontal cortex as compared to those without dementia. The same accentuation of neuronal loss is seen in brains with HIV-specific neuropathology (Weis et al, personal communication).

Neuroradiological–Neuropathological Correlation

In spite of the large number of AIDS patients, few studies have produced correlated results of modern neuro-imaging techniques with neuropathology as seen at post mortem. One patient with dementia and autopsy-proven HIVE had signs of atrophy and widespread patchy and confluent areas of high signal abnormality in the white matter at MRI (Jarvik et al, 1988). In a study of 21 patients with autopsy-proven HIVE (14 with dementia), cortical atrophy was found in

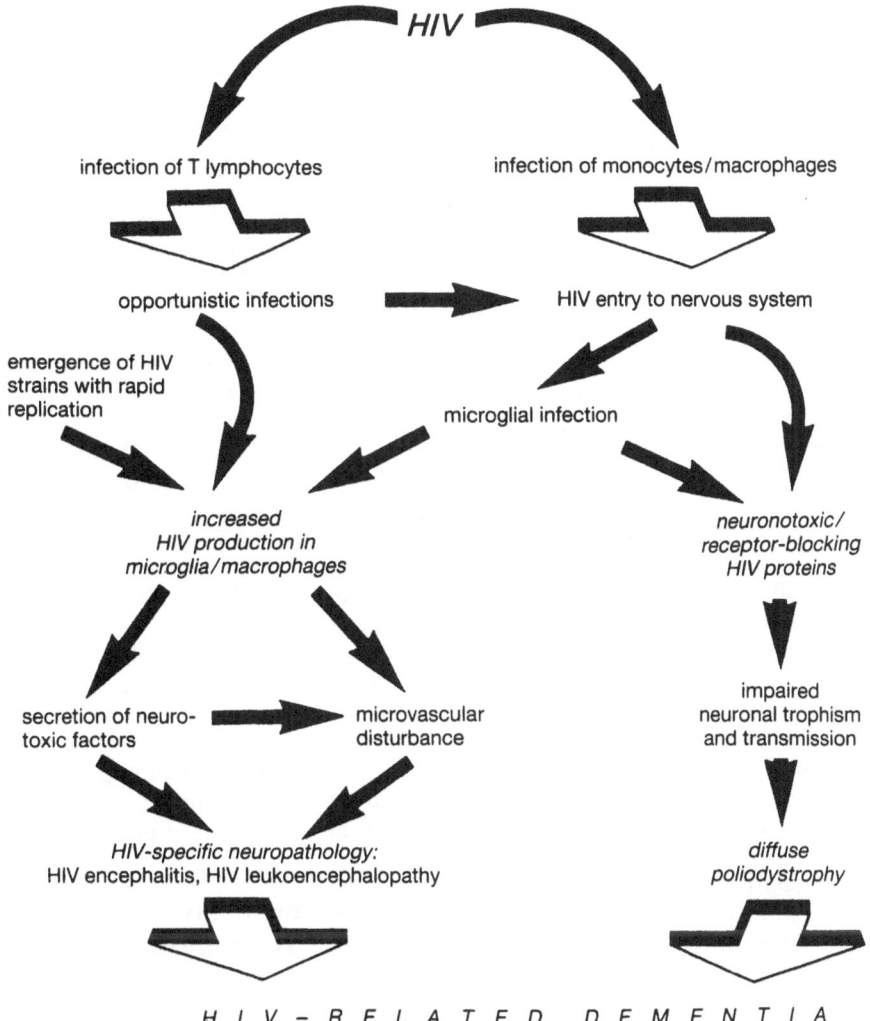

Figure 7.7. Unifying scheme illustrating the main pathogenetic pathways of the development of HIV-related dementia.

all patients, whereas white matter abnormalities, better seen with MRI than CT, were found in a minority. The authors concluded that the small HIVE lesions are not usually visualized either by CT or MRI (Post et al, 1988). The clinical diagnosis of HIVE usually antedated the neuroradiographic diagnosis (Post et al, 1988). In another study, five patients with HIV dementia showed high-signal lesions in the splenium of the corpus callosum and in the crura of the fornices using MRI. All patients had HIV-specific neuropathology at autopsy, although no specific changes were seen in the regions of the MRI abnormality (Kieburtz et al, 1990). Finally, McArthur (1987) reported that cerebral atrophy was found in 77 per cent of patients with AIDS-

related dementia using MRI. This atrophy may be due to white matter damage of HIV-specific neuropathology, as well as to grey matter involvement by DPD.

Pathogenesis

Although HIV-specific neuropathology has been sufficiently characterized in morphological terms, its pathogenesis remains to be elucidated. So far, all pathogenetic considerations must be regarded as hypotheses which urgently need to be substantiated. However, a unifying concept emerges which can make it possible to identify major factors and pathways, as summarized in Figure 7.7 and detailed in the text below.

Cellular Tropism of HIV in the Nervous System

There is general agreement that microglia and macrophages, which are considered by many to constitute a common lineage of haematogenous origin (Hickey and Kimura, 1988) (see also Chapter 11), are the essential targets of HIV in nervous tissue (see Figures 7.1–7.3, 7.5, 7.6) (Koenig et al, 1986; Wiley et al, 1986; Budka et al, 1987, 1988; Pumarola-Sune et al, 1987; Vazeux et al, 1987; Ward et al, 1987; Michaels et al, 1988b; Price et al, 1988a; Eilbott et al, 1989; Budka, 1990b). Microglia, brain macrophages and macrophage-derived multinucleated giant cells may possess 500–1500 copies of HIV RNA per cell; this amount is at least 10-fold higher than that found in blood leukocytes (Meltzer et al, 1990). The overall importance of macrophages for HIV infection of nervous tissues is paralleled by their paramount significance in the general pathogenesis of the HIV infection: macrophages are likely to be the first cell infected by HIV, are major reservoirs for HIV during all stages of the infection, regulate pace and intensity of disease progression and control virus latency in HIV-infected T-cells (Meltzer et al, 1990).

Infection by HIV of vascular endothelium within the CNS has been claimed by several groups (Wiley et al, 1986; Ward et al, 1987; Artigas et al, 1989b; Rhodes et al, 1989), but is insufficiently documented. Ultrastructural proof has not been provided. Immunohistochemical study by the author, including double-labelling with HIV and endothelial markers, did not show infection of endothelium, but suggested that labelled perivascular microglia/macrophages may be erroneously interpreted as endothelium (Budka, 1990b). Other unpublished immuno-histochemical observations by the author of the human CNS in a variety of pathological conditions demonstrated that cells labelled with anti-macrophage markers may mimic the light micro-scopic appearances of endothelial cells as far as shape and localization are concerned. Further-more, several reports claim, but do not unequi-vocally document, HIV infection of astroglia, oligodendroglia and neurons (Wiley et al, 1986; Györkey et al, 1987; Ward et al, 1987; Artigas et al, 1989b). It is clear that the shape of infected cells is not sufficient for their identification:

microglia with prominent formation of cellular processes may closely resemble glial and nerve cells (Budka, 1989, 1990b). Electron microscopic appearances of infected cells may be erroneously interpreted as neural as shown in Budka and Lassmann's discussion (1988) of the findings of Györkey et al (1987). Whenever immunocyto-logical double-labelling for HIV and specific cellular markers were used (Figure 7.1a,b), evi-dence could not be provided that cells other than microglia/macrophages were productively infected within the CNS in vivo (Budka, 1990b).

In contrast to the in vivo situation, HIV in-fection of true neuro-ectodermal cells has been demonstrated in vitro. It remains to be seen whether the infected cells of malignant gliomas (Cheng-Mayer et al, 1987; Dewhurst et al, 1987; Koyanagi et al, 1987; Harouse et al, 1989) or fetal cells (Cheng-Mayer et al, 1987; Harouse et al, 1989) are a useful model for the situation in the mature brain in vivo. It is remarkable that only a few viral transcripts and no cytopathogenetic effects were observed (Dewhurst et al, 1987). At present, it cannot be excluded that a hypothetical long-lasting latent HIV infection of glial and neuronal cells in vivo results in functional and metabolic changes including cell death. However, indirect mechanisms of nervous tissue damage seem more likely (see page 181).

Development of the Nervous Tissue Lesion by HIV: The Role of Virus Load

Since microglia/macrophages are distinctive elements in HIV neuropathology (Budka 1986, 1990b; Petito et al, 1986; Budka et al, 1987; Michaels et al, 1988b; Price et al, 1988a; Lang et al, 1989), the resulting disease can be con-sidered as macrophage-mediated (Budka, 1988, 1989). Perivascular arrangement of HIVE foci would support the possibility of transport of HIV-infected macrophages through the wall of blood vessels – the theory of the 'Trojan horse' – which may smuggle the virus into the CNS (Price et al, 1988a). Infection by HIV of vascular endothelium may be another source of viral entry into the CNS, but it remains to be proven beyond doubt (see above). Since the CNS seems to be in-fected early in the course of infection (Johnson et al, 1988), latently infected local cells, especially microglia, may represent an important local HIV reservoir until later stages of the infection, when HIV replication is activated by various stimuli. Isolated microglial cells producing HIV proteins, as immunohistochemically demonstrated in 10 per cent of AIDS brains (Budka, 1990b), are probably the initial phase of productive HIV infection within the CNS.

Subsequent development of HIV-specific neuropathology is regularly associated with prominent production of HIV (Figures 7.1–7.2, 7.5) (Budka, 1990b). Our immunohistochemical data correlate well with recent studies of HIV RNA expression by in situ hybridization in vacuolar myelopathy (Weiser et al, 1990), and of the polymerase chain reaction (PCR) technique showing a considerably higher proportion of unintegrated HIV DNA in HIVE than in non-HIVE brains (Pang et al, 1990); DNA integration is not necessary for expression of HIV protein products (Stevenson et al, 1990). It may be concluded that, at least in the terminal stages of the AIDS infection, the increasing viral load causes development of HIV-specific neuropathology on a basis of dose–effect relation (Budka, 1990b). Host factors such as decreasing anti-HIV immunity (Price et al, 1988a) and viral factors are likely to collaborate in this process. The beneficial effect of the anti-replicatory drug zidovudine in patients with encephalopathy (Pizzo et al, 1988), suggesting that neurological dysfunctions may be, to some extent, reversible (Schmitt et al, 1988), supports the important role of increased HIV production in the pathogenesis of brain damage by HIV.

Indirect Mechanisms of Damage

Secretion of Neurotoxic Factors

HIV-specific neuropathology can be interpreted as macrophage-mediated disease (see page 180). Macrophages infected by HIV may be intimately associated with damaged neural parenchyma, especially with myelin. This may involve secretion of neurotoxic factors by microglia/macrophages, including HIV proteins, enzymes, toxic oxygen metabolites or monokines (Maier et al, 1989; Meltzer et al, 1990). Interestingly, myelin damage, similar to vacuolation of CNS in AIDS patients, was recently observed in vitro after incubation of organotypic myelinated cultures with the monokine tumour necrosis factor (TNF) (Selmaj and Raine, 1988). TNF was shown to increase HIV replication in vitro (Ito et al, 1989). On the other hand, monocytes/macrophages may produce increasing amounts of TNF after infection with HIV (Molina et al, 1989). Therefore, a vicious circle of HIV production and secretion

of damaging monokines such as TNF might result in prominent HIV-tissue damage.

Microvascular Disturbance

As described above (see page 174), lesions of HIVE have a prominent perivascular location, probably following entry of HIV-infected haematogenous cells through the vessel walls, and may show microvascular pathology of various types. It was recently suggested that altered vascular permeability may be the cause of damage to myelin and axons in HIV-lep (Smith et al, 1990). It is conceivable that factors secreted by HIV-infected macrophages may influence capillary endothelium, as has been shown for cytokines including TNF (Beutler and Cerami, 1987).

Effects on Neuronal Trophism and Transmission

A few years ago, reports of a new type of tissue damage due to interference by HIV protein with neurotrophic factors or neurotransmitters gained much interest (Gray et al, 1988; Johnson et al, 1988; Michaels et al, 1988a; Price et al, 1988a; Budka, 1989). Peptide T, an octapeptide which is homologous with a conserved region of HIV envelope protein, is related to vasoactive intestinal polypeptide (VIP) (Ruff et al, 1987), which is an established neuronal transmitter and trophic factor in vitro. Thus, a scenario of competitive blocking of neuronal receptors by HIV proteins is conceivable. The HIV envelope protein was demonstrated to kill fetal nerve cells in vitro (Brenneman et al, 1988). Even the CSF of a patient with AIDS dementia had such an effect (Buzy et al, 1989). Most recently, the HIV envelope protein gp120 was found to increase intracellular free calcium, and to injure rodent retinal ganglion cells and hippocampal neurons in culture. This neuronal injury could be prevented by calcium channel antagonists (Dreyer et al, 1990). It is this context of a therapeutic perspective which attributes special importance to our recent finding of neuronal loss in the frontal cortex of AIDS patients (Ketzler et al, 1990).

Special HIV Variants in the Nervous System

Heterogeneity among HIV isolates is now a well recognized phenomenon (Cheng-Mayer and Levy, 1990), and the genes involved are discussed in Chapter 1. Since increased virus load is considered essential for the development of HIV-specific neuropathology (see page 181), it is important that HIV variants with a high replication rate dominate later in the course of the infection (Tersmette et al, 1989). This is in agreement with the development of clinical and neuropathological disease occurring mostly in the AIDS stage (Price et al, 1988a). Moreover, some isolates from the CNS (CNS tissue and/or CSF) were designated as 'neurotropic' and were distinct from isolates from the blood. There was: (1) efficient replication in blood macrophages, but (2) poor infectivity for T-cell lines; (3) reduced cytopathogenicity; (4) absent modulation of CD4 expression in infected cells; and (5) resistance against serum neutralization (Cheng-Mayer et al, 1989; Chiodi et al, 1989). Two HIV variants with different biological properties were isolated from the CNS of one patient (Koyanagi et al, 1987). Duplication of the negative regulatory gene *nef* was recently found in the brain of a patient with progressive dementia (Anand et al, 1989). Other preliminary molecular genetic data with recombinant viruses in the envelope region suggest that this gene plays a role in controlling cell tropism and cytopathology in vitro (Cheng-Mayer and Levy, 1990).

Contributory Role of Opportunistic Infections

Opportunistic infections such as those produced by CMV or JC virus may contribute to the development of HIV-associated neurological disease by transactivation of production of HIV; the reverse situation is also conceivable (Johnson et al, 1988; Nelson et al, 1990). Co-localization of CMV and HIV within the very same cell has been exceptionally described in the brain (Nelson et al, 1988) (see Figure 6.46). We have also observed co-localization of HIV and cryptococci within the same macrophage (Figure 7.3), indicating that phagocytic properties may be preserved in HIV-producing macrophages. Another mechanism by which opportunistic viruses may increase HIV dissemination to the nervous system is by attracting HIV-infected monocytes/macrophages to the site of infection (Nelson et al, 1990). How-

ever, a detailed immunocytochemical analysis of HIV expression within lesions of the CNS produced by opportunists found local HIV co-infection to be rare (Budka, 1990b), and opportunists, such as CMV, were never found in HIV-specific lesions (Budka et al, 1987; Budka, 1989, 1990b; Schmidbauer et al, 1989a,b). These results indicate that local transactivation by other viruses may occur, but does not seem to be a common event.

Immune Pathogenesis

The perivascular localization of the myelin damage in foci of HIVE is similar to the changes observed in perivenous/post-infectious encephalitis – the immune-mediated complication of many viral diseases (measles, rubella, etc). However, evidence for a humoral immune reaction is lacking in AIDS brains (Lenhardt and Wiley, 1989). In addition, there are hardly any signs for a cellular immune reaction: lymphocytes are sparse in HIV-specific neuropathology and possible effector mechanisms would have to include macrophages, which have a dominating pathogenetic role in other theories.

Metabolic Pathogenesis

Most recently, a decrease of methyl-group carriers was found in the CSF of HIV-infected children with neurological complications, and defective methylation was suggested as cause of neurological damage by HIV (Surtees et al, 1990). Since methyl-group transfer also involves folate and vitamin B_{12}, impairment of methylation might manifest in a fashion similar to folate and B_{12} deficiencies, which cause subacute combined degeneration of the spinal cord. This is an attractive new possibility for explaining the development of VM or vacuolar leukoencephalopathy. Moreover, AIDS patients may be deficient in vitamin B_{12}. Such deficiency was found most recently to contribute to impaired cognitive function in even generally asymptomatic HIV-infected individuals (Baum et al, 1990). Another metabolic pathway of HIV neuropathogenicity could involve synthesis of phospholipids, which are especially abundant in the nervous tissue. Depression of phospholipids synthesis and perturbation of the cellular membrane have been described as important mechanisms of HIV cytotoxicity in non-neuronal cell cultures (Lynn et al, 1988).

Acknowledgement. This work was supported by the Austrian Fund for the Advancement of Scientific Research (P8672-MED).

References

American Psychiatric Association. (1980) Committee on Nomenclature and Statistics: Diagnostic and statistical manual of mental disorders, 3rd edn. American Psychiatric Association: Washington, DC.

Anand R, Thayer R, Srinivasan A et al. (1989) Biological and molecular characterization of human immunodeficiency virus (HIV-1BR) from the brain of a patient with progressive dementia. Virology 168:79–89.

Artigas J, Niedobitek F, Grosse G et al. (1989a) Spongiform encephalopathy in AIDS dementia complex: report of five cases. J AIDS 2:374–81.

Artigas J, Freund K, Grosse G et al. (1989b) Immunhistochemische Darstellung von HIV-p24-Antigen in formalinfixiertem und paraffineingebettetem Hirn- und Rückenmarksgewebe. Pathologe 10:61–3.

Baum MK, Beach R, Morgan R et al. (1990) Vitamin B_{12} and cognitive function in HIV infection. VIth Internat Conf AIDS, San Francisco, CA.

Beutler B, Cerami A. (1987) Cachectin: more than a tumor necrosis factor. N Engl J Med 316:379–85.

Brenneman DE, Westbrook GL, Fitzgerald SP et al. (1988) Neuronal cell killing by the envelope protein of HIV and its prevention by vasoactive intestinal peptide. Nature 335:639–42.

Budka H. (1986) Multinucleated giant cells in brain: a hallmark of the acquired immune deficiency syndrome (AIDS). Acta Neuropathol 69:253–8.

Budka H. (1987) Das morphologische Korrelat der HIV-Infektion des Gehirns. In: Fischer P-A, Schlote W. AIDS und Nervensystem. Springer: Berlin, 117–32.

Budka H. (1988) Pathogenesis of human immunodeficiency virus (HIV)-associated brain lesions: a neuropathological evaluation. Ann NY Acad Sci 540:630–3.

Budka H. (1989) Human immunodeficiency virus (HIV)-induced disease of the central nervous system: pathology and implications for pathogenesis. Acta Neuropathol 77:225–36.

Budka H. (1990a) Neuropathology of AIDS. In: Chopra JS et al, ed. Advances in neurology, Proc XIVth World Congr Neurol, Int Congr Ser 883, Elsevier: Amsterdam 193–202.

Budka H. (1990b) Human immunodeficiency virus (HIV) envelope and core proteins in CNS tissues of patients with the acquired immune deficiency syndrome (AIDS). Acta Neuropathol 79:611–19.

Budka H, Lassmann H. (1988) Human immunodeficiency virus (HIV) in glial cells? J Infect Dis 157:203.

Budka H, Costanzi G, Cristina S et al. (1987) Brain pathology induced by infection with the human immunodeficiency virus (HIV). A histological, immunocytochemical, and electron microscopical study of 100 autopsy cases. Acta Neuropathol 75:185–98.

Budka H, Maier H, Pohl P. (1988) Human immunodeficiency virus in vacuolar myelopathy of the acquired immunodeficiency syndrome. N Engl J Med 319:1667–8.

Budka H, Wiley CA, Kleihues P et al. (1991) Consensus report. HIV-associated disease of the nervous system:

review of nomenclature and proposal for neuropathology-based terminology. Brain Pathol 1:142–50.

Buzy JM, Brenneman DE, Siegal FP et al. (1989) Cerebrospinal fluid from cognitively impaired patient with acquired immunodeficiency syndrome shows gp120-like neuronal killing in vitro. Am J Med 87:361–2.

Cheng-Mayer C, Levy JA. (1990) Human immunodeficiency virus infection of the CNS: characterization of 'neurotropic' strains. Curr Top Microbiol Immunol 160:145–56.

Cheng-Mayer C, Rutka JT, Rosenblum ML et al. (1987) Human immunodeficiency virus can productively infect cultured human glial cells. Proc Natl Acad Sci USA 84:3526–30.

Cheng-Mayer C, Weiss C, Seto D et al. (1989) Isolates of human immunodeficiency virus type 1 from the brain may constitute a special group of the AIDS virus. Proc Natl Acad Sci USA 86:8575–9.

Chiodi F, Valentin A, Keys B et al. (1989) Biological characterization of paired human immunodeficiency virus type 1 isolates from blood and cerebrospinal fluid. Virology 173:178–187.

Ciardi A, Sinclair E, Scaravilli F et al. (1990) The involvement of the cerebral cortex in HIV encephalopathy: a morphological and immunohistochemical study. Acta Neuropathol 80:92–4.

Dewhurst S, Sakai K, Bresser J et al. (1987) Persistent productive infection of human glial cells by human immunodeficiency virus (HIV) and by infectious molecular clones of HIV. J Virol 61:3774–82.

Dreyer EB, Kaiser PK, Offermann JT et al. (1990) HIV-1 coat protein neurotoxicity prevented by calcium channel antagonists. Science 248:364–7.

Eilbott DJ, Peress N, Burger H et al. (1989) Human immunodeficiency virus type 1 in spinal cords of acquired immunodeficiency syndrome patients with myelopathy: expression and replication in macrophages. Proc Natl Acad Sci USA 86:3337–41.

Gabuzda DH, Ho DD, de la Monte SM et al. (1986) Immunohistochemical identification of HTLV-III antigen in brains of patients with AIDS. Ann Neurol 20:289–95.

Giangaspero F, Scanabissi E, Baldacci MC et al. (1989) Massive neuronal destruction in human immunodeficiency virus (HIV) encephalitis. A clinico-pathological study of a pediatric case. Acta Neuropathol 78:662–5.

Gray F, Gherardi R, Scaravilli F. (1988) The neuropathology of the acquired immune deficiency syndrome (AIDS). A review. Brain 11:245–66.

Györkey F, Melnick JL, Györkey P. (1987) Human immunodeficiency virus in brain biopsies of patients with AIDS and progressive encephalopathy. J Infect Dis 155:870–6.

Harouse JM, Kunsch C, Hartle HT et al. (1989) CD4-independent infection of human neural cells by human immunodeficiency virus type 1. J Virol 63:2527–33.

Hénin D, Hauw J-J. (1989) The neuropathology of AIDS. In: McKendall RR. Handbook of clinieal neurology, vol 12(56). Elsevier: Amsterdam, 507–24.

Hickey WF, Kimura H (1988) Perivascular microglial cells of the CNS are bone marrow-derived and present antigen in vivo. Science 239:290–2.

Ito M, Baba M, Sato A et al. (1989) Tumor necrosis factor enhances replication of human immunodeficiency virus (HIV) in vitro. Biochem Biophys Res Commun 158:307–12.

Janssen RS, Cornblath DR, Epstein LG et al. (1991) Nomenclature and research case definitions for neurological manifestations of human immunodeficiency virus type-1 (HIV-1) infection. Report of a working group of the American Academy of Neurology AIDS Task Force.

Neurology 41:778–85.

Jarvik JG, Hesselink JR, Kennedy C et al. (1988) Acquired immunodeficiency syndrome. Magnetic resonance patterns of brain involvement with pathologic correlation. Arch Neurol 45:731–6.

Johnson RT, McArthur JC, Narayan O. (1988) The neurobiology of human immunodeficiency virus infections. FASEB J 2:2970–81.

Jones HR, Ho DD, Forgacs P et al. (1988) Acute fulminating fatal leukoencephalopathy as the only manifestation of human immunodeficiency virus infection. Ann Neurol 23:519–22.

Ketzler S, Weis S, Haug H et al. (1990) Loss of neurons in the frontal cortex in AIDS brains. Acta Neuropathol 80:92–5.

Kieburtz KD, Ketonen L, Zettelmaier AE et al. (1990) Magnetic resonance imaging findings in HIV cognitive impairment. Arch Neurol 47:643–5.

Kleihues P, Lang W, Burger PC et al. (1985) Progressive diffuse leukoencephalopathy in patients with acquired immune deficiency syndrome (AIDS). Acta Neuropathol 68:333–9.

Koenig S, Gendelman HE, Orenstein JM et al. (1986) Detection of AIDS virus in macrophages in brain tissue from AIDS patients with encephalopathy. Science 233:1089–93.

Koyanagi Y, Miles S, Mitsuyasu RT et al. (1987) Dual infection of the central nervous system by AIDS viruses with distinct cellular tropisms. Science 236:819–22.

Lang W, Miklossy J, Deruaz JP et al. (1989) Neuropathology of the acquired immune deficiency syndrome (AIDS): a report of 135 consecutive autopsy cases from Switzerland. Acta Neuropathol 77:379–90.

Lenhardt TM, Wiley CA. (1989) Absence of humorally mediated damage within the central nervous system of AIDS patients. Neurology 39:278–80.

Lenhardt TM, Super MA, Wiley CA. (1988) Neuropathological changes in an asymptomatic HIV seropositive man. Ann Neurol 23:209–10.

Levy RM, Bredesen DE. (1988) Central nervous system dysfunction in acquired immunodeficiency syndrome. In: Rosenblum ML, Levy RM, Bredesen DE, ed. AIDS and the nervous system. Raven Press: New York, 29–63.

Lynn WS, Tweedale A, Cloyd MW. (1988) Human immunodeficiency virus (HIV-1) cytotoxicity: perturbation of the cell membrane and depression of phospholipid synthesis. Virology 163:43–51.

Maier H, Budka H, Lassmann H et al. (1989) Vacuolar myelopathy with multinucleated giant cells in the acquired immune deficiency syndrome (AIDS). Light and electron microscopic distribution of human immunodeficiency virus (HIV) antigens. Acta Neuropathol 78:497–503.

McArthur JC. (1987) Neurologic manifestations of AIDS. Medicine (Baltimore) 66:407–37.

McArthur JC, Becker PS Parisi JE et al. (1989) Neuropathological changes in early HIV-1 dementia. Ann Neurol 26:681–4.

Meltzer MS, Skillman DR, Gomatos PJ et al. (1990) Role of mononuclear phagocytes in the pathogenesis of human immunodeficiency virus infection. Annu Rev Immunol 8:169–94.

Meyenhofer MF, Epstein LG, Cho E-S et al. (1987) Ultrastructural morphology and intracellular production of human immunodeficiency virus (HIV) in brain. J Neuropathol Exp Neurol 46:474–84.

Michaels J, Sharer LR, Epstein LG. (1988a) Human immunodeficiency virus type 1 (HIV-1) infection of the nervous system: a review. Immunodefic Rev 1:71–104.

Michaels J, Price RW, Rosenblum MK. (1988b) Microglia in the giant cell encephalitis of acquired immune deficiency

syndrome: proliferation, infection and fusion. Acta Neuropathol 76:373–9.

Miller EN, Selnes OA, McArthur J et al. (1990) Neuropsychological performance in HIV-1-infected homosexual men: the multicenter AIDS cohort study (MACS). Neurology 40:197–203.

Molina J-M, Scadden DT, Byrn R et al. (1989) Production of tumor necrosis factor alpha and interleukin 1β by monocytic cells infected with human immunodeficiency virus. J Clin Invest 84:733–7.

Monte de la SM, Ho DD, Schooley RT et al. (1987) Subacute encephalomyelitis of AIDS and its relation to HTLV-III infection. Neurology 37:562–9.

Navia BA, Price RW. (1987) The acquired immunodeficiency syndrome dementia complex as the presenting or sole manifestation of human immunodeficiency virus infection. Arch Neurol 44:65–9.

Navia BA, Cho E-S, Petito CK et al. (1986a) The AIDS dementia complex: II. Neuropathology. Ann Neurol 19:525–35.

Navia BA, Jordan BD, Price RW. (1986b) The AIDS dementia complex: I. Clinical features. Ann Neurol 19:517–24.

Nelson JA, Reynolds-Kohler C, Oldstone MBA et al. (1988) HIV and HCMV coinfect brain cells in patients with AIDS. Virology 165:286–90.

Nelson JA, Ghazal P, Wiley CA. (1990) Role of opportunistic viral infections in AIDS. AIDS 4:1–10.

Pang S, Koyanagi Y, Miles S et al. (1990) High levels of unintegrated HIV-1 DNA in brain tissue of AIDS dementia patients. Nature 343:85–9.

Petito CK, Cho E-S, Lemann W et al. (1986) Neuropathology of acquired immunodeficiency syndrome (AIDS): an autopsy review. J Neuropathol Exp Neurol 45:635–46.

Pizzo PA, Eddy J, Falloon J et al. (1988) Effect of continuous intravenous infusion of zidovudine (AZT) in children with symptomatic HIV infection. N Engl J Med 319:889–96.

Popovic M, Sarngadharan MG, Read E et al. (1984) Detection, isolation, and continuous production of cytopathic retroviruses (HTLV-III) from patients with AIDS and pre-AIDS. Science 224:497–500.

Post MJD, Tate LG, Quencer RM et al. (1988) CT, MR, and pathology in HIV encephalitis and meningitis. AJNR 9:469–76.

Price RW, Brew BJ. (1988) The AIDS dementia complex. J Infect Dis 158:1079–83.

Price RW, Brew B, Sidtis J et al. (1988a) The brain in AIDS: central nervous system HIV-1 infection and AIDS dementia complex. Science 239:586–92.

Price RW, Sidtis JJ, Navia BA et al. (1988b) The AIDS dementia complex. In: Rosenblum ML, Levy RM, Bredesen DE, ed. AIDS and the nervous system. Raven Press: New York, 203–19.

Pumarola-Sune T, Navia BA, Cordon-Cardo C et al. (1987) HIV antigen in the brains of patients with the AIDS dementia complex. Ann Neurol 21:490–6.

Rhodes RH. (1987) Histopathology of the central nervous system in the acquired immune deficiency syndrome. Hum Pathol 18:636–43.

Rhodes RH, Ward JM, Cowan RP et al. (1989) Immunohistochemical localization of human immunodeficiency viral antigens in formalin-fixed spinal cords with AIDS myelopathy. Clin Neuropathol 8:22–7.

Ruff MR, Martin BM, Ginns EI et al. (1987) CD4 receptor binding peptides that block HIV infectivity cause human monocyte chemotaxis. Relationship to vasoactive intestinal polypeptide. FEBS Lett 211:17–22.

Schmidbauer M, Budka H, Ulrich W et al. (1989a) Cytomegalovirus (CMV) disease of the brain in AIDS and connatal infection: a comparative study by histology, immunocytochemistry, and in situ DNA hybridization. Acta Neuropathol 79:286–293.

Schmidbauer M, Budka H, Ambros P. (1989b) Herpes simplex virus (HSV) DNA in microglial nodular brainstem encephalitis. J Neuropathol Exp Neurol 48:645–52.

Schmidbauer M, Budka H, Shah KV. (1990a) Progressive multifocal leukoencephalopathy (PML) in AIDS and in the pre-AIDS era. A neuropathologic comparison using immunocytochemistry and in situ DNA hybridization for virus detection. Acta Neuropathol 80:375–80.

Schmidbauer M, Budka H, Okeda R et al. (1990b) Multifocal vacuolar leukoencephalopathy: a distinct HIV-associated lesion of the brain. Neuropathol Appl Neurobiol 16: 437–43.

Schmitt FA, Bigley JW, McKinnis R et al. and the AZT Collaborative Working Group. (1988) Neuropsychological outcome of zidovudine (AZT) treatment of patients with AIDS and AIDS-related complex. N Engl J Med 319: 1573–8.

Seitelberger F. (1975) General neuropathology of the degenerative diseases of the central nervous system. In: Vinken PJ, Bruyn GW. Handbook of clinical neurology, vol 21. North Holland: Amsterdam, 43–71.

Selmaj KW, Raine CS. (1988) Tumor necrosis factor mediates myelin and oligodendrocyte damage in vitro. Ann Neurol 23:339–46.

Selnes OA, Miller E, McArthur J et al. (1990) The Multicenter AIDS Cohort Study: HIV-1 infection: no evidence of cognitive decline during the asymptomatic stages. Neurology 40:204–8.

Sharer LR, Cho E-S, Epstein LG. (1985) Multinucleated giant cells and HTLV-III in AIDS encephalopathy. Hum Pathol 16:760.

Sharer LR, Epstein LG, Cho E-S et al. (1986) Pathologic features of AIDS encephalopathy in children: evidence for LAV/HTLV-III infection of brain. Hum Pathol 17:271–84.

Shaw GM, Harper ME, Hahn BH et al. (1985) HTLV-III infection in brains of children and adults with AIDS encephalopathy. Science 227:177–82.

Sidtis JJ, Price RW. (1990) Early HIV-1 infection and the AIDS dementia complex. Neurology 40:323–6.

Smith TW, DeGirolami U, Hénin D et al. (1990) Human immunodeficiency virus (HIV) leukoencephalopathy and the microcirculation. J Neuropathol Exp Neurol 49:357–70.

Snider WD, Simpson DM, Nielsen S et al. (1983) Neurological complications of acquired immune deficiency syndrome: analysis of 50 patients. Ann Neurol 14:403–18.

Stevenson M, Haggerty S, Lamonica CA et al. (1990) Integration is not necessary for expression of human immunodeficiency virus type 1 protein products. J Virol 64:2421–5.

Surtees R, Hyland K, Smith I. (1990) Central-nervous-system methyl-group metabolism in children with neurological complications of HIV infection. Lancet 335:619–21.

Tersmette M, Gruters RA, de Wolf F et al. (1989) Evidence for a role of virulent human immunodeficiency virus (HIV) variants in the pathogenesis of acquired immunodeficiency syndrome: studies on sequential HIV isolates. J Virol 63: 2118–25.

Vazeux R, Brousse N, Jarry A et al. (1987) AIDS subacute encephalitis. Identification of HIV-infected cells. Am J Pathol 126:403–10.

Vinters HV, Anders KH. (1990) Neuropathology of AIDS. CRC Press: Boca Raton.

Ward JM, O'Leary TJ, Baskin GB et al. (1987) Immunohistochemical localization of human and simian immunodeficiency viral antigens in fixed tissue sections. Am J Pathol 127:199–205.

Weiser B, Peress N, La Neve D et al. (1990) Human immunodeficiency virus type 1 expression in the central nervous system correlates directly with extent of disease. Proc Natl Acad Sci USA 87:3997–4001.

Whitehouse P. (1986) The concept of subcortical and cortical dementia: another look. Ann Neurol 19:1–6.

Wiley CA, Schrier RD, Nelson JA et al. (1986) Cellular localization of human immunodeficiency virus infection within the brains of acquired immune deficiency syndrome patients. Proc Natl Acad Sci USA 83:7089–93.

Wilkie FL, Eisdorfer C, Morgan R et al. (1990) Cognition in early human immunodeficiency virus infection. Arch Neurol 47:433–40.

8 Myelopathies

C.K. Petito

Introduction

Spinal cord disease is frequently seen at post-mortem examinations of patients with human immunodeficiency virus (HIV) infection and the acquired immunodeficiency syndrome (AIDS) (Anders et al, 1986; Petito et al, 1986). Routine examination of the spinal cord discloses microscopic abnormalities in almost 50 per cent of all AIDS cases, whereas spinal cord abnormalities are only found in approximately 10 per cent of non-AIDS cases (Kamin and Petito, 1991). If this part of the central nervous system (CNS) is examined only when there are clinical symptoms and signs referable to the spinal cord, many lesions may be missed since clinical myelopathy may be absent, not detected in severely-ill patients, or inseparable from symptoms due to disease elsewhere in the peripheral or central nervous systems. The frequency and distribution of CNS diseases in AIDS necessitates increased numbers of sections submitted for routine microscopic examination of the spinal cord. We therefore submit sections at two thoracic and one lumbo-sacral levels and additional sections are obtained as indicated.

Diseases of the spinal cord were encountered in 48 per cent of all 178 HIV-infected patients who were autopsied at our hospital over a 9-year period, and who had a postmortem examination

of the spinal cord (Table 8.1). Vacuolar myelopathy (Petito et al, 1985) was the most common disorder and was present in 29 per cent of all spinal cords. Not surprisingly, infections were the second most common abnormality, although their incidence was lower than that found in brain, and they were rarely seen in the absence of similar infection in the brain. Routine histologic examination, exclusive of immunohistochemistry or in

Table 8.1. Spinal cord diseases in 178 patients[a] with AIDS[b]

Disease	No. of patients	%[b]
HIV-related myelopathies	53	30
Vacuolar myelopathy	51	29
Gracile tract degeneration	2	1
Microglial nodule myelitis	17	10
HIV myelitis	9	5
Herpes virus myelitis	14	8
CMV	9	5
HSV	3	2
VZV	2	1
Infectious myelitis (others)	11	7
Fungal	8	4
Bacterial	1	1
M. tuberculosis	2	1
Toxoplasmosis	1	1
Lymphoma	4	2
Normal	93	52

[a] Of whom 170 were adults and 8 were children.
[b] The total percentage is greater than 100 since many patients had more than one disease.

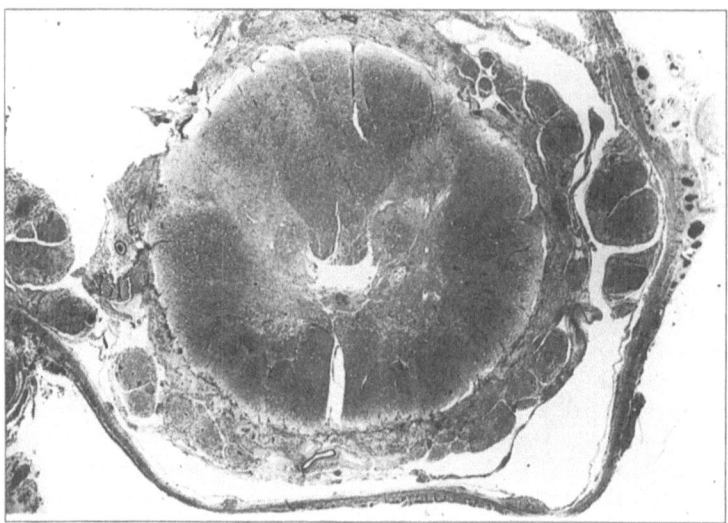

Figure 8.1. Lumbar spinal cord with leptomeningitis due to *Mycobacterium tuberculosis*. The nerve roots are encased by dense inflammation and are focally demyelinated. The cord periphery is pale and contains small foci of granulomatous inflammation and necrosis. Luxol–fast blue–hematoxylin–eosin, ×5.

situ hybridization, identified specific viral myelitis in 13 per cent of patients: HIV and cytomegalovirus (CMV) were found in 9 patients each, Herpes simplex virus (HSV) in 3 patients and Varicella zoster virus (VZV) was found in 2 patients. An additional 10 per cent of patients had scattered microglial nodules in the spinal cord, the etiology of which was not determined since characteristic micro-organisms, intranuclear inclusions and multinucleated cells were absent. Fungal, parasitic and bacterial infections were found in only 7 per cent of all spinal cords (Figure 8.1), and only sporadic case reports of these infections are found in the literature (Mehren et al, 1988; Woolsey et al, 1988).

The clinical incidence of spinal cord diseases in AIDS patients is much lower than its autopsy incidence. In large series detailing the neurological complications of AIDS, myelopathies were described in less than 10 per cent of patients. The low incidence is due, in part, to the fact that clinical symptoms and signs of spinal cord disease may not be apparent or investigated in terminally-ill bedridden patients, or may be indistinguishable from disease in the brain or the peripheral nervous system. Snider et al (1983), who first described the neurological complications of AIDS, found no cases of myelopathies in any of their 50 patients. McArthur (1987) reviewed the neurological complications in 186 HIV-infected patients and found clinical myelopathy in 13 (7 per cent). Vacuolar myelopathy was the presumed or confirmed diagnosis in 12 of these

patients, and necrotizing myelitis consistent with VZV in 1. Myelopathies have not been reported, or are rare, in HIV-positive individuals who are otherwise healthy (McArthur et al, 1989), or who have lymphadenopathy syndrome (Janssen et al, 1988) or AIDS-related complex (Janssen et al, 1989). One case report describes the onset of an acute myelopathy at the time of HIV seroconversion (Denning et al, 1987).

The majority of clinical myelopathies are caused by vacuolar myelopathy. It is the most frequent spinal cord disease in AIDS patients, and up to 60 per cent of patients with vacuolar myelopathy are symptomatic (Petito et al, 1985). Necrotizing myelitis due to herpes virus infection can also be symptomatic, and several case reports have described a rapidly progressive ascending myelopathy secondary to CMV or HSV. Although the total number of AIDS patients with clinical myelitis due to herpes virus infection is low, the relative incidence of symptomatic myelitis is high in herpes-infected spinal cords. Clinically symptomatic herpes virus myelitis was seen in only 3 per cent of the 178 AIDS patients examined at autopsy at our institution (Table 8.1). In contrast, clinical myelopathy was present in 4 of the 14 patients (29 per cent) with CMV, CMV/HSV or VZV myelitis. One of the 2 patients with VZV myelitis was reportedly asymptomatic, but would have had spinal cord signs if examined, since a partially organizing transverse necrosis of almost the entire cord at T4 level was found at postmortem examination.

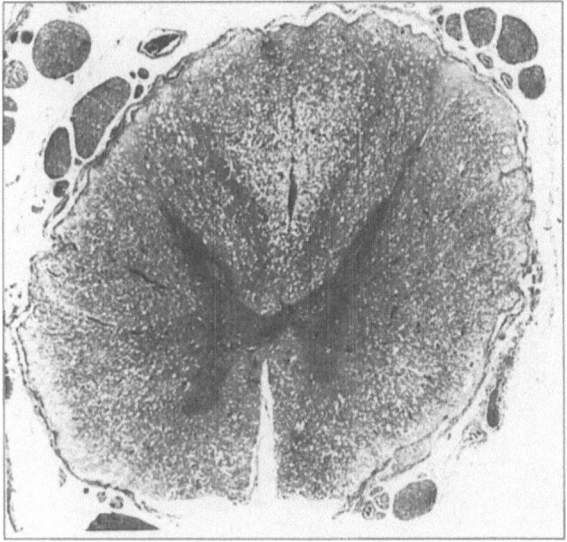

Figure 8.2. Thoracic spinal cord with severe vacuolar myelopathy. Numerous vacuoles are scattered throughout posterior, lateral and anterior columns, and are concentrated in the fasciculus gracilis. Luxol–fast blue–hematoxylin–eosin, ×10.

HIV-Related Myelopathies

The following myelopathies are closely associated with HIV infection. They are not confined to the AIDS population and rarely may occur in other patients, especially those who are immuno-compromised. However, their etiology is uncertain and may differ among AIDS patients. The relationship between these myelopathies and direct HIV infection of the spinal cord is not clear.

Vacuolar Myelopathy

Vacuolar myelopathy is the most common post-mortem abnormality of the spinal cord in AIDS patients (Goldstick et al, 1985; Petito et al, 1985). It is a disorder that is characterized by spongy degeneration of myelin sheaths in the posterior and lateral columns, and may resemble subacute combined degeneration of the spinal cord associated with vitamin B_{12} deficiency (Pant et al, 1968). It is mainly a disease of adult patients with AIDS, in whom it has an incidence at postmortem examination ranging from 18 to 30 per cent. Vacuolar myelopathy is rare in children. It was found in only 2 of 24 children with AIDS examined by Sharer et al (1990) and in none of the 15 children reviewed by Dickson et al (1989), although the oldest child in the latter series had focal vacuolation in the posterior columns.

Vacuolar myelopathy may be the sole spinal cord disease present, as depicted in Figure 8.2, which is from a 29-year-old man with severe vacuolar myelopathy, who also had a primary brain lymphoma. It may also be associated with co-existent disease in the spinal cord, as shown in Figure 8.3, which was obtained from a 45-year-old man with severe vacuolar myelopathy, who

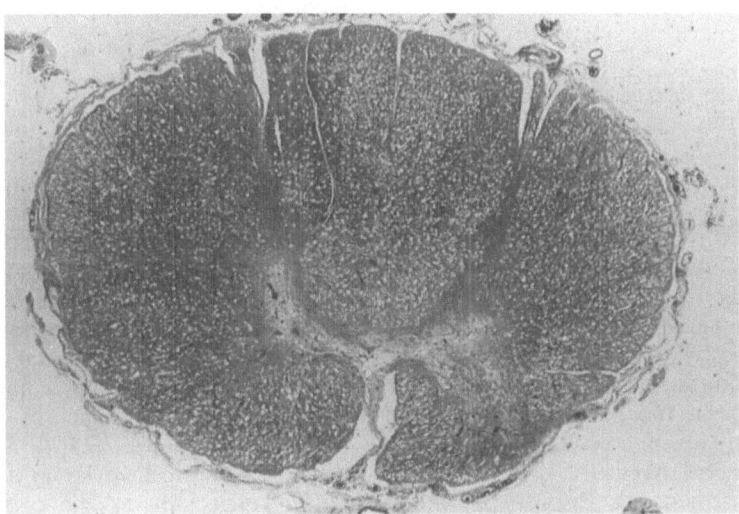

Figure 8.3. Lumbar spinal cord with severe vacuolar myelopathy and CMV myeloradiculitis. Numerous vacuoles are scattered throughout the white matter. Cystic necrosis of central gray matter and anterior horn is secondary to CMV myelitis. Luxol–fast blue–hematoxylin, ×9.

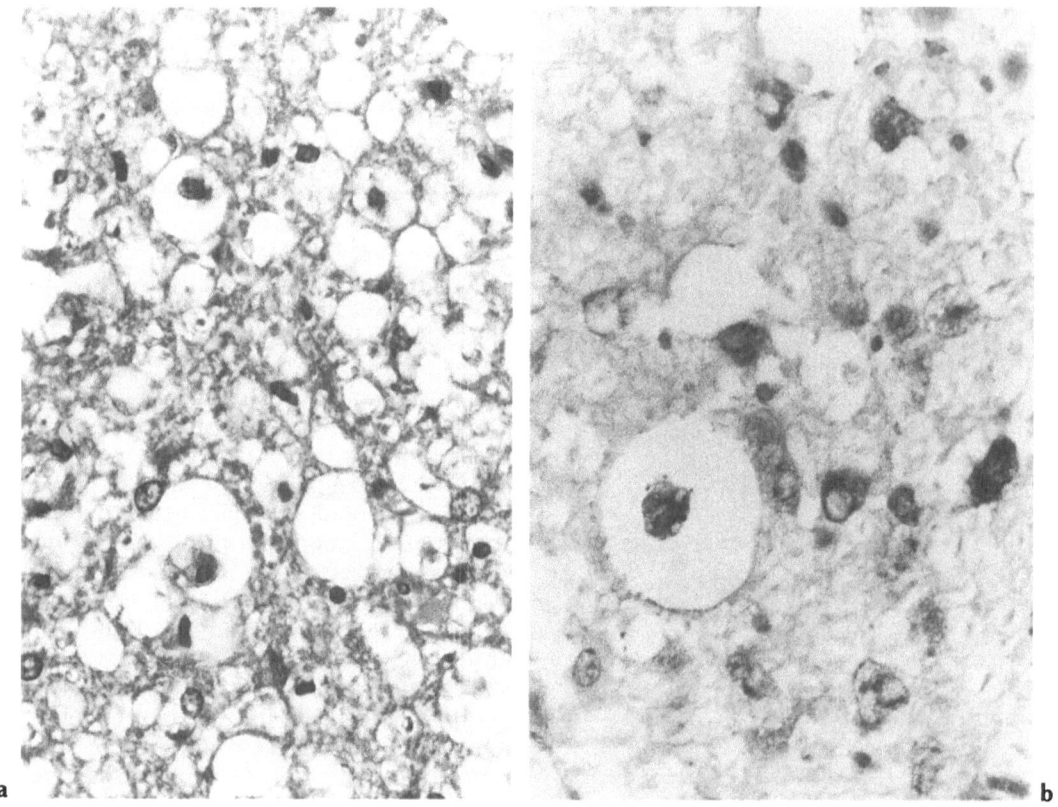

Figure 8.4. a Vacuolar myelopathy with lipid-laden macrophages in vacuoles and parenchyma. Hematoxylin– eosin, ×160. b Vacuolar myelopathy. Macrophages stained with biotinylated *Ricinus communis* agglutinin-I. Hematoxylin, ×400.

also had a CMV myeloradiculitis, as well as CMV ventriculo-encephalitis, HIV encephalitis and healed brain toxoplasmosis.

Gross examination of the spinal cord is normal in mild and moderately severe cases, whereas swelling and a white discoloration of the lateral or posterior columns can be detected in cases of severe vacuolar myelopathy. Microscopic examination reveals vacuoles within spinal cord white matter with maximum disease at the mid-to-lower thoracic cord; the posterior or lateral columns are more involved than the anterolateral or anterior columns. In some cases, the fasciculus gracilis is more severely involved (Figure 8.2) than the remaining white matter of the spinal cord. The vacuolation is confined to central myelin and, when present in the posterior nerve root entry zone, abruptly terminates at the junction of central with peripheral myelin.

We initially classified this disorder as mild, moderate or severe (Grade I, II or III) based on the greatest number of vacuoles per transverse section, in order to correlate clinical symptoms and signs with pathological changes (Petito et al, 1985). However, classification according to site of maximum damage may be more important than overall disease severity in offering important clues regarding etiology (Grafe and Wiley, 1989).

The vacuoles are sharply-outlined spaces surrounded by a thin myelin sheath and an earlier ultrastructural study suggests that they are formed by swellings within the myelin sheaths (Petito et al, 1985). Macrophage infiltration and reactive astrocytosis are proportional to the severity of disease and may be infrequent in cases of mild or even moderate severity (Figure 8.4a). Identification of monocytes or macrophages can be helped by the use of special stains, such as non-specific esterase or histochemical detection with macrophage markers such as *Ricinus communis agglutinin*-I (Figure 8.4b). Since postmortem artifacts can produce spongy changes in white matter, especially at the cord periphery, it is useful to confine the diagnosis to those cases in which at least a single macrophage can be identified.

Axons are relatively preserved in mild cases, although occasional spheroids can be found with careful examination. A preliminary report suggests that some cases of vacuolar myelopathy are accompanied by demyelination and remyelination without axonal abnormalities (Becker et al, 1989). Axonal loss can be prominent in regions of confluent vacuolation, and secondary wallerian degeneration can be found in posterior columns of upper thoracic or cervical cord and the corticospinal tracts at lower lumbosacral levels. Ultrastructural studies have suggested that the mechanism of axonal damage may be axonal compression by the intramyelin sheath swelling (Petito et al, 1985).

It is likely that vacuolar myelopathy is reversible in the early stages of disease when myelin vacuolation is present without axonal destruction. Experimental models of intramyelin sheath swelling produced by intoxicants such as lead or hexachlorophene indicate that removal of the inciting agent reverses the swelling. Likewise, clinical disorders of intramyelin sheath swelling in the spinal cord due to deficiencies of vitamins B_1, B_{12} and folic acid, or intoxication with nitrous oxide, are also reversible in the early stages of disease.

Clinical symptoms and signs of vacuolar myelopathy correlate well with the location, as well as with the severity, of the lesions. Table 8.2 summarizes the clinico-pathological correlations in vacuolar myelopathy, and is adapted from an earlier study of 20 AIDS patients with the disease (Petito et al, 1985). Myelopathic symptoms included spastic paraparesis, ataxia and incontinence, and were present in all 5 patients with Grade III disease, 4 of the 5 patients with Grade II disease, and 3 of the 8 patients with Grade I disease. Peripheral neuropathies were common in these patients.

Table 8.2. Clinical features in 20 patients with vacuolar myelopathy

Disease severity	No. of patients	Leg weakness	UMN[a] signs	Ataxia	Incontinence
Mild	8	1	–	1	3
Moderate	7	5	4	3	5
Severe	5	5	4	4	4
Total	20	11	8	8	12

Adapted from Petito et al (1985).
[a] UMN, upper motor neuron.

The etiology of vacuolar myelopathy is unclear, and may not be the same in all patients. By itself, intramyelin sheath swelling is a non-specific re-

sponse to a variety of stimuli including toxins, nutritional disturbances and metabolic dysfunction. Myelin vacuolation can accompany infection and often occurs in regions immediately adjacent to infectious foci, as illustrated later in this chapter. In our original report (Petito et al, 1985), vacuolar myelopathy was described as diffuse vacuolation of posterior and lateral columns of varying degrees of severity unaccompanied by inflammation other than occasional macrophages in areas of myelin destruction. Cases in which the vacuoles are concentrated in the lateral or posterior columns may have different etiologies (Grafe and Wiley, 1989). Cases in which vacuoles are restricted to discrete foci may be secondary to adjacent infection, as described below.

A metabolic cause for vacuolar myelopathy in some patients is an attractive hypothesis since the pathology of this disorder closely resembles the subacute combined degeneration of vitamin B_{12} deficiency, and since as many as 25 per cent of all AIDS patients have deficient absorption and reduced serum levels of this vitamin (Burkes et al, 1987; Herbert, 1988; Harriman et al, 1989). To date, however, there has been no evidence linking B_{12} deficiency to vacuolar myelopathy, although only a few series of vacuolar myelopathy have included information about serum vitamin B_{12} and folic acid levels. In our original report (Petito et al, 1985), 12 out of 20 patients were subjected to such studies and all had normal values except 1 patient with reduced serum folic acid levels.

A second major hypothesis for the etiology of vacuolar myelopathy is HIV infection of the spinal cord. Vacuolar myelopathy occurs more frequently in AIDS patients with HIV encephalitis than in those without HIV encephalitis (Petito et al, 1986), although there does not appear to be an association between vacuolar myelopathy and spinal cord infection including HIV and CMV, as shown in Table 8.3. Eilbott et al (1989) found

Table 8.3. Associated spinal cord disease in 51 patients with vacuolar myelopathy

Disease severity	No. of patients	HIV	CMV	MGN[a]	VZV	None
Mild	27	1	2	7	1	16
Moderate	16	–	–	2	–	14
Severe	8	1	3	2	–	4
Total	51	2	5	11	1	34
(%)[b]		(4)	(10)	(22)	(2)	(67)

[a] MGN = microglial nodules.
[b] The total percentage is greater than 100 since some patients had more than one disease.

HIV RNA in macrophages and multinucleated cells in regions of vacuolar myelopathy in 3 AIDS patients using in situ hybridization in formalin-fixed spinal cords. HIV RNA was not detected in macrophages from areas of other CNS infections in the brains of the same patients. Subsequently these authors have reported similar findings in 10 AIDS patients with vacuolar myelopathy (Weisner et al, 1990). Others have also identified HIV antigen by immunohistochemistry in spinal cord macrophages in 3 AIDS patients with vacuolar myelopathy (Maier et al, 1989; Rhodes et al, 1989; Budka, 1990).

In contrast, Rosenblum et al (1989), using immunohistochemistry, viral culture and in situ hybridization with radiolabeled probes in frozen spinal cord, detected HIV only in 3 out of 15 AIDS patients with vacuolar myelopathy. In those 3, HIV was associated with microglial nodules and multinucleated cells, rather than with areas of vacuolation. Grafe and Wiley (1989) studied the spinal cords of 26 AIDS patients, of whom 8 had vacuolar myelopathy. They detected HIV antigen by immunohistochemistry in monocytes only in the 2 with HIV myelitis. Our laboratory has found HIV antigen in rare macrophages in only 1 of 36 cases with moderate or severe vacuolar myelopathy and PCR studies have failed to correlate the presence of HIV genomic material in spinal cord with vacuolar myelopathy.

A third mechanism that may cause vacuolar myelopathy is infection by an opportunistic virus. Sharer et al (1986) have suggested that the rarity of vacuolar myelopathy in children with HIV encephalitis may indicate that reactivation of a previously acquired opportunistic infection is more likely to be responsible for vacuolar myelopathy than HIV. The identification of HIV-related myelopathies such as vacuolar myelopathy (Kamin and Petito, 1991) and gracile tract degeneration (Geller et al, 1977; Davis et al, 1988) in immunosuppressed non-AIDS patients further suggests that opportunistic viruses may be responsible in some cases.

Finally, infection of the peripheral nervous system, including the dorsal root ganglia, can result in secondary changes in the cord and may be responsible for some of the lesions in cases of vacuolar myelopathy. In our experience, some cases with mild diffuse vacuolation can also have severe vacuolation and axonal loss of the fasciculus gracilis that is out of proportion to that seen in the rest of the cord. These cases resemble the gracile tract degeneration described by Rance et al (1988). Possible candidates for an infectious agent

include reactivation of latent viruses such as VZV and HSV, or direct infection of peripheral ganglia by HIV.

At the present time, there is no evidence that implicates retroviruses other than HIV as the cause of vacuolar myelopathy in AIDS patients, and serum antibodies directed against human T-lymphotropic virus Type I (HTLV-I) have not been found in AIDS patients with vacuolar myelopathy (Brew et al, 1989). However, several non-HIV retroviruses in animals and humans including murine leukemia virus and HTLV-I can produce spinal cord diseases. (For review, see Johnson and McArthur, 1987 and Roman, 1987).

Murine leukemia virus produces a spongiform polioencephalomyelopathy in wild mice, which results clinically in a flaccid paraparesis (Gardner et al, 1973). Productive viral replication is concentrated in endothelia and pericytes (Brooks et al, 1980; Swarz et al, 1981; Pitts et al, 1987). This disease is mainly located in spinal cord gray matter rather than white matter, and the vacuolation is secondary to intracellular swellings in neuronal and astrocytic cytoplasm rather than in myelin sheaths (Gardner et al, 1973; Brooks et al, 1980; Swarz et al, 1981; Pitts et al, 1987). HTLV-I is closely linked to spinal cord disease in man, and is now considered the etiological agent of Japanese and tropical spastic paraparesis (Gardner et al, 1973; Osame et al, 1987). The neuropathology of HTLV-I-associated myelopathy (HAM) is distinct from vacuolar myelopathy, and is characterized by inflammation, necrosis and demyelination in the gray and white matter of the spinal cord (Vernant et al, 1987; Akizuki et al, 1988; Piccardo et al, 1988; Moore et al, 1989).

Gracile Tract Degeneration

Rance and colleagues (1988) have described 4 AIDS patients who, at autopsy, had selective degeneration of the fasciculus gracilis. Microscopically, loss of myelin and axons in the fasciculus gracilis was the most prominent finding, as shown in Figure 8.5. Of their 4 cases, one had a focal area of necrosis in the anterior columns, and one had myelin vacuolation in the fasciculus cuneatus and lateral columns. Degeneration of the gracile tracts was prominent at thoracic levels, but only mild at lumbosacral levels.

Diffuse myelin vacuolation and degeneration of the gracile tract may be found together. In addition, vacuolar myelopathy may involve the gracile tract. The etiology may be similar in both and, if

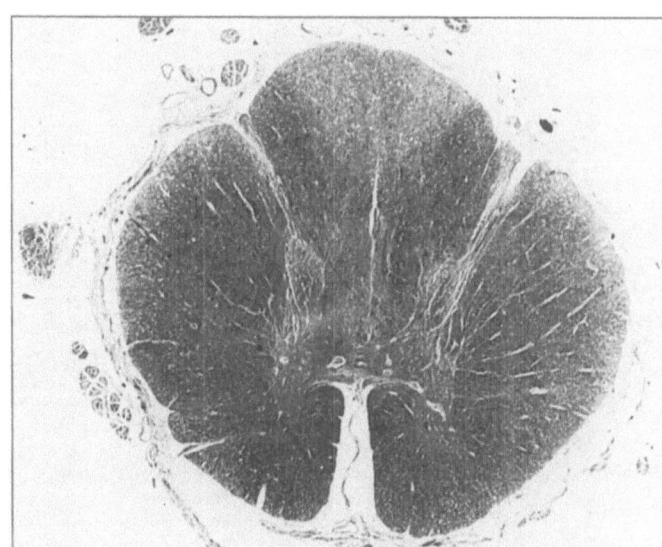

Figure 8.5. Thoracic spinal cord with gracile tract degeneration. Luxol–fast blue–hematoxylin–eosin, ×10.

infectious, may be due to disease in the lumbosacral dorsal root ganglia (see also page 106).

Delayed Myelination of Corticospinal Tracts

Pediatric patients with AIDS rarely have vacuolar myelopathy of the CNS, and opportunistic infections in general are infrequent in the brains and spinal cords of these children. HIV myelitis, with its characteristic microglial nodules and multinucleated cells, can accompany HIV encephalitis and thus may be encountered at postmortem examination in pediatric patients with AIDS.

An AIDS-related spinal cord disorder that may be unique to pediatric patients is delayed myelination of the corticospinal tracts in the thoracic spinal cord. Dickson and his colleagues (1989) examined in detail the spinal cords of 15 children with AIDS. The lateral corticospinal tracts in the spinal cords of 5 of these children were poorly myelinated, but had a normal number of axons as compared with age-matched controls. This change was not accompanied by myelin breakdown, demyelination or macrophage infiltration. Inflammatory cells, including multinucleated cells characteristic of HIV infection, were not present and immunohistochemical studies failed to detect HIV antigens. The mechanism of delayed myelination is unclear, but the authors postulate that its localization to regions still undergoing active myelination suggests a specific inhibition of oligodendroglial function. The absence of demonstrable HIV and inflammation in these regions may indicate that this disorder is related to metabolic or nutritional deficiencies in these severely-ill children, rather than to disturbances in glial function secondary to HIV or to cytokine-induced injury from inflammatory cells.

Human Immunodeficiency Virus Myelitis

The microscopic features of HIV infection of the CNS can occasionally be seen in the spinal cord as well as in the brain, although the spinal cord rarely, if ever, is the only site of the characteristic HIV-related changes. HIV myelitis was present in 5 per cent of the 178 spinal cords examined at our institution (Table 8.1), and in 8 per cent of the 26 spinal cords reported by Grafe and Wiley (1989).

As described elsewhere in this book, HIV infection of the CNS is usually associated with microglial nodules and multinucleated cells in the centrum semi-ovale and subcortical gray matter, but with severe infection, the spinal cord can also be involved. The characteristic microglial nodules and multinucleated cells can be found in both the gray and white matter of the spinal cord (Figure

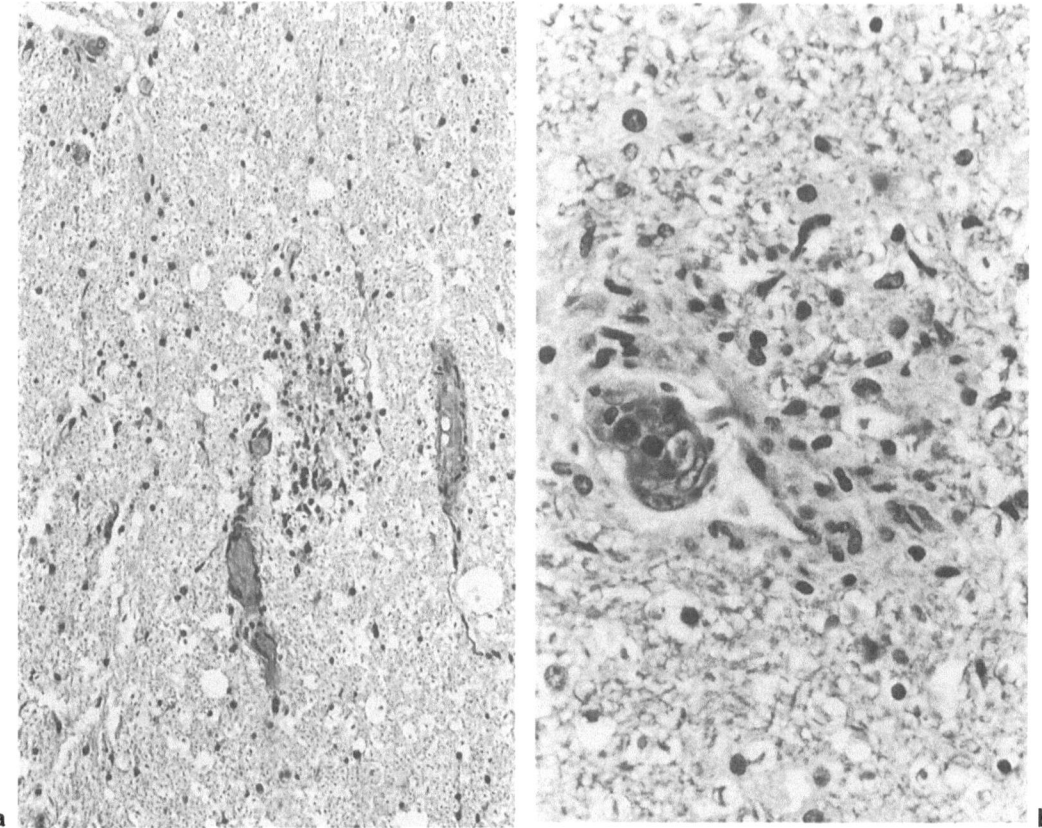

Figure 8.6a,b. HIV myelitis. **a** An inflammatory focus in lateral column with scattered vacuoles in adjacent white matter. Hematoxylin–eosin, ×220. **b** Sparse paravascular inflammation with focal loss of myelin and axons. Hematoxylin–eosin, ×400.

8.6a), and HIV can be identified within these inflammatory foci by immunohistochemistry or in situ hybridization (Grafe and Wiley, 1989; Rosenblum et al, 1989). The inflammatory infiltrates (Figure 8.6b) can result in small foci of demyelination or focal axonal or neuronal loss and gliosis. White matter adjacent to these foci can be vacuolated; however, unless serial sections are performed, the inflammatory infiltrates may not be apparent.

Herpes Virus Myelitis

Cytomegalovirus

Cytomegalovirus is a frequent cause of opportunistic infection in the CNS of adult AIDS patients. It usually presents as a diffuse microglial nodule encephalitis (Morgello et al, 1987; Vinters

et al, 1989), although necrotizing lesions occur in 10 per cent of all cases (Morgello et al, 1987). The spinal cord is not usually involved in CMV infection, and, in our series of 178 autopsies, CMV myelitis was found in only 5 per cent (Table 8.1). When present, however, CMV myelitis may be clinically symptomatic and may show several pathological patterns.

Necrotizing myelitis due to CMV characteristically is seen as diffuse or multifocal subpial regions of necrosis, scattered foci of demyelination (Figure 8.7a), or axonal spheroid collections (Figure 8.7b). The typical enlarged inclusion-bearing cells of CMV are usually concentrated in the subpial white matter in these cases (Figures 8.8a,b). The specific viral identification can be aided by antisera or cDNA probes directed against CMV antigens or genomic material. An accompanying leptomeningitis is seen, and occasionally CMV vasculitis may also be encountered. The inflammatory response is variable, but may be sparse and predominantly mononuclear

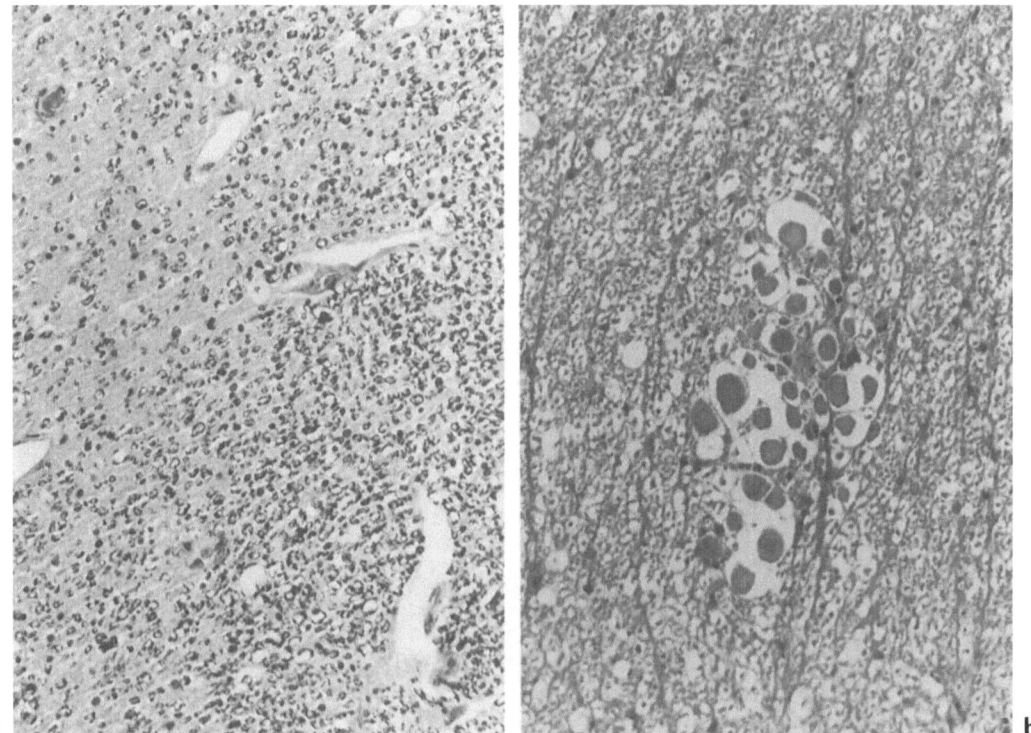

Figure 8.7a,b. CMV myelitis. **a** Focus of demyelination in posterior column. Luxol–fast blue–hematoxylin–eosin, ×160. **b** Small collection of axonal spheroids in lateral column. Hematoxylin–eosin, ×400.

(Booss et al, 1989). In regions of recent necrosis, either acute or chronic inflammation may be seen. In chronic cases, regions of cystic necrosis (Figure 8.3) contain scattered lipid-laden macrophages and blood vessels, and are surrounded by reactive astrocytes with prominent fibrillary processes.

Necrotizing CMV myelitis is often associated with a diffuse ventriculo-encephalitis of the brain, suggesting possible spread to the spinal cord via the cerebrospinal fluid. This pattern was found in 3 of 30 cases of CMV infection of the CNS reported by Morgello et al (1987), and in 1 case each reported by Tucker et al (1985) and Behar et al (1987).

Radiculitis usually accompanies CMV myelitis, or may be the predominant pattern of spinal cord infection (Eidelberg et al, 1986a; Singh et al, 1986; Behar et al, 1987; Morgello et al, 1987). Enlarged CMV-inclusion-bearing cells may be numerous (Figure 8.9), and their immuno-reactivity for S-100 protein identifies them as Schwann cells (Grafe and Wiley, 1989). Infection of nerve roots is accompanied by loss of myelin and axons, although inflammation may be inconspicuous.

CMV infection of the spinal cord can also be seen as scattered microglial nodules containing the characteristic enlarged inclusion-bearing cell of CMV. This was the most frequent pattern in our autopsy series (Table 8.1), and was present in 6 of the 9 cases of CMV myelitis. There is no evidence that the microglial pattern of infection is symptomatic.

Herpes Simplex Virus

Herpes simplex virus myelitis is rare, but when present is usually symptomatic and is associated with co-infection by CMV. Britton et al (1985) described an AIDS patient with a progressive thoracic myelopathy who, at autopsy, had a large region of cystic necrosis and hemorrhage in the spinal cord at the T7 level. Cowdry Type A intra-nuclear inclusions were present in the regions of organizing necrosis in the spinal cord, as well as in the anterior spinal artery. Immunohistochemical studies identified both HSV-2 and CMV in the spinal cord, although they did not identify the viral origin of Cowdry Type A inclusions within

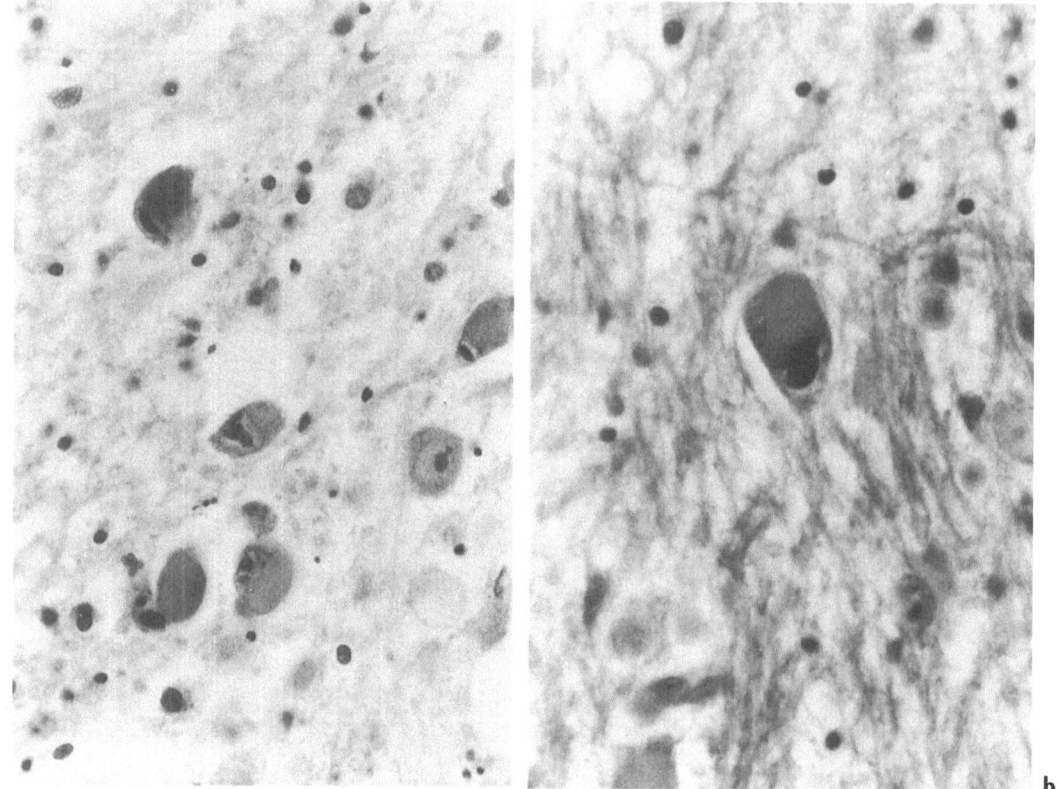

Figure 8.8a,b. CMV myelitis. **a** Subpial collection of enlarged CMV-inclusion-bearing cells in areas of necrosis. Hematoxylin–eosin, ×160. **b** Solitary CMV-inclusion-bearing cell in subpial white matter. Hematoxylin–eosin, ×400.

the inflamed artery. A second patient described by Tucker et al (1985) had a rapidly-progressive ascending myelitis which, at autopsy, was due to diffuse subpial regions of necrosis in the spinal cord, again due to co-infection with CMV and HSV.

In our series, HSV encephalomyelitis was identified in 2 cases, and in both there was an associated CMV encephalomyelitis. Large wedge-shaped regions of acute necrosis were present in the lower medulla and spinal cord (Figure 8.10). Enlarged CMV-inclusion-bearing cells, as well as the smaller Cowdry Type A intranuclear inclusions of Herpes simplex, were found in or at the periphery of the necrotic regions.

Varicella Zoster Virus

Varicella zoster virus myelitis is an infrequent complication of herpes zoster, and almost always occurs with immunosuppression. In AIDS patients, VZV myelitis is rare, and much less common than CMV or HSV myelitis. The prob-

able diagnosis of VZV myelitis was made in only 1 out of a total of 586 AIDS patients in 3 clinical series examining the neurological complications of AIDS (Snider et al, 1983; Levy et al, 1985; McArthur, 1987). In the present review of all AIDS autopsies performed at our institution over an 8-year period, we encountered VZV myelitis in 2 out of the 178 spinal cords examined. A further 156 AIDS patients have been reported whose spinal cords were studied at postmortem examination (Anders et al, 1986; Rhodes, 1987; Grafe and Wiley, 1989), and none was found to have VZV myelitis.

The pathology of VZV myelitis in AIDS is similar to that described in immunocompromised non-AIDS patients (Rose et al, 1964; Devinsky et al, 1991). Hemorrhage and necrosis in the posterior horns are common. In addition, focal or segmental necrosis, demyelination, or vasculitis with or without secondary infarction can also be seen. In a recent study of VZV myelitis in 13 patients (Devinsky et al, 1991), 3 of whom had AIDS, posterior horn abnormalities were seen in all of the 9 patients examined at autopsy, wedge-

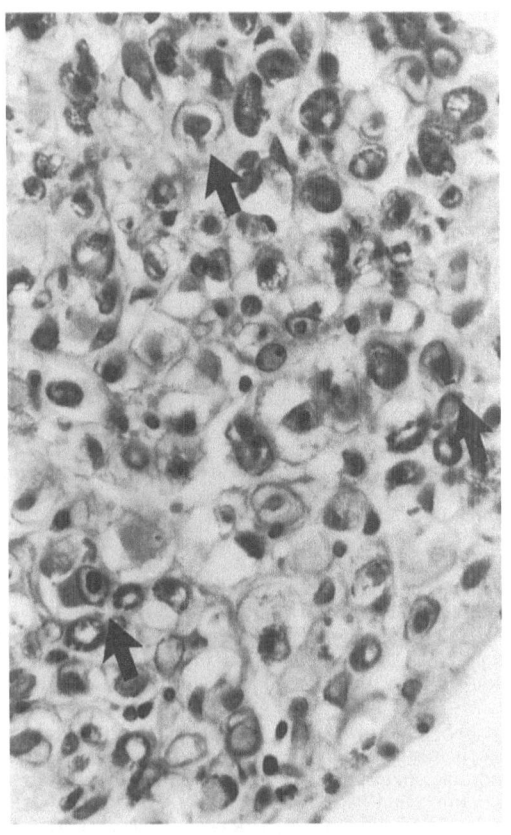

Figure 8.9. Anterior root from spinal cord in Figure 8.3 showing loss of myelinated axons and scattered CMV-inclusion-bearing cells (*arrows*). Hematoxylin–eosin, ×400.

shaped areas of demyelination were seen in 6 of the 9 patients, and acute necrotizing vasculitis was seen in 4 patients. Cowdry Type A intranuclear inclusions, characteristic of VZV, were usually associated with the intraparenchymal lesions, but were uncommon in the inflamed leptomeningeal arteries. Immunohistochemical studies can be used to identify the viral origin of the inclusions, and have also identified VZV antigen in non-inclusion-bearing cells, including those within the intima of VZV arteritis (Eidelberg et al, 1986b; Morgello et al, 1988).

Acknowledgements. The technical skills of Mrs Darleen Vecchio and the secretarial help of Ms Geraldine Winfrey are gratefully acknowledged. This chapter is supported in part by a grant from the National Institutes of Health, NINCDS R01-NS27416.

References

Akizuki S, Setoguchi M, Nakazato O et al. (1988) An autopsy case of human T-lymphotrophic virus Type 1-associated myelopathy. Hum Pathol 19:988–90.

Anders KH, Guerra WF, Tomiyasu U et al. (1986) The neuropathology of AIDS, UCLA experience and review. Am J Pathol 124:537–58.

Becker PS, Griffin JW, McArthur JC et al. (1989) Vacuolar

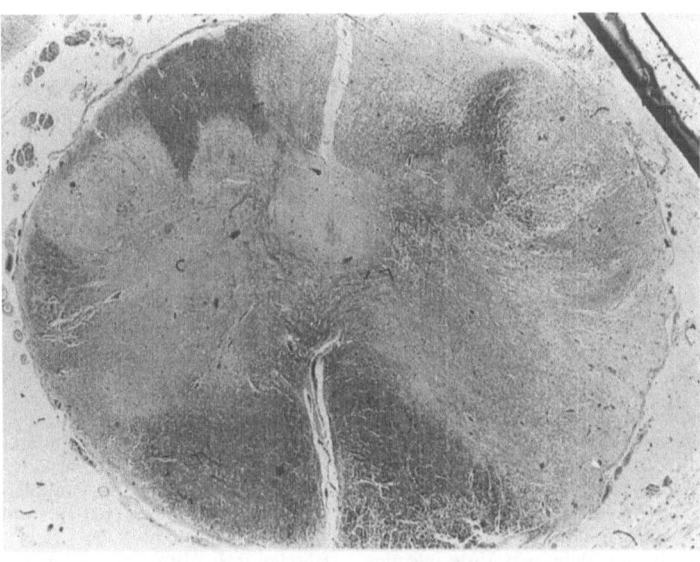

Figure 8.10. Lower medulla with combined CMV and HSV encephalomyelitis and large subpial regions of necrosis. Luxol–fast blue–hematoxylin–eosin, ×7.5.

myelopathy in Human Immunodeficiency Virus (HIV) infection: central remyelination. J Neuropathol Exp Neurol 48:383.

Behar R, Wiley C, McCutchan JA. (1987) Cytomegalovirus polyradiculoneuropathy in acquired immune deficiency syndrome. Neurology 37:557–61.

Booss J, Dann PR, Griffith BP et al. (1989) Host defense response to cytomegalovirus in the central nervous system: predominance of the monocyte. Am J Pathol 134:71–8.

Brew BJ, Hardy W, Zuckerman E et al. (1989) AIDS-related vacuolar myelopathy is not associated with coinfection by human T-lymphotropic virus Type 1. Ann Neurol 26:679–81.

Britton CB, Mesa-Tejada R, Fenoglia CM et al. (1985) A new complication of AIDS: thoracic myelitis caused by herpes simplex virus. Neurology 35:1071–4.

Brooks BR, Swarz JR, Johnson RT. (1980) Spongiform polioencephalomyelopathy: I. Pathogenesis of infection in newborn mice. Lab Invest 43:480–6.

Budka H. (1990) Human immunodeficiency virus (HIV) envelope and core proteins in CNS tissue of patients with the acquired immune deficiency syndrome (AIDS). Acta Neuropathol 79:611–19.

Burkes RL, Cohen H, Krailo M et al. (1987) Low serum cobalamin levels occur frequently in the acquired immuno-deficiency syndrome and its related disorders. Eur J Haematol 38:141–7.

Davis DG, Wiley CA, Markesbury WR. (1988) Spinal cord posterior column degeneration in bone marrow transplant patients. J Neuropathol Exp Neurol 47:303.

Denning DW, Anderson J, Rudge P et al. (1987) Acute myelopathy associated with primary infection with Human Immunodeficiency Virus. Br Med J 294:143–4.

Devinsky O, Cho E-S, Petito CK et al. (1991) Herpes zoster myelitis. Brain, in press.

Dickson DW, Belman AL, Kim TS et al. (1989) Spinal cord pathology in pediatric acquired immunodeficiency syndrome. Neurology 39:227–35.

Eidelberg D, Sotrel A, Vogel H et al. (1986a) Progressive polyradiculopathy in acquired immunodeficiency syndrome. Neurology 36:912–16.

Eidelberg D, Sotrel A, Horoupian DS et al. (1986b) Thrombotic cerebral vasculopathy associated with herpes zoster. Ann Neurol 19:7–14.

Eilbott DJ, Peress N, Burger H et al. (1989) Human immunodeficiency virus Type 1 in spinal cords of acquired immunodeficiency syndrome patients with myelopathy: expression and replication in macrophages. Proc Natl Acad Sci USA 86:3337–41.

Gardner MB, Henderson BE, Officer JE et al. (1973) A spontaneous lower motor neuron disease apparently caused by indigenous Type-C RNA virus in wild mice. J Natl Cancer Inst 51:1243–54.

Geller A, Gilles F, Schwachman H. (1977) Degeneration of fasciculus gracilis in cystic fibrosis. Neurology 27:185–7.

Goldstick L, Mandybur TI, Bode R. (1985) Spinal cord degeneration in AIDS. Neurology 35:103–6.

Grafe MR, Wiley CA. (1989) Spinal cord and peripheral nerve pathology in AIDS: The roles of cytomegalovirus and Human Immunodeficiency Virus. Ann Neurol 25:561–6.

Harriman GR, Smith PD, Horne MK et al. (1989) Vitamin B_{12} malabsorption in patients with acquired immunodeficiency syndrome. Arch Intern Med 149:2039–41.

Herbert V. (1988) B_{12} deficiency in AIDS. JAMA 260:2837.

Janssen RS, Saykin AJ, Kaplan JE et al. (1988) Neurological complications of Human Immunodeficiency Virus infection in patients with lymphadenopathy syndrome. Ann Neurol 23:49–55.

Janssen RS, Saykin AJ, Cannon L et al. (1989) Neurological and neuropsychological manifestations of HIV-1 infection: association with AIDS-related complex but not asymptomatic HIV-1 infection. Ann Neurol 26:592–600.

Johnson RT, McArthur JC. (1987) Editorial: myelopathies and retroviruses. Ann Neurol 21:113–16.

Kamin SS, Petito CK. (1991) Idiopathic myelopathies in non-AIDS patients. Hum Pathol 22:816–24.

Levy RM, Bredesen DE, Rosenblum ML. (1985) Neurological manifestations of the acquired immunodeficiency syndrome (AIDS): experience at USCF and review of the literature. J Neurosurg 62:475–95.

Maier H, Budka H, Lassmann H, Pohl P. (1989) Vacuolar myelopathy with multinucleated giant cells in the acquired immune deficiency syndrome (AIDS). Acta Neuropathol 78:497–503.

McArthur JC. (1987) Neurologic manifestations of AIDS, Medicine (Baltimore) 66:407–37.

McArthur JC, Cohen BA, Selnes OA et al. (1989) Low prevalence of neurological and neuropsychological abnormalities in otherwise healthy HIV-1-infected individuals: results from the multicenter AIDS cohort study. Ann Neurol 26:601–11.

Mehren M, Burns PJ, Mamani F et al. (1988) Toxoplasmic myelitis mimicking intramedullary spinal cord tumor, Neurology 38:1648–50.

Moore GRW, Traugott U, Scheinberg LC et al. (1989) Tropical spastic paraparesis: a model of virus-induced, cytotoxic T-cell-mediated demyelination. Ann Neurol 26:523–30.

Morgello S, Cho E-S, Nielsen S et al. (1987) Cytomegalovirus encephalitis in patients with acquired immunodeficiency syndrome: an autopsy study of 30 cases and a review of the literature. Hum Pathol 18:289–97.

Morgello S, Block GA, Price RW et al. (1988) Varicella-Zoster virus leucoencephalopathy and cerebral vasculopathy. Arch Pathol Lab Med 112:173–7.

Osame M, Matsumoto M, Usuko K et al. (1987) Chronic progressive myelopathy associated with elevated antibodies to human T-lymphotropic virus Type 1 and adult T-cell leukemia-like cells.Ann Neurol 21:117–22.

Pant SS, Asbury AK, Richardson EP Jr. (1968) The myelopathy of pernicious anemia. Acta Neurol Scand Suppl 44:4–36.

Peress N, Burger H, LaNeve D et al. (1990) HIV-1 expression-AIDS vacuolar myelopathy correlates with extent of disease (abstr). J Neuropathol Exp Neurol 49:351.

Petito CK, Navia BA, Cho E-S et al. (1985) Vacuolar myelopathy pathologically resembling subacute combined degeneration in patients with the acquired immunodeficiency syndrome. N Engl J Med 312:874–9.

Petito CK, Cho E-S, Lemann W et al. (1986) Neuropathology of acquired immunodeficiency syndrome (AIDS): an autopsy review. J Neuropathol Exp Neurol 45:635–46.

Piccardo P, Ceroni M, Rodgers-Johnson P. (1988) Pathological and immunological observations on tropical spastic paraparesis in patients from Jamaica. Ann Neurol 23:s156–60.

Pitts OM, Powers JM, Bilello JA. (1987) Ultrastructural changes associated with retroviral replication in central nervous system capillary endothelial cells. Lab Invest 56:401–9.

Rance NE, McArthur JC, Cornblath DR et al. (1988) Gracile tract degeneration in patients with sensory neuropathy and AIDS. Neurology 38:265–71.

Rhodes RH. (1987) Histopathology of the central nervous system in the acquired immunodeficiency syndrome. Hum Pathol 18:636–43.

Rhodes RH, Ward JM, Cowan RP, Moore PT. (1989) Immunohistochemical localization of human immunodeficiency viral antigens in formalin-fixed spinal cords with AIDS myelopathy. Clin Neuropathol 8:22–7.

Roman GC. (1987) Retrovirus-associated myelopathies. Arch Neurol 44:659–63.

Rose FC, Brett EM, Burston J. (1964) Zoster encephalomyelitis. Arch Neurol 11:155–72.

Rosenblum M, Scheck AC, Cronin K et al. (1989) Disassociation of AIDS-related vacuolar myelopathy and productive HIV infection of the spinal cord. Neurology 39:892–6.

Sharer LR, Epstein LG, Cho E-S, et al. (1986) HTLV-III and vacuolar myelopathy. N Engl J Med 315:62–3.

Sharer LR, Dowling PC, Michaels J et al. (1990) Spinal cord disease in children with HIV infection: a combined molecular and neuropathological study. Neuropathol Appl Neurobiol 16:317–31.

Singh BM, Levine S, Yarrish RL et al. (1986) Spinal cord syndromes in the acquired immune deficiency syndrome. Acta Neurol Scand 73:590–8.

Snider WD, Simpson DM, Nielsen S et al. (1983) Neurological complications of the acquired immunodeficiency syndrome: analysis of 50 patients, Ann Neurol 14:403–18.

Swarz JR, Brooks BR, Johnson RT. (1981) Spongiform polioencephalomyelopathy caused by a murine retrovirus. II. Ultrastructural localization of virus replication and spongiform changes in the central nervous system. Neuropathol Appl Neurobiol 7:365–80.

Tucker T, Dix RD, Katzen C et al. (1985) Cytomegalovirus and herpes simplex virus ascending myelitis in a patient with acquired immune deficiency syndrome. Ann Neurol 18: 74–9.

Vernant JC, Maurs L, Gessain A et al. (1987) Endemic tropical spastic paraparesis associated with human T-lymphotropic virus Type 1: a clinical and seroepidemiological study of 25 cases. Ann Neurol 21:123–30.

Vinters HV, Kwok MK, Ho HW et al. (1989) Cytomegalovirus in the nervous system of patients with acquired immune deficiency syndrome. Brain 112:245–68.

Weisner B, Peress N, LaNeve D et al. (1990) Human immunodeficiency virus Type 1 expression in the central nervous system correlates directly with extent of disease. Proc Natl Acad Sci USA 87:3997–4001.

Woolsey RM, Chambers TJ, Chung HD et al. (1988) Mycobacterial meningomyelitis associated with human immunodeficiency virus infection. Arch Neurol 45:691–3.

9 Neuropathology of AIDS in Children

L.R. Sharer and M. Mintz

Introduction

The acquired immunodeficiency syndrome (AIDS) was first recognized in 1981, when reports appeared that described the unexplained occurrence of opportunistic infections in homosexual men in Los Angeles and New York City (Gottlieb et al, 1981; Masur et al, 1981). A similar syndrome was soon recognized in persons who used illicit drugs by the intravenous route and in patients who had received blood products, including hemophiliacs (Davis et al, 1983; Small et al, 1983). By 1983 it became apparent that AIDS could also occur in children, many of whose parents either had AIDS or were at risk of developing AIDS (Oleske et al, 1983; Rubinstein et al, 1983). That same year and the following year, the retrovirus that causes AIDS, initially termed lymphadenopathy virus (LAV), human T-cell lymphotropic virus Type III (HTLV-III), and AIDS retrovirus (ARV), now called human immunodeficiency virus Type 1 (HIV-1), was isolated and characterized (Barré-Sinoussi et al, 1983; Gallo et al, 1984; Levy et al, 1984). These discoveries led to improved understanding of AIDS and its transmission from one person to another, and also facilitated the study of the disease in children.

Following the description of intellectual decline in adults with AIDS (Snider et al, 1983), reports appeared of a new, progressive encephalopathy in children with the disorder (Epstein et al, 1984, 1985; Belman et al, 1985). It was suggested early on that this encephalopathy might be related to the newly-described retrovirus that causes AIDS (Epstein et al, 1985). This suggestion was fueled in part by demonstration of the transmission of HIV-1 to chimpanzees using brain tissue from children and adults who had died of AIDS (Gajdusek et al, 1985), as well as by the finding of HIV-1 nucleic acid sequences in the brains of AIDS patients, including children, who had encephalopathy (Shaw et al, 1985). This particular encephalopathy did not seem to resemble any that had been encountered previously, including those seen in children with other immunodeficiencies. Certain pathologic features, particularly the presence of multinucleated giant cells in brain tissue, appeared to be unique to the disorder (Sharer et al, 1985, 1986). In retrospect, it was easier to study progressive encephalopathy in children than in adults, who more often had other complicating factors, particularly opportunistic infections of the central nervous system (CNS) (Petito et al, 1986).

It is now generally accepted that HIV-1 infects the CNS in a substantial number of affected persons, and that this infection is directly or indirectly related to the progressive encephalopathy, also termed HIV encephalopathy or AIDS-dementia complex (Price et al, 1988; Janssen et al, 1989), although the exact pathogenesis is unknown. The hypothesis that primary CNS infection by HIV-1 causes the encephalopathy is strengthened by reports of dramatic improvements in children treated with the anti-retroviral compound zidovudine even when their immune deficiency

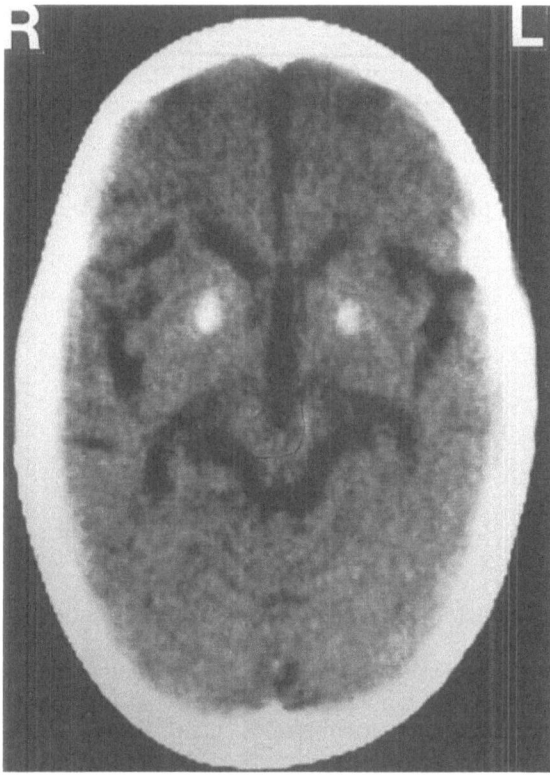

Figure 9.1. Computerized tomogram scan of the brain of a 10-year-old girl with HIV-1 infection and progressive encephalopathy. There are characteristic hyperdensities in the region of the basal ganglia, as well as enlargement of the sulci. (Reprinted from Mintz (1992) © Yearbook Medical Publishers, with permission.)

has otherwise progressed (Pizzo et al, 1988; Sharer and Epstein, 1989; Mintz, 1992).

or evidence of brain atrophy on imaging studies; (2) progressive motor dysfunction; and (3) loss of neurodevelopmental milestones (Mintz et al, 1989a; Mintz, 1992). Children are usually normal at birth, with normal head circumference. With time in younger affected children, head circumference fails to increase at a normal rate. The natural history of the progressive encephalopathy can be quite variable, with occasional, spontaneous 'honeymoon' or plateau periods, but the overall course is one of relentless, downhill neurological deterioration (Epstein et al, 1986; Diamond, 1989; Mintz et al, 1989a; Mintz, 1992). In the terminal stages, children with encephalopathy are apathetic, weak and spastic, with global loss of communicative and language skills. Brain imaging studies at this point frequently disclose severe atrophy, as well as mineralizations in the basal ganglia (Figure 9.1) (Belman et al, 1986).

At the Children's Hospital AIDS Program (CHAP) in Newark, New Jersey, 159 children with HIV-1 infection have been assessed for neurological dysfunction since 1982. Three-quarters of these children have met criteria for a diagnosis of AIDS, as defined by the Centers for Disease Control (Centers for Disease Control, 1987). About 75 per cent of the children with AIDS, or half of the entire HIV-1 infected cohort, have had progressive encephalopathy. This encephalopathy has proved to be one of the most devastating complications of HIV-1 infection in children (Falloon et al, 1989; Scott et al, 1989) – the occurrence of progressive encephalopathy in a child with AIDS predicts a poor prognosis (Epstein et al, 1986; Scott et al, 1989).

Clinical Manifestations

Clinically, children with HIV-1 infection may fall into one of three groups: those who manifest progressive encephalopathy, a progressive neurological deterioration secondary to primary HIV-1 infection of the CNS; those with evidence of static encephalopathy, a non-progressive neurological insult not directly related to HIV-1 infection; and those with a normal neurological examination (Epstein et al, 1986; Mintz et al, 1989a; Mintz, 1992). Neurological findings in children with HIV-1-related progressive encephalopathy comprise a well-defined clinical triad: (1) impaired brain growth, with either acquired microcephaly

Pathologic Features

Our own experience is derived from an autopsy series of 36 children who died with HIV-1 infection. These children all lived in the greater metropolitan region of New York City; most of them resided in Newark, New Jersey, or its environs. The ages at death of the children in this series ranged from 4 months to 13 years; most of them were below the age of 5 years. Of these children, 26 had suffered perinatal, or vertical, transmission of HIV-1 infection; 5 had been infected by transfusion of blood or blood products; 3 children at risk of perinatal transmission had also received transfusions; and in 2 patients the risk factor was unknown. At least 19 of the 26 children with perinatal transmission had mothers who were intravenous drug users, the largest single risk factor for transmission of HIV-1 in this series. None of the transfusion cases received blood products after 1985, the year in which mandatory donor screening for antibody to HIV-1 was instituted in the USA. Detailed observations of the first 11 patients in this series were reported earlier (Sharer et al, 1986).

General autopsy findings in this series revealed a variety of disorders, including *Pneumocystis carinii* pneumonia, disseminated cytomegalovirus (CMV) infection, disseminated *Mycobacterium avium–intracellulare* complex (MAI) infection, lymphoid interstitial pneumonitis (Joshi et al, 1985), polyclonal lymphoproliferation (Joshi et al, 1987a) and systemic arteriopathy (Joshi et al, 1987b). These findings were recently reviewed by Joshi (1990).

Gross Neuropathologic Findings

Brain weights were almost invariably less than expected for age and were usually more than one standard deviation below published norms (Schulz et al, 1962). In the younger children, diminished brain weight correlated directly with small head circumference. Mild-to-moderate atrophy of the brain was often apparent, with moderate dilatation of the lateral ventricles and, occasionally, mild enlargement of the sulci. In a few cases of children who died under 1 year of age, there was apparent retardation of brain development, with poor gray–white differentiation (Epstein et al, 1988a).

Microscopical Findings

A summary of the microscopical findings in the 36 cases is given in Table 9.1.

Table 9.1. Microscopical neuropathologic features in autopsies of 36 children with HIV-1 infection and AIDS

Feature	No. of cases	%
Vascular/juxtavascular mineralization	33	92
White matter abnormality (pallor and/or gliosis)	28	78
Inflammatory cell infiltrates	27	75
Multinucleated cells	22	61
Multinucleated giant cells	13	36
Vascular inflammation	11	30.5
Opportunistic infections	5	14

Mineralizations

The most common CNS finding in this autopsy series was mineralization, present to at least a minimal degree in over 90 per cent of the brains. This was also the most common neuropathological finding in a series of 26 autopsies of children with HIV-1 infection reported by Dickson et al (1989a). Mineralizations were usually basophilic, consisting of calcium and iron, possibly with other minerals as well (Perl et al, 1989). They were extracellular and were most common in the putamen and globus pallidus, where they were usually adjacent to small vessels (Figure 9.2a). Mineralization of the walls of larger blood vessels was also noted (Figure 9.2b). In several cases the basal ganglia mineralizations were severe, with confluent masses of dense, basophilic material. Mineralizations also occurred in the frontal white matter, where they were less often juxtavascular (Figure 9.2c). There were fewer cases with mineralizations in white matter than in basal ganglia, but all the cases with mineralizations in the frontal white matter also had them in the basal ganglia. A few cases also had focal juxtavascular mineralizations in the cerebral cortex, particularly at the depths of sulci (Figure 9.2d). Mineralizations in all sites were generally unassociated with inflammatory cell changes, although there was on occasion inflammation of the adventitia of larger, mineralized vessels in the basal ganglia (Sharer et al, 1986). Mineralizations were never noted in the caudate nuclei, despite their high frequency in the putamen. In addition, typical juxtavascular mineralizations did not occur outside the cerebral hemispheres.

Mineralizations were illustrated in one of the earliest pathological descriptions of the brain in children with AIDS (Church and Isaacs, 1984). Other investigators have stressed the frequency and importance of these mineralizations (Belman et al, 1986, 1988), and sequential radiographic studies have demonstrated the development of basal ganglia mineralizations with time (Epstein et al, 1987). On pathological examination they were often severe in children who died before the age of 1 year, a finding at variance with that of Dickson et al (1989a). We have also observed typical juxtavascular mineralizations of the basal ganglia in the oldest case of presumed perinatal transmission that we have examined, a child who was symptomatic from the age of 5 years, and who was 13 years old at the time of his death. This patient is thought to be one of the oldest cases of perinatally acquired HIV-1 infection

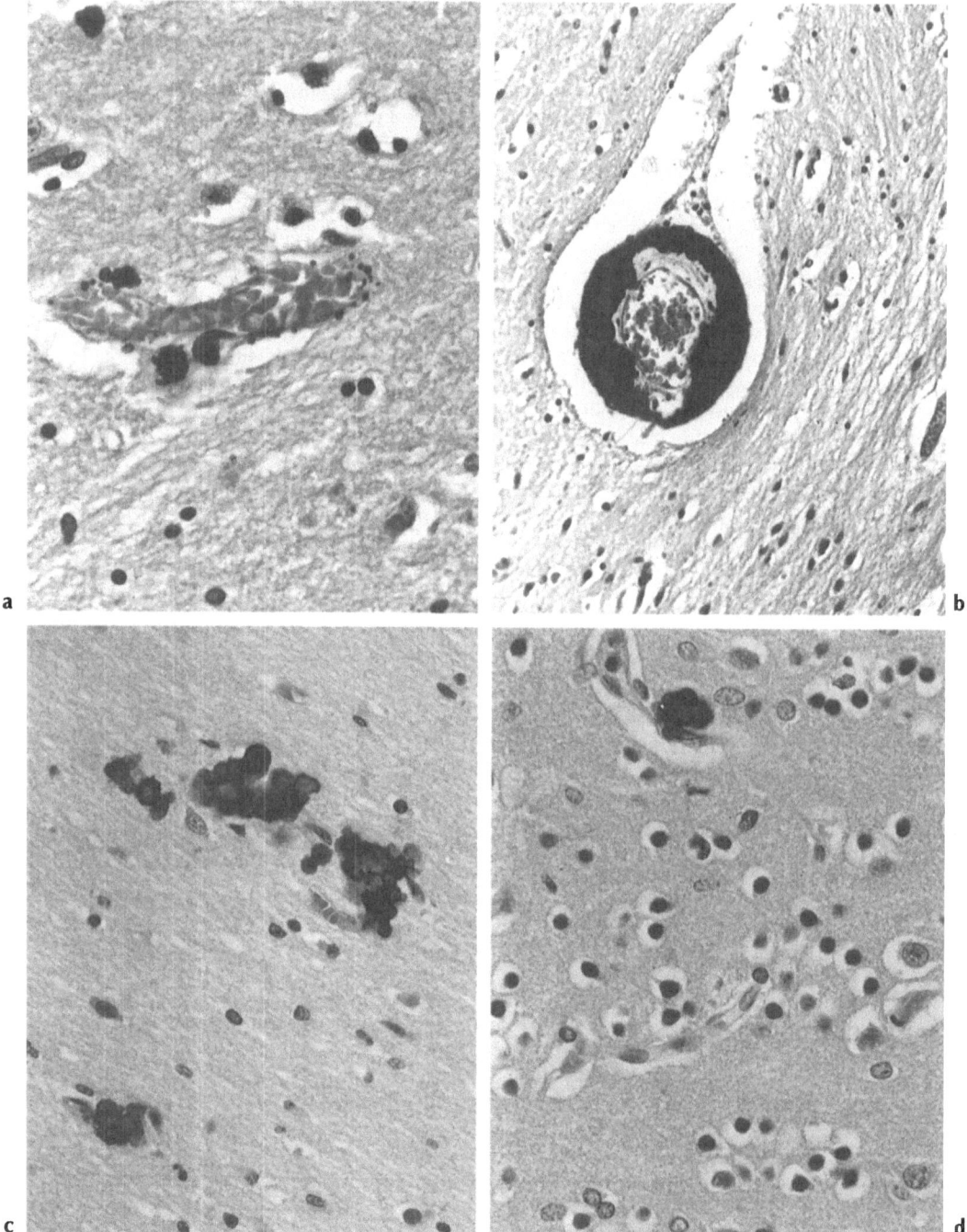

Figure 9.2. a Vascular and juxtavascular mineralizations, putamen, 12-month-old girl. Note fine mineralizations in vessel walls, as well as larger, basophilic concretions. **b** Basophilic mineralization of wall of larger vessel, putamen, same patient as in **a**. Note smaller mineralizations as well. **c** Mineralizations in frontal white matter, 6-month-old boy. Mineralizations were also present in the basal ganglia in this case. **d** Juxtavascular mineralizations in frontal cortex at depth of a sulcus, same patient as in **c**. (Hematoxylin and eosin. Original magnifications: **a** ×500; **b** ×200; **c** and **d** ×400.)

on record (J Oleske and E Connor, personal communication).*

White Matter Changes

A high proportion of cases (78 per cent) had white matter changes, consisting of either pallor, gliosis, or both. Pallor was usually generalized and was accompanied by diffuse astrocytosis. In a few instances pallor was more focal, surrounding regions of inflammatory change. Pallor was apparent on special stains for myelin, although in brains from younger subjects it was often difficult to determine if the degree of staining was reduced from what was expected for age. Immunocytochemistry for glial fibrillary acidic protein (GFAP) was employed to assess the degree of astrocytosis (Figure 9.3, facing p. 208). The intensity of the astrocytic reaction by this technique was often greater than would have been expected from examination of routinely (hematoxylin and eosin) stained sections, a phenomenon also commented upon in the white matter of adult subjects with HIV-1 infection (McArthur et al, 1989).

The myelin pallor was in some instances accompanied by a comparable reduction in the intensity of axonal staining, while in other cases there appeared to be relative sparing of axons, suggesting demyelination. Myelin breakdown with lipid-laden macrophages was not apparent, except in rare instances. In some of the younger subjects, there was a suggestion of delayed myelination. Myelin pallor was also apparent in the corticospinal tracts in the spinal cord, where its occurrence correlated directly with the degree of pallor of the white matter of the cerebral hemispheres (Sharer et al, 1990). Dickson et al (1989b) have also reported apparent wallerian degeneration of the corticospinal tracts in the cord in several of the children they examined.

Inflammatory Cell Infiltrates

Loose collections of cells, consisting of variable numbers of lymphocytes, plasma cells, microglial cells, astrocytes and macrophages, in some cases with multinucleated cells, were present in the CNS of 27 of the 36 cases (Figure 9.4). These inflammatory cell infiltrates bore some resemblance to classical glial–microglial or Babès nodules, but they were less tightly cellular (Cho and Sharer, 1990). They were most commonly located in the following regions: the central white matter of the cerebral hemispheres; the basal ganglia,

especially the putamen and the external capsule–claustrum region; and the pons, particularly the basis pontis. However, examples of these cell collections were found in all parts of the CNS, from the cerebral cortex to the gray and white matter of the spinal cord (Figure 9.4). The lesions were rarely associated with neuronal necrosis or neuronophagia, even when they occurred in regions that had a dense population of nerve cells. In some instances, inflammatory cell infiltrates in the white matter were accompanied by focal myelin pallor. The numbers of infiltrates varied considerably from case to case, being quite numerous in some cases and sparse in others. It was often possible to detect the presence of HIV-1 in these lesions, by either immunocytochemistry or in situ hybridization. Despite the high frequency of inflammatory infiltrates in the pons, none of the patients in this series had a clinical brainstem syndrome, although at least one such case has been reported (Raphael et al, 1989).

Multinucleated Cells and Multinucleated Giant Cells

Multinucleated cells were found in the CNS of 22 of the 36 cases, with multinucleated giant cells (MGC) in 13 of these. All the cases that had multinucleated cells also had inflammatory cell infiltrates. Many of the cases had cells with clusterings of multiple, dark nuclei, with scanty cytoplasm. These cells have been named 'nuclear cluster cells' (Figure 9.5) (Sharer et al, 1986). The scanty cytoplasm of these cells could be stained with the lectin *Ricinus communis* agglutinin-1 (RCA-1), a marker for macrophages, microglia and endothelial cells (Michaels et al, 1988a), suggesting that they were of monocyte/macrophage origin (Figure 9.6, facing p. 208). Nuclear cluster cells almost invariably occurred as part of inflammatory cell infiltrates, rather than by themselves in the neuropil.

Multinucleated giant cells with copious cytoplasm were a conspicuous feature of the cases in

---

* We believe that this 13-year-old boy, who died in 1989, was infected in 1976. It is now generally accepted that HIV-1 infection entered the United States in 1975 or earlier, and the New York metropolitan area, including Newark, would appear to have been a locus of involvement from the beginning of the epidemic. There is always a possibility that this child was infected by some other route, such as through sexual abuse, but he was suspected of having an AIDS-like illness as early as age 5 years. Both his parents had AIDS, and they were intravenous drug users.

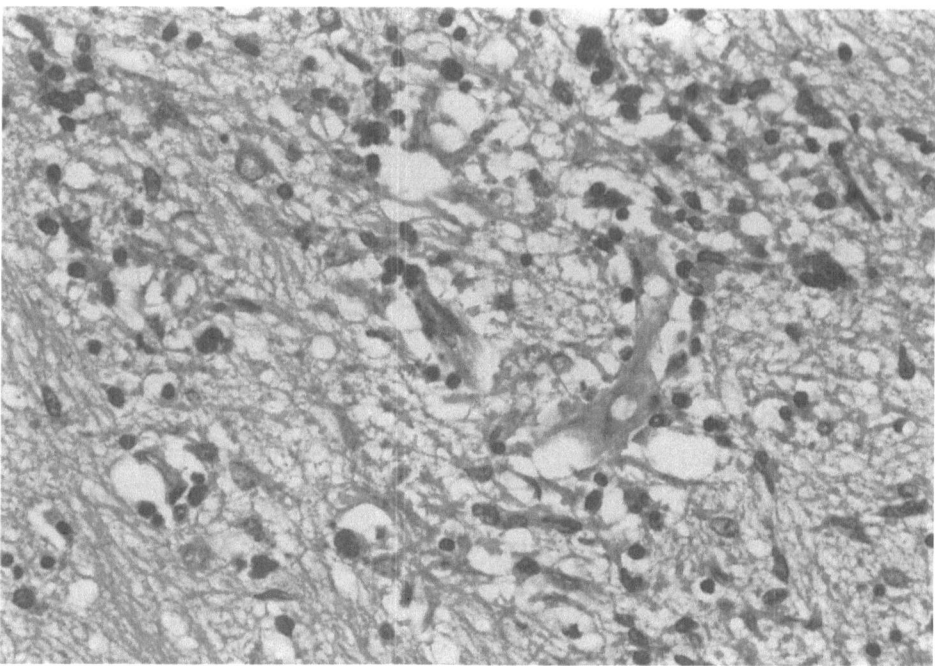

Figure 9.4. Inflammatory cell infiltrate, cerebral white matter, 12-month-old girl (same patient as in Figures 2 **a** and **b**). A variety of cells is present, including cells with multiple nuclei. The background white matter is pale. H&E, original magnification ×400.

which they occurred. The cells had several forms, including Langhans-like cells with peripherally-arrayed nuclei and pale granular cytoplasm with an eosinophilic center, and large macrophage-like cells with nuclei dispersed throughout the cytoplasm (Figure 9.7a). Some of the nuclei had fine tapering processes or bridges of nucleoplasm between them, similar to those seen in adult cases (see Chapter 6) (Figure 9.7b) (Mizusawa et al, 1988). The MGC were generally located in inflammatory cell infiltrates, although single MGC could also be found. In some cases there were scattered perivascular MGC with only a few macrophages. The MGC stained strongly with the RCA-1 lectin and on immunohistochemistry were negative for GFAP. With frozen sections, they could be readily demonstrated to contain either HIV-1 protein or nucleic acid sequences, by in situ hybridization or immunocytochemistry (Figure 9.8, facing p. 208). In a few cases, large numbers of HIV-1 particles could readily be found within MGC, by electron microscopy (Figure 9.9) (Meyenhofer et al, 1987). Some of the MGC had fine tapering cytoplasmic processes that were both RCA-1 positive and HIV-1 antigen positive, suggesting that the cells had either arisen by fusion of microglia or had incorporated microglia within them (Michaels et al, 1988a).

Vascular Inflammation

Inflammation of the walls of small- to medium-sized blood vessels occurred in less than one-third of the cases (Figure 9.10). Inflammatory cells in the vessel walls included lymphocytes, plasma cells, mononuclear macrophages, and, in some instances, MGC. Vascular inflammation was especially prominent in the cerebral white matter. Fibrinoid necrosis was rare (Epstein et al, 1985). Vessel thrombosis and infarction of CNS tissue did not occur as complications of this inflammation, which in some respects resembled a vasculitis. HIV-1 antigen or nucleic acid sequences were infrequently detected in association with these vascular lesions.

Opportunistic Infections

Opportunistic or reactivated latent infections were infrequent within the CNS of the subjects in this series, and were documented in only four, or possibly five, cases. This observation has been confirmed by other investigators examining cases of children who died with HIV-1 infection (Dickson et al, 1989a; Kozlowski et al, 1990). This low incidence contrasts with the high

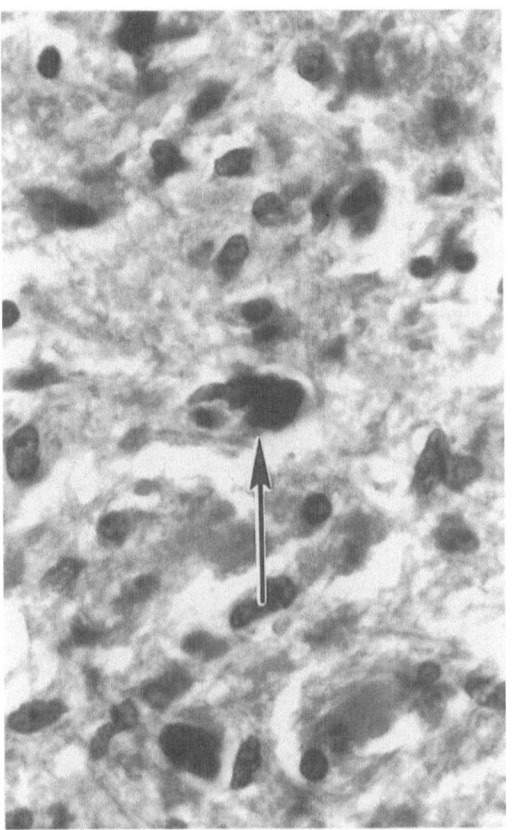

Figure 9.5. Nuclear cluster cell (*arrow*) within an inflammatory cell infiltrate, base of pons, 6-month-old boy. H&E, original magnification ×800.

incidence of infectious complications seen on neuropathological examination of adults with AIDS (Levy et al, 1985; Petito et al, 1986). Of the children in our series, two had evidence of CMV infection of the CNS. One was an 11-year-old boy with hemophilia and transfusion-associated AIDS, who had evidence of disseminated CMV at post mortem. The other was a 5-year-old boy who had elevated serum antibody titers to CMV during life and had been treated with intravenous acyclovir on several occasions; no CMV lesions were detected outside the CNS at autopsy. There was also a third case in which CMV was suspected in the CNS but could not be confirmed. One child had disseminated *Candida* infection, with involvement of multiple organs, including the brain, by microabscesses containing pseudohyphae.

The fifth patient was a 30-month-old boy with a rapidly progressive spinal cord syndrome. At autopsy he had measles virus infection, largely confined to the gray matter of the spinal cord, with chronic inflammation, loss of nerve cells

including anterior horn cells, and Cowdry type A intranuclear inclusions (Figure 9.11, facing p. 208). The onset of this spinal cord disorder followed an episode of acute, generalized measles virus infection by an interval of 4 months. Electron microscopical examination of the gray matter of the cord revealed intranuclear paramyxovirus particles with a diameter of approximately 18 nm (Figure 9.12). The diagnosis was confirmed by in situ hybridization, using a highly specific measles virus probe. It is of interest that the spinal cord in this case also demonstrated vacuolar myelopathy (Sharer et al, 1990). A few measles type lesions with Cowdry type A intranuclear inclusions were also seen in the brain, which in addition had inflammatory cell infiltrates and HIV-1 type MGC.

None of the cases in this series had documented involvement of the CNS by either *Toxoplasma gondii* or *Cryptococcus neoformans*, two of the most common opportunistic infections of the brain in adults with AIDS (see Chapter 6). At least two children with HIV-1 infection have been described as having had CNS toxoplasmosis (Biggemann et al, 1987; Shanks et al, 1987), but its occurrence in children with AIDS is distinctly unusual, with no cases documented in the series compiled by Kozlowski et al (1990) from four institutions, including our own (Kozlowski et al, 1990). We have observed CNS toxoplasmosis at post mortem in at least one case, a 3-month-old infant with congenital *Toxoplasma gondii* and presumed HIV-1 infection not included in the current series (Cohen-Addad et al, 1988). The case with disseminated *Candida* referred to above, a 4-year-old girl, was also found to have a large, necrotizing, focally-calcified mass adjacent to the right lateral ventricle. Although the lesion resembled a burned-out or treated *Toxoplasma* abscess (Navia et al, 1986), no organisms could be recognized in this lesion, even with the use of immunocytochemistry for *T. gondii* antigens.

A case of progressive multifocal leukoencephalopathy (PML) due to JC papovavirus has recently been documented in a 7-year-old boy with AIDS (Vandersteenhoven et al, 1992). This is the first report of a case of PML in a child with HIV-1 infection.

Miscellaneous Findings

There were several other findings that were either directly or indirectly related to HIV-1 infection. A 6-year-old boy, previously described (Sharer et al, 1986), had severe nerve cell loss throughout the cerebral cortex, with many MGC and

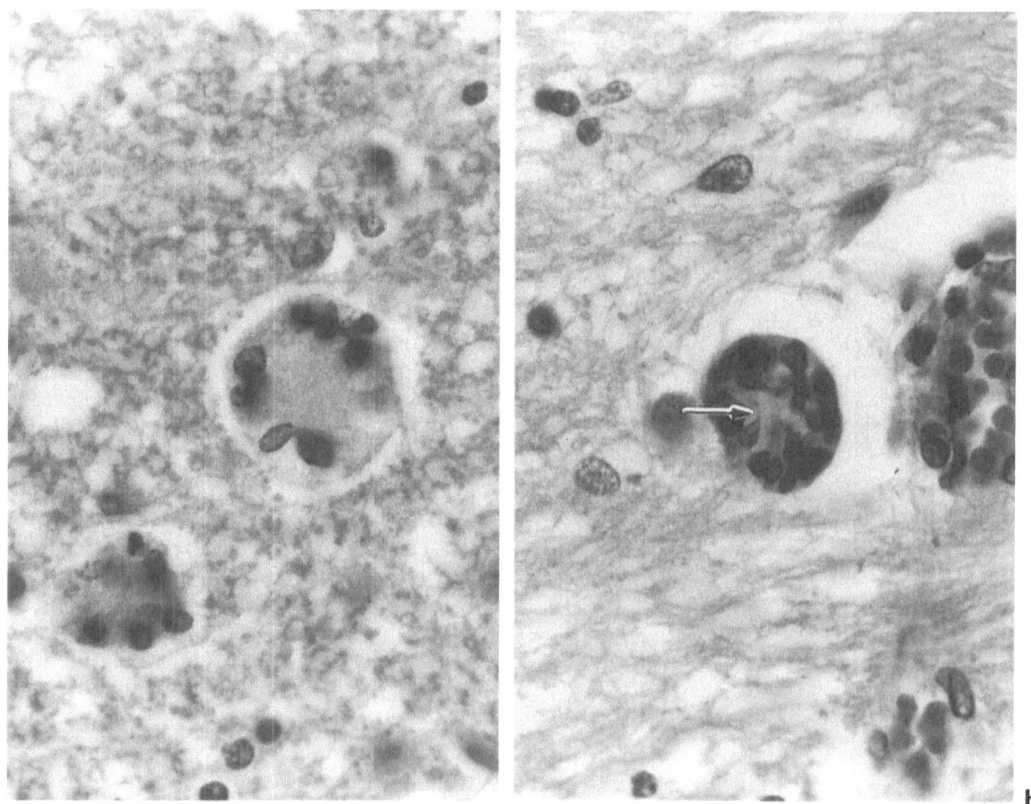

Figure 9.7. a MGC with varying morphology, cerebral white matter, 13-year-old boy. The cell on the upper right has peripherally arrayed nuclei, while nuclei in the other cell are dispersed in the cytoplasm. **b** MGC in corpus callosum, same patient as in **a**. Note tapering nuclear processes or nuclear bridges (*arrow*). H&E, original magnification, ×800.

numerous HIV-1 particles on electron microscopy (Figure 9.9), similar to the rare, necrotizing HIV-1 encephalopathy described by other authors (Clague et al, 1988; Giangaspero et al, 1989). A 2-year-old girl, also reported previously, had a multicentric primary CNS lymphoma of the B-cell type (Epstein et al, 1988b).

Two children had vacuolar myelopathy. One was the 30-month-old boy with measles infection described above, and the other was a 9-year-old girl who also had severe HIV-1 infection of the cord (Sharer et al, 1990). The spinal cord lesions in these cases were identical to those seen in adult subjects with vacuolar myelopathy (Petito et al, 1986) (see Chapter 8), with varying degrees of vacuolar change of the lateral and posterior columns, with the most prominent involvement in the thoracic (dorsal) segments of the cord. For reasons that are unknown (see Chapter 8), vacuolar myelopathy occurs much more rarely in children than in adults with HIV-1 infection; these two cases are the first to be reported in the literature (Sharer et al, 1990).

Four cases had lesions that may have been related to AIDS. One child, a 12-month-old girl, had necrotizing lesions with basophilic mineralization of axons confined to the pons (Figure 9.13), similar to the pontine leukoencephalopathy described by Vinters et al (1987) in patients with various immune deficiency disorders, including AIDS. One 18-month-old girl had an acute necrotizing lesion, with hemorrhage, in the basal ganglia on one side, thought possibly to be either an acute infarct or acute viral encephalitis. Attempts to identify herpes virus within the lesion were unsuccessful. A 7-year-old boy had marked intimal hyperplasia involving large arteries at the base of the brain and in the leptomeninges, with recent infarction in the basal ganglia (Cho et al, 1987). No MGC were seen in either the involved vessels or the brain as described by Yankner et al (1986) in an adult. A similar case has been described by Dickson et al (1989a). Another child, the 13-year-old boy described earlier, also had intimal hyperplasia of the arteries of the circle of Willis, without

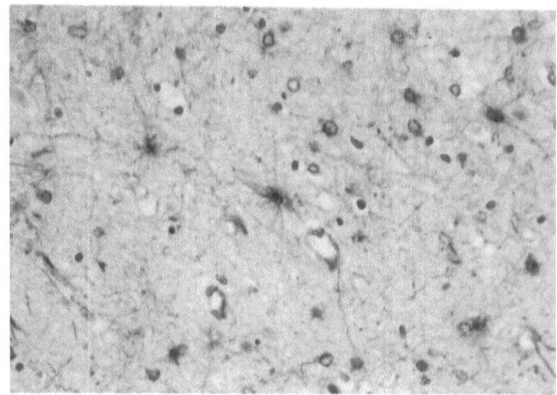

Figure 9.3. Reactive astrocytes in cerebral white matter, 6-month-old boy. Avidin-biotin immunocytochemistry with polyclonal antiserum to glial fibrillary acidic protein (GFAP), hematoxylin counterstain. Original magnification ×100.

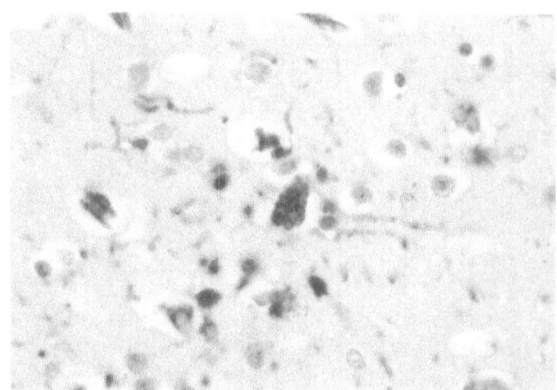

Figure 9.6. Nuclear cluster cell in center of the field, with reaction product in scanty cytoplasm, cerebral cortex, 12-month-old girl (same patient as in Figure 9.4). Other positive cells are present in the field, some of which are microglia with fine processes. Lectin histochemistry using biotinylated *Ricinus communis* agglutinin-1, hematoxylin counterstain. Original magnification ×160.

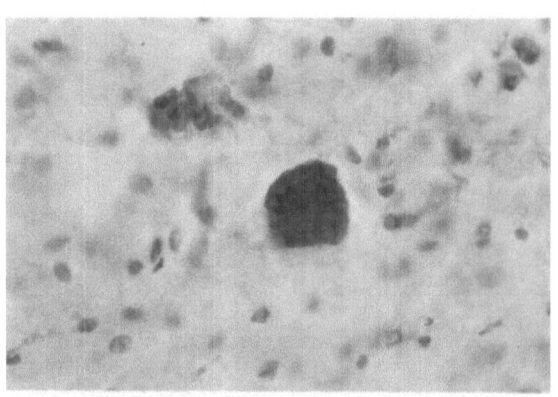

Figure 9.8. Cells stained for HIV-1 p25 core protein, hippocampus, 9-year-old girl. A positively-stained MGC is present in the center of the field, with other positive cells, including macrophages and cells with processes, consistent with microglia. Frozen section, avidin-biotin immunocyto-chemistry using monoclonal antibody to p25, hematoxylin counterstain. Original magnification ×160.

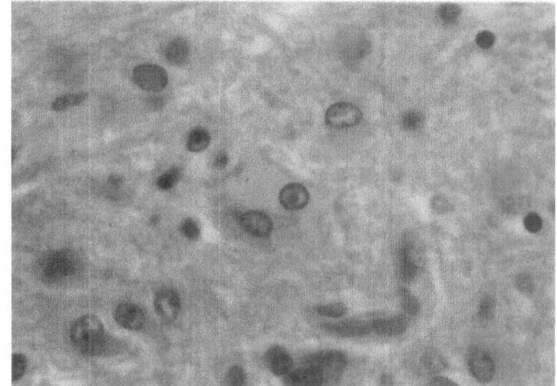

Figure 9.11. Eosinophilic, Cowdry type A intranuclear inclusion, apparently in an astrocyte, gray matter of lumbar spinal cord, 30-month-old boy with measles myelitis. H&E, oil immersion. Original magnification ×250.

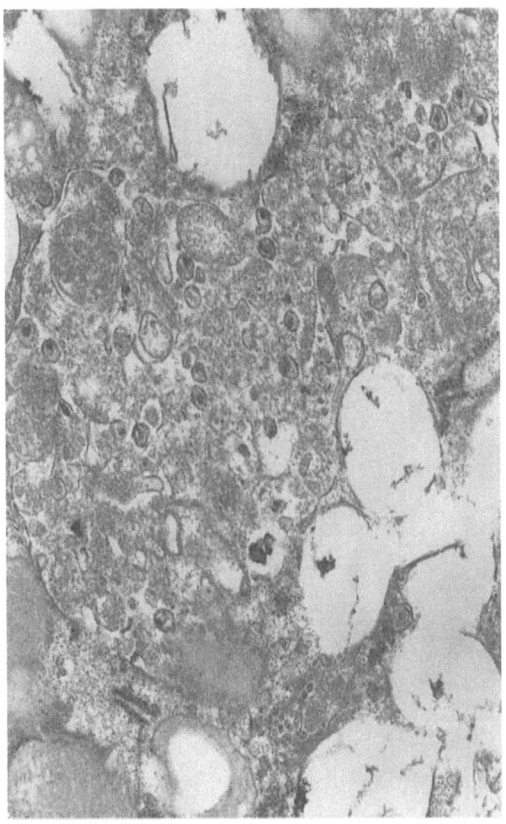

Figure 9.9. Lentiviral particles with eccentric dense cores within the cytoplasm of a MGC, cerebral cortex of 6-year-old boy with HIV-1 infection and severe encephalopathy. This same patient had severe cortical neuronal loss and gliosis. Original magnification ×28 000.

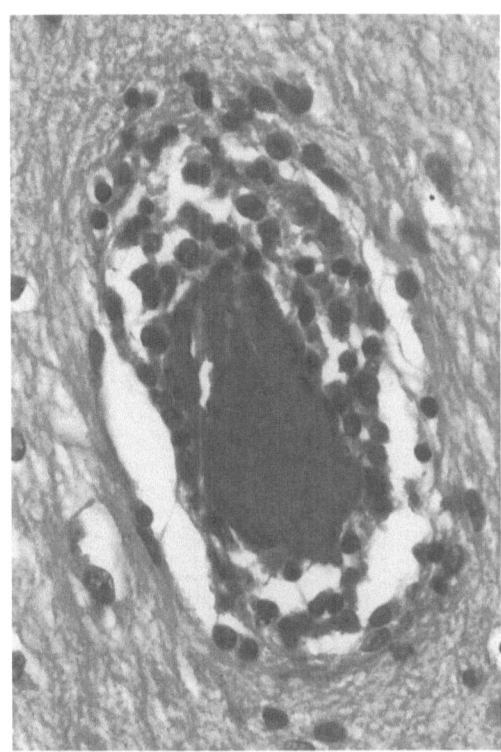

Figure 9.10. Inflammation of wall of vein, cerebral white matter, 8-month-old girl. H&E, original magnification ×500.

evidence of either MGC in the vessels or cerebral infarction. Many MGC were present in the brain in this case (Figure 9.7). One child, a 4-year-old girl with a primary leiomyosarcoma in the abdomen, had microscopic metastases in the brain.

Several patients had general neuropathologic abnormalities felt to be unrelated to HIV-1 infection and AIDS. Three had evidence of recent (two cases) or old (one case) anoxic damage, with nerve cell change in Sommer's sector of the hippocampus. One additional patient, a 2½-year-old boy, had severe anoxic damage of the entire cerebral cortex, basal ganglia and cerebellum, with early organization. There was evidence of old, non-hemorrhagic Wernicke's encephalopathy involving the mammillary bodies in one patient, a 5-year-old boy. An 18-month-old boy had diffuse cerebral edema of unknown cause, with central transtentorial herniation and Duret hemorrhages in the upper brainstem. The oldest

child, the 13-year-old boy mentioned above, had organizing contusions of the left temporal lobe, following a traumatic head injury that occurred 5 months prior to death. Finally, two patients had small vascular malformations in the cerebral hemispheres, without evidence of hemorrhage. One of these had dystrophic, basophilic mineralization of axons in the region of the malformation.

Review of Reported Cases

There have been other series of CNS findings in children with HIV-1 infection, but only that of Dickson and his colleagues at the Albert Einstein College of Medicine in the Bronx, New York, has been reported in detail (Dickson et al, 1989a,b; Wiley et al, 1990). The findings of this group of investigators are similar to those in our series, with the added description of cases with extensive vascular pathology, with stroke in 6 out of 25

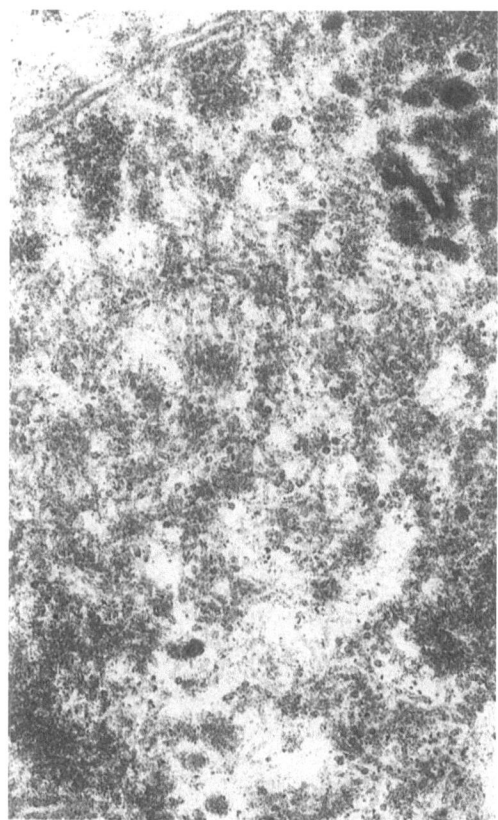

Figure 9.12. Eosinophilic, Cowdry type A intranuclear inclusion, apparently in an astrocyte, gray matter of lumbar spinal cord, 30-month-old boy with measles myelitis. H&E, oil immersion. Original magnification ×250.

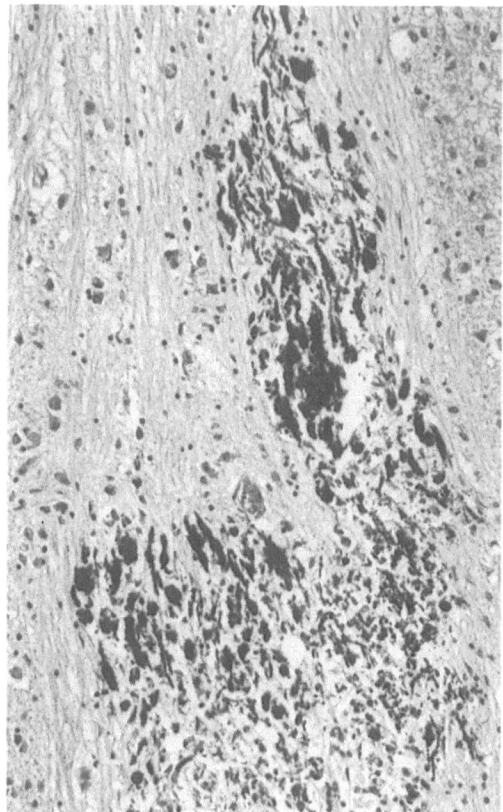

Figure 9.13. Paramyxovirus-like nucleocapsids in nucleus of unidentified cell, lumbar gray matter, same patient as Figure 9.11. Numerous elongated and cross-sectional profiles are present. Original magnification ×47 500.

cases (Park et al, 1990). In this series and in the larger compilation of cases by Kozlowski et al (1990), there was also a paucity of opportunistic infections, with CMV as the most common agent. Sporadic reports of other findings have also appeared, including one Golgi study in which an abnormality of dendritic branching was detected in the brain of a 2-year-old child (Mervis et al, 1987).

Conclusions

Pathological study of the brains of children with HIV-1 infection has helped to shed light on the pathogenesis of the encephalopathy that occurs in AIDS patients, since children, when compared with adults, infrequently have recognizable opportunistic infections of the CNS that can cloud the clinical and pathological picture. It is difficult to invoke a role for co-factor infections, such as CMV, in the pathogenesis of HIV-1 infection of the CNS, since these other types of infection are so rare in the childhood cases. In a sense, HIV-1 infection of the CNS in children is a 'cleaner' system, and it is easier to accept a direct or indirect role for HIV-1 as the cause of the progressive encephalopathy seen in this population of patients.

The low incidence of opportunistic infections in the CNS of children with AIDS in the USA is best explained by the limited time of exposure that these children have had to the common pathogens during their short lifetimes. There have been no published reports of neuropathological findings in children with AIDS from developing countries that have a high incidence of perinatal HIV-1 infection (Mann et al, 1986). It is thus unknown whether the low incidence of CNS opportunistic and reactivated, latent infections observed in children with HIV-1 infection in the USA would be recapitulated in other countries where the incidence of opportunistic infections outside the CNS is likely to be high (Boccon-Gibod, 1990).

Many questions regarding AIDS and the nervous system remain to be answered, the most important of which is whether or not HIV-1 is neurotropic (see also Chapter 7). Current information suggests only limited tropism of the virus for cells of neuroectodermal origin in vivo, although there is persuasive evidence for in vitro infection of cells of neuronal and glial origin (Michaels et al, 1988b; Li et al, 1990). Infection of oligodendrocytes or neurons would do much to explain the progression of the encephalopathy, but to date neither has been convincingly demonstrated. It is more difficult to explain the pathogenesis of the disorder on the basis of infection of monocytes, macrophages and MGC, although infection of microglia could disturb the normal function, admittedly unknown, of these cells in their resting state. As a consequence of this difficulty, it has been postulated that indirect mechanisms involving cytokines, including tumor necrosis factor alpha (TNFα), produced by infected macrophages and MGC or by activated astrocytes, play a role (Michaels et al, 1988b; Mintz et al, 1989b).

Several authors have presented evidence of infection of CNS capillary endothelial cells by HIV-1 (Wiley et al, 1986; Rostad et al, 1987). In our own material, we have been unable to document that there is infection of these cells, and other investigators have either had difficulty distinguishing endothelial cell infection from pericyte infection by light microscopy (Budka, 1990), or have seen endothelial cell infection only rarely (Kure et al, 1989). It has been proposed that infection of CNS capillary endothelial cells could cause alteration of the blood–brain barrier. In simplistic fashion, one might imagine that blood–brain barrier infection would result in cerebral edema, but by and large brain edema has not been a major finding in adults and children who have died with HIV-1 infection.

Newer issues have surfaced recently, with the advent of therapy for HIV-1 infection with anti-retroviral agents. Several investigators have described dramatic clinical improvements in the progressive encephalopathy in children, including decrease in the severity of cerebral atrophy on imaging studies, following institution of zidovudine therapy (Matthes et al, 1988; Pizzo et al, 1988). We have observed a similar effect in a number of our own patients (Mintz, 1992). This dramatic near-reversal of CNS deterioration with anti-retroviral therapy has not been reported

in adult subjects with HIV encephalopathy, although improvements in cognitive testing have been described (Schmitt et al, 1988). The pathological basis for the improvement in children has not been reported and so far remains unknown. We have not, as yet, had autopsies performed on any of our own patients who were treated with zidovudine. Because of the wide variability of the degree of CNS involvement by HIV type lesions in children with HIV-1 infection, it is unlikely that any useful pathological information about the effects on HIV-1 encephalitis of zidovudine or other anti-retroviral agents can be obtained from the study of one or even a few treated cases. Furthermore, we have recently observed a number of children on long-term zidovudine therapy who have, after initial improvement, experienced later neurological deterioration (Mintz et al, 1990).

Lastly, a key question remains, and it has been the most difficult to address. The timing of infection of the CNS by HIV-1 in children who have acquired the virus perinatally, or vertically, from their mothers is currently unknown. Preliminary evidence has been presented suggesting infection of the fetal CNS during gestation (Jovaisas et al, 1985; Lyman et al, 1990), but more solid information is needed. Indeed, the timing of infection of the fetus itself is uncertain, and currently the terms 'perinatal' and 'vertical' are preferred to 'in utero' and 'transplacental' (Peckham et al, 1988; Plotkin et al, 1988). Nevertheless, protocols for experimental anti-retroviral therapy of HIV-1 infected pregnant women and their fetuses are already in use.

The stimulus for HIV-1 to enter the CNS, both in perinatal and non-perinatal cases, is unknown. Some authors have favored a 'Trojan horse' theory, with virus entering as a passenger in monocytes/macrophages attracted to the CNS for some other reason (Haase, 1986). In children it is difficult to accept this hypothesis, since evidence for the 'other reason' is usually lacking. Like other lentiviruses, HIV-1 may naturally enter the CNS because it is neurotropic, but this has yet to be proved.

Acknowledgements. Dr Leon G. Epstein initiated the clinical neurological studies, and Dr James M. Oleske and Dr Edward M. Connor took care of the patients. Dr Vijay Joshi, Dr Frederick DiCarlo and Dr Gloria Kohut performed the autopsies. Dr Jennifer Michaels performed the immunocytochemistry. Ms Dorothea Joppe, Mr Markus Meyenhofer and Miss Tracy Roscoe assisted in the preparation of the electron micrographs. The p25 monoclonal antibody was a gift of Dr Kathy Shriver and Dr Lynn Goldstein, of Genetic Systems. This study was supported by US Public Health Service Grants NS 25121 and NS 28754.

References

Barré-Sinoussi F, Chermann JC, Rey F et al. (1983) Isolation of a T-lymphotropic retrovirus from a patient at risk for acquired immune deficiency syndrome (AIDS). Science 220:868–71.

Belman AL, Ultmann MH, Horoupian D et al. (1985) Neurological complications in infants and children with acquired immune deficiency syndrome. Ann Neurol 18: 560–6.

Belman AL, Lantos G, Horoupian D et al. (1986) Calcification of the basal ganglia in infants and children with AIDS. Neurology 36:1129–99.

Belman AL, Diamond G, Dickson D et al. (1988) Pediatric acquired immunodeficiency syndrome: neurologic syndromes. Am J Dis Child 142:29–35.

Biggemann B, Voit T, Neuen E et al. (1987) Neurologic manifestations in three German children with AIDS. Neuropediatrics 18:99–106.

Boccon-Gibod L. (1990) Pathology of AIDS in African patients. In: Joshi VV, ed. Pathology of AIDS and other manifestations of HIV infection. Igaku Shoin: New York, 329–46.

Budka H. (1990) Human immunodeficiency virus (HIV) envelope and core proteins in CNS tissues of patients with the acquired immune deficiency syndrome (AIDS). Acta Neuropathol 79:611–19.

Centers for Disease Control. (1987) Classification system for human immunodeficiency virus (HIV) infection in children under 13 years of age. MMWR 36:225–36.

Cho E-S, Sharer LR. (1990) Central nervous system in HIV infection. In: Joshi VV, ed. Pathology of AIDS and other manifestations of HIV infection. Igaku Shoin: New York, 43–63.

Cho E-S, Sharer LR, Peress NS et al. (1987) Intimal proliferation of leptomeningeal arteries and brain infarcts in subjects with AIDS (Abstr). J Neuropathol Exp Neurol 46:385.

Church JA, Isaacs H. (1984) Transfusion-associated acquired immune deficiency syndrome in infants. J Pediatr 105: 731–7.

Clague CPT, Ostrowski MA, Deck JHN et al. (1988) Severe diffuse necrotizing cortical encephalopathy in acquired immune deficiency syndrome (AIDS): an immunocytochemical and ultrastructural study (Abstr). J Neuropathol Exp Neurol 47:346.

Cohen-Addad NE, Joshi VV, Sharer LR et al. (1988) Congenital acquired immunodeficiency syndrome and congenital toxoplasmosis: pathologic support for a chronology of events. J Perinatol 8:328–31.

Davis KC, Horsburgh CR, Hasiba U et al. (1983) Acquired immunodeficiency syndrome in a patient with hemophilia. Ann Intern Med 98:284–6.

Diamond GW. (1989) Developmental problems in children with HIV infection. Ment Retard 27:213–17.

Dickson DW, Belman AL, Park YD et al. (1989a) Central nervous system pathology in pediatric AIDS: an autopsy study. APMIS 8 (suppl):40–57.

Dickson DW, Belman AL, Kim TS et al. (1989b) Spinal cord pathology in pediatric acquired immunodeficiency syndrome. Neurology 39:227–35.

Epstein LG, Sharer LR, Joshi VV et al. (1984) Progressive encephalopathy in children with acquired immune deficiency syndrome: clinical and neuropathological findings. Ann Neurol 16:414 (abstr).

Epstein LG, Sharer LR, Joshi VV et al. (1985) Progressive encephalopathy in children with acquired immune deficiency syndrome. Ann Neurol 17:488–96.

Epstein LG, Sharer LR, Oleske JM et al. (1986) Neurologic manifestations of HIV infection in children. Pediatrics 78: 678–87.

Epstein LG, Berman CZ, Sharer LR et al. (1987) Unilateral calcification and contrast enhancement of the basal ganglia in a child with AIDS encephalopathy. AJNR 8:163–5.

Epstein LG, Sharer LR, Goudsmit J. (1988a) Neurological and neuropathological features of human immunodeficiency virus infection in children. Ann Neurol 23 (suppl):19–23.

Epstein LG, DiCarlo FJ, Joshi VV et al. (1988b) Primary lymphoma of the central nervous system in children with acquired immunodeficiency syndrome. Pediatrics 82: 355–63.

Falloon J, Eddy J, Wiener L et al. (1989) Human immunodeficiency virus infection in children. J Pediatr 114:1–30.

Gajdusek DC, Amyx HL, Gibbs CJ et al. (1985) Infection of chimpanzees by human T-lymphotropic retroviruses in brain and other tissues from AIDS patients. Lancet 1:55–6.

Gallo RC, Salahuddin SZ, Popovic M et al. (1984) Frequent detection and isolation of cytopathic retroviruses (HTLV-III) from patients with AIDS and at risk for AIDS. Science 224:500–2.

Giangaspero F, Scanabissi E, Baldacci MC et al. (1989) Massive neuronal destruction in human immunodeficiency virus (HIV) encephalitis: a clinico-pathological study of a pediatric case. Acta Neuropathol 78:662–5.

Gottlieb MS, Schroff R, Schanker RM et al. (1981) *Pneumocystic carinii* pneumonia and mucosal candidiasis in previously healthy homosexual men. Evidence of a new acquired cellular immunodeficiency. N Engl J Med 305: 1425–31.

Haase A. (1986) Pathogenesis of lentivirus infections. Nature 322:130–6.

Janssen RS, Cornblath DR, Epstein LG et al. (1989) Human immunodeficiency virus (HIV) infection and the nervous system: report from the American Academy of Neurology AIDS Task Force. Neurology 39:119–22.

Joshi VV. (1990) Pathology of acquired immunodeficiency syndrome (AIDS) in children. In: Joshi VV, ed. Pathology of AIDS and other manifestations of HIV infection. Igaku Shoin: New York, 239–69.

Joshi VV, Oleske JM, Minnefor AB et al. (1985) Pathologic pulmonary findings in children with the acquired immunodeficiency syndrome: a study of ten cases. Hum Pathol 16:241–6.

Joshi VV, Kauffman S, Oleske JM et al. (1987a) Polyclonal polymorphic B-cell lymphoproliferative disorder with prominent pulmonary involvement in children with acquired immune deficiency syndrome. Cancer 59:1455–62.

Joshi VV, Pawel B, Connor E et al. (1987b) Arteriopathy in children with acquired immune deficiency syndrome. Pediatr Pathol 7:261–75.

Jovaisas E, Koch MA, Schafer A et al. (1985) LAV/HTLV-III in 20-week fetus (Letter). Lancet 2:1129.

Kozlowski PB, Sher JH, Dickson DW et al. (1990) Central nervous system in pediatric HIV infection: experience from a multicenter study. In: Kozlowski PB, Snider DA, Vietze PM et al, eds. Brain in pediatric AIDS. Karger: Basel,

132–46.

Kure K, Lyman WD, Weidenheim KM et al. (1989) Subacute AIDS encephalitis: identification of human immunodeficiency virus (HIV)-infected cells by an improved double-immunostaining method (Abstr). J Neuropathol Exp Neurol 48:385.

Levy JA, Hoffman AD, Kramer SM et al. (1984) Isolation of lymphocytopathic retroviruses from San Francisco patients with AIDS. Science 225:840–84.

Levy RM, Bredesen DE, Rosenblum ML. (1985) Neurological manifestations of the acquired immunodeficiency syndrome (AIDS): experience at UCSF and review of the literature. J Neurosurg 62:475–95.

Li XL, Moudgil T, Vinters HV et al. (1990) CD4-independent, productive infection of a neuronal cell line by human immunodeficiency virus type 1. J Virol 64:1383–7.

Lyman WD, Kress Y, Kure K et al. (1990) Detection of HIV in fetal central nervous system tissue. AIDS 4:917–20.

Mann JM, Francis H, Davachi F et al. (1986) Human immunodeficiency virus seroprevalence in pediatric patients 2 to 14 years of age at Mama Yemo Hospital, Kinshasa, Zaire. Pediatrics 78:673–7.

Masur HM, Michelis MA, Greene JB et al. (1981) An outbreak of community-acquired Pneumocystis carinii pneumonia: manifestations of cellular immune dysfunction. N Engl J Med 305:1431–8.

Matthes J, Walker LA, Watson JG et al. (1988) AIDS encephalopathy with response to treatment. Arch Dis Child 63:545–7.

McArthur JC, Becker PS, Parisi JE et al. (1989) Neuropathological changes in early HIV-1 dementia. Ann Neurol 26:681–4.

Mervis RF, Boesel C, Rosenblum M et al. (1987) Abnormal neuronal morphology in the brain of an AIDS-infected child: a Golgi study (Abstr). Abstracts of the Society for Neuroscience, 17th Annual Meeting, New Orleans, LA, USA, November 21, 1987. Abstract no. 348.1, p. 1252.

Meyenhofer MF, Epstein LG, Cho E-S et al. (1987) Ultrastructural morphology and intracellular production of human immunodeficiency virus (HIV) in brain. J Neuropathol Exp Neurol 46:474–84.

Michaels J, Price RW, Rosenblum MK. (1988a) Microglia in the giant cell encephalitis of acquired immune deficiency syndrome: proliferation, infection, and fusion. Acta Neuropathol 76:373–9.

Michaels J, Sharer LR, Epstein LG. (1988b) Human immunodeficiency virus type 1 (HIV-1) infection of the nervous system: a review. Immunodeficiency Rev 1:71–104.

Mintz M. (1992) Neurological abnormalities. In: Yogev R, Connor EM, ed. Management of HIV infection in infants and children. Mosby Yearbook: St. Louis, 247–85.

Mintz M, Epstein LG, Koenigsberger MR. (1989a) Neurological manifestations of acquired immunodeficiency syndrome in children. Int Pediatr 4:161–71.

Mintz M, Rapaport R, Oleske JM et al. (1989b) Elevated serum levels of tumor necrosis factor are associated with progressive encephalopathy in children with acquired immunodeficiency syndrome. Am J Dis Child 143:771–4.

Mintz M, Connor EM, Oleske JM et al. (1990) Neurologic deterioration in children on long-term zidovudine therapy (Abstr). Presented at the Ninth AIDS Clinical Trials Group Meeting, Bethesda, MD, USA, July 10–13, 1990.

Mizusawa H, Hirano A, Llena F. (1988) Nuclear bridges within multinucleated giant cells in subacute encephalitis of acquired immunodeficiency syndrome (AIDS). Acta Neuropathol 76:166–9.

Navia BA, Petito CK, Gold JWM et al. (1986) Cerebral toxoplasmosis complicating the acquired immune deficiency syndrome: clinical and neuropathological findings in 27 patients. Ann Neurol 19:224–38.

Oleske J, Minnefor A, Cooper RJ et al. (1983) Immune deficiency in children. JAMA 249:2345–9.

Park YD, Belman AL, Kim T-S et al. (1990) Stroke in pediatric acquired immunodeficiency syndrome. Ann Neurol 28:303–11.

Peckham CS, Senturia YD, Ades AE et al. (1988) The European Collaborative Study, mother-to-child transmission of HIV infection. Lancet 22:1039–43.

Perl DP, Good PF, Jabbar G. (1989) Basal ganglia mineralization in AIDS: evidence of massive aluminum deposition (Abstr). J Neuropathol Exp Neurol 48:384.

Petito CK, Cho E-S, Lemann W et al. (1986) Neuropathology of acquired immunodeficiency syndrome (AIDS): an autopsy review. J Neuropathol Exp Neurol 45:635–46.

Pizzo PA, Eddy J, Falloon J et al. (1988) Effects of continuous infusion of zidovudine (AZT) in children with symptomatic HIV infection. N Engl J Med 319:889–96.

Plotkin SA, Evans HE, Fost NC et al. (1988) The Task Force on Pediatric AIDS, perinatal human immunodeficiency virus infection. Pediatrics 82:941–4.

Price RW, Brew B, Sidtis J et al. (1988) The brain in AIDS: central nervous system HIV-1 infection and AIDS dementia complex. Science 239:586–92.

Raphael SA, de Leon G, Sapin J. (1989) Symptomatic primary human immunodeficiency virus infection of the brain stem in a child, Pediatr Infect Dis J 8:654–6.

Rostad SW, Sumi SM, Shaw C-M et al. (1987) Human immunodeficiency virus (HIV) infection in brains with AIDS-related leukoencephalopathy. AIDS Res Hum Retroviruses 3:363–73.

Rubinstein AM, Sicklick M, Gupta A et al. (1983) Acquired immunodeficiency with reversed T4/T8 ratios in infants born to promiscuous and drug-addicted mothers. JAMA 249:2350–6.

Schmitt FA, Bigley JW, McKinnis R et al. (1988) Neuropsychological outcome of zidovudine (AZT) treatment of patients with AIDS and AIDS-related complex. N Engl J Med 319:1573–8.

Schulz DM, Giordano DA, Schulz DH. (1962) Weights of organs of fetuses and infants. Arch Pathol 74:244–50.

Scott GB, Hutto C, Makuch R et al. (1989) Survival in children with perinatally acquired human immunodeficiency virus type 1 infection. N Engl J Med 321:1791–6.

Shanks GD, Redfield RR, Fischer GW. (1987) Toxoplasma encephalitis in an infant with acquired immunodeficiency syndrome, Pediatr Infect Dis J 6:70–1.

Sharer LR, Epstein LG. (1989) Intravenous infusion of zidovudine (AZT) in children with HIV infection (Letter). N Engl J Med 320:806.

Sharer LR, Cho E-S, Epstein LG. (1985) Multinucleated giant cells and HTLV-III in AIDS encephalopathy, Hum Pathol 16:760.

Sharer LR, Epstein LG, Cho E-S et al. (1986) Pathologic features of AIDS encephalopathy in children: evidence for LAV/HTLV-III infection of brain. Hum Pathol 17:271–84.

Sharer LR, Dowling PC, Michaels J et al. (1990) Spinal cord disease in children with HIV-1 infection: a combined molecular and neuropathological study, Neuropathol Appl Neurobiol 16:317–31.

Shaw GM, Harper ME, Hahn BH et al. (1985) HTLV-III infection in brains of children and adults with AIDS encephalopathy. Science 227:177–82.

Small CB, Klein RS, Friedland GH et al. (1983) Community-acquired opportunistic infections and defective cellular immunity in heterosexual drug abusers and homosexual men. Am J Med 74:433–41.

Snider WD, Simpson DM, Nielsen SL et al. (1983) Neurological complications of acquired immune deficiency
syndrome: analysis of 50 patients. Ann Neurol 14:403–
18.

Vandersteenhoven JJ, Dbaibo G, Boyko OB et al. (1992)
Progressive multifocal leukoencephalopathy in pediatric
acquired immunodeficiency syndrome. Pediatr Infect Dis J
11:232–37.

Vinters HV, Anders KH, Barach P. (1987) Focal pontine
leukoencephalopathy in immunosuppressed patients. Arch
Pathol Lab Med 111:192–6.

Wiley CA, Schrier RD, Nelson JA et al. (1986) Cellular
localization of human immunodeficiency virus infection
within the brains of acquired immune deficiency syndrome
patients. Proc Natl Acad Sci USA 83:7089–93.

Wiley CA, Belman AL, Dickson DW et al. (1990) Human
immunodeficiency virus within the brains of children with
AIDS. Clin Neuropathol 9:1–6.

Yankner BA, Skolnick PR, Shoukimas GM et al. (1986)
Cerebral granulomatous angiitis associated with isolation
of human T-lymphotropic virus type III from the central
nervous system. Ann Neurol 20:362–4.

10 Peripheral Nerve and Muscle Disease in HIV Infection

G.N. Fuller and J.M. Jacobs

Introduction

Peripheral neuropathies were among the first neurological conditions to be described in AIDS (Snider et al, 1983). In the intervening 8 years, much has been learned of the pathogenesis of AIDS and many of its neurological complications. The demonstration of the direct involvement of HIV in central nervous system (CNS) conditions (discussed in Chapters 6 and 7) has led many workers to seek a similar role for HIV in the peripheral nervous system (PNS). No conclusive evidence has yet been found for this. The neuropathology of AIDS in the PNS is less well documented than in the CNS, and peripheral nerve syndromes are frequently without neuropathological correlates.

The study of peripheral neuropathies presents specific problems related to the particular organization of the PNS. Few postmortem studies include adequate sampling of the cells of origin of peripheral nerve fibres, or of the fibres themselves. In most cases neuropathological examination is limited to biopsied or postmortem sural nerves.

A further problem concerns the method of examination of peripheral nerves. Paraffin sections are invaluable for the identification of histological changes within nerves, for example,

the presence of inflammatory cells, or evidence of vasculitis. However, changes to nerve fibres themselves, which may be of a relatively subtle nature, can only be assessed accurately by the use of semithin resin sections and of teased fibre preparations. Few pathological reports include full examination by these methods.

Classification

Classifications have evolved based upon the recognition that different clinical syndromes occur at successive stages of HIV infection (McArthur, 1987; Dalakas and Pezeshkpour, 1988; Parry 1988); Figure 10.1 shows a scheme derived from these. During early HIV-seropositivity there is no immunosuppression, though patients appear more liable to abnormalities of immune function. As immunosuppression becomes more marked, opportunistic infections and tumours may cause different peripheral nerve syndromes. At the same time patients become increasingly likely to develop peripheral neuropathy as a result of the neurotoxic side-effects of drug treatment, or from malnutrition.

The stages of HIV infection have been categorized (CDC, 1986), and the simplified system used here comprises the stages of seroconversion (CDC stage 1), HIV-seropositive well (CDC

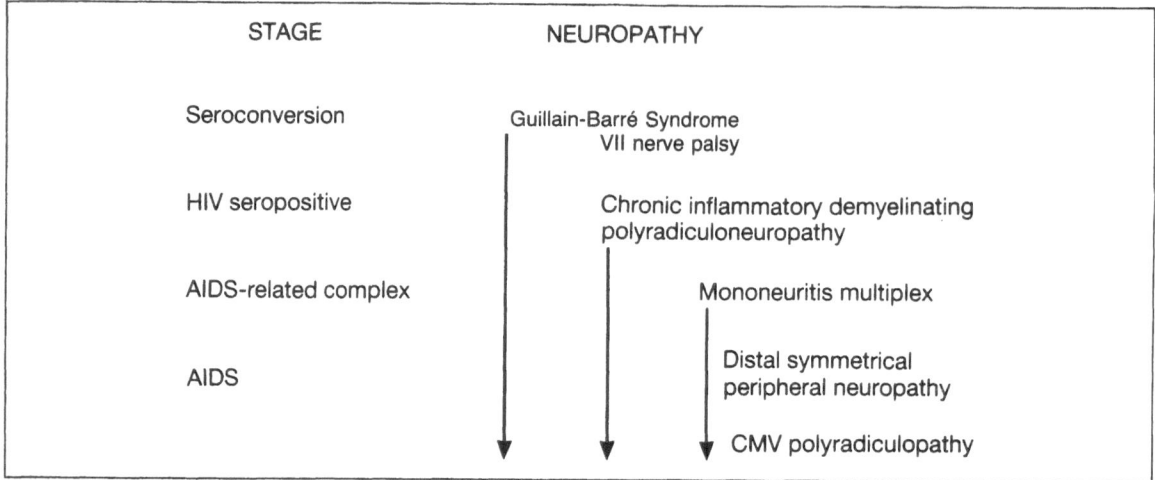

Figure 10.1. Classification of peripheral nerve pathology in HIV infection.

stage 2), and ARC and AIDS combined (CDC stages 3 and 4).

The rate of occurrence of the various neuropathies is difficult to ascertain and each will be discussed separately. HIV-seropositive well patients very rarely develop neuropathy. In ARC and AIDS, mononeuritis multiplex and polyradiculopathies are rare, while the distal sensory peripheral neuropathies are common, occurring in about one-third of patients.

Seroconversion

A seroconversion illness was recognized by Cooper et al (1985), and Ho et al (1985a). Peripheral nerve syndromes reported to occur in association with seroconversion and seroconversion illnesses include: acute inflammatory demyelinating polyradiculopathy (AIDP) (Hagberg et al, 1986; Piette et al, 1986); lower motor neuron (LMN) facial palsy, often associated with aseptic meningitis (Piette et al, 1986; Wiselka et al, 1987); and bilateral brachial neuritis (Calabrese et al, 1987). All these syndromes have been self-limiting and no neuropathology is available.

These neuropathies have been interpreted as evidence of early neurotropism of the HIV, particularly since HIV has been isolated in the cerebrospinal fluid (CSF) in some cases (Hagberg et al, 1986; Piette et al, 1986). However, these PNS syndromes have also been associated with seroconversion to other viruses, for example

CMV in AIDP (Winer et al, 1988). Thus, while these neuropathies possibly reflect a specific effect of HIV, alternative explanations include contemporaneous infection and seroconversion for another agent, or a non-specific mechanism common to HIV and other viruses.

HIV-Seropositive

Inflammatory Neuropathies

Both acute (AIDP) and chronic inflammatory demyelinating polyradiculopathy (CIDP) have been described in HIV-seropositive patients (Cornblath et al, 1987; Lange et al, 1988; Miller et al, 1988; Léger et al, 1989). At this stage of the disease, patients are otherwise well, and perhaps not previously recognized to be HIV-positive. It has therefore been possible to study these syndromes in the absence of other HIV-related diseases, which can complicate investigation. However, the use of number of publications as an index has led to an inflated perception of their frequency, and it is noteworthy that in a London population of 1500 HIV-positive patients, Fuller et al (1990a) found no cases over a 2 year period.

AIDP (Guillain–Barré Syndrome): Clinical Features

These were originally described by Cornblath et al (1987) and later by others (Léger et al, 1989).

As in HIV-seronegative patients, AIDP presents with a rapidly-evolving flaccid areflexic paralysis and mild sensory symptoms. Respiratory muscles may be involved, and assisted ventilation may be required. Cerebrospinal fluid protein is usually elevated, up to 4 g/l. Unlike HIV-negative AIDP, there is often a mild lymphocytic pleocytosis of up to 300 cells/mm^3. Immunological studies show reversal of the CD4/CD8 ratio and a polyclonal elevation of gammaglobulin. Electrophysiological findings are similar to those reported in sero-negative AIDP, which may be normal initially but later show slowing of conduction, prolongation of distal motor latencies and conduction block. Reduction of compound muscle action potentials with evidence of denervation on EMG consistent with axonal loss may follow. The prognosis appears to be good, although recovery is slow. Patients are reported to improve following plasmapheresis (Cornblath et al, 1987), although recovery may occur spontaneously.

CIDP: Clinical Features

Patients usually present with chronic weakness, absent reflexes in arms and legs, and relatively minor sensory loss (Cornblath et al, 1987; Miller et al, 1988). The onset may be progressive or relapsing. Cerebrospinal fluid protein is usually elevated, and there may be a mild lymphocytic pleocytosis of up to 50 cells/mm^3. Immunological abnormalities are similar to those found in AIDP, though the CD4/CD8 reversal may be more marked. Electrophysiological studies show changes similar to, though less marked than, those in AIDP. Most patients recover, either spontaneously or following steroids or plasmapheresis.

AIDP and CIDP: Pathological Features

Neuropathological reports are limited to biopsies of sural or superficial peroneal nerves (Cornblath et al, 1987; Riggs et al, 1987; Lange et al, 1988; Miller et al, 1988; Chaunu et al, 1989; Léger et al, 1989). One of the most consistent findings is mononuclear cell infiltration around epi- and endoneurial vessels. Léger et al (1989) noted the affected vessels to be <80 μm in diameter. The degree of infiltration is variable, and in a few cases no inflammatory cells were identified. In two cases reported by Cornblath et al (1987), immunostaining with monoclonal Leu 4 showed that many infiltrating cells were T-lymphocytes.

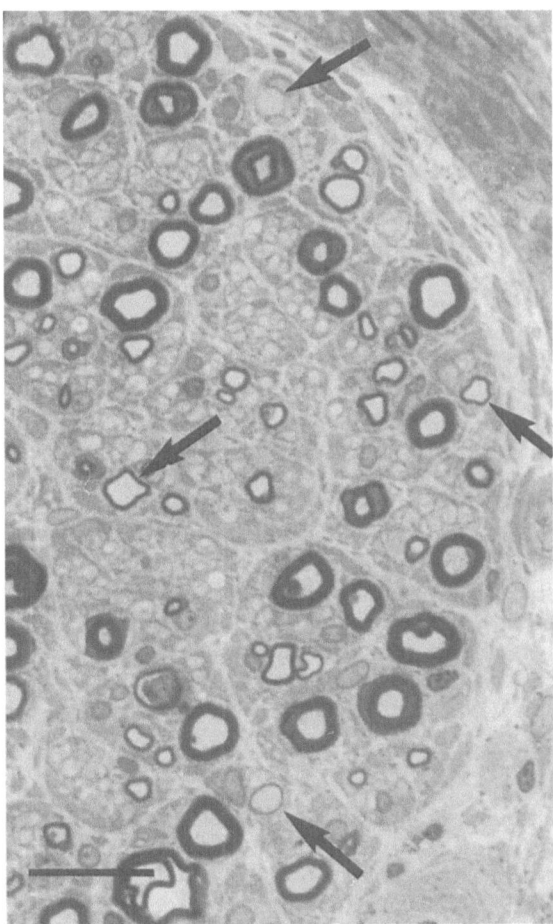

Figure 10.2. One-micrometre Araldite section of sural nerve from a patient with AIDP. Several demyelinated or thinly remyelinating fibres (*arrows*) are seen. Bar represents 20 μm.

Subperineurial oedema was noted in some of the cases studied by Cornblath et al (1987).

Demyelination is reported in many cases (Figure 10.2), and is most reliably identified in teased fibres (Chaunu et al, 1989; Léger et al, 1989). Electron microscopy of nerves from a few cases reported by Cornblath et al (1987) showed myelin stripping by macrophages.

Axonal degeneration is a common finding, although variable in degree. Quantitative studies (Chaunu et al, 1989; Léger et al, 1989) showed markedly-reduced myelinated fibre densities in a few cases, although in many, densities were within normal limits. In two cases seen by Cornblath et al (1987), myelinated fibre degeneration was associated with the presence of lipid-containing cells in the perineurium. Miller et al (1988) describe positive immunostaining for IgM in the perineurium, which is likely to be a non-specific change, probably related to increased vascular permeability associated with endoneurial pathology.

In situ hybridization for HIV has been uniformly negative, although HIV was cultured from one nerve (Ho et al, 1985b; Cornblath et al, 1987).

AIDP and CIDP: Pathogenesis

AIDP and CIDP have similar features in both HIV-positive and HIV-negative patients, and in both cases there may be a common immunopathogenic aetiology, a suggestion supported by reports of other immune-mediated abnormalities in HIV-positive patients, notably idiopathic thrombocytopenic purpura and the possible response to immunosuppressive agents. A direct role for HIV in AIDP and CIDP seems unlikely in view of the negative in situ hybridization findings and the uniformly good prognosis.

VII Nerve Palsy

Lower motor neuron facial palsies, unilateral or bilateral, have been reported as Bell's palsy where no aetiology is suggested (Brown et al, 1988). Not infrequently, an associated aseptic meningitis is found (Bélec et al, 1989). The cause of this apparent increase in frequency of LMN facial paralysis in HIV infected patients is not known.

ARC and AIDS

Subclinical

A number of electrophysiological and neuropathological studies have shown that subclinical peripheral nerve abnormalities are common in the later stages of HIV infection (de la Monte et al, 1988; Mah et al, 1988; So et al, 1988; Vishnubhakat and Beresford 1988; Carne et al, 1989; Chaunu et al, 1989). In a group of 30 patients with AIDS, but without symptoms or signs of peripheral neuropathy, Fuller et al (1991a) found a reduction of 37 per cent in sural sensory action potential, of 34 per cent in common peroneal compound muscle action potential, and reduced sensory and motor velocities of between

2 and 12 per cent compared with HIV-negative controls. These deficits represent changes throughout the population, since the standard deviations were similar to those of normal controls. They showed no correlation with parameters of HIV infection, for example, duration of AIDS or CD4 count, or with nutritional parameters such as Body Mass Index or serum albumin.

Pathological Features

Sural nerves were examined from five asymptomatic cases (Fuller et al, 1991a). No inflammatory cells were seen in paraffin sections. In semithin resin sections, one or two degenerating myelinated fibres were seen in every fascicle and examination of teased fibres confirmed the presence of ongoing myelinated fibre degeneration (Figure 10.3). Quantitative studies showed a reduction of about 30 per cent of myelinated fibres; no other qualitative abnormalities of nerve fibres were found. Unmyelinated fibre densities were consistently lower than those of non-AIDS controls, but because of the wide range of normal values these differences were not statistically significant. Electron microscopic examination also revealed the presence of tuboreticular bodies in endothelial cells of epi- and endoneurial endothelial cells (see below).

Subclinical peripheral nerve involvement is suggested to be evidence of an effect of HIV on the PNS (Chaunu et al, 1989). However, the changes are similar in degree to those found in other chronic diseases such as lymphoma (Walsh, 1971), carcinoma of the lung (Campbell and Paty, 1974) and other tumours (Paul et al, 1978), and probably reflect factors common to many disease states rather than the specific effect of HIV.

In view of these findings, it is important that asymptomatic AIDS or ARC patients are used as electrophysiological or pathological controls when assessing patients with peripheral nerve symptoms.

Mononeuritis Multiplex

Mononeuritis multiplex is the characteristic presentation for a number of specific multifocal neuropathies, and sometimes for more generalized conditions. In ARC and AIDS it is relatively rare, and is associated with three specific pathologies.

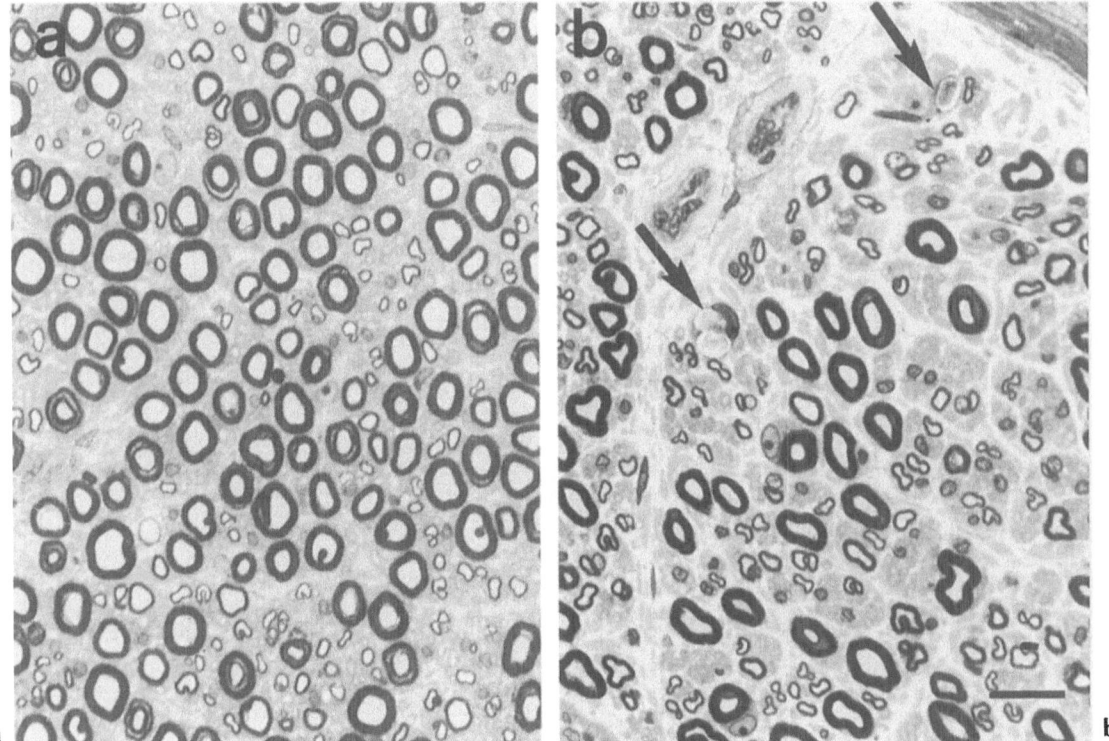

Figure 10.3a,b. One-micrometre Araldite sections of sural nerve from: **a** a sudden-death control (postmortem) of similar age to **b**; and **b** a case of AIDS without symptoms of peripheral neuropathy. In **b** the density of myelinated fibres is mildly decreased. Lipid debris (*arrows*) indicates degeneration of two myelinated fibres. Bar represents 20 μm.

Vasculitis

Examination of nerves from some patients who have presented with mononeuritis, usually of sudden onset, has shown a necrotizing vasculitis affecting medium-sized arterioles. In some cases, there are systemic symptoms (Weber et al, 1987), in others only the PNS appears to be affected (Gherardi et al, 1989). The aetiology is not known, but the suggestion has been made (Said et al, 1988) that necrotizing arteritis may result from a deposition of immune complexes from HIV infection by a process similar to that occurring in hepatitis B infection. The demonstration of HIV by in situ hybridization in perivascular CD8 lymphocytes in the epi- and endoneurium in two cases (Gherardi et al, 1989) could suggest a direct involvement of HIV in the pathogenesis of peripheral nerve vasculitis. However, the frequency of HIV infection of CD8 lymphocytes in other tissues was not examined in their patients, so the significance of these findings is uncertain. Acute degeneration of the majority of myelinated fibres was a common finding in one series of cases of necrotizing vasculitis, including three with

AIDS (Said et al, 1988); unequal fascicular distribution of fibre degeneration was sometimes a striking feature. Figure 10.4 illustrates necrotizing vasculitis in the femoral nerve of a patient with ARC who had had an episode of hepatitis B a year previously. The very short history (<24 hours) of clinical symptoms accounts for the absence of nerve fibre degeneration.

Lymphoma

Systemic B-cell non-Hodgkin's lymphoma occurs in about 3 per cent of patients with AIDS (Beral et al, 1989), a similar rate to that seen in other immunosuppressed states (Kinlen et al, 1979). This type of lymphoma infiltrates the peripheral nerves in up to 40 per cent of HIV-negative cases (Jellinger and Radiaszkiewicz, 1976) and has been reported as the presentation of lymphoma in AIDS (Gold et al, 1988). It is also a frequent cause of cranial neuropathies (Snider et al, 1983), when there may be an associated lymphomatous meningitis. The prognosis of these lymphomas in AIDS is poor, but there is frequently good

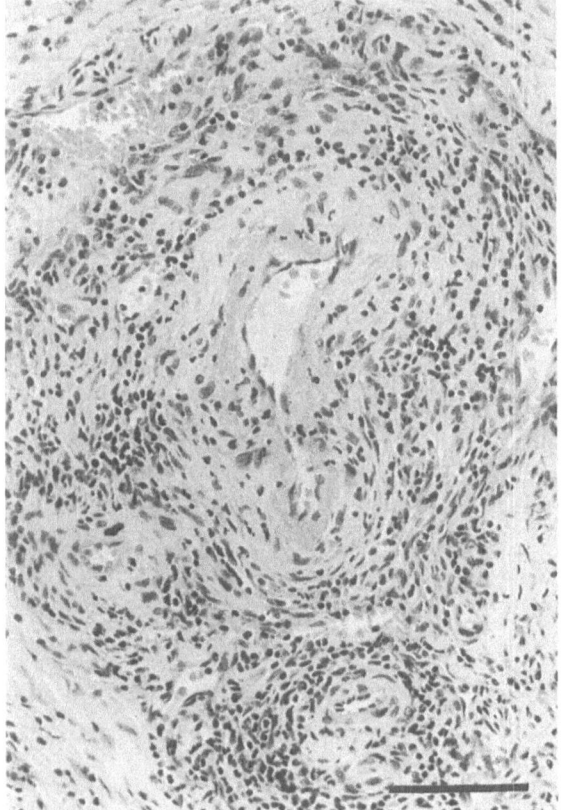

Figure 10.4. Paraffin section of the femoral nerve showing fibrinoid necrosis of an epineurial arteriole. Inflammatory cells infiltrate and surround the vessel. Bar represents 100 μm.

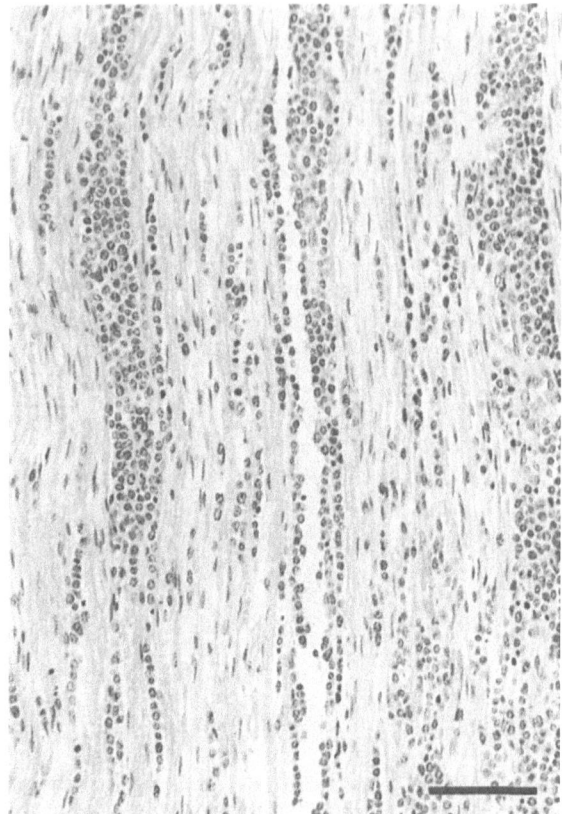

Figure 10.5. Paraffin section, stained with haematoxylin and eosin, of sural nerve showing infiltration of the fascicle by cells which were found to be predominantly B-lymphocytes. Bar represents 100 μm.

symptomatic relief from chemotherapy, often with resolution of the peripheral and cranial nerve signs.

Figure 10.5 shows a nerve taken post mortem from a case of lymphoma. There is infiltration of one fascicle of the femoral nerve by B-lymphocytes. The remaining fascicles, and several other peripheral nerves examined, showed no infiltration. Only occasional acute fibre degeneration was seen in the affected nerve, presumably because the pathology was of very short duration.

Cytomegalovirus

CMV has occasionally been identified in the peripheral nerves of patients who present with mononeuritis multiplex (Small et al, 1989; Fuller et al, 1990b; Griffin JW, personal communication). Inclusions are found in Schwann cells and endothelial cells, and their presence is associated

with acute inflammatory cell infiltration. CMV has been reported in peripheral nerves in the absence of clinical correlates (Grafe and Wiley, 1989). The virus is also an important cause of a polyradiculopathy (see below). In view of the small size of the peripheral nerve biopsies, it seems likely that multifocal pathology will be frequently missed.

Other Forms of Mononeuritis Multiplex

Lipkin et al (1985) reported a series of patients with mononeuritis multiplex, which frequently involved cranial nerves, with an associated aseptic meningitis in five out of six cases where the CSF was examined. Electrophysiological studies showed evidence of multifocal neuropathy in three cases. Perivascular inflammatory infiltrates, without stated evidence of vasculitis, were found in biopsies from several cases, and axonal degeneration and demyelination were identified from paraffin sections. These cases could represent a mononeuritis multiplex presentation of a diffuse inflammatory neuropathy or a variant of the vasculitis.

Cranial Neuropathies

As mentioned above, cranial neuropathies, particularly those affecting the facial nerve, can be associated with meningitis, most frequently with aseptic and lymphomatous meningitis, though occasionally with cryptococcal and tuberculous types. CMV meningo-encephalitis has been reported to cause multiple cranial neuropathies (Behar et al, 1987).

Cranial neuropathies can also result from Varicella zoster virus reactivation, as exemplified by the LMN facial weakness that is associated with vesicles on the hard palate and external ear (the Ramsey Hunt syndrome) (Bélec et al, 1989).

Central nervous system disease may simulate peripheral cranial nerve involvement when cranial nuclei are damaged within the substance of the brain.

Polyradiculopathies

Cytomegalovirus

Initial reports have variously referred to this syndrome as 'CMV-induced demyelination' (Moscowitz et al, 1984), 'Guillain–Barré syndrome' (Bishopric et al, 1985), a 'spinal cord syndrome' (Singh et al, 1986) and 'polyneuropathy' (Jeantils et al, 1986). Full postmortems were performed on these and other cases (Tucker et al, 1985; Eidelberg et al, 1986; Behar et al, 1987; Jacobson et al, 1988; Mahieux et al, 1989), so that CMV polyradiculopathy is the best characterized clinico-pathological syndrome in HIV infection. It occurs in less than 1 per cent of AIDS patients (Guiloff et al, 1988), usually in those patients with a prior AIDS-defining opportunistic infection or tumour, though it may be the first presentation of AIDS (Mahieux et al, 1989).

Clinical Features. The clinical onset is stereotyped, usually beginning with low back pain or lumbosacral radicular pain, followed by sphincter disturbance. Then, usually within 3 weeks of the onset of signs, a rapidly-progressive flaccid areflexic paraparesis develops, sometimes with patchy sensory loss. The arms remain neurologically normal. Lumbar puncture usually reveals CSF neutrophil pleocytosis with reduced glucose, although in the early stages the CSF may be normal. CMV has been isolated from the CSF, but a negative virology finding could still be consistent with this diagnosis.

Untreated, the syndrome progresses, with ascending paralysis, leading to death usually within 6 weeks of onset. Anti-CMV treatment was reported as unsuccessful by Jacobson et al (1988), but more recent reports (Miller et al, 1989; Novak et al, 1989; Fuller et al, 1990c) suggest that the anti-CMV drug, ganciclovir, prolongs survival and may improve the neurological disability.

Neuropathological Features. Neuropathological changes are found predominantly in the cauda equina; the spinal cord is involved in about half of the cases. Both dorsal and ventral lumbar and sacral roots, and affected areas of the spinal cord, show multiple foci of necrosis, with loss of axons and myelin. Mahieux et al (1989) noted sparing of axons at the periphery of these lesions. Inflammatory infiltrates, mainly containing polymorphonuclear leucocytes and lymphocytes, were seen in areas of necrosis, around blood vessels, and in the leptomeninges. CMV inclusions were found in Schwann cells (Figure 10.6).

Singh et al (1986) and Eidelberg et al (1986) also described focal necrosis of small blood vessels in the nerve roots, with inflammatory cell infiltration and thrombosis. Endothelial cells in affected roots contained CMV inclusions. Eidelberg et al (1986) examined dorsal root ganglia which they found to be unaffected; spinal motor cells often showed chromatolysis as a result of ventral root damage.

In some cases there was evidence of ascending progress of the infection with involvement of roots at higher levels of the cord. Behar et al (1987) described focal changes, particularly at entry zones of the motor cranial nerves and ventral roots.

CMV appears to have an affinity for ependymal surfaces, and ependymal cells containing CMV inclusions are sometimes seen protruding into the ventricular lumen (Eidelberg et al, 1986; Mahieux et al, 1989). Eidelberg et al (1986) have suggested that the particular susceptibility of lumbar and sacral roots to CMV infection might be due to the transport of released infected ependymal cells along CSF pathways to the caudal regions, where further cell-to-cell spread of CMV occurs in thecal and spinal subarachnoid spaces.

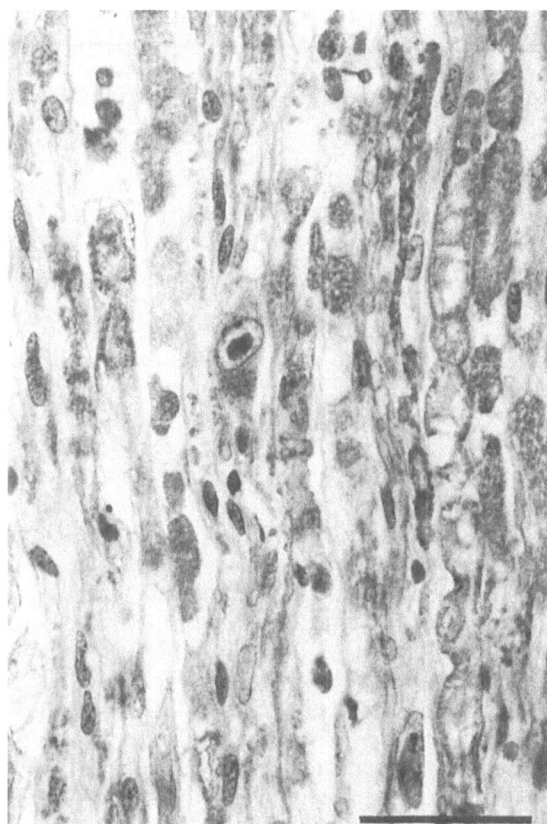

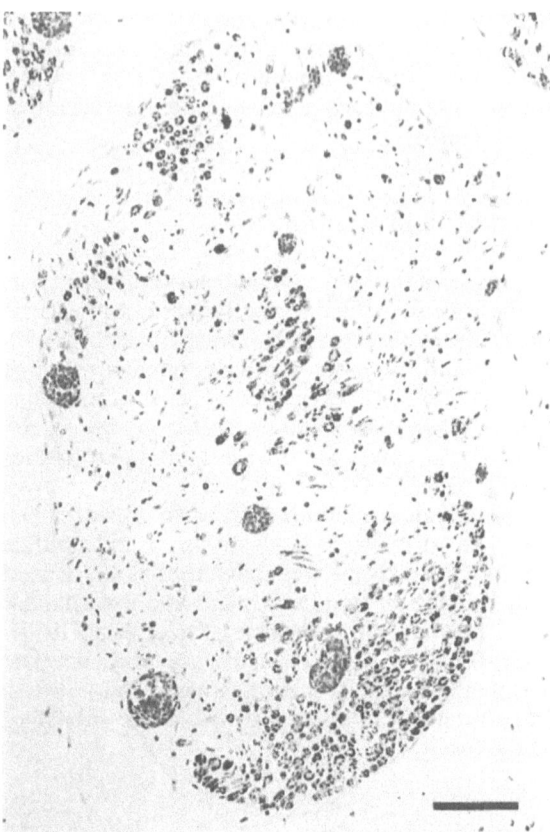

Figure 10.6. Paraffin section of lumbar root from a case of CMV polyradiculopathy. A nuclear inclusion body is seen in a Schwann cell. Bar represents 50 µm.

Figure 10.7. Paraffin section of root from cauda equina, stained with luxol-fast blue to show myelin. There is patchy loss of myelinated fibres. Bar represents 200 µm.

In a case where there was an apparent clinical response to ganciclovir, patchy axon loss was found in the cauda equina, but without the inflammatory infiltrate and with only occasional CMV inclusions (Figure 10.7) (Fuller et al, 1990c). The absence of inflammatory cells and rare CMV inclusions were features similar to those seen in CMV retinitis following a clinical response to treatment (Pepose et al, 1987).

It seems clear that CMV is the cause of this syndrome, infecting both Schwann cells and endothelial cells. The mechanism of tissue damage is less certain, but may involve a direct cytopathic effect, an immune-mediated bystander effect, or an endarteritis.

Other Polyradiculopathies

There are several aetiologies that should be considered in the differential diagnosis of the CMV polyradiculopathy. Single case reports have described syphilitic (Lanska et al, 1988), tuberculous (Woolsey, 1989), Herpes simplex virus (Dahan et al, 1988) and lymphomatous (Gold et al, 1988) polyradiculopathy in AIDS patients. These need to be distinguished either on serological grounds or by the predominant lymphocytic pleocytosis. Other agents have been described in patients with other causes of immunosuppression, such as zygomycosis, or with malignancies.

Distal Symmetrical Peripheral Neuropathies

The distal symmetrical peripheral neuropathies (DSPN) were among the first PNS complications to be described in AIDS when Snider et al (1983) reported 8 patients with this disorder, associated in 7 with painful dysaesthesiae. Since then it has become clear that DSPN are the most common form of peripheral neuropathy in HIV infection and AIDS (McArthur, 1987; Cornblath and McArthur, 1988; Dalakas and Pezeshkpour, 1988). Most reports have suggested a multifactorial aetiology, citing nutritional deficiencies, infections, and drugs as possible factors (Snider et al, 1983; McArthur, 1987; Cornblath and McArthur, 1988). Others have suggested that some peripheral nerve abnormalities are caused by HIV itself, the evidence being based on the finding of a positive HIV culture from peripheral nerve tissue (despite negative results for in situ hybridization) (Ho et al, 1985b; Cornblath et al, 1987), and on the presence of an 'HIV-like' particle in the axoplasm of a myelinated fibre (Bailey et al, 1988), a rather unlikely site for the virus since it does not seem to infect neurons. In addition, protagonists for this view draw a parallel with the role of HIV in the CNS.

The distal symmetrical peripheral neuropathies are a heterogeneous group, suggesting that different aetiologies are involved. In approximately 60 per cent of patients with a DSPN, pain in the feet is a prominent feature (Cornblath and McArthur, 1988), and this has been the basis of characterizing a clinically distinct subgroup (Lange et al, 1988; Fuller et al, 1989).

Painful Peripheral Neuropathy

Clinical Features. The clinical features of painful peripheral neuropathy (PPN) have been described by Bailey et al (1988), Cornblath and McArthur (1988), Lange et al (1988), Chaunu et al (1989), Léger et al (1989), and Fuller et al (1989).

Painful peripheral neuropathy usually presents in AIDS, although it may occur in patients with severe ARC. It usually begins with a tight feeling in the feet which progresses over 1–4 weeks to pain which is described as burning or dysaesthesiae, is usually exacerbated by contact, and is almost always limited to the feet. It is associated with numbness and paraesthesiae. Patients rarely complain of weakness. Examination soon after the pain has begun may reveal few abnormalities. Mild sensory loss, particularly of superficial modalities, is found a month after the pain begins and may be associated with impaired or reduced ankle reflexes. Weakness and wasting are rarely found. The pain often resolves or improves spontaneously, and may be helped by amitriptyline or carbemazepine. There is little progression of the sensory deficit but ankle reflexes are gradually lost.

Nerve conduction studies reveal a marked reduction in sensory action potentials, especially in the legs, and mildly-reduced motor conduction velocities and compound muscle action potentials. There is some evidence of chronic partial denervation on EMG.

Neuropathological Features. Few full postmortems have been reported and most pathology is based on studies of sural and superficial peroneal nerves. Fuller et al (1989) examined sural nerves from 8 patients with PPN, 4 biopsies and 4 taken post mortem. There were no inflammatory cells in paraffin sections of any of the 8 nerves. In 1-μm resin sections, the appearances were remarkably consistent, with moderate loss of myelinated fibres and occasional evidence of ongoing fibre degeneration (Figure 10.8). Subperineurial and endoneurial oedema were present in each nerve to varying degrees. A common feature was the presence of myelinated fibres with axons which were disproportionately small for myelin sheath thickness (Fuller et al, 1990d) (Figures 10.8 and 10.9). This axonal atrophy was most marked in the larger fibres. In some nerves all of the larger fibres were atrophic, in others, a smaller proportion of fibres were affected. There was an occasional thinly-myelinated fibre. No regeneration clusters were seen.

Quantitatively, the mean fibre density (4396 ± 931/mm^2, n = 8) was found to be about 60 per cent of the sudden-death controls of similar age (7845 ± 1414/mm^2, n = 8). Histograms of fibre diameter distribution showed an apparently-reduced peak of large fibres (Figure 10.10). This was found to be due not to a selective loss of large fibres, but to a decrease in fibre size consequent upon the atrophic axonal change, illustrated by the reduction of 'g' ratios (axon diameter/fibre diameter) in many fibres (Figure 10.11).

The atrophic fibres varied considerably in shape, from a circular to a highly irregular profile. To overcome problems in morphometry associated with changes in shape, measurements were made of axonal area from electron micrographs, and the numbers of myelin lamellae were counted. Results showed (Figure 10.12) that whilst the distribution of numbers of myelin lamellae was similar to that in the control nerves, the axonal areas in the PPN nerves were markedly reduced (Fuller et al, 1990d).

Examination of teased fibres showed occasional degenerating myelinated fibres in all nerves. In fibres from some nerves, the myelin sheaths had a very lumpy outline, an appearance often associated with axonal atrophy. There was occasional

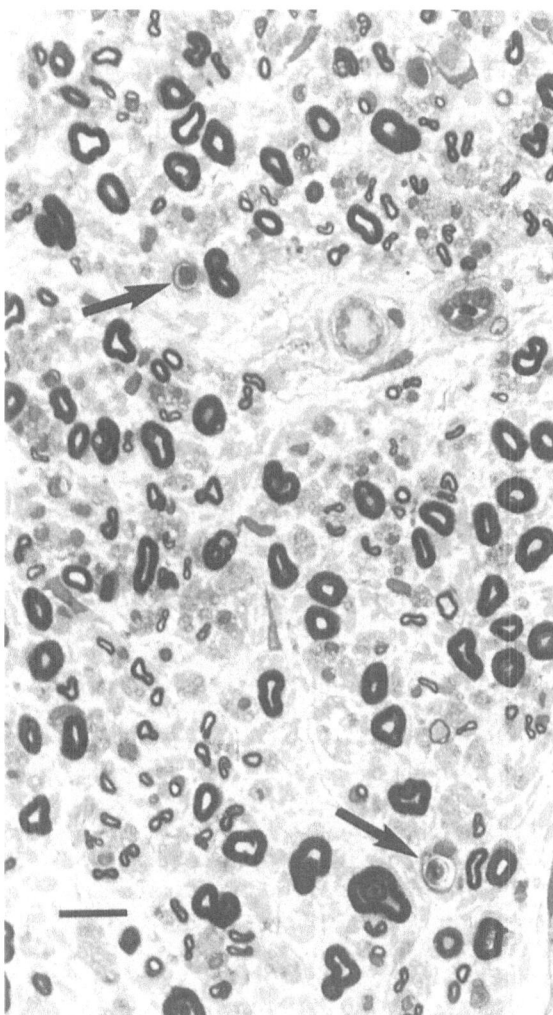

Figure 10.8. One-micrometre Araldite section of sural nerve from a case of PPN. The density of myelinated fibres is moderately reduced. Two degenerating fibres are indicated by *arrows*. Many fibres appear atrophic with axons small relative to myelin sheath thickness. There is considerable endoneurial oedema. Bar represents 20 μm.

evidence of demyelination, most probably secondary to the axonal changes.

There was evidence of some loss of unmyelinated axons (Figure 10.13), and although the densities (mean $22\,260 \pm 3022$/mm^2, n = 6) were consistently lower than the sudden-death controls of similar age ($30\,900 \pm 7117$/mm^2, n = 8), this difference did not reach statistically significant levels.

Other studies of the painful neuropathies have shown some other features, although axonal loss is common to all cases. Perivascular inflammatory cells were seen by Bailey et al (1988), de la Monte et al (1988), and in one case by Léger et al (1989). Chaunu et al (1989) described widespread subacute microvasculitis in four cases. These patients with microvasculitis and PPN may suffer from multifocal inflammatory neuropathies presenting as distal symmetrical painful neuropathy. Thus, they should perhaps be distinguished nosologically from the PPN described above, which lacks inflammation, and be considered with other multifocal microvasculitides, such as those which present as mononeuritis multiplex (see above).

Reliable evidence of demyelination can only be obtained from semithin resin sections and/or teased fibres. Where these methods were used, segmental demyelination was found quite frequently, but it is not clear whether it was primary, or secondary to axonal changes.

The findings in peripheral nerve biopsies give little indication as to the site of the pathological insult, but they would be in keeping with either an axonopathy or a neuronopathy. The absence of regenerating fibres could suggest neuronopathy, but does not exclude an axonopathy.

Association of PPN with CMV. Painful peripheral neuropathy has been associated with CMV infection at other sites, notably in the retina. This association, which is particularly striking at the onset of the syndrome, has led to the suggestion that PPN may be caused by a CMV dorsal root ganglionitis (Fuller et al, 1989). Circumstantial evidence offers further support: the dorsal root ganglion is a site of predilection for other herpes virus (H. simplex, H. zoster) infections; the changes in the peripheral nerve are similar to those found following H. zoster radiculitis (Zacks et al, 1964); the dorsal root ganglion lacks a blood–nerve barrier, and CMV infections of the CNS in AIDS occur at sites with reduced blood barriers, for example, the retina, in association with HIV-related retinal infarcts (cotton wool spots); and CMV is an established pathogen in the peripheral nervous system in AIDS causing lumbosacral polyradiculopathy. The proximal nerve roots have been identified as sites of CMV infection by neuropathologists (Grafe and Wiley, 1989; Vinters et al, 1989), although clinical correlates for the findings were not available.

CMV may not be the only cause of a PPN in AIDS. However, there is increasing evidence that the PPN is part of the spectrum of CMV-related peripheral nerve disease in AIDS.

Rance et al (1988) describe autopsy findings in four men with AIDS and distal paraesthesiae and dysaesthesiae affecting the lower limbs. There

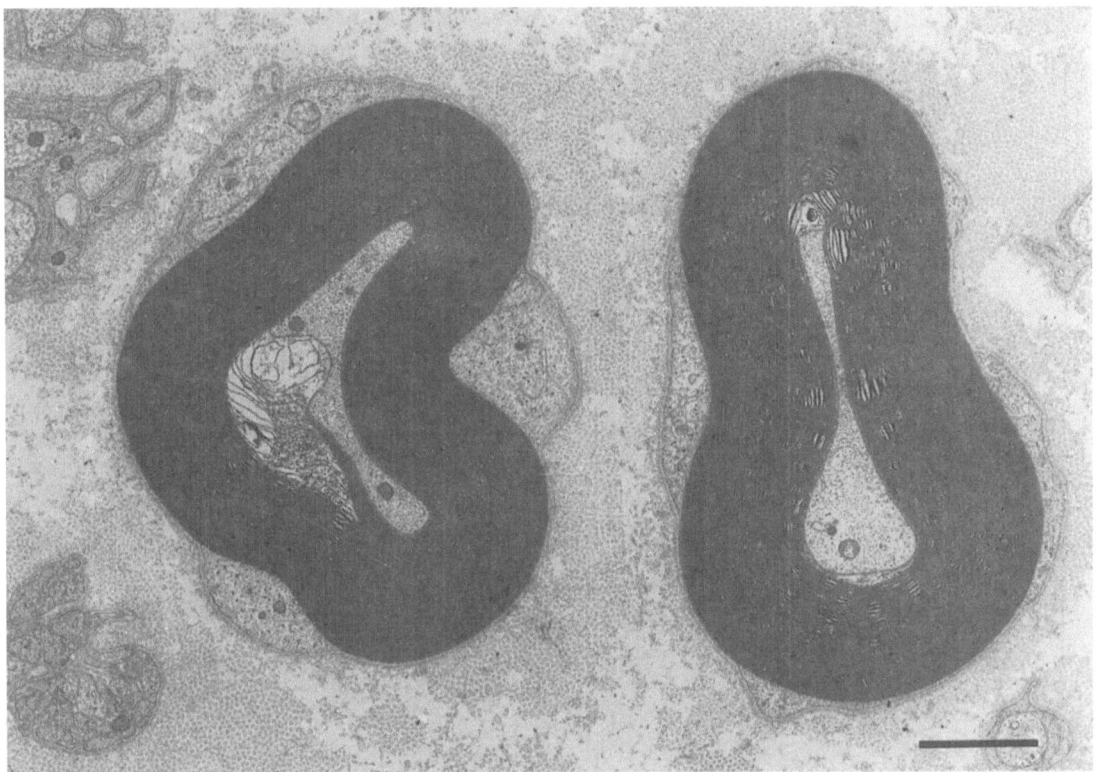

Figure 10.9. Electron micrograph of two myelinated fibres from the same case of PPN as shown in Figure 10.8. The axons are small relative to the thickness of the myelin sheaths. Bar represents 2 μm.

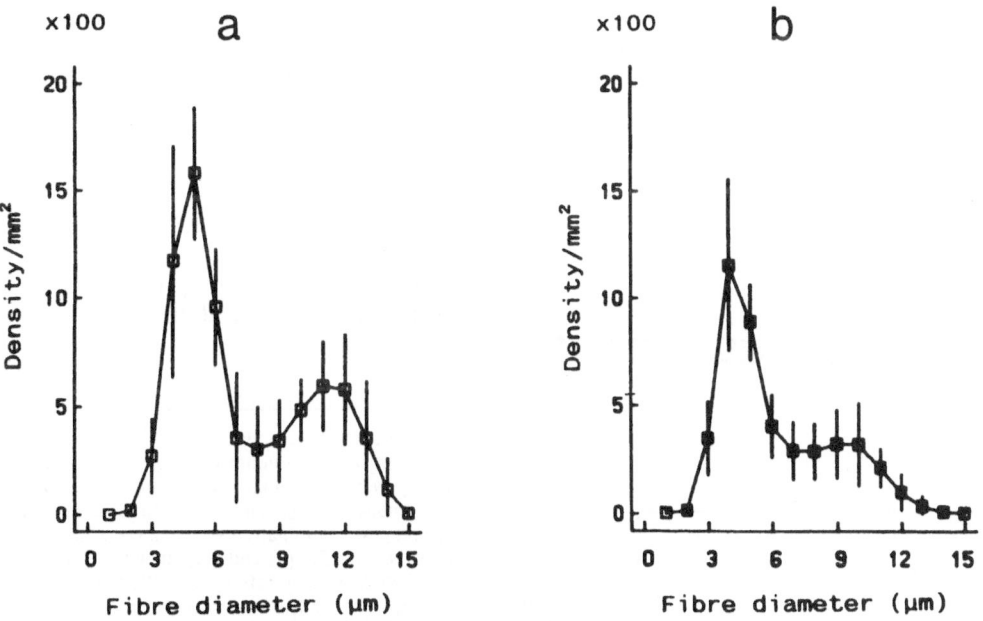

Figure 10.10a,b. Plots of myelinated fibre density/mm^2 classified by fibre diameter, of: a the means of 5 sudden-death controls; and b the means of nerves from 8 patients with PPN. The *vertical bar* indicates the standard deviation for the mean. There is apparent loss of larger fibres, and a slight overall shift to the left. The plot of asymptomatic control nerve fibres (not shown) was almost identical to that for the sudden-death controls.

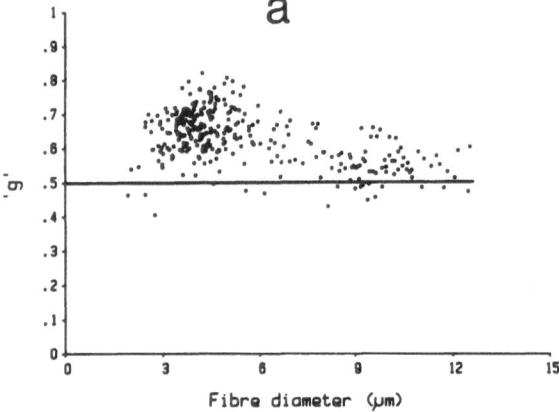

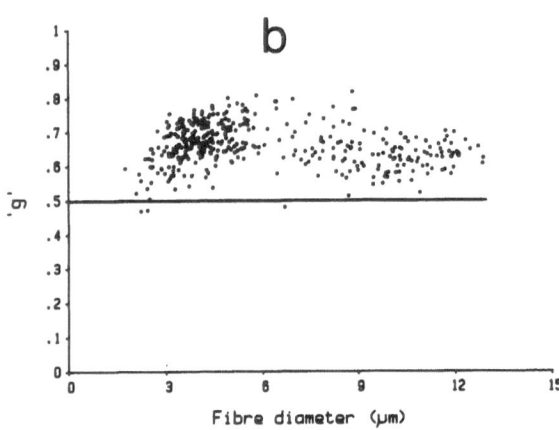

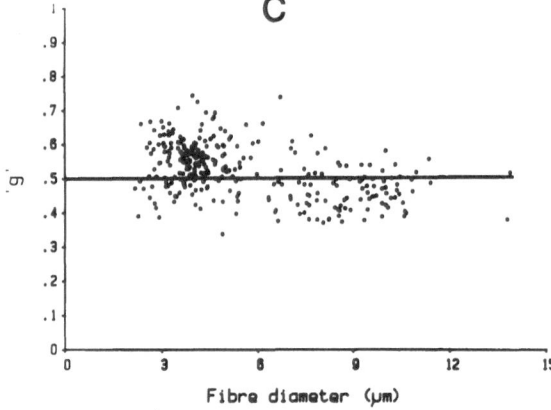

Figure 10.11a–c. The distribution of 'g' ratios plotted against fibre diameters from nerves of: **a** a sudden death control; **b** an asymptomatic AIDS patient; and **c** a patient with PPN. An arbitrary line is drawn at 0.5, to show that many more PPN fibres lie below this value than either the sudden-death control or the asymptomatic AIDS nerve fibres.

was degeneration of the gracile tract with loss of axons and myelin sheaths, most prominent in the cervical and upper thoracic cord. Dorsal root ganglia were available from only one patient, who had mild gracile cord degeneration. There was a modest loss of sensory neurons, indicated by the presence of nodules of Nageotte, and associated with some lymphocytic infiltration. The presence of some giant neurofilamentous axonal swellings within the ganglia was noted. There was no evidence of viral inclusions or opportunistic infection. No peripheral nerves were available from any of the cases, although electrophysiological studies in one case had shown absent sural sensory action potentials. These findings suggest a distal axonal degeneration affecting both peripheral and central processes of sensory ganglion cells. The authors speculate that this could result from infection by retroviruses, herpes viruses or CMV.

There is a single full postmortem report of a patient who had a PPN and later developed lower motor neuron weakness (Robert et al, 1989). CMV inclusions were found in endothelial cells of small vessels in dorsal root ganglia, peripheral nerves and also in the epimysium of muscle. Schwann cells did not appear to be infected. In situ hybridization showed no wider distribution of CMV than that suggested by the demonstration of inclusions. There was loss of approximately 50 per cent of myelinated fibres in the deep peroneal and sural nerves, large fibres being the most affected. Occasional degenerating fibres were seen in teased preparations, but there was no evidence of demyelination.

Non-painful Distal Symmetrical Peripheral Neuropathies

This is a heterogeneous group, in which many factors may be implicated, and other infective

Figure 10.12a–c. Lamellae counts and axonal areas. The distribution of the cumulative frequency of lamellae counts (*upper panel*) is similar in: **a** sudden-death controls, **b** asymptomatic controls, and **c** PPN. In the *lower panel*, there is a difference in cumulative frequency distribution of axonal areas between PPN (**c**) and the other two groups (**a,b**) – the areas of PPN fibres rarely exceed 30 µm in area, whilst they reach 60 µm in the two control groups.

Figure 10.13. Electron micrograph of a sural nerve from a patient with PPN. The presence of bands of Schwann cell processes (*arrows*) suggests loss of some unmyelinated axons. Bar represents 5 µm.

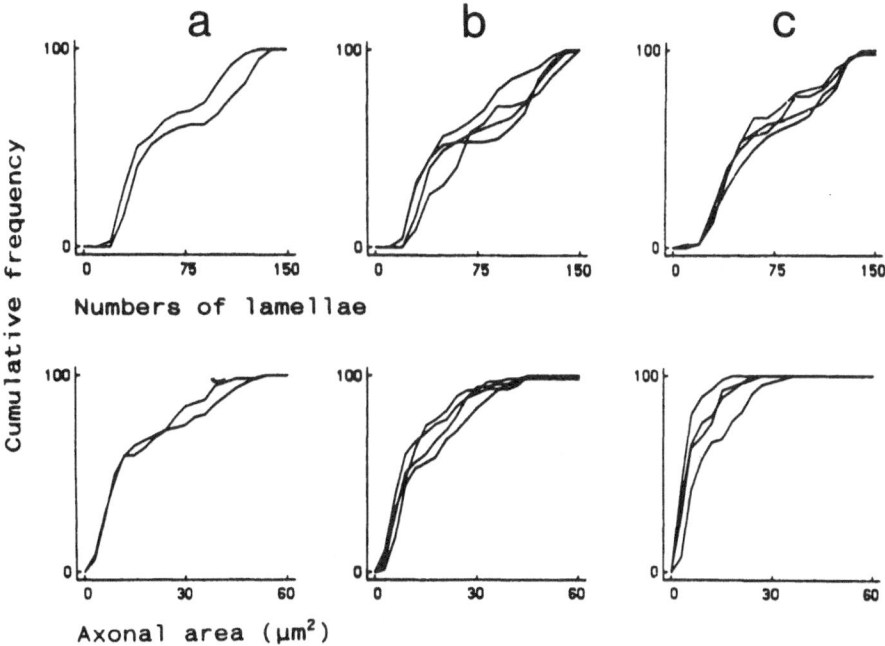

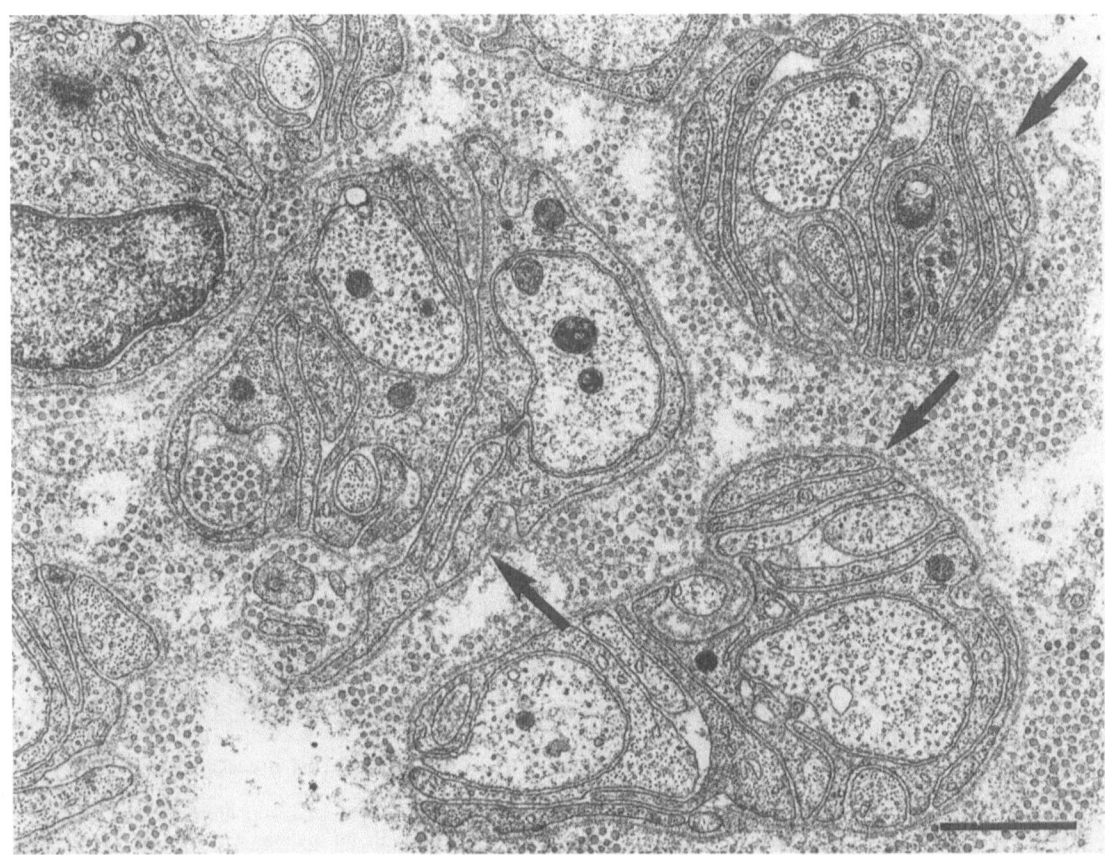

Figure 10.13.

agents, such as *Mycobacterium avium–intracellulare* (JW Griffin, personal communication) or herpes viruses may be involved. Nutritional factors may also be important, notably vitamin B_{12}, which is frequently abnormal in AIDS (Harriman et al, 1989). There is some evidence that weight loss alone is associated with peripheral nerve abnormalities (Sotaniemi, 1984). AIDS patients are also exposed to multiple drug therapies and to many novel and relatively-untested or alternative therapies, and peripheral neuropathy associated with these may be hard to distinguish from other causes. In particular, peripheral neuropathies are the dose-limiting adverse effect of two new anti-retroviral agents, dideoxycytidine (ddC) (Merigan et al, 1989; Dubinsky et al, 1989) and dideoxyinosine (ddI).

Neuropathological Features. Fuller et al (1990e) examined sural nerves from three cases of non-painful distal symmetrical peripheral neuropathy. No inflammatory cells were seen in paraffin sections. There was evidence of loss of myelinated fibres both in transverse sections and in teased fibre preparations. Quantitative studies showed the mean myelinated fibre density to be $4796/mm^2$; the control mean density was $7845 \pm 1414/mm^2$ (n = 8). Qualitatively, no abnormalities were identified in the myelinated fibres. By electron microscopy, there was also evidence of loss of unmyelinated axons (mean $22\,475/mm^2$), but this was not statistically significantly different from control values.

Toxic Neuropathy

The multiple therapies received by AIDS patients put them at risk of toxic neuropathies. Frequently, it will be difficult to distinguish toxic neuropathies from the DSPN seen in AIDS. However, it is occasionally possible to make such a distinction as exemplified by the case of a man with AIDS, treated with thalidomide (100 mg daily for 22 months) for aphthous ulcers. He developed a distal painless peripheral neuropathy affecting the feet, with absence of light touch and pinprick in the toes and impairment up to the ankles. Vibration sense was lost in the toes, but proprioception was normal. Examination of paraffin sections of his sural nerve biopsy showed no inflammatory cells. In resin sections there was a moderately-reduced density of myelinated fibres (4342/mm²) (Figure 10.14), and some of the larger fibres appeared atrophic. The fibre diameter histogram shows a selective loss of the

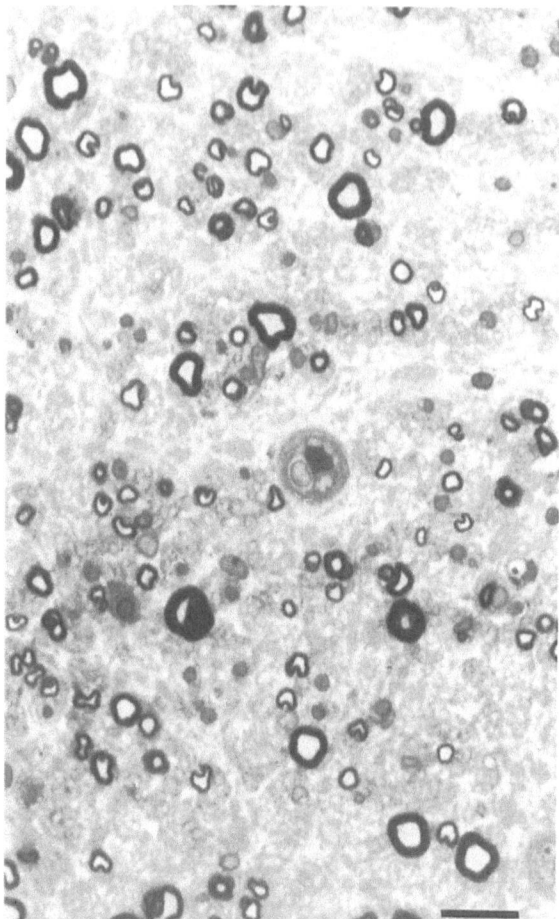

Figure 10.14. One-micrometre Araldite section from an AIDS patient with a neuropathy associated with thalidomide treatment. The density of myelinated fibres is reduced, with loss of the larger fibres. Bar represents 20 µm.

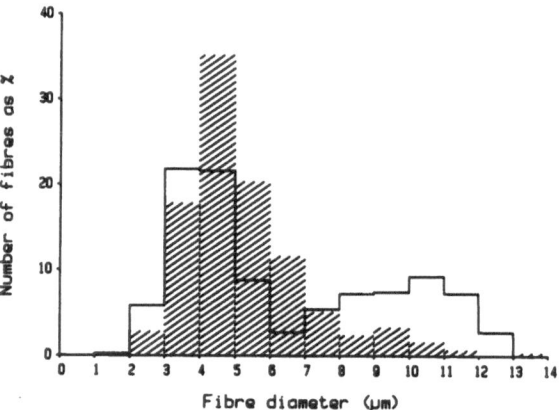

Figure 10.15. Histogram showing fibre diameter distributions (as percentages) in the patient with thalidomide-associated neuropathy (*hatched*) and an age-matched control (*open*). There is loss of large diameter fibres.

larger fibres (Figure 10.15). After the discontinuation of treatment with thalidomide, the symptoms were subjectively improved although there was little change in the sensory impairment (Fuller et al, 1991b). In previous descriptions of thalidomide neuropathy (Le Quesne, 1984; Lagueny et al, 1986), it has been noted that symptoms may continue for some time after treatment has stopped. It was of interest that the patient later developed red hands, another symptom of thalidomide neuropathy (Le Quesne, 1984). These clinical features, and the clear demonstration of a selective loss of large diameter fibres, strongly suggest that the neuropathy is associated with thalidomide treatment.

Incidental and Isolated Case Reports

There are several peripheral nerve syndromes which have been associated with HIV infection in single case reports, for example, motor neuron disease (amyotrophic lateral sclerosis) (Hoffman et al, 1985). These probably represent chance co-occurrences, rather than true associations.

Autonomic Neuropathies

Following the sudden death of a patient undergoing a transcutaneous needle lung biopsy a series of patients were identified who had markedly abnormal autonomic function tests (Craddock et al, 1987). This report was followed by prospective studies showing marked subclinical autonomic dysfunction (Freeman et al, 1990). However, it was soon noted that the autonomic function tests usually relied on the cardiovascular reflexes, and that these were frequently abnormal in patients with other systemic diseases – notably those recovering from pneumonia, and patients with diarrhoea or anaemia (Lohmoller et al, 1987). The situation is further complicated because of the frequent pathological involvement of the adrenals, particularly with CMV, which is usually unrecognized clinically.

While there are a few case reports that suggest autonomic involvement in AIDS, its clinical features and frequency have yet to be determined.

Tubuloreticular Inclusions

Tubuloreticular inclusions (TRI) are structures arising from cytoplasmic membrane systems. In AIDS, TRI have been found in lymphocytes,

endothelial cells and histiocytes in many tissues. In the CNS they have been described, together with cylindrical confronting cisternae (see Chapter 6). The significance of TRI is not known. They were thought at one time to be viral inclusions. Experimental studies show that they can be induced by α- and β-interferon, but not by γ-interferon.

In the PNS, TRI were identified in endothelial cells in nerve biopsies from two patients with AIDS-related peripheral neuropathy in studies by Dalakas and Pezeshkpour (1988) and Gastaut et al (1987) respectively. In both cases it was suggested that these structures might be implicated in the pathogenesis of the neuropathy. Their presence in peripheral nerve was also noted by Grafe and Wiley (1989). Fuller and Jacobs (1989) found TRI in endothelial cells in nerve biopsies from 11 patients with AIDS, 9 of whom had neuropathies of different types, and 2 of whom had no clinical evidence of peripheral neuropathy (Figure 10.16). It therefore seems unlikely that TRI are associated with the pathogenesis of peripheral neuropathies in AIDS, but are instead a manifestation of a more generalized response, probably related to the high interferon levels that occur in AIDS.

Myopathies

Myopathy is now well recognized as a complication of AIDS. Generalized weakness due to the systemic disease may be difficult to differentiate from a primary neuromyopathic disorder. Elevated creatine phosphokinase (CPK) levels and electromyographic studies can help in identifying some myopathies, but muscle biopsy may be essential for diagnosis.

Polymyositis

Polymyositis is the most common of the AIDS-associated myopathies, and may be the presenting sign of HIV infection (Dalakas et al, 1986). Muscle biopsies show inflammatory infiltrates of lymphocytes, macrophages and atypical histiocytes (Dalakas et al, 1986). Multinucleated giant cells were seen in a case reported by Bailey et al (1987), and their presence was interpreted as a possible sign of direct HIV infection of muscle. Active phagocytosis and muscle fibre necrosis were evident. In three cases reported by Marolda

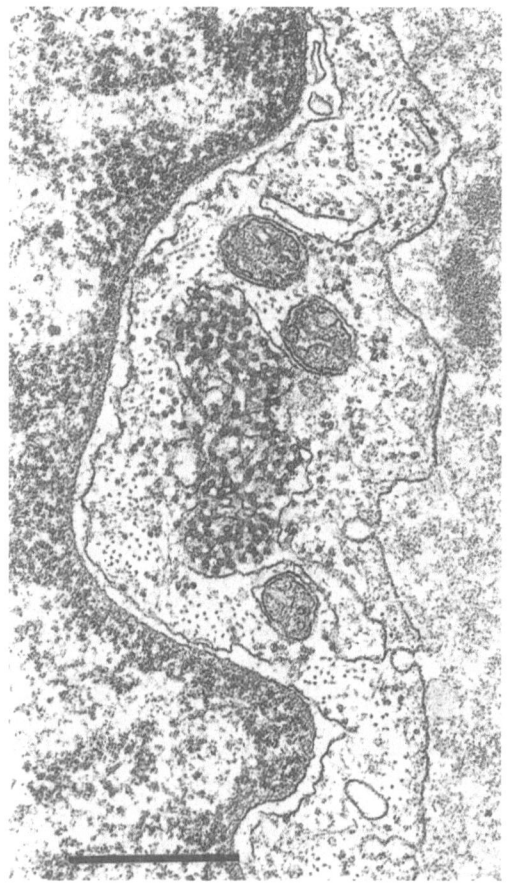

Figure 10.16. Electron micrograph of an endoneurial endothelial cell in the sural nerve of a patient with AIDS showing a tubuloreticular inclusion. Bar represents 0.5 μm.

et al (1989), histochemical staining showed atrophy and increased numbers of type I fibres. HIV has not been identified in muscle fibres, but viral antigens were shown immunocytochemically, using specific antibodies to HIV, in CD4-positive lymphoid cells surrounding or invading muscle cells (Dalakas et al, 1986). HIV major core protein (p24) has been detected in macrophages and damaged muscle cells. However, in situ hybridization techniques for HIV were negative (Dalakas and Pezeshkpour, 1988; Simpson and Bender, 1988).

Recent studies indicate that the inflammatory myopathy is suppressor-cytotoxic T-cell mediated and restricted to class 1 major histocompatibility complex (MHC-1) antigens in the muscle fibres (Dalakas et al, 1990). Factors leading to the expression of this antigen are not yet known.

Polymyositis develops in simian AIDS when induced by type D retrovirus, which is related to

Mason–Pfizer monkey virus (see Dalakas and Pezeshkpour, 1988).

Non-inflammatory Myopathy

A few cases of non-inflammatory myopathy with focal necrosis and regeneration have been described (Lange et al, 1988; Simpson and Bender, 1988; Dalakas et al, 1989). Stern et al (1987) reported a case of non-inflammatory myopathy with extensive degeneration, but no regeneration. In two patients in the advanced stages of AIDS, Panegyres et al (1990) described necrotizing myopathy with an unusual segmental vesicular change of myofibres. Electron microscopy showed the vacuoles to correspond to dilated cisternae of the sarcoplasmic reticulum.

Nemaline Rod Myopathy

This myopathy is characterized histologically by the presence of atrophic, mainly type I fibres with central nuclei that contain granular, basophilic and fuchsinophilic material. Electron microscopy of these cells shows distortion, thickening and misalignment of Z-bands, forming nemaline bodies 0.5–1.0 μm in length (Dalakas and Pezeshkpour, 1988; Gonzalez et al, 1988; Simpson and Bender, 1988; Cabello et al, 1990). Loss of sarcomeric thick filaments has also been described (Gonzalez et al, 1988; Cabello et al, 1990). In most cases there is no inflammatory infiltration. The pathological features of this myopathy are similar to those of adult-onset nemaline myopathy, first described by Engel and Resnick (1966). In both, CPK levels are normal, or only slightly elevated.

Some muscle biopsies in AIDS-related myopathies show both inflammatory and nemaline changes, leading to the suggestion that the two patterns of myopathy may represent ends of a continuous spectrum (Simpson and Bender, 1988; Cabello et al, 1990).

Type II Muscle Fibre Atrophy

Type II muscle fibre atrophy may occur in AIDS patients suffering from rapid weight loss, undernutrition, or the remote effects of malignancy, who develop proximal muscle weakness with normal CPK levels (Dalakas and Pezeshkpour, 1988).

Zidovudine-Induced Myopathy

Zidovudine, used in the treatment of AIDS, has been recognized as a cause of myalgia in up to 8 per cent of patients (Richman et al, 1987). Bessen et al (1988) described four patients who developed a polymyositis-like syndrome whilst being treated with zidovudine. Muscle biopsy showed scattered 'dissolution' of myofibres, but little inflammation. In a case reported by Gorard et al (1988), muscle biopsy showed widespread focal necrosis affecting short segments of muscle fibres. Within 7 days of withdrawal of zidovudine, clinical symptoms had improved significantly.

Dalakas et al (1990) reported myopathy in patients receiving long-term zidovudine. In general the symptoms improved after discontinuing the drug. Muscle biopsies from 20 cases of HIV-related myopathy were studied, of whom 15 had received zidovudine. All 20 cases had a T-cell-mediated inflammatory myopathy, but those given zidovudine also had ragged red fibres. Electron microscopy confirmed the presence of abnormal mitochondria containing paracrystalline inclusions, of which ragged red fibres are indicative. No ragged red fibres were found in biopsies of patients not given zidovudine. Following the discontinuation of the drug in 8 patients, clinical improvement of myopathy was noted, and repeat muscle biopsies from 2 cases showed marked histological improvement.

Four patients of Panegyres et al (1990) developed myopathy while receiving zidovudine. One case had an inflammatory myopathy but all four cases showed a vesicular change in muscle fibres, identified by electron microscopy as being due to mitochondrial abnormalities including increase in size, vacuolation, and changes in numbers and arrangement of cristae. These changes were only seen in fibres showing other evidence of injury. Muscle symptoms resolved when treatment was discontinued, and a repeat biopsy in one case showed no abnormalities.

Acknowledgements. We wish to acknowledge help received in relation to our publications cited in this chapter. We are most grateful to Dr R.J. Guiloff for advice and helpful discussion, to Dr B.G. Gazzard for allowing patients in his care to be assessed, and to Dr J.N. Harcourt-Webster for postmortem material. We wish to thank the technical staff of the Neuropathology Department, Institute of Neurology, and particularly Andrew Beckett, for preparation of nerves for microscopy; and Steve Durr for the photography. GF was in receipt of a MRC Research Training Fellowship. The work was supported in part by the Special Trustees of Westminster and Roehampton Hospitals. We are grateful to the Joint Research Advisory Committee of the Institute of Neurology for computer equipment used in these studies.

References

Bailey RO, Turok DI, Jaufmann BP et al. (1987) Myositis and acquired immunodeficiency syndrome. Hum Pathol 18:749–51.

Bailey RO, Baltch AL, Venkatesh R et al. (1988) Sensory motor neuropathy associated with AIDS. Neurology 38:886–91.

Behar R, Wiley C, McCutchan JA. (1987) Cytomegalovirus polyradiculoneuropathy in acquired immune deficiency syndrome. Neurology 37:557–61.

Bélec L, Gherardi R, Georges AJ et al. (1989) Peripheral facial paralysis and HIV infection: report of four African cases and review of the literature. J Neurol 236:411–14.

Beral V, Peterman T, Berkelman R. (1989) Epidemiology of AIDS-associated lymphomas: United States. In: Fifth International Conference on AIDS, Montreal, Canada (1989), Abstract WAP 17, IDRC: Ottowa, p. 17.

Bessen LJ, Greene JB, Louie E et al. (1988) Severe polymyositis-like syndrome associated with zidovudine therapy of AIDS and ARC. N Engl J Med 318:708.

Bishopric G, Bruner J, Butler J et al. (1985) Guillain–Barré syndrome with cytomegalovirus infection of peripheral nerves. Arch Pathol Lab Med 109:1106–8.

Brown MM, Thompson A, Goh BT et al. (1988) Bell's palsy and HIV infection. J Neurol Neurosurg Psychiatry 51:425–6.

Cabello A, Martinez-Martin P, Guierrez-Rivas E et al. (1990) Myopathy with nemaline structures associated with HIV infection. J Neurol 237:64–8.

Calabrese LH, Proffitt MR, Levin Yen-Lieberman B et al. (1987) Acute infection with the human immunodeficiency virus (HIV) associated with acute brachial neuritis and erythematous rash. Ann Intern Med 107:849–51.

Campbell MJ, Paty DW. (1974) Carcinomatous neuromyopathy: 1. Electrophysiological studies. J Neurol Neurosurg Psychiatry 37:131–41.

Carne CA, Stibe C, Brokenhurst A et al. (1989) Subclinical neurological and neuropsychological effect of infection with HIV. Genitourin Med 65:151–6.

Centers for Disease Control. (1986) CDC classification system for human T-lymphotropic virus type III/lymphadenopathy-associated virus infections. MMWR 35:334–9.

Chaunu MP, Ratinahirana H, Raphael M et al. (1989) The spectrum of changes on 20 nerve biopsies in patients with HIV infection. Muscle Nerve 12:452–9.

Cooper DA, Gold J, Maclean P. (1985) Acute AIDS retrovirus infection. Definition of a clinical illness associated with seroconversion. Lancet i:537–40.

Cornblath DR, McArthur JC. (1988) Predominantly sensory neuropathy in patients with AIDS and AIDS related complex. Neurology 38:794–6.

Cornblath DR, McArthur JC, Kennedy PGE. (1987) Inflammatory demyelinating peripheral neuropathies associated with human T-cell lymphotropic virus type III infection. Ann Neurol 21:32–40.

Craddock C, Pasrol G, Bull R et al. (1987) Cardiorespiratory arrest and autonomic neuropathy in AIDS. Lancet 2:16–18.

Dahan P, Haettich B, Le Parc JM. (1988) Meningoradiculitis due to herpes simplex virus disclosing HIV infection. Ann Rheum Dis 47:440.

Dalakas MC, Pezeshkpour GH. (1988) Neuromuscular diseases associated with human immunodeficiency virus infection. Ann Neurol 23 (suppl):s38–48.

Dalakas MC, Pezeshkpour GH, Gravell M et al. (1986) Polymyositis associated with AIDS retrovirus. JAMA 256:2381–3.

Dalakas MC, Wichman A, Sever JL. (1989) AIDS and the nervous system. JAMA 261:2396–2399.

Dalakas MC, Illa I, Pezeshkpour GH et al. (1990) Mitochondrial myopathy caused by long-term zidovudine therapy. N Engl J Med 322:1095–1105.

Dubinsky RM, Yarchoan R, Dalakas M et al. (1989) Reversible axonal neuropathy from the treatment of AIDS and related disorders with 2'3'-dideoxycytidine (ddC). Muscle Nerve 12:856–60.

Eidelberg D, Sotrel A, Vogel H et al. (1986) Progressive polyradiculopathy in acquired immune deficiency syndrome. Neurology 36:912–16.

Engel WK, Resnick JS. (1966) Late-onset rod myopathy: a newly recognised, acquired and progressive disease. Neurology 16:308–9.

Freeman MS, Friedman LS, Broadbridge C. (1990) Autonomic function in human immunodeficiency virus infection. Neurology 40:575–80.

Fuller GN, Jacobs JM. (1989) Cytomembranous inclusions in the peripheral nerve in AIDS. Acta Neuropathol (Berl) 79:336–9.

Fuller GN, Jacobs JM, Guiloff RJ. (1989) Association of painful peripheral neuropathy in AIDS with cytomegalovirus infection. Lancet ii:937–41.

Fuller GN, Jacobs JM, Guiloff RJ. (1990a) Incidence and classification of peripheral nerve syndromes in HIV infection. A prospective study. J Neurol Neurosurg Psychiatry 54:751.

Fuller GN, Greco C, Miller RG. (1990b) Cytomegalovirus and mononeuropathy multiplex in AIDS. Neurology 40 (suppl 1):301.

Fuller GN, Gill SK, Guiloff RJ et al. (1990c) Ganciclovir for lumbosacral polyradiculopathy in AIDS. Lancet i:48–9.

Fuller GN, Jacobs JM, Guiloff RJ. (1990d) Axonal atrophy in the painful neuropathy in AIDS. Acta Neuropathol (Berl) 81:198–203.

Fuller GN, Jacobs JM, Guiloff RJ. (1990e) Non-painful distal symmetrical peripheral neuropathies in AIDS. J Neurol Neurosurg Psychiatry 54:755.

Fuller GN, Jacobs JM, Guiloff RJ. (1991a) Subclinical peripheral nerve involvement in AIDS: an electrophysiological and pathological study. J Neurol Neurosurg Psychiatry 54:318–24.

Fuller GN, Jacobs JM, Guiloff RJ. (1991b) Thalidomide, peripheral neuropathy and AIDS. Int J STD AIDS 2:369–370.

Gastaut JA, Gastaut JL, Pellissier JF et al. (1987) Neuropathie périphérique et infection par le retrovirus HIV. Presse Méd 16:1057.

Gherardi R, Lebargy F, Gaulard P et al. (1989) Necrotizing vasculitis and HIV replication in peripheral nerves. N Engl J Med 321:685–6.

Gold JE, Jimenez E, Zalusky R. (1988) Human immunodeficiency virus-related lymphoreticular malignancies and peripheral neurologic disease: a report of four cases. Cancer 61:2318–24.

Gonzales MF, Olney RK, Yuen TS et al. (1988) Subacute structural myopathy associated with human immunodeficiency virus infection. Arch Neurol 45:585–7.

Gorard DA, Henry K, Guiloff RJ. (1988) Necrotising myopathy and zidovudine. Lancet i:1050.

Grafe MR, Wiley CA. (1989) Spinal cord and peripheral nerve pathology in AIDS: the roles of cytomegalovirus and human immunodeficiency virus. Ann Neurol 25:561–6.

Guiloff RJ, Fuller GN, Roberts A et al. (1988) Nature, incidence and prognosis of neurological involvement in the acquired immunodeficiency syndrome in Central London. Postgrad Med J 64:919–25.

Hagberg L, Malmvall B-E, Svennerholm L et al. (1986) Guillain–Barré syndrome as an early manifestation of HIV central nervous system infection. Scand J Infect Dis 18:591–2.

Harriman GR, Smith PD, Horne MK. (1989) Vitamin B12 malabsorption in patients with acquired immunodeficiency syndrome. Arch Intern Med 149:2039–41.

Ho DD, Sarngadharan MG, Resnick L et al. (1985a) Primary human T-lymphotropic virus type III infection. Ann Intern Med 103:880–3.

Ho DD, Rota TR, Schooley RT et al. (1985b) Isolation of HTLV-III from cerebrospinal fluid and neural tissue of patients with neurological syndromes related to the acquired immunodeficiency syndrome. N Engl J Med 313:1493–8.

Hoffman PM, Festoff BW, Giron LT et al. (1985) Isolation of LAV/HTLV-III from a patient with amyotrophic lateral sclerosis. N Engl J Med 313:324–5.

Jacobson MA, Mills J, Rush J et al. (1988) Failure of antiviral therapy for acquired immunodeficiency syndrome-related cytomegalovirus myelitis, Arch Neurol 45:1090–2.

Jeantils V, Lemaitre MO, Robert J et al. (1986) Subacute polyneuropathy with encephalopathy in AIDS with human cytomegalovirus pathogenicity. Lancet ii:1039.

Jellinger K, Radiaszkiewicz T. (1976) Involvement of the central nervous system in malignant lymphomas. Virchows Arch[A] 370:345–62.

Kinlen LJ, Sheil AGR, Peto J et al. (1979) Collaborative United Kingdom–Australian study of cancer in patients treated with immunosuppressive drugs. Br Med J 2:1461–6.

Lagueny A, Rommel A, Vignolly et al. (1986) Thalidomide neuropathy: an electrophysiologic study. Muscle Nerve 9:837–44.

Lange DJ, Britton CB, Younger DS et al. (1988) The neuromuscular manifestations of human immunodeficiency virus infections. Arch Neurol 45:1084–8.

Lanska MJ, Lanska DJ, Schmidley JW. (1988) Syphilitic polyradiculopathy in an HIV-positive man. Neurology 38:1297–1301.

Léger JM, Bouche P, Bolgert F et al. (1989) The spectrum of polyneuropathies in patients infected with HIV. J Neurol Neurosurg Psychiatry 52:1369–74.

Le Quesne PM. (1984) Neuropathy due to drugs. In: Dyck PJ, Thomas PK, Lambert EH et al, ed. Peripheral Neuropathy, 2nd edn. WB Saunders: Philadelphia, 2162–79.

Lipkin WI, Parry G, Kiprov D et al. (1985) Inflammatory neuropathy in homosexual men with lymphadenopathy. Neurology 35:1479–83.

Lohmoller G, Matuschke A, Goebel FD. (1987) Testing for neurological involvement in HIV infection. Lancet ii:1532.

Mah V, Vartavarian LM, Akers MA et al. (1988) Abnormalities of peripheral nerve in patients with human immunodeficiency virus infection. Ann Neurol 24:713–17.

Mahieux F, Gray F, Fenelon G et al. (1989) Acute myeloradiculitis due to cytomegalovirus as the initial manifestation of AIDS. J Neurol Neurosurg Psychiatry 52:270–4.

Marolda M, De Mercato R, Camporeale FS et al. (1989) Myopathy associated with AIDS. Ital J Neurol Sci 10:423–7.

McArthur JC. (1987) Neurologic manifestations of AIDS. Medicine 66:407–37.

Merigan TC, Skowron G, Bozzette SA et al. (1989) Circulating p24 antigen levels and responses to dideoxycytidine in human immunodeficiency virus (HIV) infections. Ann Intern Med 110:189–94.

Miller RG, Parry GJ, Pfaeffl W et al. (1988) The spectrum of peripheral neuropathy associated with ARC and AIDS. Muscle Nerve 11:857–63.

Miller RG, Storey RG, Greco C. (1989) Successful treatment of progressive polyradiculopathy in AIDS patients. Neurology 39 (suppl 1):271.

Monte de la SM, Gabuzda DH, Ho DD et al. (1988) Peripheral neuropathy in the acquired immunodeficiency syndrome. Ann Neurol 23:485–92.

Moskowitz LB, Gregorios JB, Hensley GT et al. (1984) CMV induced demyelination associated with AIDS. Arch Pathol Lab Med 108:873–7.

Novak IS, Trujillo JR, Rivera VM et al. (1989) Ganciclovir in the treatment of CMV infection in AIDS patients with neurologic complications. Neurology 39 (suppl 1):379–80.

Panegyres PK, Papadimitriou JM, Hollingsworth JA et al. (1990) Vesicular change in the myopathies of AIDS. Ultrastructural observations and their relationship to zidovudine treatment. J Neurol Neurosurg Psychiatry 53:649–55.

Parry GJ. (1988) Peripheral neuropathies associated with human immunodeficiency virus infection. Ann Neurol 23 (suppl):s49–53.

Paul T, Katiyar BC, Misra S. (1978) Carcinomatous neuromuscular syndromes: a clinical and quantitative electrophysiological study. Brain 101:53–63.

Pepose JS, Newman C, Bach MC et al. (1987) Pathologic features of cytomegalovirus retinopathy after treatment with the agent ganciclovir. Ophthalmology 94:414.

Piette AM, Tusseau F, Vignon D et al. (1986) Acute neuropathy coincident with seroconversion for anti-LAV/HTLV-III. Lancet i:852.

Rance NE, McArthur JC, Cornblath DR et al. (1988) Gracile tract degeneration in patients with sensory neuropathy in AIDS. Neurology 38:265–71.

Richman DD, Fischl MA, Grieco MH et al. (1987) The toxicity of azidothymidine (AZT) in the treatment of patients with AIDS and AIDS-related complex. A double-blind, placebo-controlled trial. N Engl J Med 317:192–7.

Riggs JE, Rogers JS II, Schochet SS Jr et al. (1987) AIDS-related neuropathy. W V Med J 83:167–9.

Robert ME, Geraghty JJ, Miles SA et al. (1989) Severe neuropathy in a patient with acquired immune deficiency syndrome: evidence for widespread cytomegalovirus infection of peripheral nerve and human immunodeficiency

virus-like immunoreactivity of anterior horn cells. Acta Neuropathol (Berl) 79:255–61.

Said G, Lacroix-Ciaudo C, Fujimura H et al. (1988) The peripheral neuropathy of necrotizing arteritis; a clinicopathological study. Ann Neurol 23:461–5.

Simpson DM, Bender AN. (1988) Human immunodeficiency virus-associated myopathy: analysis of 11 cases. Ann Neurol 24:79–84.

Singh BM, Levine S, Yarrish RL et al. (1986) Spinal cord syndromes in the acquired immune deficiency syndrome. Acta Neurol Scand 73:590–8.

Small PM, McPhaul LW, Sooy CD et al. (1989) Cytomegalovirus infection of the laryngeal nerve presenting as hoarseness in patients with acquired immunodeficiency syndrome. Am J Med 86:108–10.

Snider WD, Simpson DM, Nielsen S et al. (1983) Neurological complications of acquired immune deficiency syndrome: analysis of 50 patients. Ann Neurol 14:403–18.

So YT, Holtzman DM, Abrams DI et al. (1988) Peripheral neuropathy associated with acquired immunodeficiency syndrome: prevalence and clinical features from a population based survey. Arch Neurol 45:945–8.

Sotaniemi KA. (1984) Slimmer's paralysis – peroneal neuropathy during weight reduction. J Neurol Neurosurg Psychiatry 47:564–6.

Stern R, Gold J, Dicarlo EF. (1987) Myopathy complicating the acquired immune deficiency syndrome. Muscle Nerve 10:318–22.

Tucker T, Dix RD, Katzen C et al. (1985) Cytomegalovirus and herpes simplex virus ascending myelitis in a patient with acquired immune deficiency syndrome. Ann Neurol 18:74–9.

Vinters HV, Kwok MK, Ho HW et al. (1989) Cytomegalovirus in the nervous system of patients with the acquired immune deficiency syndrome. Brain 112:245–68.

Vishnubhakat SM, Beresford R. (1988) Prevalence of peripheral neuropathy in HIV disease: prospective study of 40 patients. Neurology 38 (suppl 1):350.

Walsh JC. (1971) Neuropathy associated with lymphoma. J Neurol Neurosurg Psychiatry 34:42–50.

Weber CA, Figueroa JP, Calabro JJ et al. (1987) Co-occurrence of the Reiter syndrome and acquired immunodeficiency. Arch Intern Med 107:112–13.

Winer JB, Hughes RAC, Anderson MJ et al. (1988) A prospective study of acute idiopathic neuropathy. II Antecedent events. J Neurol Neurosurg Psychiatry 51:613–18.

Wiselka MJ, Nicholson KG, Ward SC et al. (1987) Acute infection with human immunodeficiency virus associated with facial nerve palsy and neuralgia. J Infect Dis 15:189–94.

Woolsey RM. (1989) Treatable causes of meningomyelitis in individuals infected with human immunodeficiency virus. Arch Neurol 46:723.

Zacks SI, Langfitt TW, Elliot FA. (1964) Herpetic neuritis: a light and electron microscopic study. Neurology 14:744–50.

11 Role of the Macrophage in HIV Encephalitis

M.M. Esiri

AIDS has caused a resurgence of interest in the macrophages of the central nervous system (CNS). The reason is that in the CNS HIV predominantly infects macrophages. This is in contrast to the lymphoreticular tissues, the other main site of attack, in which CD4 (helper) lymphocytes appear to be the main target of infection and damage, and macrophages and monocytes more subsidiary ones. In the CNS, macrophage infection is associated with pathological changes that, in some cases, can eventually become clinically overwhelming. Yet how these changes are brought about remains obscure. To understand the possible effects that macrophage infection with HIV may have on the rest of the CNS, we need to know what influences macrophages normally exert, and whether these break down or become pathologically altered when HIV infection occurs. Unfortunately, very little is known at present about normal functions of macrophages in the nervous system. Even the question of which cells in the nervous system belong to the macrophage lineage has long been, and to a lesser extent remains, a matter of controversy. The aims of this chapter are first to outline recent views on the microscopic anatomy, immunocytochemistry and function of macrophages in the CNS, and second to summarize the effect that HIV has on this cell population and the ways in which this may lead to CNS damage.

Mononuclear Phagocytes of the Nervous System: Macrophages and Microglia

Tissue macrophages are heterogeneous components of a widely dispersed system of cells – mononuclear phagocytes – principally concerned with ingesting and killing foreign organisms and assisting lymphocytes in mounting immune responses. Tissue macrophages also remove effete cellular constituents of the body and are increasingly recognized as a source of secreted molecules that may influence the behaviour of neighbouring cells in the host tissue. Precursors of tissue macrophages are produced after early fetal life in the bone marrow, circulate in the blood as monocytes, and migrate, apparently at random, into normal host tissues and, in larger numbers, under chemotactic stimuli, into inflammatory lesions (Gordon, 1986; Johnston, 1988). Once they have entered their host tissue they differentiate into a variety of forms, some with detectably different functions, apparently at the dictate of the tissue environment in which they find themselves. Once they have reached their destination they are thought to have a life span of several months. Thus, in patients who had received a bone marrow transplant after ablation of

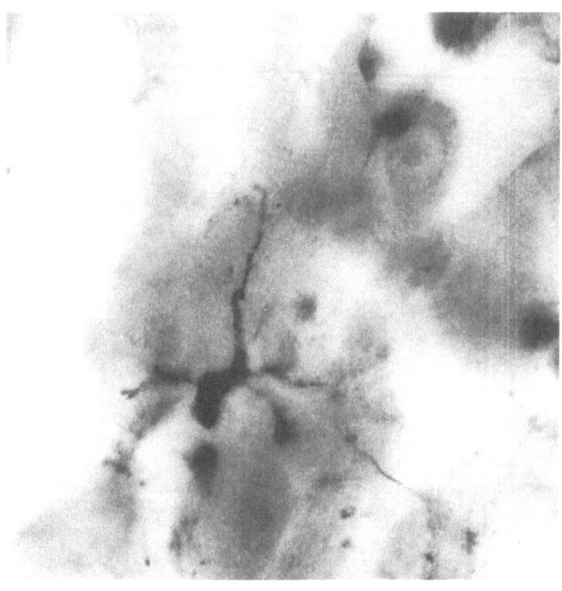

Figure 11.1. Resting microglial cell in human cerebral cortex, in a frozen, formalin-fixed section stained with silver carbonate. Note multiple slender processes and small cell body. The nucleus is obscured by the stain.

their own marrow, donor macrophages took about 3 months to replace host alveolar macrophages (Thomas et al, 1976).

Hortega (1919, 1932) was the first to describe microglia, and the first of many investigators to turn his attention to the migration and maturation of blood monocytes into macrophages in the CNS. He made use of a new silver carbonate stain, with which he claimed that mesodermal elements from the blood enter the nervous system and leptomeninges around the time of birth and, in the brain, form amoeboid microglial cells which eventually differentiate into resting microglia. Resting microglia, which, according to Hayes et al (1987), make up about 13 per cent of normal human white matter glial cells, have a small oval, elongated or comma-shaped nucleus, with dispersed chromatin and fine branching crenellated cytoplasmic processes. In white matter they are predominantly bipolar, with their long axes in the plane of the fibre tracts they occupy. In cortical grey matter they are bipolar or stellate in shape (Figure 11.1). Processes of adjacent cells rarely overlap with one another. Hortega considered that a resting microglia remains quiescent until a local disturbance occurs which stimulates it to enlarge, divide and form reactive microglial cells. Although this view was fairly generally accepted for a while (with a few

dissentions), it fell into some disfavour when newer techniques – electron microscopy, autoradiography and immunocytochemistry – were applied to try to confirm it.

Early electron microscopists who studied the nervous system had initial difficulty in identifying microglial cells at all (Maxwell and Kruger, 1965; Kruger and Maxwell, 1966; King, 1968). This was partly due to the small size of the resting microglial cell population, and the inconspicuous nature of the cells themselves. Further study led to the recognition of scattered bipolar cells with fine ramifying processes and various ultrastructural features, such as relatively electron-dense cytoplasm containing mitochondria, ribosomes, rough endoplasmic reticulum, Golgi apparatus, dense bodies, vesicles and a few intermediate filaments; all features consistent with mononuclear phagocytes. These cells were identified as resting microglial cells (Vaughn and Peters, 1968; Mori and Leblond, 1969; Barön and Gallego, 1972; Blakemore, 1975) (Figure 11.2). Similar, slightly larger cells, containing more cytoplasmic organelles, were found after applying various lesions to the CNS, and identified as reactive microglial cells (Matthews and Kruger, 1973; Persson, 1976; Fujita et al, 1981; Brierley and Brown, 1982).

Autoradiography has been used in numerous experimental studies in animals to investigate the origin of resting and reactive microglial cells. The earlier studies were reviewed by Oemichen (1978, 1982). In many studies tritiated thymidine was injected before, or just after, infliction of focal injury to the CNS, and labelled cells were sought in and beyond the lesions with autoradiography, often combined with electron microscopy (Adrian and Walker, 1962; Konigsmark and Sidman, 1963; Huntington and Terry, 1966; Adrian, 1968; Kitamura et al, 1972; Kitamura, 1973; Kitamura and Fujita, 1975; Fujita and Kitamura, 1976; Adrian et al, 1978; Schelper and Adrian, 1980). Finding labelled cells in the lesions was generally taken as evidence for a vascular origin of the cells, since the uninjured CNS contained hardly any detectable (dividing) labelled cells under these conditions. Results were variable and, according to Schelper and Adrian (1986), their interpretation may in some instances have been confused by the fact that tritiated thymidine could have been re-utilized at the injury site and taken up by locally-dividing cells, not necessarily of blood origin. However, in general, findings suggested that CNS damage from trauma or inflammation that involved a local breakdown in the blood–brain barrier caused local accumulations of labelled phagocytes derived from the blood,

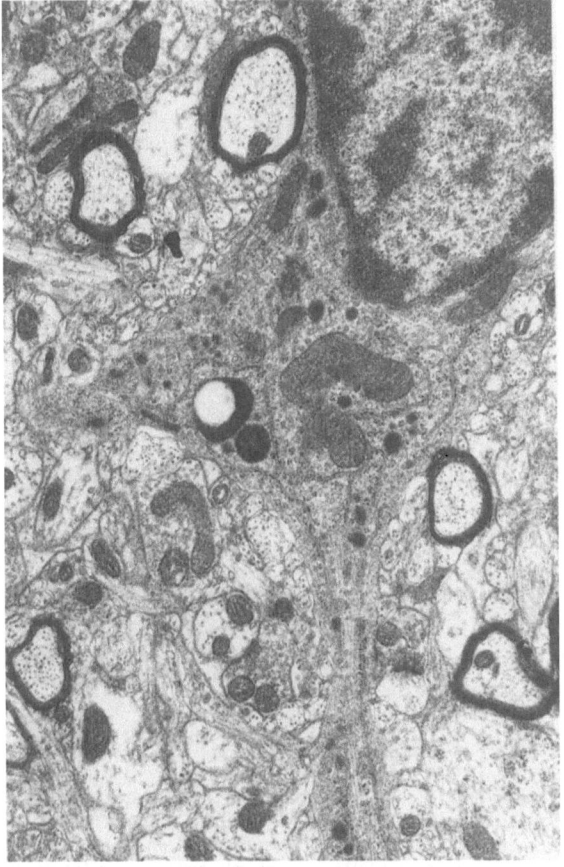

Figure 11.2. Electron micrograph of part of a resting micro-glial cell. The nucleus is at top right. A narrow process is seen extending to the bottom of the photograph. The cytoplasm contains dense bodies and mitochondria. Uranyl acetate and lead citrate stain, ×91 000. (Kindly supplied by Dr W. Blakemore.)

circulating cells, though the situation was different in newborns (Imamoto and Leblond, 1978; Ling, 1980).

When it was first introduced, immunocyto-chemistry seemed to offer a clue to the origin and nature of microglia. If marker antigens could be used to identify all cells of the mononuclear phagocyte series, they could theoretically settle the question of the origin of microglia. Unfortunately the blood monocytes, against which such antibodies were first raised, lack some of the structural and functional attributes of differentiated macrophages. Some of these antibodies reacted with tissue macrophages in other organs, and with some perivascular and leptomeningeal macrophages in the CNS, but not with microglia (Oemichen et al, 1979; Wood et al, 1979; Tsuchihashi et al, 1981; Esiri and Booss, 1984; Matsuomoto et al, 1985). The conclusion drawn by some was that microglia were therefore not members of the mononuclear phagocyte system. More recent immunocytochemical studies using antibodies raised against tissue macrophages have, however, come to different conclusions, and firmly support Hortega's original view that microglial cells are indeed members of the mononuclear phagocyte lineage of cells, albeit exceptionally specialized ones (Perry et al, 1985; Franklin et al, 1986; Davey et al, 1988; Adams et al, 1989). Independent histochemical evidence also supports this view (Murabe and Sano, 1982a; Vorbrodt and Wisniewski, 1982).

As far as their functional capacities are concerned, resting microglial cells lack many of the phagocytic properties of their extraneural tissue macrophage counterparts. For example, they show no acid phosphatase, lysozyme or non-specific esterase activity (Oemichen, 1978; Kitamura, 1980). Conversely, they show thiamine pyrophosphatase activity, which is lacking in other macrophages and monocytes, and which suggests that they may have a role in thiamine metabolism in the nervous system (Murabe and Sano, 1982a). In grey matter, resting microglia show a close association with neurons and synaptic endings (Murabe and Sano, 1981, 1982b), and in response to axotomy, their processes become insinuated between the reactive neuronal surface and presynaptic endings on it (Blinzinger and Kreutzberg, 1968; Sumner, 1975; Torvik and Soreide, 1975; Sumner, 1977). On the basis of such observations, it has been suggested that microglia may play a role in the remodelling of synapses. Recent studies of microglia in the neurohypophysis of rats also support this notion (Pow et al, 1989). Microglia and leptomeningeal

whereas other changes, not associated with blood–brain barrier breakdown, were marked by unlabelled reactive microglia, presumably of local origin. Similar conclusions were reached in the recently reported autoradiographic study from Schelper and Adrian (1986), who used ^{125}I-labelled iodo-deoxyuridine to overcome the problem of thymidine re-utilisation. In stab injuries in mice, they found many labelled mono-cytes and macrophages but no labelled microglia, although there were increased numbers of un-labelled reactive microglial cells. Monocytes, they firmly stated, enter the injured CNS to become macrophages but not microglia. In adult animals, using tracer techniques, it was difficult or impossible to find any resting microglial cells in the normal, uninjured CNS that were derived from

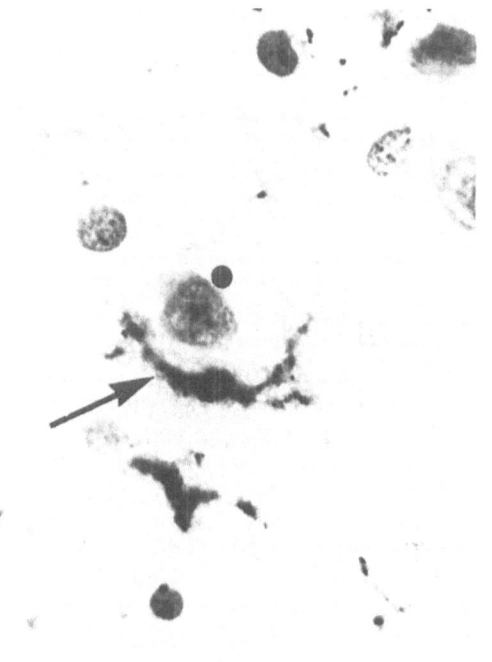

Figure 11.3. Resting microglial cell (*arrow*) in human cerebral cortex demonstrated by means of RCA-1 lectin binding to a paraffin section. The cell is closely juxtaposed to a neuron, the nucleus of which is visible. Counterstained with haematoxylin. (Preparation by Dr C.S. Morris.)

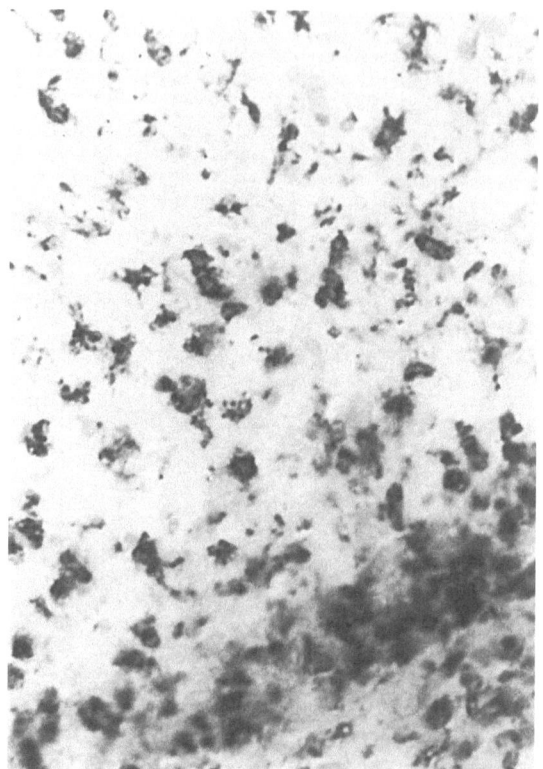

Figure 11.4. Reactive microglia in human cerebral white matter from a case of multiple sclerosis, demonstrated in a frozen section treated with a monoclonal antibody to MHC Class II antigens (RFDRI). Counterstained with haematoxylin.

and perivascular macrophages can also undergo pinocytosis and take up tracer proteins, and therefore may play a role in removal of soluble substances from the extracellular space (Wagner et al, 1974). Resting and reactive microglia and macrophages all show histochemical evidence of nucleoside diphosphatase activity and strong binding of the lectin *Ricinus communis* agglutinin-1 (Murabe and Sano, 1982a; Vorbrodt and Wisniewski, 1982; Mannoji et al, 1986) (Figure 11.3), but the functional significance of these reactions, which they share with capillary endothelium, is not clear. Regarding immune functions of microglia, reactive forms and, according to some studies, resting forms at a lower level, as well as macrophages, express MHC Class II antigens (Hauser et al, 1983; Daar et al, 1984; Lampson and Hickey, 1986; Woodroofe et al, 1986; Esiri and Reading, 1987; Hayes et al, 1987; Sobel and Ames, 1988). This suggests that microglia may have the potential to present processed antigens on their surface to helper T-lymphocytes. IgG Fc-receptors and complement type 3 receptors have been described on resting microglia in mice (Perry et al, 1985),

though not in humans (Wood et al, 1979). Histochemically, and in their antigenic reactions with marker antibodies, reactive microglia bear much closer resemblances to macrophages than do resting microglia. Thus, they possess lysozyme, acid phosphatase and non-specific esterase activity, and carry surface IgG Fc-receptors (Oemichen and Huber, 1976; Oemichen et al, 1980) and readily-detectable MHC Class II antigens (Figure 11.4).

Turning to morphological aspects of microglial cell reactions to disease in the CNS (Duchen, 1984; Dolman, 1985), there is probably no pathological condition in the immunocompetent individual which does not involve a microglial cell reaction. In this respect, microglia are similar to astrocytes which, at least in white matter, will take on a reactive appearance under almost any pathological stimulus. However, it is of interest that the reactions of these two cell types do not invariably take place on the same scale and at the

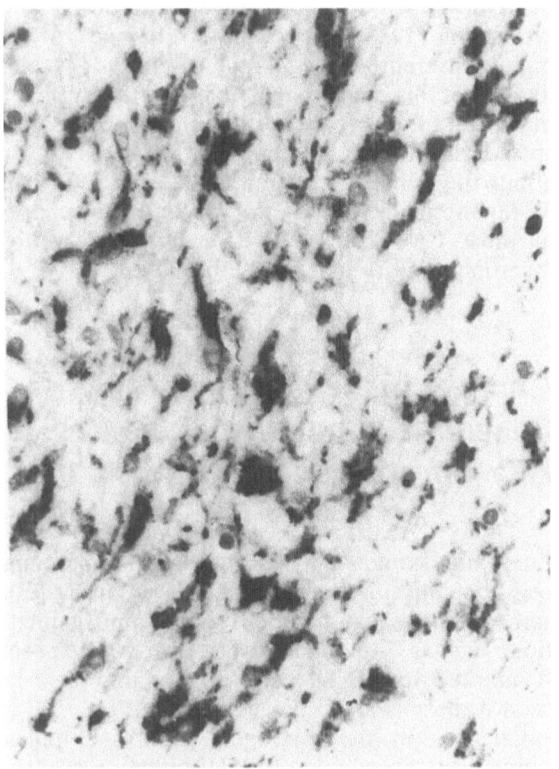

Figure 11.5. Reactive microglia in the pons from a case of subacute sclerosing panencephalitis demonstrated in a paraffin section treated with RCA-1 lectin. Counterstained with haematoxylin. (Preparation by Dr C.S. Morris.)

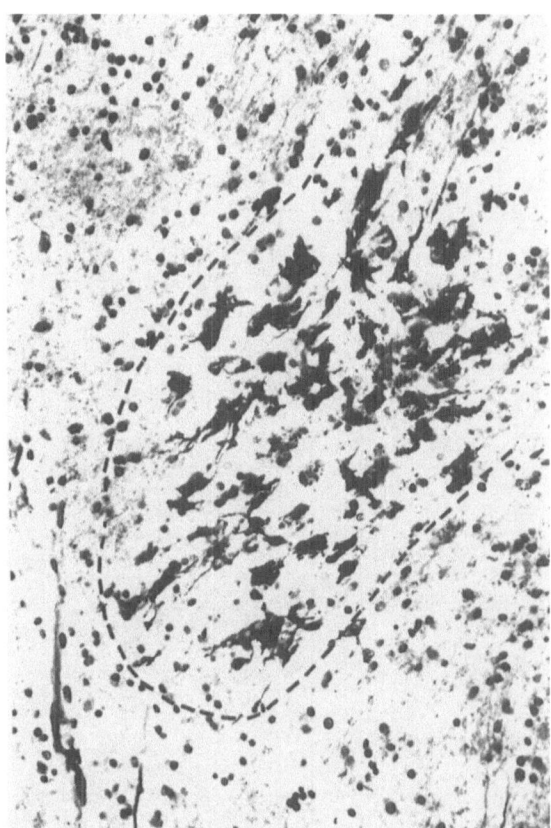

Figure 11.6. Frozen formalin-fixed section from a case of severe head injury, showing a cluster of reactive microglial cells (*outlined*) demonstrated using silver carbonate stain.

same pace. Thus, there may be a florid reactive astrocytosis in longstanding cerebral oedema with only a mild microglial cell reaction and, conversely, in some infective conditions the microglial reaction tends to be more prominent than the astrocytic response, especially at an early stage. A microglial cell response consists of an increase in size of the cells and, in some contexts, an increase in number and length of the processes of each cell, so that they come to overlap. The packing density of the cells is also increased. Some of the most striking reactions of microglial cells in human pathology are seen in subacute and chronic infective conditions, such as subacute sclerosing panencephalitis, due to measles infection, and general paralysis of the insane, due to syphilis (Figure 11.5). Focal collections of plumper cells with fewer, shorter processes are found chiefly in white matter in closed head injuries (Figure 11.6), and in grey matter in cytomegalovirus encephalitis. In tumours, infarcts and plaques of multiple sclerosis, the mononuclear phagocytes that accumulate in the lesions resemble macrophages rather than microglia. Characteristically, they develop into lipid-filled, round-bodied cells. Reactive microglia can be found beyond the margins of the lesions (Figure 11.7). These variations suggest that for the full development of the microglial phenotype, relatively-intact neural tissue – grey or white – is required. Conversely, it is not microglia but macrophages and conventional alternative forms such as epithelioid cells and multinucleated giant cells (MGC) that accumulate in the meninges and superficial Virchow–Robin spaces in chronic meningitis.

In summary, Hortega's view of the origin and nature of microglial cells are broadly held to be valid by many current authors (Perry and Gordon, 1989). Macrophages are widely believed to enter the developing nervous system as blood monocytes. The stimulus for their migration may be programmed cell death in the nervous system. Some become perivascular and subarachnoidal,

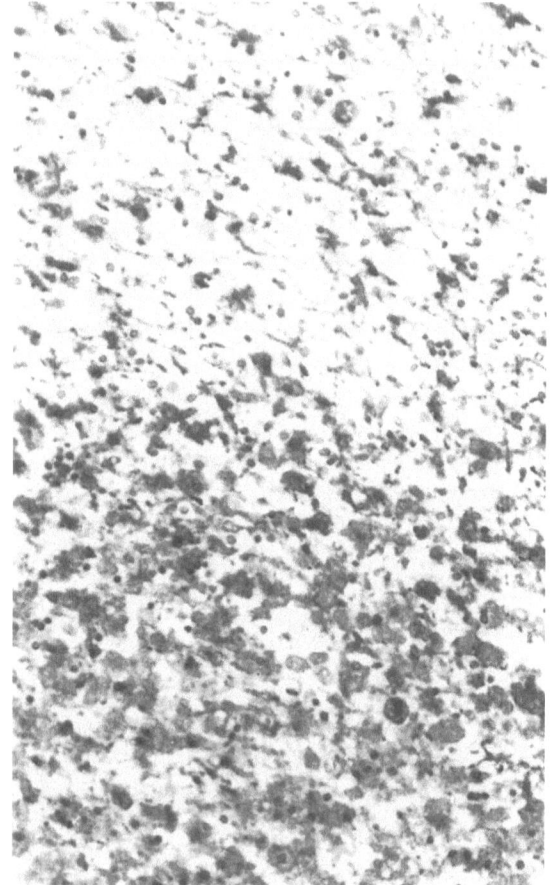

Figure 11.7. Acute multiple sclerosis plaque (*bottom*) containing plump, round-bodied macrophages. At the *top*, in surrounding myelinated white matter there are numerous process-bearing reactive microglia. Frozen section treated with antimacrophage monoclonal antibody EBMII. Counterstained with haematoxylin.

been described for macrophages at other sites (Johnston, 1988) may well be barred to the highly-differentiated microglial cell. It seems likely that this sacrifice will be found to have been made in the interests of taking on additional specialized functions when more becomes known about the behaviour of microglia. In any event, by far the main source of activated macrophages in most CNS disease is from newly-recruited macrophages brought by the blood.

Changes in the Mononuclear Phagocyte Population of the CNS in HIV Infection and AIDS

Dramatic changes may be seen in the microglial cell population in the brain in AIDS, but at least some of these may be due to opportunistic infections such as CMV encephalitis or toxoplasmosis. A marked microglial cell reaction may also be seen if there is a cerebral lymphoma.

Changes in the microglial and macrophage population directly due to HIV infection are best evaluated in cases that lack these additional, complicating features. Early changes could be expected in cases of HIV infection that die before the onset of AIDS. Death may be the result of suicide or incidental illness, but more commonly occurs among HIV-positive haemophiliacs who frequently die without developing AIDS (Esiri et al, 1989). These patients often show fatal intracranial haemorrhage, but the effects of this on the overall evaluation of the changes, particularly if it is no more than a few hours old, will be relatively slight. Older haemorrhages can be avoided when assessing the microglial reaction. Almost all such HIV-infected cases show some degree of microglial and macrophage alteration. In most cases, microglial cells show a slight, patchy, but relatively diffuse increase both in numbers, and in size and prominence of their processes (Figure 11.8). This is seen both in grey and white matter. Superimposed on this slight background of reactive change, some cases show well-defined nodules or clusters of plump, reactive microglial cells with shorter processes (Figure 11.9). In grey matter and periventricular tissue, these microglial nodules may be indistinguishable from microglial nodules due to CMV infection except that, unlike the latter, CMV inclusions or other evidence for the presence of CMV from nucleic acid hybridization or immunocyto-

or choroid plexus macrophages. Others differentiate into resting microglial cells over the course of a few days to a few weeks, depending on the species. Well-differentiated microglial cells can be detected in human fetuses by the second trimester (Esiri et al, 1991a). Resting microglial cells probably have only a limited capacity to convert to reactive microglia, perhaps proliferating in the process, and respond with very restricted phagocytic activity to a pathological insult to the nervous system. However, they may be able to carry out immune functions quite effectively by increasing their expression of MHC Class II products, and presenting antigens locally to activated lymphocytes (Woodroofe et al, 1986; Esiri and Reading, 1987). Nevertheless, the major effects of macrophage activation that have

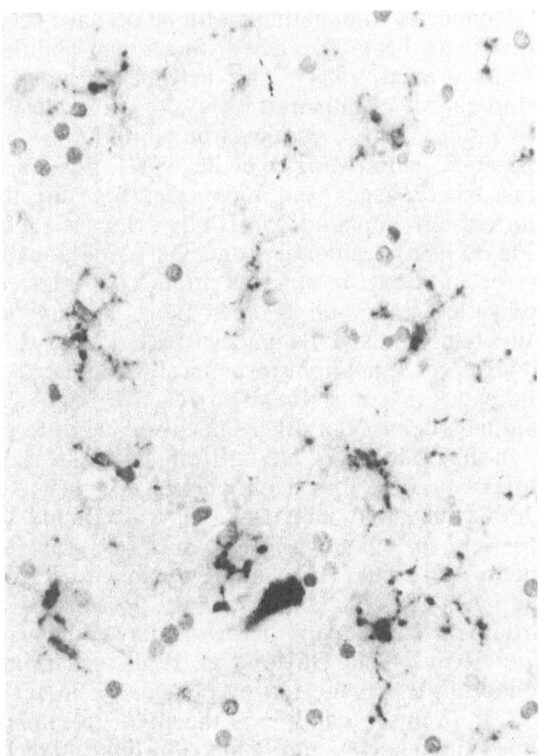

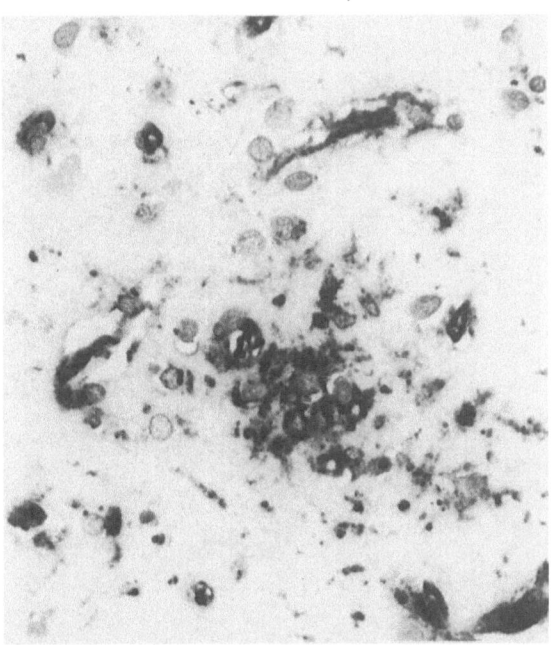

Figure 11.9. Cerebral cortex from an HIV-positive haemophiliac showing a focal collection of microglia. Such microglial nodules may contain CMV antigens. RCA-1 lectin binding on paraffin section counterstained with haematoxylin. (Preparation by Dr C.S. Morris.)

Figure 11.8. Low power view of cerebral white matter from an HIV-positive haemophiliac. No specific pathology was present but there is a diffuse increase in size and prominence of process-bearing microglia. RCA-1 lectin binding on paraffin section counterstained with haematoxylin. (Preparation by Dr C.S. Morris.)

chemistry is lacking, and HIV may be detectable (see below).

A further feature that is common in brains of cases with HIV infection, but not full-blown AIDS, is the presence of sparse cuffs of perivascular lymphocytes and macrophages, usually surrounding veins, and occurring in grey or white matter and in the leptomeninges (personal observations). These do not harbour demonstrable viruses. These features – mild, diffuse and/or focal reactive microglial cells and mononuclear perivascular infiltrates – may occur in the absence of any other changes, but there may be diffuse pallor or myelin staining in the central cerebral white matter of such cases. If the microglial reaction is more marked and widespread, it is usually accompanied by diffuse reactive astrocytosis with myelin pallor or rarefaction, or cerebral oedema.

In 20–30 per cent of cases of AIDS, but rarely in those without fully-developed AIDS, the microglial cell changes outlined above are accompanied by MGC, which have been repeatedly

shown in immunocytochemical studies to carry macrophage-specific markers (Sharer et al, 1985; Budka, 1986; Dickson, 1986; Gartner et al, 1986; Koenig et al, 1986; Stoler et al, 1986; Wiley et al, 1986a; Budka et al, 1987; Gendelman et al, 1989). They vary in size from that of a large macrophage to 4–6 times greater, ie 50–60 µm across. They may have only 2–3 nuclei, but the larger ones have 20–30 or even more, usually peripherally situated (Figure 11.10). The cytoplasm may contain lipofuscin pigment. These cells are found in perivascular spaces and in parenchymal microglial/macrophage clusters bordering these spaces. Less commonly, they are solitary and scattered or in the company of a few macrophages, in grey or white matter. Central cerebral white matter, deep grey matter of basal ganglia and thalamus, or the brainstem are the sites where they are most frequently found. They are usually associated with diffuse or focal myelin pallor, oedema or rarefaction of white matter. Less commonly, they are associated with damage to axons as well. Multinucleated giant cells have been found in the lesions of HIV-myelopathy of the spinal cord, as well as in HIV encephalopathy (Maier et al, 1989). However, they have not been associated so far with peripheral neuropathy

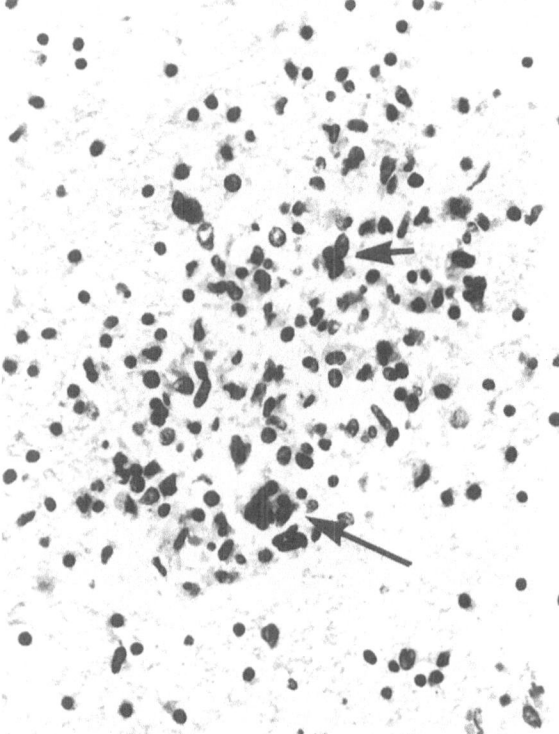

Figure 11.10. White matter from an HIV-positive haemophiliac containing a focal collection of inflammatory cells including a few multinucleated cells (*arrows*). Haematoxylin and eosin stain.

Monocytes from patients with AIDS have been shown to have depressed functional abilities (Murray et al, 1984). This may be an indirect consequence of impaired release of macrophage-activating factors, particularly γ-interferon, by T-lymphocytes (Murray et al, 1987). Evidence that macrophages and monocytes are directly functionally impaired by HIV infection is slight (Fauci, 1988; Lechtenberg and Sher, 1988), but in a recent study monocytes from HIV-infected individuals showed a decreased capacity to stimulate a mixed lymphocyte reaction (Twigg et al, 1991). Immunohistochemically, they display the same reactions found in reactive microglia and macrophages in other infective conditions.

In the CNS, MGC are far from unique to HIV infection. As well as being present in fungal and chronic bacterial infections, they are found in diseases of unknown origin such as granulomatous arteritis (Cravioto and Feigin, 1959; Nurick et al, 1972; Sandlin et al, 1979), and also in arteritis accompanying herpes zoster (Linneman and Alvira, 1980; Doyle et al, 1983) and in immunosuppressive measles encephalitis (Esiri et al, 1982). In these conditions the MGC resemble those found in HIV encephalopathy but they have not been as fully investigated for a macrophage origin.

In summary, the HIV-infected mononuclear phagocyte population in the CNS survives, and indeed enlarges, as the course of the infection progresses. This enables these cells to act as a reservoir of the virus, and is in contrast to the situation with respect to the CD4 helper lymphocyte population, which progressively contracts as AIDS progresses.

in HIV infection. The evidence that these cells are infected by HIV is discussed below.

Morphologically, the MGC in the brain and spinal cord in AIDS resemble in some respects those found in chronic infective granulomas. It seems likely that they are produced under the influence of an HIV-coded protein. Cells resembling these macrophage-derived MGC are seen in HIV-infected T-cell cultures in vitro (Popovic et al, 1984), and the HIV envelope protein gp120 is known to effect fusion of cells by a direct action on their cell membranes. In classical infective granulomas, on the other hand, MGC may be produced under the influence of a soluble product released from activated T-lymphocytes, possibly γ-interferon (Weinberg et al, 1984). The resulting MGC, regardless of their precise mode of formation, may well have similar properties. Formal investigation of the function and biochemistry of MGC in culture suggests that they closely resemble macrophages (Schlesinger et al, 1984).

Infection of Mononuclear Phagocytes of the Nervous System in HIV Infection and AIDS

Numerous studies using in situ hybridization histochemistry and immunocytochemistry have shown that HIV antigens and nucleic acid are found in macrophages, microglial cells and MGC in HIV encephalopathy (Epstein et al, 1985; Budka, 1986; Gabuzda et al, 1986; Koenig et al, 1986; Wiley et al, 1986b; Budka et al, 1987; Pumarola-Sune et al, 1987; Vazeux et al, 1987; Gendelman et al, 1988). Retroviral particles have also been identified in association with MGC and mononuclear cells having ultrastructural

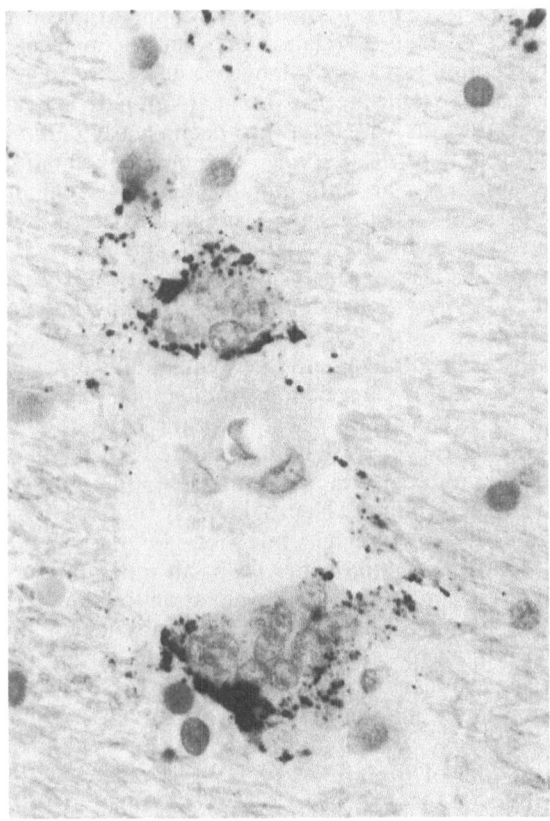

Figure 11.11. Perivascular multinucleated giant cells in the white matter of a case of HIV-1 encephalopathy. The section has been treated to demonstrate p24 antigen of HIV-1 (black). Viral antigen is seen particularly around the margins of the cells. Counterstained with haematoxylin. (Immunostained preparation by Dr C.S. Morris.)

appearances consistent with macrophages in HIV encephalopathy (Epstein et al, 1985; Sharer et al, 1985; Budka et al, 1987; Koenig et al, 1986; Meyenhofer et al, 1987; Mirra and del Rio, 1989). Viral particles were found in the endoplasmic reticulum and budding from, or immediately outside, the plasma membrane. HIV has also been cultured from mononuclear phagocytes derived from a brain biopsy from a case of AIDS with neurological manifestations (Gartner et al, 1986). In the spinal cord, HIV antigens (Maier et al, 1989) and nucleic acid (Weiser et al, 1990) have been found in MGC in white matter showing a vacuolar myelopathy, though the association of HIV-infected MGC with this myelopathy is more controversial than with encephalitis (Grafe and Wiley, 1989; Rosenblum et al, 1989). In particular, MGC are the cells most regularly shown to contain abundant HIV, with indications of proliferation of the virus (Figure 11.11). If MGC are present, the proportion of them, and of mononuclear phagocytes that can be shown to contain HIV (by immunocytochemistry) is quite high, particularly if frozen material, in which the antigens are well preserved, is examined. In the absence of MGC, the chance of detecting any HIV in the brain by immunocytochemistry, nucleic acid hybridization or electron microscopy is greatly reduced.

If the evidence for HIV infection of mononuclear phagocytes and MGC is strong, there is much less certain evidence about how damage to the nervous system is produced, and almost no evidence about the localization of HIV in the CNS in the asymptomatic early stages following infection, even though the virus can be cultured from the cerebrospinal fluid (Resnick et al, 1985) and specific local antibody production to HIV occurs (Goudsmit et al, 1986; Elovaara et al, 1987; Andersson et al, 1988; Resnick et al, 1988). These more puzzling aspects of CNS HIV infection are discussed below.

Evidence for Infection of Other CNS Cell Types by HIV

One possible explanation for damage to the CNS in AIDS is infection of other cell types in addition to mononuclear phagocytes. For example, oligodendrocyte infection might lead to myelin damage as it does in progressive multifocal leukoencephalopathy with JC papovavirus infection; and endothelial cell infection might cause cerebral oedema. There is some tentative support for this hypothesis, but it is not, at present, firmly based. A few of the immunocytochemical and in situ hybridization studies have claimed to detect HIV in endothelial cells (Wiley et al, 1986b; Ward et al, 1987), neuroglia (Stoler et al, 1986; Pumarola-Sune et al, 1987; Ward et al, 1987; Esiri et al, 1991b) and neurons (Wiley et al, 1986a; Pumarola-Sune et al, 1987). Gyorkey et al (1987) claim to have identified cells containing retrovirus particles in biopsies from AIDS cases as oligodendrocytes, and Mirra and del Rio (1989) found an infected astrocyte, also in ultrastructurally-examined biopsy material. Some of these reports, with their photographic evidence, have not been regarded as entirely convincing (see Budka et al, 1987 and Chapter 7, for example). Endothelial cells can be hard to distinguish from perivascular macrophages, particularly in frozen sections, and reactive microglia can display long processes that mimic those of astrocytes. Other authors have searched carefully for evidence of infection in cells other than mononuclear phagocytes and MGC, and failed to find it. On the other hand, there are interesting in vitro studies of HIV infection in glial cells, which indicate that under the culture conditions employed, normal and neoplastic neuroglia can be infected with HIV (Cheng-Mayer et al, 1987; Chiodi et al, 1987; Popovic et al, 1988; Clapham et al, 1989; Watabe et al, 1989). Some HIV isolates seemed to be more effective than others at infecting neuroglial cells in culture, though the percentage of cells infected was always low and the yield of progeny virus also low. CD4 membrane antigen was apparently not involved in the entry of HIV into neuroglial cell lines, in contrast to T-cells and macrophages (Clapham et al, 1989). Growth and multiplication of cultures of neoplastic glial cells infected with HIV was considerably reduced compared with uninfected controls, despite only a few cells being infected (Popovic et al, 1988). Taken together, this evidence suggests that it is premature at this stage to dismiss the possibility of neuroglial cell infection with HIV occurring in vivo.

Comparison of HIV with Other Lentiviruses: Inflicted Damage to the CNS

The lentivirus responsible for visna—maedi in sheep offers interesting comparisons with HIV and AIDS. Like HIV, visna virus infects macrophages and produces demyelinating disease of the CNS (Narayan et al, 1983; Haase, 1986; Kennedy et al, 1989). However, the visna CNS lesions, unlike those of AIDS, are intensely inflammatory. This difference may well reflect the immunosuppressive effects of HIV which limit the capacity of the host to mount an inflammatory response. Emphasis has been placed on increased MHC Class II molecule expression by macrophages by visna virus in the pathogenesis of nervous system, joint and lung disease in visna—maedi (Kennedy et al, 1985). It has been suggested that this may lead to auto-immune disease, including CNS demyelination, through excess immune stimulation by auto-antigens (Johnson, 1982). However, the increased MHC Class II expression in visna is entirely to be expected as part of the activation of macrophages that accompanies any viral infection, and it can be readily demonstrated in a variety of human encephalitides. The more unusual feature of visna (as of HIV) is the prolonged, life-long, low-grade infection it causes in macrophages. The crucial importance of persistent viral infection as the

driving force in provoking CNS damage in visna is indicated by a correlation between sheep species susceptibility to CNS damage and ability of the virus to replicate in brain (Kennedy et al, 1989). Moreover, it has not been clearly demonstrated that the nervous system demyelination in visna has an auto-immune basis. It remains equally possible that the visna lesions are due to infected macrophages in the CNS (and elsewhere) acting as a reservoir of infection that can spread to involve oligodendrocytes. Experimental intracerebral inoculation of visna virus in sheep can result in oligodendrocyte infection (Stowring et al, 1985), thus demonstrating that oligodendrocytes can act as host cells for the virus, though whether they are killed, or rendered incapable of maintaining myelin when infected, has not yet been determined. Neuroglial cell infection (or, indeed, macrophage infection in the CNS), by attracting into the CNS an inflammatory infiltrate, may enable demyelination to take place as a 'bystander' effect, even in the absence of any viral-induced damage to the neuroglia. Studies of visna that can cast light on which of these various alternative mechanisms of CNS damage is significant may contribute indirectly to an understanding of the effects of HIV infection on the CNS.

Pathogenesis of CNS Damage by HIV and its Relation to Mononuclear Phagocyte Infection: Possibilities and Remaining Uncertainties

HIV infection of the CNS is the first human viral infection in which the major site of infection, at least at the stage when tissue damage occurs, has been clearly shown to be in cells of the mononuclear phagocyte lineage. We do not know if the same cell type harbours the virus during the asymptomatic stage of the infection or whether, at this stage, CD4 lymphocytes are involved. It has been suggested that HIV enters the nervous system in infected monocytes, which then differentiate into infected macrophages and microglial cells (Stowring et al, 1985; Haase, 1986; Ho et al, 1987), and the widespread distribution of the lesions is consistent with this hypothesis. Initial meningeal infection followed by spread through Virchow—Robin spaces (Esiri and Gay, 1990) could also probably explain the distribution of lesions. Initially, it seems likely that infection is latent or expressed at only a very low level. Since symptomatic infection of the CNS, in most patients, occurs relatively late in the course of AIDS when T-cell function has become ineffective, T-cells (in the perivascular infiltrates that can be seen in asymptomatic subjects in the brain and meninges?) may earlier have been involved in keeping the infection in the nervous system quiescent. Symptomatic infection would then be seen as emerging when T-cell control is lost. Another possible means by which infection might spread to CNS macrophages would be the development of new genetic variants of the virus that are particularly well suited to proliferation in macrophages. There are indications from viral-culture studies that variants with tropism for macrophages can indeed be found in the CNS in AIDS (Gartner et al, 1986; Rübsamen-Waigman et al, 1986; Evans et al, 1987; Koyanagi et al, 1987). Increase in the available macrophage and reactive microglial cell population, caused by extra recruitment of monocytes in response to opportunistic infections, would also promote HIV proliferation in the CNS by providing more host cells. Other viruses may also act to promote HIV replication in the CNS. CMV is of particular interest in this context, as it is capable of infecting monocytes and macrophages (Schrier et al, 1985; Wiley et al, 1986b). In HIV-positive subjects, systemic CMV infection may predispose to the progression to AIDS (Webster et al, 1989), and CMV and HIV infection in the same brain, even in the same microglial nodules and individual cells, has been reported on several occasions (Wiley and Nelson, 1990; Budka et al, 1987; Vinters et al, 1989). Furthermore, murine CMV infection, when administered to transgenic mice containing integrated copies of the HIV long terminal repeat (LTR) linked to the bacterial gene for chloramphenicol acetyl-transferase (CAT), caused a 3–5-fold increase in CAT activity under conditions of macrophage stimulation by macrophage colony-stimulating factor or phorbol myristate acetate (Gendelman et al, 1988). These two agents, which are capable of inducing cellular differentiation, on their own augmented spontaneous CAT expression 2–3-fold. These experiments show that macrophage differentiation and superinfection with CMV are both possible means by which HIV production may be increased to produce symptomatic CNS infection in humans. Other DNA viruses can act in a similar manner to increase HIV-LTR-directed gene expression (Gendelman et al, 1986). Infection of CD4 lymphocytes by HIV entails interaction with the part of the CD4 antigen which acts as a receptor for the virus (Dalgliesh et al, 1984; Klatzmann et al, 1984). Cell-to-cell spread of HIV infection in the CNS might similarly be expected to be dependent on expression of CD4 antigen on the surface of recipient cells. Low levels of CD4 expression have been described on macrophages (but not microglia) (Vazeux et al, 1987), and also on endothelium (Rhodes and Ward, 1989) in HIV infection. However, there are indications from in vitro studies that infection of neuroglial cell lines may occur independently of CD4 expression (Cheng-Mayer et al, 1987; Chiodi et al, 1987; Clapham et al, 1989; Harouse et al, 1989).

The final question to be considered is how infection of macrophages, MGC and microglial cells leads to the pathological changes, particularly prominent in white matter, of HIV encephalopathy (see Chapter 7). Infection of neuroglia is one possibility which has already been referred to. Another is that HIV-infected mononuclear phagocytes produce either cell- or virus-coded products which are deleterious to the nervous system. One such factor might be tumour necrosis factor. Alternatively, infected mononuclear phagocytes may be hampered in performance of their normal functions in the nervous system, or macrophages may utilize, for their own metabolism, scarce substances that are required for normal neuron or neuroglial function. Thus, one suggestion advanced to explain the similarity of HIV myelopathy to subacute combined degeneration of the cord due to vitamin B_{12} or folate deficiency, was that macrophages in the lesions may use up all locally-available folate, resulting in a shortage of that available to other cells (Smith et al, 1987). Yet another possibility is that, as suggested in visna, prolonged increase in MHC Class II expression on infected and uninfected mononuclear phagocytes in the CNS may precipitate the development of auto-immune damage to myelin. Humoral factors, such as antibody and complement, could have a role in such a process. It is noteworthy that auto-immunity has been suggested as having a role in other tissue damage in AIDS, including thrombocytopenia (Morris et al, 1982; Abrams et al, 1986), and even damage to CD4 lymphocytes (Klatzmann and Gluckman, 1986). Finally, macrophages are known to be essential for providing a secretory product (interleukin-1) which increases production of nerve growth factor following axotomy in peripheral nerves (Heumann et al, 1987; Lindholm et al, 1987). This factor is essential for repair of damaged peripheral nerve. If mild degrees of damage to the CNS were found to require similar participation by macrophages in repair, and CNS macrophages in AIDS are deficient at secreting such factors, this might provide a basis for the type of damage that is found in the CNS in AIDS. There are preliminary indications from trials of zidovudine therapy in AIDS that neurological symptoms can be improved with this therapy (Yarchoan et al, 1987; Portegies et al, 1989). It is plausible that this treatment, by controlling HIV replication, could act favourably to reverse many of the possible processes outlined above. It is also feasible that additional therapeutic strategies could be devised if investigation can demonstrate which of these possible pathogenetic pathways are actually taken in human HIV infection.

system for HIV production, which allows small-scale infection of other CNS constituents. At present, although there is evidence that neuroglia can be infected by HIV in vitro, the evidence regarding their infection in vivo remains inconclusive. The discovery that macrophages in the CNS harbour HIV infection has re-awakened interest in the mononuclear phagocytes of the nervous system, and should act as a stimulus for the investigation of the still enigmatic functions of normal microglia.

Acknowledgement. This work was supported by the Medical Research Council, UK.

Summary

Macrophages, MGC and reactive microglia, all now widely recognized as belonging to the mononuclear phagocyte system, act as host cells for HIV in advanced stages of infection in the CNS. In about one-quarter to one-third of AIDS cases, they can be shown to carry a heavy viral load by the time of death. The infected cell population seems to increase sharply as CNS disease progresses. Whether infected cells are damaged, die, and are replaced by further monocyte recruitment is not known. Infection of macrophages probably depends on their possession of surface CD4 molecules, which act as receptors for the virus. At an early asymptomatic stage of infection, it is less certain that macrophages carry the virus in the CNS. Whatever the host cell type, at this stage T-cells may exert a suppressive effect on viral replication in the CNS, an effect which is lost in the terminal stages of AIDS. Other factors may also allow macrophage infection to flourish in the later stages of infection. These include generation of virus variants particularly well suited for surviving and proliferating in macrophages; assistance of HIV replication by other virus infections, particularly CMV; and increased availability of host cells as other pathogens attract more monocytes into the CNS. No defects of macrophage/microglial cell function have been demonstrated in the CNS in AIDS, but this is an aspect of the neuropathology that deserves attention, since it remains unclear how macrophage infection with HIV produces the damage observed. Possibly, secreted products of infected and uninfected macrophages act in a non-specific cytotoxic manner. An alternative explanation for this damage might be provided if macrophages act as a reservoir and amplification

References

Abrams DI, Kiprov DD, Goedert JJ et al. (1986) Antibodies to human T-lymphotropic virus type III and development of acquired immune deficiency syndrome in homosexual men presenting with immune thrombocytopenia. Ann Intern Med 104:47–50.

Adams CWM, Poston RN, Buk SJ. (1989) Pathology, histochemistry and immunocytochemistry of lesions in acute multiple sclerosis. J Neurol Sci 92:291–306.

Adrian EK Jr. (1968) Cell division in injured spinal cord. Am J Anat T23:501–20.

Adrian EK Jr, Walker BE. (1962) Incorporation of thymidine H³ by cells in normal and injured mouse spinal cord. J Neuropathol Exp Neurol 21:597–609.

Adrian EK Jr, Williams MG, George FC. (1978) Fine structure of reactive cells in injured nervous tissue labelled with ³H-thymidine injected before injury. J Comp Neurol 180:815–39.

Andersson MA, Bergstrom TB, Blomstrand C et al. (1988) Increasing intrathecal lymphocytosis and immunoglobulin G production in neurologically asymptomatic HIV-I infection. J Neuroimmunol 19:291–304.

Barõn M, Gallego A. (1972) The relation of the microglia with the pericytes in the rat cerebral cortex. Z Zellforsch 128:42–57.

Blakemore WF. (1975) The ultrastructure of normal and reactive microglia. Acta Neuropath (Berl) 6 (suppl):273–8.

Blinzinger K, Kreutzberg G. (1968) Displacement of synaptic terminal from regenerating motoneurons by microglial cells. Z Zellforsch 85:145–57.

Brierley JB, Brown AW. (1982) The origin of lipid phagocytes in the central nervous system 1. The intrinsic microglia. J Comp Neurol 211:397–406.

Budka H. (1986) Multinucleated giant cells in brain: a hallmark of the acquired immune deficiency syndrome (AIDS). Acta Neuropathol (Berl) 69:253–8.

Budka H, Costanzi G, Cristina S et al. (1987) Brain pathology induced by infection with the human immunodeficiency virus (HIV). A histological, immunocytochemical and electron microscopical study of 100 autopsy cases. Acta Neuropathol (Berl) 75:185–98.

Cheng-Mayer C, Rutka Y, Rosenblum M et al. (1987) Infection of cultured human brain cells with the human immunodeficiency virus (HIV). Proc Natl Acad Sci USA 84:3526–30.

Chiodi F, Fuerstenberg S, Gildlund M et al. (1987) Infection of brain-derived cells with the human immunodeficiency virus. J Virol 61:1244–7.

Clapham P, Weber J, Whitby D et al. (1989) Soluble CD4 blocks the infectivity of diverse strains on HIV and SIV for T cells and monocytes but not for brain and muscle cells. Nature 337:368–70.

Cravioto H, Feigin I. (1959) Non-infectious granulomatous angiitis with a predilection for the nervous system. Neurology 9:599–609.

Daar AS, Fuggle SV, Fabre JW et al. (1984) The detailed distribution of MHC Class II antigens in normal human organs. Transplantation 38:293–8.

Dalgliesh A, Beverley P, Clapham P et al. (1984) The CD4 (T4) antigen is an essential component of the receptor for the AIDS retrovirus. Nature 312:763–7.

Davey FR, Cordell JL, Erber WN et al. (1988) A monoclonal antibody (Y1/82A) with specificity towards peripheral blood monocytes and tissue macrophages. J Clin Pathol 41:753–8.

Dickson DW. (1986) Multinucleated giant cells in acquired immunodeficiency syndrome encephalopathy. Origin from endogenous microglia? Arch Pathol Lab Med 110:967–8.

Dolman CL. (1985) Microglia. In: Davis RL, Robertson DM, ed. Textbook of neuropathology. Williams and Wilkins: Baltimore, 117–37.

Doyle PW, Gibson G, Dolman CL. (1983) Herpes zoster ophthalamicus with contralateral hemiplegia: identification of cause. Ann Neurol 14:84–85.

Duchen LW. (1984) General pathology of neurons and neuroglia. In: Adams JH, Corsellis JAN, Duchen LW, ed. Greenfield's neuropathology, 4th edn. Arnold: London, 1–52.

Elovaara I, Iiavanainen M, Valle S-L et al. (1987) CSF protein and cellular profiles in various stages of HIV infection related to neurological manifestations. J Neurol Sci 78:331–42.

Epstein LG, Sharer LR, Cho E-S et al. (1985) HTLV-III/LAV-like retrovirus particles in the brains of patients with AIDS encephalopathy. AIDS Res Hum Retroviruses 1:447–54.

Esiri MM, Booss J. (1984) Comparison of methods to identify microglial cells and macrophages in the human central nervous system. J Clin Pathol 37:150–6.

Esiri MM, Gay D. (1990) Annotation: immunological and neuropathological significance of the Virchow–Robin space. J Neurol Sci 100:3–8.

Esiri MM, Reading MC. (1987) Macrophage populations associated with multiple sclerosis plaques. Neuropathol Appl Neurobiol 13:451–65.

Esiri MM, Oppenheimer DR, Brownell B et al. (1982) Distribution of measles antigen and immunoglobulin-containing cells in the CNS in subacute sclerosing panencephalitis and atypical measles encephalitis. J Neurol Sci 53:29–43.

Esiri MM, Scaravilli F, Millard PM et al. (1989) Neuropathology of HIV infection in haemophiliacs: comparative necropsy study. Br Med J 299:1312–15.

Esiri MM, Al-Izzi MS, Reading MC. (1991a) Macrophages, microglial cells and HLA-DR antigens in foetal and infant human brain. J Clin Pathol 101:59–72.

Esiri MM, Morris CS, Millard PR. (1991b) Fate of oligodendrocytes in HIV-1 infection. AIDS 5:1081–8.

Evans LA, McHugh TM, Stites DP et al. (1987) Differential ability of human immunodeficiency virus isolates to productively infect human cells. J Immunol 138:3415–18.

Fauci AS. (1988) The human immunodeficiency virus:

infectivity and mechanisms of pathogenesis. Science 239:617–22.

Franklin WA, Mason DY, Pulford K et al. (1986) Immuno-histological analysis of human mononuclear phagocytes and dendritic cells using monoclonal antibodies. Lab Invest 54:322–35.

Fujita S, Kitamura T. (1976) Origin of brain macrophages and the nature of microglia. In: Zimmerman HM, ed. Progress in neuropathology, vol 3. Grune and Stratton: London, 1–50.

Fujita S, Tsuchihashi Y, Kitamura T. (1981) Origin, morphology and function of the microglia. Prog Clin Biol Res 59A:141–50.

Gabuzda DH, Ho DD, de la Monte SM et al. (1986) Immunohistochemical identification of HTLV-III antigen in brains of patients with AIDS. Ann Neurol 20:289–95.

Gartner S, Markovitz P, Markovitz D et al. (1986) Virus isolation from and identification of HTLV-III/LAV-producing cells in brain tissue from a patient with AIDS. JAMA 256:2365–71.

Gendelman HE, Phelps W, Feigenbaum L et al. (1986) Transactivation of the human immunodeficiency virus long terminal repeat sequence by DNA viruses. Proc Natl Acad Sci USA 83:9759–63.

Gendelman HE, Leonard JM, Dutko FJ et al. (1988) Immunopathogenesis of human immunodeficiency virus infection in the central nervous system. Ann Neurol 23 (suppl):578–81.

Gendelman HE, Orenstein JM, Baca LM et al. (1989) The macrophage in the persistence and pathogenesis of HIV infection. AIDS 3:475–95.

Gordon S. (1986) Biology of the marcrophage. J Cell Sci Suppl 4:267–86.

Goudsmit J, Wolters EC, Bakker M et al. (1986) Intrathecal synthesis of antibodies to HTLV-III in patients without AIDS or AIDS-related complex. Br Med J 292:1231–4.

Grafe MR, Wiley CA. (1989) Spinal cord and peripheral nerve pathology in AIDS: the roles of cytomegalovirus and human immunodeficiency virus. Ann Neurol 25:561–6.

Gyorkey F, Melnick JL, Gyorkey P. (1987) Human immuno-deficiency virus in brain biopsies of patients with AIDS and progressive encephalopathy. J Infect Dis 155:870–6.

Haase AT. (1986) Pathogenesis of lentivirus infections. Nature 322:130–6.

Harouse JM, Kunsch C, Hartle HT et al. (1989) CD4-independent infection of human neural cells by human immunodeficiency virus type I. J Virol 63:2527–33.

Hauser SL, Bhan AK, Gilles FH et al. (1983) Immuno-histochemical staining of human brain with monoclonal antibodies that identify lymphocytes, monocytes and the Ia antigen. J Immunol 5:197–205.

Hayes GM, Woodroofe MN, Cuzner ML. (1987) Microglia are the major cell type expressing MHC Class II in human white matter. J Neurol Sci 80:25–37.

Heumann R, Lindholm D, Bandtlow C et al. (1987) Differential regulation of mRNA encoding nerve growth factor and its receptor in rat sciatic nerve during development, degeneration and regeneration: role of macrophages. Proc Natl Acad Sci USA 84:8735–9.

Ho DD, Pomerantz RJ, Kaplan JC. (1987) Pathogenesis of infection with human immunodeficiency virus. N Engl J Med 317:278–86.

Hortega P del Rio. (1919) 'Tracer elemento' de los centros nervosos. I: La microglia en estado normal. II: Intervencion de la microglia en los procesos patologicos. III: Naturaleza probable de la microglia. Biol Sci Exp Biol 9:69–120.

Hortega P del Rio. (1932) Microglia. In: Penfield W, ed.

Cytology and cellular pathology of the nervous system. Hoeber: New York, 482–534.

Huntington HW, Terry RD. (1966) The origin of reactive cells in cerebral stab wounds. J Neuropathol Exp Neurol 25:646–53.

Imamoto K, Leblond CP. (1978) Radioautographic investigation of gliogenesis in the corpus callosum of young rats. II Origin of microglial cells. J Comp Neurol 180:139–63.

Johnson RT. (1982) Viral infections of the nervous system. Raven Press: New York.

Johnston RB. (1988) Monocytes and macrophages. N Engl J Med 318:747–52.

Kennedy PGE, Narayan O, Ghotbi Z et al. (1985) Persistent expression of Ia antigen and viral genome in visna–maedi virus-induced inflammatory cells. J Exp Med 162:1970–28.

Kennedy PGE, Narayan O, Zink MC et al. (1989) The pathogenesis of visna, a lentivirus induced immuno-pathologic disease of the central nervous system. In: Gilden DH, Lipton HLK, ed. Clinical and molecular aspects of neurotropic virus infection. Kluwer Academic Publishers: Amsterdam, 393–421.

King JS. (1968) A light electron microscopic study of perineuronal glial cells and processes in the rabbit neocortex. Anat Rec 161:111–24.

Kitamura T. (1973) The origin of brain macrophages: some considerations on the microglia theory of del Rio Hortega, Acta Pathol Jpn 23:11–26.

Kitamura T. (1980) Dynamic aspects of glial reactions in altered brain. Path Res Pract 168:301–43.

Kitamura T, Fujita S. (1975) The role of haematogenous cells in alterations of mouse brains following stab wounding, inoculation of Japanese encephalitis virus and retrograde degeneration of facial nucleus. In: Kornyey ST, Tariska ST, Gosztomyc G, ed. VII Int Congr Neuropathol Exc Med 1:37–40.

Kitamura T, Hattori H, Fujita S. (1972) Autoradiographic studies on histogenesis of brain macrophages in the mouse. J Neuropathol Exp Neurol 31:502–18.

Klatzmann D, Gluckman JC. (1986) HIV infection: facts and hypotheses. Immunol Today 7:291–6.

Klatzmann D, Champagne E, Clamaret S et al. (1984) T-lymphocyte T4 molecule behaves as a receptor for human retrovirus LAV. Nature 312:767–8.

Koenig S, Gendelman HE, Orenstein JM et al. (1986) Detection of AIDS virus in macrophages in brain tissue from AIDS patients with encephalopathy. Science 233:1089–93.

Konigsmark BW, Sidman RL. (1963) Origin of brain macrophages in the mouse. J Neuropathol Exp Neurol 22:643–76.

Koyanagi Y, Miles S, Mitsuyasu RT et al. (1987) Dual infection of the central nervous system by AIDS viruses with distinct cellular tropisms. Science 236:819–22.

Kruger L, Maxwell DS. (1966) Electron microscopy of oligodendrocytes in normal rat cerebrum. Am J Anat 118:411–36.

Lampson LA, Hickey WF. (1986) Monoclonal antibody analysis of MHC expression in human biopsies: tissue ranging from 'histologically normal' to that showing different levels of glial tumour involvement. J Immunol 136:4054–62.

Lechtenberg R, Sher JH. (1988) AIDS in the nervous system. Churchill Livingstone: New York.

Lindholm D, Heumann R, Meyer M et al. (1987) Interleukin-1 regulates synthesis of nerve growth factor in non-neuronal cells of rat sciatic nerve. Nature 330:658–9.

Ling EA. (1980) Transformation of monocytes into amoeboid

microglia and into microglia in the corpus callosum of postnatal rats, as shown by labelling monocytes by carbon particles. J Anat 128:847–58.

Linneman CC, Alvira MM. (1980) Pathogenesis of varicella-zoster angiitis in the CNS. Arch Neurol 37:239–40.

Maier H, Budka H, Lassmann H et al. (1989) Vacuolar myelopathy with multinucleated giant cells in the acquired immune deficiency syndrome. Acta Neuropathol 78:497–503.

Mannoji H, Vegeer H, Becker LE. (1986) A specific histochemical marker (lectin Ricinus communis agglutinin-1) for normal human microglia, and application to routine histopathology. Acta Neuropathol (Berl) 71:341–3.

Matsuomoto Y, Watabe K, Ikuta F. (1985) Immunohisto-chemical study on neuroglia identified by the monoclonal antibody against a macrophage differentiation antigen (Macl). J Neuroimmunol 9:379–89.

Matthews MA, Kruger L. (1973) Electron microscopy of non-neuronal cellular changes accompanying neural degeneration in thalamic nuclei of the rabbit II reactive elements within the neuropil. J Comp Neurol 148:313–46.

Maxwell DS, Kruger L. (1965) Small blood vessels and the origin of phagocytes in the rat following heavy particle irradiation. Exp Neurol 12:33–54.

Meyenhofer MF, Epstein LG, Cho E-S et al. (1987) Ultra-structural morphology and intracellular production of human immunodeficiency virus (HIV) in brain. J Neuro-pathol Exp Neurol 46:474–84.

Mirra SS, del Rio C. (1989) The fine structure of acquired immunodeficiency syndrome encephalopathy. Arch Pathol Lab Med 113:858–65.

Mori S, Leblond CP. (1969) Identification of microglia in light and electron microscopy. J Comp Neurol 135:57–79.

Morris L, Distenfeld A, Amorosi E et al. (1982) Autoimmune thrombocytopenic purpura in homosexual men. Ann Intern Med 96:714–17.

Murabe Y, Sano Y. (1981) Thiamine pyrophosphatase activity in the plasma membrane of microglia. Histochemistry 71:45–52.

Murabe Y, Sano Y. (1982a) Morphological studies on neuroglia. VI Postnatal development of microglial cells. Cell Tissue Res 225:469–85.

Murabe Y, Sano Y. (1982b) Morphological studies on neuroglia. V Microglial cells in the cerebral cortex of the rat with special reference to their possible involvement in synaptic function. Cell Tissue Res 223:493–506.

Murray HW, Rubin BY, Masur H et al. (1984) Impaired production of lymphocytes and immune (gamma) interferon in the acquired immunodeficiency syndrome. N Engl J Med 310:883–9.

Murray HW, Scavuzzo D, Jacobs JL et al. (1987) In vitro and in vivo activation of human mononuclear phagocytes by interferon – γ: studies with normal and AIDS monocytes. J Immunol 138:2457–62.

Narayan O, Strandberg JD, Griffin DE et al. (1983) Aspects of the pathogenesis of visna in sheep. In: Mims CA, Cuzner MC, Kelly RE, ed. Viruses and demyelinating disease. Academic Press: London, 125–40.

Nurick S, Blackwood W, Mair WGP. (1972) Giant cell granu-lomatous angiitis of the central nervous system. Brain 95:133–42.

Oemichen M. (1978) Mononuclear phagocytes in the central nervous system. Springer-Verlag: Berlin.

Oemichen M. (1982) Functional properties of microglia. In: Smith WT, Cavanagh JB, ed. Recent advances in neuro-pathology, vol 2. Churchill Livingstone: London, 83–107.

Oemichen M, Huber H. (1976) Reactive microglia with membrane features of mononuclear phagocytes. J Neuropathol Exp Neurol 35:30–9.

Oemichen M, Wietholter H, Greaves MF. (1979) Immunological analysis of human microglia: lack of monocytic and lymphoid membrane differentiation antigens. J Neuropathol Exp Neurol 38:99–103.

Oemichen M, Wietholter H, Gencic M. (1980) Cytochemical markers for mononuclear phagocytes as demonstrated in reactive microglia and globoid cells. Acta Histochem 66: 243–52.

Perry VH, Gordon S. (1989) Resident macrophages of the central nervous system: modulation of phenotype in relation to a specialised microenvironment. In: Goetzl EJ, Spector NH, ed. Neuroimmune networks: physiology and diseases. Alan Liss: New York, 119–25.

Perry VH, Hume DA, Gordon S. (1985) Immunohistochemical localisation of macrophages and microglia in the adult and developing mouse brain. Neuroscience 15: 313–26.

Persson L. (1976) Cellular reactions to small cerebral wounds in the rat frontal lobe. An ultrastructural study. Virch Arch [B] 22:21–37.

Popovic M, Sarngadharan MG, Read E. (1984) Detection, isolation and continuous production of cytopathic retroviruses (HTLV-III) from patients with AIDS and pre-AIDS. Science 224:497–500.

Popovic M, Mellert W, Eyfle V et al. (1988) Role of mononuclear phagocytes and accessory cells in human immunodeficiency virus type I infection of the brain. Ann Neurol 23 (suppl):574–7.

Portegies P, de Gans T, Lange JMA et al. (1989) Declining incidence of AIDS dementia complex after introduction of zidovudine treatment. Br Med J 299:819–21.

Pow DV, Perry VH, Morris JF et al. (1989) Microglia in the neurohypophysis associate with and endocytose terminal portions of neurosecretory neurons. Neuroscience 33:567–78.

Pumarola-Sune T, Navia BA, Cordon-Cardo C et al. (1987) HIV antigen in the brains of patients with the AIDS dementia complex. Ann Neurol 21:490–6.

Resnick L, di Marzo-Veronese F, Schupbach J et al. (1985) Intra-blood–brain-barrier synthesis of HTLV-III-specific IgG in patients with neurologic symptoms associated with AIDS or AIDS-related complex. N Engl J Med 313:1498–1504.

Resnick L, Berger JR, Shapshak P et al. (1988) Early penetration of the blood–brain-barrier by HIV. Neurology 38:9–14.

Rhodes RH, Ward JM. (1989) Immunohistochemistry of human immunodeficiency virus in the central nervous system and an hypothesis concerning the pathogenesis of AIDS meningo-encephalomyelitis. In: Rotterdam H, Sommers SC, ed. Progress in AIDS pathology. vol 1. Field and Wood: Philadelphia, 167–79.

Rosenblum M, Scheck AC, Cronin K et al. (1989) Dissociation of AIDS-related vacuolar myelopathy and productive HIV-1 infection of the spinal cord. Neurology 39:892–6.

Rübsamen-Waigmann H, Becker WB, Helm EB et al. (1986) Isolation of variants of lymphocytopathic retroviruses from the peripheral blood and cerebrospinal fluid of patients with ARC or AIDS. J Med Virol 19:335–44.

Sandlin R, Alexander WS, Hornabrook RW et al. (1979) Granulomatous angiitis of the CNS. Arch Neurol 36:433–5.

Schelper RL, Adrian EK Jr. (1980) Non-specific esterase activity in reactive cells in injured nervous tissue labelled with ^{3}H-thymidine or ^{125}Iododeoxyuridine injected before injury. J Comp Neurol 194:829–44.

Schelper RL, Adrian EK Jr. (1986) Monocytes become macrophages; they do not become microglia; a light and electron microscopic autoradiographic study using 125-Iododeoxyuridine. J Neuropathol Exp Neurol 45:1–19.

Schlesinger L, Musson RA, Johnston RB Jr. (1984) Functional and biochemical studies of multinucleated giant cells derived from the culture of human monocytes. J Exp Med 159:1289–94.

Schrier RD, Nelson JA, Oldstone MD. (1985) Detection of human cytomegalovirus in peripheral blood lymphocytes in a natural infection. Science 230:1048–51.

Sharer LR, Cho E-S, Epstein LG. (1985) Multinucleated giant cells and HIV-III in AIDS encephalopathy. Hum Pathol 16:760.

Smith I, Howells DW, Kendall B et al. (1987) Folate deficiency and demyelination in AIDS. Lancet ii:215.

Sobel RA, Ames MB. (1988) Major histocompatibility complex molecular expression in the human central nervous system; immunohistochemical analysis of 40 patients. J Neuropathol Exp Neurol 41:19–28.

Stoler MH, Eskin TA, Benn S et al. (1986) Human T-cell lymphotropic virus type III infection of the central nervous system. A preliminary in situ analysis. JAMA 256:2360–4.

Stowring L, Haase AT, Petursson G et al. (1985) Detection of visna virus antigens and RNA in glial cells in foci of demyelination. Virology 141:311–18.

Sumner BEH. (1975) A quantitative analysis of boutons of different types of synapses in normal and injured hypoglossal nuclei. Exp Neurol 49:406–17.

Sumner BEH. (1977) Responses in the hypoglossal nucleus to delayed regeneration of the transected hypoglossal nerve: a quantitative ultrastructural study. Exp Brain Res 29:219–31.

Thomas ED, Ramberg RE, Sale GE et al. (1976) Direct evidence for a bone marrow origin of the alveolar macrophage in man. Science 192:1016–18.

Torvik A, Soreide AJ. (1975) The perineuronal glial reaction after axotomy. Brain Res 95:519–29.

Tsuchihashi Y, Kitamura T, Fujita S. (1981) Immunofluorescence studies of monocytes in the injured rat brain. Acta Neuropathol (Berl) 53:213–19.

Twigg HL, Weissler JC, Yoffe B, Ball EJ, Lipscomb MA. (1991) Monocyte accessory cell formation in patients infected with the human immunodeficiency virus. Clin Immunol Immunopathol 59:436–48.

Vaughn JE, Peters A. (1968) A third neuroglial cell type. An electron microscopic study. J Comp Neurol 133:269–88.

Vazeux R, Brousse N, Jarry A et al. (1987) AIDS subacute encephalitis; identification of HIV-infected cells. Am J Pathol 126:403–10.

Vinters HV, Kwok MK, Ho HW et al. (1989) Cytomegalovirus in the nervous system of patients with the acquired immunodeficiency syndrome. Brain 112:245–68.

Vorbrodt AW, Wisniewski HM. (1982) Plasmalemma-bound nucleoside diphosphatase as a cytochemical marker of central nervous system (CNS) mesodermal cells. J Histochem Cytochem 30:418–24.

Wagner HJ, Pilgrim C, Brandl J. (1974) Penetration and removal of horseradish peroxidase injected into cerebrospinal fluid. Role of cerebral perivascular spaces, endothelium and microglia. Acta Neuropathol (Berl) 27:299–315.

Ward JM, O'Leary TJ, Baskin GB et al. (1987) Immunohistochemical localisation of human and simian immunodeficiency viral antigens in fixed tissue sections. Am J Pathol 127:199–205.

Watabe K, Saida T, Kim SU. (1989) Human and simian glial cells infected by human T-lymphotropic virus type I in culture. J Neuropathol Exp Neurol 48:610–19.

Webster A, Lee CA, Cook DG et al. (1989) Cytomegalovirus

infection and progression towards AIDS in haemophiliacs with human immunodeficiency virus infection. Lancet ii: 63–6.

Weinberg JB, Hobbs MM, Misukomis MA. (1984) Recombinant human γ-interferon induces human monocyte polykaryon formation. Proc Natl Acad Sci (USA) 81:4554–7.

Weiser B, Peress N, La Neve et al. (1990) Human immunodeficiency virus type 1 expression in the central nervous system correlates directly with extent of disease. Proc Natl Acad Sci USA 87:3997–4001.

Wiley CA, Nelson JA. (1990) Role of human immunodeficiency virus and cytomegalovirus in AIDS encephalitis. Am J Pathol 133:73–81.

Wiley CA, Schrier RD, Nelson JA et al. (1986a) Cellular localisation of human immunodeficiency virus infection within the brains of acquired immune deficiency syndrome patients. Proc Natl Acad Sci USA 83:7089–93.

Wiley CA, Schrier RD, Denaro FJ et al. (1986b) Localisation within the CNS of cytomegalovirus proteins and genome during fulminant infection in an AIDS patient. J Neuropathol Exp Neurol 45:127–39.

Wood GW, Gollahon KA, Tilzer SA et al. (1979) The failure of microglia in normal brain to exhibit mononuclear phagocyte markers. J Neuropathol Exp Neurol 38:369–76.

Woodroofe MN, Bellamy AS, Feldman M et al. (1986) Immunocytochemical characterisation of the immune reaction in the nervous system in multiple sclerosis: possible role for microglia in lesion growth. J Neurol Sci 74:135–52.

Yarchoan R, Berg G, Bronwers P et al. (1987) Response of human immunodeficiency-virus-associated neurological disease to 3¹ azido 3¹ μ deoxythymidine. Lancet i:132–5.

12 Immunohistochemical and In Situ Hybridization Study of HIV Encephalitis

C.A. Wiley

Introduction

The discovery of HIV as the etiologic agent of AIDS has already been described in this book. Suffice it to say that by the middle of 1984 it was accepted that a third human retrovirus had been isolated, which was associated with AIDS. While some still doubt that HIV causes AIDS (Duesberg, 1988), it would appear that much of the immunosuppression of AIDS can be attributed directly to HIV infection (Blattner et al, 1988).

In January of 1985, Shaw et al demonstrated by Southern blot analysis that HIV was within the central nervous system (CNS) of 5 out of 15 of the patients they studied (Shaw et al, 1985). The importance of this observation was suggested by the finding that in some of the demented patients, HIV nucleic acids were detected at higher levels in the CNS than in lymphoid organs like the spleen and lymph nodes. This led these investigators to propose that the brain might be the reservoir of HIV, protecting it from immune surveillance within the immunologically-privileged site of the CNS. In retrospect, these were rather prescient assertions, given the size of the sample studied. Following this observation, the virus was labelled 'neurotropic' and an intensive hunt for HIV in the CNS was begun.

Studies of CSF in HIV-seropositive patients demonstrated numerous abnormal findings. A mild pleocytosis and elevated level of protein were almost universally observed. Oligoclonal bands of immunoglobulin were commonly observed, and some of these were found exclusively in the CNS (consistent with intrathecal production) and were specific to HIV. Ho et al (1985) and Levy et al (1985) showed that HIV could frequently be recovered from CSF. In fact, later studies showed that HIV could be recovered at about the time of seroconversion when a subacute meningitis was present (Carne et al, 1985; Cooper et al, 1985; Goudsmidt et al, 1986). This has been interpreted by some to mean that HIV enters the CNS early in the course of infection (Price et al, 1988). Resnick et al (1985) and Goudsmidt et al (1986) demonstrated intrathecal production of antibody to HIV, further supporting the chronicity of CSF infection.

However, since HIV is also readily recovered from cerebrospinal fluid (CSF) samples of asymptomatic HIV-1-seropositive patients, the relationship between HIV within the CSF and its involvement in the CNS parenchyma is to date unclear. If one extrapolates from analogies with other viral infections of the CNS, for example Herpes simplex virus (HSV), the presence of virus within the CSF is rarely associated with extension into the CNS parenchyma. In fact, in the case of HSV, encephalitis occurs by a separate mechanism (axonal extension through nasal or pelvic nerves) from that leading to CSF infection (Stroop and Schaefer, 1987). Nevertheless, frequent recovery of HIV from the CSF fuelled the search for HIV in the CNS.

Visualization of HIV in the Nervous System

Electron Microscopy

Ultrastructural studies of HIV encephalitis have been extremely difficult. The first study by Epstein et al (1984) identified HIV particles within multinucleated giant cells. However, particles were so rare, and human tissue so inadequately fixed, that it was difficult to identify the lineage of infected cells. Several groups have now claimed identification of HIV particles budding from astrocytes and oligodendroglia, but as noted below, the rarity of these observations has clouded their significance (Gyorkey et al, 1987; Mirra and del Rio, 1989).

Immunocytochemistry

Immunocytochemistry has several advantages over electron microscopy in the study of viral infections of the CNS. Since immunocytochemistry is usually performed at the light microscopy level, tissue sampling is not as much of a problem as it is in electron microscopy. Even viral infections known for their high level of virion production (for example, JC virus in progressive multifocal leukoencephalopathy, or HSV in acute encephalitis of animals or man), sometimes require long painstaking searches to find morphologic virion at the ultrastructural level. In the case of retroviral infections, electron microscopy is disadvantaged by the relative rarity of virions and by the virion pleomorphism which makes study of less than ideally fixed tissues of human studies particularly suspect. By permitting detection of viral antigens as opposed to virions, immunocytochemistry has the additional advantage of permitting detection of viral infections where intact virions are not produced. If there is some arrest in the viral production cycle such that only certain viral proteins are made, these may be detected by immunocytochemistry. As most viral infections have regulatory proteins that govern 'latent' and active viral production, it is not surprising that immunocytochemistry is a more sensitive technique than electron microscopy.

It was not until 1986 that HIV antigens could be demonstrated in situ (Gabuzda et al, 1986; Koenig et al, 1986; Stoler et al, 1986; Wiley et al, 1986; Pomarola-Sune et al, 1987; Vazeux et al, 1987; Ward et al, 1987). There are many explanations for why this hunt took so long. First was the absence of good reagents. Readily-available human polyclonal anti-sera were notoriously plagued with non-specific binding. Rabbit polyclonal sera were similarly plagued, and murine monoclonal antibodies required time to produce and screen. Second was the absence of well-preserved human tissue. After AIDS was proven to be caused by an infectious agent for which inactivation data were scarce, autopsies were avoided. The diffuse and deep nature of the CNS lesions made biopsy material almost non-existent. Third, and most importantly, when the cellular localization was performed, it did not make any 'sense'.

Drawing from analogies with other acute and chronic viral encephalitides (for example, HSV and measles), significant infection of neurons and glia was expected. However, HIV antigens were detected uncommonly. Not all AIDS cases had HIV in the brain, and those that did had only occasional infected cells. How could such severe clinical changes be explained by so little virus? When virus was present, it appeared in the areas showing the severe neuropathological changes, consistent with subcortical clinical findings. However, it was not in neuroglial cells; rather, HIV was predominantly detected in some macrophages and occasional endothelial cells (Figure 12.1). How could CNS symptoms be explained by HIV infection of macrophages and endothelial cells? What were the macrophages doing in the CNS? The most likely explanation, that macrophages were consuming lysed nervous system cells, did not fit since only rare glial and neuronal elements were infected.

In Situ Hybridization

Fortunately, the immunocytochemistry results could be verified with the then-new technique of in situ hybridization (Koenig et al, 1986; Stoler et al, 1986; Wiley et al, 1986). In theory the sensitivity of in situ hybridization should be to immunocytochemistry what immunocytochemistry is to electron microscopy. In situ hybridization allows detection of viral nucleic acids (both genomic and message) using either radioactive or enzymatic techniques. Because viral replication could be 'blocked' after viral

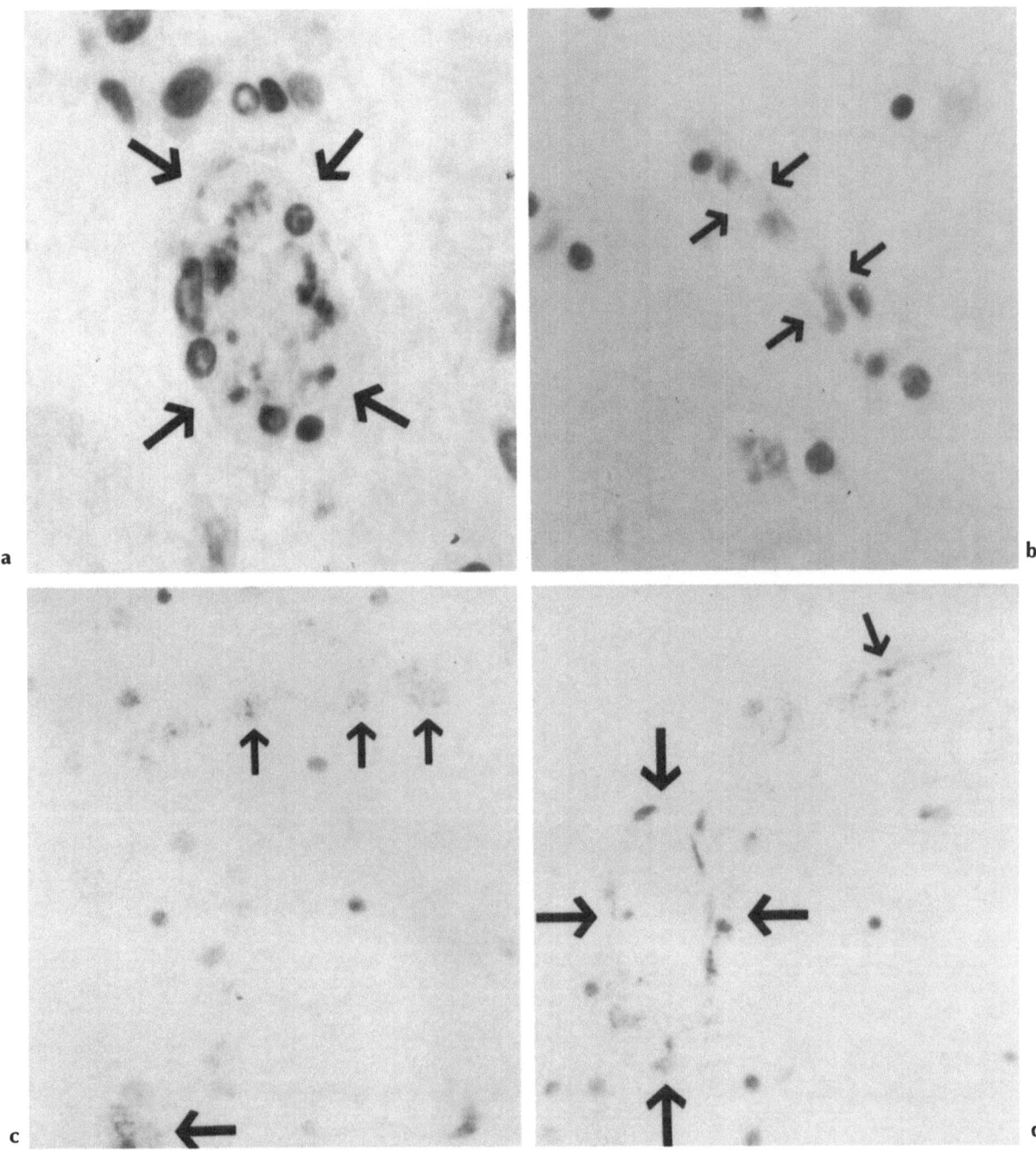

Figure 12.1a–d. Paraffin-embedded tissues from an AIDS patient with HIV encephalitis. **a** Immunocytochemical stain for HIV p24 antigen. A large multinucleated giant cell (surrounded by *arrows*) contains discrete pockets of staining (*brown*). Surrounding brain parenchyma is not immunostained. Counterstained with hematoxylin, ×500. **b** Immunocyto-chemical stain for HIV p24 antigen. A small capillary within the center of the field (surrounded by *arrows*) shows light staining along the endothelial cell margins (*light brown*). Counter-stained with hematoxylin, ×300. **c** Autoradiogram of in situ hybridization for HIV nucleic acids using Dupont ^{35}S labeled DNA probe. Numerous mononuclear cells (*arrows*) within regions of affected basal ganglia are covered with radiographic grains (*blue dots*) indicating positive hybridization for HIV nucleic acids. Counterstained with hematoxylin, ×200. **d** Autoradiogram of in situ hybridization for HIV nucleic acids using Dupont ^{35}S labeled DNA probe. Vessels (*arrows*) within regions of affected basal ganglia have endothelial cells covered with radiographic grains (*blue dots*) indicating positive hybridization for HIV nucleic acids. Counterstained with hematoxylin, ×200.

transcription but before translation of viral proteins, in theory, in situ hybridization should detect cells at an early state of infection or detect cells that have undergone an 'abortive' infection (that is, without virion production). In practice, there is significant question about whether such a sensitivity is possible using human tissues.

Identical cellular localization of HIV by in situ hybridization meant that the immunocytochemistry results were not artifacts of cross-reactivity (Figure 12.1). But this confirmation was not accompanied by the demonstration of a greater number of infected cells. Depending upon the fixation and autolysis, it was hoped that in situ hybridization would be more sensitive than immunocytochemistry, but in situ hybridization demonstrated the same small number of infected cells. Several investigators have attempted to quantify the intracellular level of infection; however, it is difficult to be confident of estimates of viral copy numbers when the tissue preservation parameters are so capricious. It has become clear that only moderately high levels of provirus or viral messages are detected by this technique, so the extent of true 'latently' infected cells remains unclear.

Polymerase Chain Reaction

Perhaps the most sensitive technique for detection of any nucleic acid is polymerase chain reaction (PCR). This technique employs a repeated enzymatic amplification of nucleic acids, such that single copies of nucleic acid molecules can be detected. This theoretical prediction of sensitivity has been borne out in several systems. Unfortunately, with this incredible sensitivity comes high specificity and potential contamination problems. While the technique works well in solution, there is little on the horizon that will permit it to work in situ. Therefore, identification of positive samples will not identify which cellular components are infected. Given that all human tissues are 'contaminated' with blood, a single infected white blood cell contained within a sample could generate a positive signal.

Despite these drawbacks, recent PCR studies by Pang et al (1990) have begun to elucidate some of the characteristics of HIV in brain tissue. Using a newly-designed quantitative PCR technique, these workers have shown that HIV in CNS tissue exists not as integrated provirus, but rather as unintegrated linear DNA. These data would suggest that cyclical re-infection is an important mechanism in CNS infection (see below). Perhaps

in situ hybridization detection is only possible in those cells that have undergone this amplification by cyclical re-infection.

Potential Mechanisms of HIV-Associated Neurological Diseases

Indirect Causes of CNS Damage

Perhaps the leading theory regarding the immunopathogenesis of HIV-mediated CNS damage centers around the idea that infected and non-infected macrophages (the principal immune cell present in cases of HIV encephalitis) release factors that mediate CNS damage. Given the litany of agents that potentially can be released by macrophages, the importance of one or a combination of these agents will be difficult to dissect. Certainly tumor necrosis factor (TNF, also known as cachectin) is an important agent to consider, given its prominent role in the cachexia associated with AIDS. Indeed, some studies have shown a correlation between *serum* TNF levels and CNS damage. However, those same studies have shown no association between *CSF* TNF levels and CNS damage. It is difficult to explain this discrepancy as being due to a degradation phenomenon; however, TNF is known to have a short biologic half-life and most of its effects are presumed to be mediated at the site of secretion. Support for the importance of macrophage-secreted factors will await further in situ localization of the myriad of macrophage-derived cytokines, enzymes and their messages.

Having received less attention because it is less commonly infected by HIV, the endothelial cell sits at a critical juncture between the brain and the systemic circulation. Disruption of this cell's function would be expected to have non-specific but deleterious effects on the CNS microenvironment. Small perturbations could be expected to lead to diffuse injury with widespread effects. Even if endothelial cell cytolysis did not occur, compromise of the cell's 'luxury functions' in maintaining the blood–brain barrier would be expected to result in amplified brain damage. In fact, the extensive gliosis and vacuolation described in the AIDS brain is very similar to that mediated by radiation damage to small blood vessels (Haymaker, 1968). Further evaluation of the importance of this route of CNS injury awaits elucidation of the degree of endothelial cell infection, and whether a non-lytic, perhaps non-productive, infection of this cell can lead to compromise of its function.

Direct Viral Causes of CNS Damage

Toxic Viral Proteins

Peptide T. Pert and associates (1986) identified a 60 kD brain protein that bound monoclonal antibodies to CD4 and also bound HIV gp120. They further identified an octapeptide sequence within the gp120 protein (termed 'peptide T') that inhibited gp120 binding and T-cell infection. These investigators proposed that interaction of gp120 with the peptide T receptor might account for memory disturbances in AIDS patients. The significance of this observation awaits reproduction by other laboratories. Nevertheless, the distribution of the receptor (predominantly within the hippocampus and hypothalamus), and its conservation in many different species that are not affected by HIV proteins, make it difficult to reconcile receptor competition with known pathologic properties of HIV.

Neuroleukin. In 1986 Gurney and co-workers (Gurney et al, 1986a,b; Lee et al, 1987) described a trophic factor that supported B-cell immunoglobulin synthesis and promoted in vitro survival of central and peripheral neurons. These investigators molecularly cloned the factor and named it neuroleukin. The deduced amino acid sequence of neuroleukin showed significant homology with a highly-conserved region of the HIV envelope protein gp120. In vitro studies demonstrated that HIV protein extracts and recombinant gp120 competitively inhibited the neurotrophic properties of neuroleukin. This inhibition could be eliminated by immunoprecipitation of HIV gp120. If a similar trophic function were postulated for mature neurons, then loss of neuronal function in AIDS patients could be explained by competitive action of gp120 for neuroleukin binding sites. This theory was complicated by two subsequent observations. When the neuroleukin gene was compared to other CNS genes, it was found to be equivalent to a neuronal enolase whose function would not have been considered to be trophic (Chaput et al, 1988; Faik et al, 1988). Second, the homology between neuroleukin and HIV-1 gp120 did not extend to HIV-2. Since HIV-2 has also been shown to cause neurologic disease, a separate mechanism would have to be proposed.

Toxic Viral Replication

Children demonstrate a spectrum of neurologic disease associated with HIV infection of the CNS (Belman et al, 1985). Some display a failure to acquire additional developmental skills, while others have a more progressive loss of previously-acquired skills. Those children who develop a progressive neurologic course have readily-detectable HIV antigen within the CNS, while those with a plateau course show little or no detectable HIV (Wiley et al, 1990). Recent studies, documenting reversal of ADC with zidovudine treatment (Yarchoan et al, 1987; Fiala et al, 1988; Yarchoan et al, 1988), would be consistent with these findings, suggesting that continued viral expression is important for progression of HIV-mediated CNS damage.

Silent Infection of Glia

Present levels of detection have suggested that glia and neurons are only rarely infected in AIDS brains in vivo. Presumably, infection of these cells is mediated through some mechanism other than CD4 since, despite one study finding message for CD4 in the CNS (FunRe et al, 1987), no studies have convincingly demonstrated CD4 antigen (Maddon et al, 1986). In vitro, virus can be 'rescued' from glial cell line cultures by co-cultivation with peripheral blood monocytes, but whether this means that viral genes are active in the glial cultures is unclear. If our inability to identify HIV in glial and neuronal cells is an artifact due to insensitivity of present detection techniques, then one could entertain a number of possible ways HIV could cause direct damage to neuronal and glial elements. With low levels of expression of viral proteins, such as transactivating and repressing proteins, significant dysfunction of neuronal cells might be expected. Alternatively, even without viral protein production, simple synthesis of large amounts of unintegrated viral DNA might incapacitate many cells (Pang et al, 1990). Low levels of glial HIV infection (below that detectable by any in situ technique) are consistent with the low levels of peripheral blood monocyte infection. We have recently used a plaque assay to dramatize the insensitivity of present technology in detecting these 'latently' infected cells (Schrier et al, unpublished observations). In these studies, immunocytochemical staining of early macrophage cultures (in vitro for less than 1 week) identified little retroviral antigen. However, after the macrophages had matured for several weeks in culture ('activated'), numerous viral antigen positive plaques could be identified. Such observations leave open the possibility that the cells we are detecting by immunocytochemistry, in situ hybridization and electron microscopy are simply the aberrant macrophage acting as a viral factory, possibly exacerbating the viral production, or alternatively simply acting as indicator cells for a more common CNS cell infection.

The prominent white matter abnormalities observed in AIDS brains has suggested that oligodendroglia may be a prime target for low-level viral infection. Some investigators have even claimed to identify significant numbers of productively HIV-infected oligodendroglia (Gyorkey et al, 1987); however, this has been challenged by others (Mirra and del Rio, 1989). Consistent with a 'silent' oligodendroglial infection, we have recently shown increased expression of MHC I on glial cells in regions of patchy white matter changes. This could represent direct viral infection, or alternatively it could represent exposure to γ-interferon produced locally, or systemically leaking through a blood–brain barrier. The pronounced gliosis observed, even in early cases of AIDS, has suggested that astrocytes could be a direct or indirect target of infection. Several groups have identified a low level of infection of these cells (Mirra and del Rio, 1989). However, gliosis is a non-specific response to a variety of pathological situations, which suggests a normal reactive process rather than a pathologic effect of direct viral infection.

Silent Infection of Neurons

In any dementing type of illness, the natural inclination would be to expect direct neuronal infection to account for the symptomatology. In fact, this bias was so strong that early localizations of the virus principally to macrophage elements in the CNS were met with scepticism. There is little evidence for significant neuronal infection; however, if this proves to be a detection artifact, a litany of theories could be raised to explain the clinical symptoms on the basis of low-level neuronal infection. Most attention has been focused on deep white matter and grey matter pathology. However, CT and MRI studies have also identified cerebral cortical atrophy with both sulcal widening and ventricular dilatation. While ventricular enlargement could be explained by deep gray matter changes, sulcal widening is more consistent with atrophy of the cortical gray

matter. Studies characterizing cortical changes associated with HIV infection have been initiated (see Chapters 6 and 7). Recently we initiated studies with Dr Robert Terry to evaluate quantitatively cortical thickness and cell counts in three different regions of the neocortex. Perpendicular measurements of cortical thickness demonstrated differences between HIV-seropositive and age-matched control brains. These differences were not significant for some cortical regions; however, the inferior parietal cortex of brains with HIV encephalitis was significantly thinner than that of age-matched controls. Substantial shifts were also noted in cortical cell size profiles in different regions of the cortex.

Conclusions

HIV infection is tightly associated with a clinical condition called AIDS dementia complex (ADC). In cases of ADC, pathologic examination of the brain demonstrates a subacute encephalitis. Immunocytochemistry and in situ hybridization have localized HIV to the pathologically-perturbed regions. However, the quantity of virus detected is incommensurate with the degree of clinical and pathologic findings. This suggests either that our present means of viral detection are too insensitive or that HIV-associated encephalopathy is an indirect effect of infection.

Unanswered Questions

There are numerous unanswered questions surrounding the pathogenesis of HIV-associated encephalopathy. Why do only a subset of HIV-seropositive patients get neurologic disease? Are these patients infected by a unique neurotropic strain of HIV? Or is there something peculiar about the host immune response that either does not control the virus or predisposes the brain to damage?

What is the mechanism of brain injury? Direct mechanisms like viral lysis are readily eliminated, but what about less obvious direct and indirect mechanisms of injury? Are viral proteins toxic? Does gp120 bind to neuroleukin receptors and deprive CNS cells of trophic factors? Or does gp120 interfere with peptide neurotransmitter action? Is damage mediated by macrophage factors? Why are macrophages in the brain? Are HIV-positive CNS macrophages the result of local infection (thus acting as indicator cells of abortively infected neuronal and glial elements), or were bone marrow monocytes latently infected and only after differentiation in tissues do they express viral proteins? Is damage mediated by infection of CNS endothelial cells? Do productively- or non-productively-infected endothelial cells maintain a blood–brain barrier? Is damage to the blood–brain barrier the initiating stimulus for macrophage ingress? Are CNS cells infected at a presently undetectable level? Can retroviruses replicate in non-dividing cells? Do HIV transcripts or regulatory proteins interfere with normal cellular function?

Answers to these questions will substantially advance our understanding of retroviral-mediated CNS damage. The utility of this knowledge will not be limited to AIDS: of the five known human retroviruses, three are associated with nervous system damage. The knowledge we gain from studies of HIV-mediated CNS infection will undoubtedly aid our understanding of these other known infections, in addition to helping to look for retroviral causes of other nervous system diseases (Wiley 1987).

Acknowledgements. This chapter was supported by a grant from the University of California AIDS Task Force, NIH grants NS-25178, MH45294, MH43298, NS27810 and NS-00928 to CAW.

References

Belman AL, Ultmann MH, Horoupian D et al. (1985) Neurological complications in infants and children with acquired immune deficiency syndrome. Ann Neurol 18:560–6.

Blattner W, Gallo RC, Temin HM. (1988) HIV causes AIDS. Science 241:515–16.

Carne CA, Tedder RS, Smith A et al. (1985) Acute encephalopathy coincident with seroconversion for anti-HLTV-III. Lancet ii:1206–8.

Chaput M, Claes V, Portetelle D et al. (1988) The neurotrophic factor neuroleukin is 90% homologous with phosphohexose isomerase. Nature 332:454–5.

Cooper DA, Gold J, Maclean P et al. (1985) Acute AIDS retrovirus infection. Definition of a clinical illness associated with seroconversion. Lancet i:537–40.

Duesberg P. (1988) HIV is not the cause of AIDS. Science 241:514, 517.

Epstein LG, Sharer LR, Cho ES et al. (1984) HTLV-III/LAV-like retrovirus particles in the brains of patients with AIDS encephalopathy. AIDS Res 1:447–54.

Faik P, Walker JI, Redmill AA et al. (1988) Mouse glucose-6-phosphate isomerase and neuroleukin have identical 3' sequences. Nature 332:455–7.

Fiala M, Cone LA, Cohen N et al. (1988) Responses of neurologic complications of AIDS to 3'-azido-3'-deoxythymidine and 9-(1,3-dihydroxy-2-propoxymethyl) guanine I. Clinical features. Rev Infect Dis 10:250–86.

FunRe I, Hahn A, Rieber EP et al. (1987) The cellular receptor (CD4) of the human immunodeficiency virus is expressed on neurons and glial cells in human brain. J Exp Med 165:1230–5.

Gabuzda DH, Ho DD, de la Monte SM et al. (1986) Immunohistochemical identification of HTLV-III antigen in brains of patients with AIDS. Ann Neurol 20:289–95.

Goudsmidt J, Wolters EC, Bakker M et al. (1986) Intrathecal synthesis of antibodies of HTLV-III in patients without AIDS and AIDS related complex. Br Med J [Clin Res] 292:1231–4.

Gurney ME, Apatoff BR, Spear GT et al. (1986a) Neuroleukin: a lymphokine product of lectin-stimulated T cells. Science 234:574–81.

Gurney ME, Heinrich SP, Lee MR et al. (1986b) Molecular cloning and expression of neuroleukin, a neurotrophic factor for spinal and sensory neurons. Science 234:566–74.

Gyorkey F, Melnick JL, Gyorkey P. (1987) Human immunodeficiency virus in brain biopsies of patients with AIDS and progressive encephalopathy. J Infect Dis 155:870–6.

Haymaker W, Ibrahim MZ, Miquel J et al. (1968) Delayed radiation effects in the brains of monkeys exposed to x- and gamma-rays. J Neuropathol Exp Neurol 27:50–79.

Ho DD, Rota TR, Schooley RT et al. (1985) Isolation of HTLV-III from cerebrospinal fluid and neural tissues of patients with neurologic syndromes related to the acquired immunodeficiency syndrome. N Engl J Med 313:1493–7.

Koenig S, Gendelman HE, Orenstein JM et al. (1986) Detection of AIDS virus in macrophages in brain tissue from AIDS patients with encephalopathy. Science 233:1089–93.

Lee MR, Ho DD, Gurney ME. (1987) Functional interaction and partial homology between human immunodeficiency virus and neuroleukin. Science 237:1047–51.

Levy JA, Shimabukuro J, Hollander H et al. (1985) Isolation of AIDS-associated retroviruses from cerebrospinal fluid and brain of patients with neurological symptoms. Lancet 2:586–8.

Maddon PJ, Dalgleish AG, McDougal JS et al. (1986) The T4 gene encodes the AIDS virus receptor and is expressed in the immune system and the brain. Cell 47:333–48.

Mirra SS, del Rio C. (1989) The fine structure of acquired immunodeficiency syndrome encephalopathy. Arch Pathol Lab Med 113:858–65.

Pang S, Koyanagi Y, Miles S et al. (1990) The structure of HIV-1 DNA in brain and blood of AIDS patients. Nature 343:85–9.

Pert CB, Hill JM, Ruff MR et al. (1986) Octapeptides deduced from the neuropeptide receptor-like pattern of antigen T4 in brain potently inhibit human immunodeficiency virus receptor binding and T-cell infectivity. Proc Natl Acad Sci USA 83:9254–8.

Price RW, Brew B, Sidtis J et al. (1988) The brain in AIDS: central nervous system HIV-1 infection and AIDS dementia complex. Science 239:586–92.

Pumarola-Sune T, Navia BA, Cordon-Cardo C et al. (1987) HIV antigen in the brains of patients with the AIDS dementia complex. Ann Neurol 21:490–6.

Resnick L, diMarzo-Veronese F, Schupbach J et al. (1985) Intra-blood–brain-barrier synthesis of HTLV-III-specific IgG in patients with neurologic symptoms associated with AIDS or AIDS-related complex. N Engl J Med 313:1498–1504.

Shaw GM, Harper ME, Hahn BH et al. (1985) HTLV-III infection in brains of children and adults with AIDS encephalopathy. Science 227:177–82.

Stoler MH, Eskin TA, Benn S et al. (1986) Human T-cell

lymphotropic virus type III infection of the central nervous system. A preliminary in situ analysis. JAMA 256:2360–4.

Stroop WG, Schaefer DC. (1987) Herpes simplex virus, type I invasion of the rabbit and mouse nervous systems revealed by in situ hybridization. Acta Neuropathol (Berl) 74:124–32.

Vazeux R, Brousse N, Jarry A et al. (1987) AIDS subacute encephalitis. Identification of HIV-infected cells. Am J Pathol 126:403–10.

Ward JM, O'Leary TJ, Baskin GB et al. (1987) Immunohistochemical localization of human and simian immunodeficiency viral antigens in fixed tissue sections. Am J Pathol 127:199–205.

Wiley CA. (1987) Implications of the neuropathology of HIV encephalitis for the pathogenesis of Alzheimer disease.

Alzheimer Dis Assoc Disord 1:236–50.

Wiley CA, Schrier RD, Nelson JA et al. (1986) Cellular localization of human immunodeficiency virus infection within the brains of acquired immune deficiency syndrome patients. Proc Natl Acad Sci USA 83:7089–93.

Wiley CA, Belman AL, Rubinstein A et al. (1990) Human immunodeficiency virus within brains of children with AIDS. Clin Neuropathol 9:1–6.

Yarchoan R, Berg G, Brouwers P et al. (1987) Response of human-immunodeficiency-virus-associated neurological disease to 3'-azido-3'-deoxythymidine. Lancet 1:132–5.

Yarchoan R, Thomas RV, Grafman J et al. (1988) Long-term administration of 3'-azido-2',3'-dideoxythymidine to patients with AIDS-related neurological disease. Ann Neurol 23:s82–7.

13 Overview of HIV in the Nervous System

P.G.E. Kennedy

General Considerations Concerning Neurological Involvement

My intention in this chapter is to summarize some of the salient features of HIV in the nervous system, and in particular to highlight important areas of uncertainty which require further investigation. Although precise figures for nervous system involvement in HIV infection are not available, nevertheless it is abundantly clear that neurologists are seeing an increasing number of HIV-related neurological conditions as the number of individuals infected with HIV steadily increases. The pace of this increase is difficult to predict, but it is a sobering thought that the actual worldwide total of AIDS cases may well be in the region of 500 000. At the end of December 1990 there were 4098 cases of AIDS in the UK, with 2256 deaths (Latest AIDS figures, 1990). Statistical predictions suggest that there will be 279 000 cases of AIDS with 179 000 deaths in the USA by 1991, but the long-term effects of changes in sexual practices on statistical predictions remain to be seen.

It is clear that the neurological complications associated with HIV infection occur at all stages of the disease, ranging from individuals who are seropositive alone to those with AIDS-related complex (ARC) – to the full-blown AIDS syndrome. Early observations from studies in both New York and San Francisco suggested that 10 per cent of AIDS patients presented with neurological symptoms, and more recent studies are certainly compatible with this view (Snider et al, 1983; Levy et al, 1985). The true incidence of neurological disease in patients with AIDS is difficult to assess, but is probably in the order of 50–60 per cent, with over 80 per cent of individuals involved on the basis of postmortem data (Kennedy 1988a,b). It is clear that certain neurological syndromes are more likely to occur at particular stages of the infection; for example, demyelinating neuropathy tends to occur early in the infection, especially in asymptomatic seropositive individuals, whereas encephalopathy and myelopathy tend to occur later in the disease.

The various complications are protean and can be divided into infections (viral, bacterial, fungal and protozoal), malignancies, cerebrovascular complications, and a variety of peripheral syndromes (Table 13.1). In some cases, the neurological complications appear to be due to direct involvement of HIV, and in others to the effect of opportunistic infection resulting from the immunosuppression. It is clear that HIV is neurotropic (Shaw et al, 1985; Price et al, 1988a) as well as lymphotropic (Lane et al, 1983), and that the nervous system is involved early in the course of the infection. Several lines of evidence are consistent with this view, notably observations that HIV may be isolated with high frequency from the cerebrospinal fluid (CSF) of patients at various stages of infection. For example, HIV has

Table 13.1. Neurological complications of HIV infection

1. CNS complications

Infections
AIDS-dementia complex
CMV encephalitis
HSV and VZV encephalitis
Cerebral toxoplasmosis
Progressive multifocal leukoencephalopathy
Vacuolar myelopathy and myelitis
Aseptic meningitis
Fungal infections
Mycobacterial infections
Treponema infection
CMV retinitis

Neoplasms
Primary CNS lymphoma
Systemic CNS lymphoma
Kaposi's sarcoma

Vascular
Cerebral haemorrhage
Cerebral infarction

2. PNS complications

Distal sensorimotor neuropathy
Acute, subacute or chronic demyelinating inflammatory
neuropathies
Mononeuritis multiplex
Herpes zoster radiculitis and cranial neuropathy
Myopathy and polymyositis

Reproduced from Kennedy (1988b) with permission.

been isolated with greater frequency from the CSF of seropositive individuals without AIDS compared to those with AIDS (Resnick et al, 1988), and in another study the rate of CSF HIV isolation was equally high in individuals irrespective of the presence of neurological symptoms (Sonnerborg et al, 1988). Moreover, the frequency of virus isolation does not appear to correlate with the severity of the immunodeficiency.

There is now general agreement that the most important of the neurological complications is an HIV encephalitis (also known as AIDS-related dementia and AIDS dementia complex, and formerly known as subacute encephalitis (Navia et al, 1986a; McArthur, 1987) (see also Chapter 7). The precise incidence of this syndrome is not known for certain, but it probably occurs in 30–40 per cent of patients with AIDS, and in one retrospective study it was reported as being the sole manifestation in 25 per cent of patients with HIV infection (Navia and Price, 1987). Some of the previous figures for the incidence of the dementia were probably overestimates. The subcortical nature of this dementia is now well documented (Tross et al, 1988). Nevertheless, encephalitis and dementia are not synonymous

since not all individuals with encephalitis are demented. In assessing the timing of the dementia in relationship to the development of HIV infection, it should be appreciated that there is no hard evidence for asymptomatic seropositive individuals having significant cognitive impairment compared with seronegative controls. It seems likely that there must be a degree of immunological impairment for the HIV-associated dementia to occur (see Chapter 3). However, there is no evidence to date to suggest a correlation between the degree of neurological involvement in HIV infection and the presence of the virus within the CNS. Because of these contradictions and uncertainties, it is critical to carry out rigorous prospective longitudinal studies of asymptomatic seropositive individuals in following the development of dementia and other complications. The various parameters used must be standardized, and the data obtained from various populations must be compared meaningfully. There is good evidence to show that neuroimaging can provide useful correlations between the degree of cerebral atrophy and the dementia in such patients (Price and Navia, 1987), although longitudinal studies are again required to define these correlations in detail. Some degree of cortical atrophy and ventricular enlargement is usually visible on CT scanning in adults presenting with mild dementia, and may actually precede the clinical symptoms (Price and Navia, 1987). MRI, however, is likely to provide a more sensitive measure of such changes (Price and Navia, 1987).

With regard to the other complications, it is noteworthy that toxoplasmosis is the commonest cause of focal neurological disease in HIV infection and has a characteristic clinical presentation, being particularly common in Haitian patients. It has also emerged that AIDS is now the most important cause of CNS lymphoma, which is the second commonest cause of intracranial mass lesions after toxoplasmosis. Progressive multifocal leukoencephalopathy (PML) may present atypically when associated with HIV, and it is quite remarkable that HIV infection is now known to be the single greatest risk factor for PML (Berger et al, 1987).

The characteristic vacuolar myelopathy which occurs in HIV infection remains somewhat of an enigma. It tends to occur late in infection, is often associated with the dementia, and has many clinical and pathological features in common with subacute combined degeneration of the cord (Petito et al, 1985). It occurs in up to 30 per cent of AIDS cases, and some authors include it in the definition of the HIV dementia (Price et al,

1988a). Moreover, there is some firm evidence for direct viral involvement of the spinal cord, since HIV antigens, HIV RNA, multinucleate giant cells and HIV isolation have been demonstrated in association with vacuolar myelopathy (Ho et al, 1985; Eilbott et al, 1989; Maier et al, 1989).

Pathology and Pathogenesis

Several studies based on large numbers of cases (70–153) have consistently shown pathological abnormalities in a very high proportion of patients with AIDS (Petito et al, 1986; Budka et al, 1987; Gray et al, 1988). A somewhat smaller survey (Lantos et al, 1989) reported cerebral pathological abnormalities in nearly 90 per cent of 26 patients who had died with HIV infection. Interestingly, only 22 of these fulfilled the CDC criteria for AIDS. Multiple cerebral abnormalities are frequently evident. Such lesions include those typical of HIV encephalitis, as well as others resulting from opportunistic infections, lymphomas and vascular lesions. The pattern of opportunistic infections in HIV infection is somewhat unusual. For example, it is not known why fungal infections (in particular aspergillosis) are less frequently found in HIV-positive individuals than in other immunocompromised patients. It is conceivable that this clinical heterogeneity is determined at least in part by the specific alterations in T-cell function seen in these various conditions.

The question of when to perform a brain biopsy (see Chapter 6) in patients with mass lesions is difficult and controversial, and each case needs to be judged on its individual merits. In the case of suspected toxoplasmosis, I favour medical treatment initially and only advocate brain biopsy if such treatment fails. A particular problem with fungal infections such as *Candida* is that the CT scan appearances are indistinguishable from those of other mass lesions. In cases where there is considerable diagnostic doubt, biopsy of the lesion may be the easiest method of obtaining a diagnosis which can then lead to effective therapy. In the case of PML, the only definitive method of diagnosis is brain biopsy. Cerebral lymphoma is also difficult to diagnose without a biopsy, although the presence of a solitary cerebral mass lesion is very suggestive of this diagnosis.

In HIV encephalitis, the lesions are most prominent in subcortical structures, in keeping with the nature of the dementia often associated with this condition. The three main types of neuropathological abnormality are a diffuse white matter pallor and/or small foci of inflammatory cells, multinucleate giant cell encephalitis, and vacuolar myelopathy (Petito et al, 1986; Price et al, 1988a; Kennedy, 1989; Lantos, 1989). The white matter pallor, associated with astrocytic proliferation, may be found in neurologically asymptomatic individuals and HIV is not often isolated in these patients. Multinucleate giant cells tend to be found in the brains of individuals with established neurological disease from whom HIV can be isolated. They are often associated with scanty inflammatory changes, such as perivascular and parenchymal lymphocytic and macrophage infiltrations (Price et al, 1988a; Kennedy, 1989; Lantos, 1989). These findings are described extensively in Chapters 6 and 7. Reactive astrocytosis is usually present and the inflammation primarily affects the deep grey and white matter, although it may also be found in the cortex. However, in many cases the process is diffuse, rather than multifocal. There is a lack of correlation between the pathological findings in HIV infection and the clinical neurological features. This is reflected by observations that only mild pathological changes have been detected in up to one-third of demented cases of AIDS and in over 50 per cent of non-demented patients dying with AIDS (Navia et al, 1986b). Moreover, in another study 90 per cent of patients with AIDS or ARC had neuropathological changes of HIV encephalitis, but only about one-third of those with moderate or severe encephalitis were demented (de la Monte et al, 1987) (see also Chapter 7).

Vacuolar myelopathy is the most frequent spinal cord disease in HIV infection, and is present in about 30 per cent of all autopsied cases (see Chapter 8). Its presence usually correlates with the characteristic brain abnormalities described, although the severity of these two pathologies is not always equivalent (Price et al, 1988a; Kennedy, 1989). Although the presence of HIV has been demonstrated within the affected cords in some cases with this condition (see above), its pathogenesis still remains unclear. The possibilities include metabolic disturbances and direct HIV involvement, as well as opportunistic viral infections. Several co-factors may also be implicated.

Although there appears to be a consensus regarding the pathological findings in HIV encephalitis, the pathogenesis of the neurological lesions is not well understood (see Chapters 7, 11 and 12). It is clear, however, that the virus is found in macrophages in the CNS. The virus is also found in microglial cells and multinucleate giant cells (Budka et al, 1987; Price et al, 1988a). There is some unconfirmed evidence for its presence in endothelial cells (Wiley et al, 1986), but to date there is no convincing evidence that HIV is present within astrocytes, oligodendrocytes or neurons (Budka et al, 1987; Kennedy, 1989). Although macrophages are likely to act as a reservoir of HIV, there is no confirmed evidence of actual impairment of macrophage function in the CNS in AIDS patients (see Chapter 11).

Any explanation of HIV pathogenesis in the nervous system must take into account the fact that only a small amount of virus has been demonstrated by a variety of techniques in areas which show the maximum pathology (see Chapter 12). In view of the paucity of demonstrated HIV in the brain and its probable absence in glial cells, it seems most unlikely that direct viral lysis plays a significant role, and indirect mechanisms of HIV-induced brain damage seem far more likely. For example, infected macrophages and lymphocytes may release toxic substances such as enzymes or lymphokines (McArthur, 1987; Price et al, 1988a; Kennedy, 1989), or HIV gene products may have a harmful effect on adjacent normal cells. It is thus possible that oligodendrocyte function may be compromised through such mechanisms. Macrophages and lymphocytes have been viewed as 'amplifiers' of brain infection (Price et al, 1988b), and a widely-held view at present is that HIV enters the brain through infected macrophages. A further possibility (which is not mutually exclusive) is that simultaneous infection of brain cells with HIV and an opportunistic virus may be important, for example, CMV, which is known to upregulate HIV expression in macrophages (Gendelman et al, 1988). Moreover, combined infection with HIV and CMV has been demonstrated in vivo in brain cells from infected patients (Nelson et al, 1988).

The mechanism of attachment of HIV to neural cells is not completely understood. Although it is well established that the CD4 molecule is the cellular receptor for HIV (Dalgleish et al. 1984), additional receptors may be important in HIV attachment to neural cells. However, in one study the CD4 molecule has been demonstrated on human neurons and glial cells in vivo (Funke et al, 1987), which does suggest a mechanism whereby HIV could be targeted to specific neural cell types. These observations have yet to be confirmed. The identity and role of alternative receptors for HIV attachment to neural cells is an important area for future research. It should also

be appreciated that functional changes in, for example, neuronal cells induced by HIV, may be important in the pathogenesis of the neurological symptoms. Such changes may be induced by a variety of mechanisms, including release of lymphokines.

Further questions which need to be resolved include the possible role of HIV variants in producing heterogeneous patterns of neurological disease, the unequivocal demonstration of the possible glial and neuronal cell specificity or otherwise of HIV, the precise role of viral gene products in modulating glial and neuronal function, the possible role of infected endothelial cells, and the relationship between blood–brain-barrier damage and macrophage trafficking in the CNS in this condition.

Neuropsychological and Neurophysiological Assessment in HIV Infection

Reference has already been made to the frequently observed lack of correlation between the neuropathology and neurological symptoms, and also the variation in the reported incidence of neuropsychological and neurological features in HIV infection. This must be due, at least in part, to geographical variations in the various studies, as well as to the lack of uniformity in the test procedures. It is absolutely clear that longitudinal, rather than cross-sectional, studies of various HIV patient cohorts must be carried out using rigorous controls. These should include not only asymptomatic seropositive patients, but also sero-negative individuals with the same risk factors as the HIV patients. Many of the current studies have been short-term so long-term studies on large numbers of patients are now required (McDonald and Weber, 1989). The importance of using standardized neuropsychological testing cannot be overstated.

Clearly it is important to develop markers of early nervous system involvement in HIV infection. This is not only important for facilitating studies of the natural history of the disease, but also for monitoring the response to treatment. Although a variety of neuropsychological, neuro-physiological and immunological parameters are available, there is no clear consensus at present as to which is the most sensitive. Neurophysiological investigations have included visual, auditory and somatosensory evoked potentials, as well as standard peripheral electromyography and nerve conduction studies. With all these procedures it is important to consider the degree of discomfort involved since patient compliance is a continuing problem, especially in those individuals who are drug abusers.

Paediatric AIDS

Children with HIV infection, acquired congenitally or otherwise, show a pattern of neurological involvement which is different from that observed in adults (see Chapter 9). For example, characteristic features of HIV encephalitis in children consist of seizures, impaired intellectual and motor function, delayed or lost milestones, microcephaly and calcification of the basal ganglia (Epstein et al, 1986; Epstein and Sharer, 1988). They may also suffer from ataxia and extrapyramidal symptoms, and the progressive encephalopathy carries a poor prognosis. There are two features of particular interest in paediatric AIDS. The first is that opportunistic infections of the nervous system occur only rarely, although the reason for this is not understood. Possibly, the primary infection with the relevant organism may not have occurred in these children (or is of shorter duration compared with adults) so that the conditions are not appropriate for such opportunistic infections to occur. Second, vacuolar myelopathies, as well as neuropathies, are very rare (see Chapter 9). Significant differences in the neuropathological findings in children compared with adults include a high incidence of HIV encephalitis with more severe and widespread lesions in children. Sharer and Mintz (see Chapter 9) have adduced these various neuropathological differences as evidence that in children the progressive HIV encephalopathy is likely to be due to the local presence of HIV. An interesting recent observation may shed some light on the pathogenesis of neurological lesions in paediatric HIV infection. It was found that in six children with congenital HIV infection and neurological complications there was defective methyl group metabolism based on assays of CSF (Surtees et al, 1990). Defective methylation may, therefore, play a part in the pathogenesis of neurological damage in such patients, and more attention should probably be paid to other possible metabolic abnormalities in all groups of patients with HIV infection.

Treatment Considerations

Currently, the most effective therapeutic agent for HIV infection is zidovudine. The beneficial effects of this drug in HIV infection are now fairly well established, although it is still unclear whether zidovudine should be started very early during HIV infection or at later stages in the disease (Anonymous, 1990). Studies of the efficacy of zidovudine in the treatment of HIV-related neurological disease are still at a relatively early stage, but there are certainly some preliminary data to suggest that it produces definite benefit. For example, administration of the drug to seven patients with HIV-related neurological disease (three with dementia, two with peripheral neuropathy, one with combined dementia and peripheral neuropathy, and one with paraplegia due to thoracic cord disease) resulted in improvement in neurological function in six (Yarchoan et al, 1988). Furthermore, three of these patients showed sustained improvement 5–18 months after therapy was initiated (Yarchoan et al, 1988). There has also been some more recent evidence to suggest that patients with the AIDS-dementia complex show improvement with this drug. For example, a recent study from the Netherlands reported a declining incidence of the AIDS-dementia complex in patients with AIDS following the introduction of systemic treatment with zidovudine (Portegies et al, 1989). Further detailed studies on large numbers of patients over relatively-long time periods will be required to establish unequivocally the precise role of zidovudine and other treatments. In view of the paucity of viral replication (as determined by currently-available molecular techniques) within the CNS in patients with HIV encephalitis, the question arises as to how exactly the drug may work in these patients. While it is possible that the inhibition of viral replication within the CNS may be of importance, it is also possible that the main effect of the drug is peripheral and that its effectiveness is dependent on diminishing the amount of virus which enters the CNS via the blood–brain barrier.

Treatment is clearly available for the various opportunistic infections, and the combination of pyrimethamine and sulphadiazine is particularly effective against toxoplasmosis, especially if treatment is started early. It is now generally agreed that anti-*Toxoplasma* therapy should be lifelong in order to prevent the high incidence of relapses. It is less clear whether patients with serological evidence of *Toxoplasma* infection should be treated prophylactically. One of the reasons for this uncertainty is that serological tests for *Toxoplasma* in patients with HIV infection are often misleading. Likewise, successful treatment with antifungal agents is possible for fungal conditions such as CNS cryptococcus, but here the mortality is much greater. Again lifelong maintenance therapy is required.

A variety of possible strategies for developing vaccines against HIV infection are the focus of much current research (De Clecq, 1990), but the ability of such vaccines to ameliorate or prevent the development of neurological complications will remain to be seen.

Animal Models for HIV Infection

There are a number of interesting animal diseases which have similarities to HIV infection and such models may prove to be useful not only in terms of vaccine development, but also in understanding the pathogenesis of HIV infection. Simian immunodeficiency virus (SIV) infection of macaques has similarities to AIDS (Letvin et al, 1986), and is likely to be particularly useful in the development of vaccines. Another disease with similarities to AIDS is feline immunodeficiency virus (FIV) infection (Hosie et al, 1989), although studies of this condition are only in the early stages and the nature of any neurological lesions has not yet been clarified. There are many intriguing similarities between HIV and the visna–maedi retrovirus of sheep. Both of these are members of the lentivirus group (as is SIV), and there are close similarities in their virion morphology as well as the nucleotide sequences in the *gag-pol* region of the genome (Gonda et al, 1985). Both HIV and visna are associated with restricted viral replication in vivo and also infect cells of the immune system, producing lymphoproliferation. However, there are also a number of differences between the two viruses in that severe immunosuppression is not seen in visna and the target immune cells are different in that visna primarily infects the macrophage, whereas the T4 cell and the macrophage are the targets for HIV infection (Dalgleish et al, 1984; Gendelman et al, 1985). Also, direct oligodendrocyte infection occurs in visna, whereas it almost certainly does not in HIV. Nevertheless, visna is a useful model system which has some definite relevance to HIV infection.

In summary, knowledge of HIV infection of the nervous system is accruing at a considerable rate, but we still lack a clear understanding of several fundamental aspects of the HIV–neural cell interaction both in vitro and in patients. Some of the key unanswered questions are summarized in Table 13.2.

Table 13.2. Some questions regarding HIV in the nervous system which require clarification

1. At what stage in the natural history of the disease is the CNS infected in HIV infection?
2. How frequently are the CNS and PNS infected in HIV infection?
3. What is the true incidence of HIV encephalitis and to what extent are geographical differences determinants of this incidence?
4. What is the pathogenesis of HIV encephalitis? Direct or indirect?
5. What is the pathogenesis of HIV-associated myelopathy? – metabolic? direct HIV? immune-mediated?
6. What is the pathogenesis of HIV-associated neuropathy and myopathy? Are direct or indirect mechanisms more important?
7. What factors underlie the lack of correlation between the pathological changes in the CNS and the clinical neurological features in HIV-associated neurological disease?
8. What role does zidovudine have in the management of the various neurological syndromes, and at what stage should it be given?
9. If zidovudine is effective, what is the main mechanism of action in view of the paucity of viral replication within the CNS?
10. What are the most sensitive neuropsychological and electrophysiological markers for the detection of early neurological disease?
11. What are the factors which underlie the marked differences between HIV infection in paediatric neurological cases and adults?
12. To what extent are studies in animal models relevant to HIV infection in humans?

References

Anonymous. (1990) Zidovudine for symptomless HIV infection. Lancet 335:821–2.

Berger JR, Kaszovitz B, Post JD et al. (1987) Progressive multifocal leukoencephalopathy associated with human immunodeficiency virus infection: a review of the literature with a report of sixteen cases. Ann Intern Med 107:78–87.

Budka H, Costanzi G, Cristina S et al. (1987) Brain pathology induced by infection with the human immunodeficiency virus (HIV). A histological, immunocytochemical, and electron microscopical study of 100 autopsy cases. Acta Neuropathol 75:185–98.

Dalgleish AG, Beverley PCL, Clapham PR et al. (1984) The CD4 (T4) antigen is an essential component of the receptor for the AIDS retrovirus. Nature 312:763–7.

De Clecq E. (1990) Targets and strategies for the antiviral

chemotherapy of AIDS. TiPS 1990; 11:198–205.

Eilbott DJ, Peress N, Burger H et al. (1989) Human immuno-deficiency virus Type I in spinal cords of acquired immuno-deficiency syndrome patients with myelopathy: expression and replication in macrophages. Proc Natl Acad Sci USA 86:3337–41.

Epstein LG, Sharer LR. (1988) Neurology of human immuno-deficiency virus infection in children. In: Rosenblum ML, Levy RM, Bredesen DE, ed. AIDS and the nervous system. Raven Press: New York, 79–101.

Epstein LG, Sharer LR, Oleske JM et al. (1986) Neurologic manifestations of human immunodeficiency virus infection in children. Pediatrics 78:678–87.

Funke I, Hahn A, Rieber EP et al. (1987) The cellular receptor (CD4) of the human immunodeficiency virus is expressed on neurons and glial cells in human brain. J Exp Med 165:1230–5.

Gendelman HE, Narayan O, Molineaux S et al. (1985) Slow persistent replication of lentiviruses: role of macrophages and macrophage-precursors in bone marrow. Proc Natl Acad Sci USA 82:7086–90.

Gendelman HE, Leonard JM, Dutko FJ et al. (1988) Immunopathogenesis of human immunodeficiency virus infection in the central nervous system. Ann Neurol 23 (suppl):s78–81.

Gonda MA, Brown MJ, Clements JE et al. (1985) Sequence homology and morphologic similarity of HTLV-III and visna virus, a pathogenic lentivirus. Science 227:173–7.

Gray F, Gherardi R, Scaravilli F. (1988) The neuropathology of the acquired immune deficiency syndrome (AIDS). Brain 111:245–66.

Ho D, Rota T, Schooley R et al. (1985) Isolation of HTLV-III from cerebrospinal fluid and neural tissues of patients with neurological syndromes related to the acquired immunodeficiency syndrome. N Engl J Med 313:1493–503.

Hosie MJ, Robertson C, Jarrett O. (1989) Prevalence of feline leukaemia virus and antibodies to feline immunodeficiency virus in cats in the United Kingdom. Vet Rec 128:293–7.

Kennedy PGE. (1988a) Human immunodeficiency virus infections of the nervous system. Curr Opin Neurol Neuro-surg 1:147–52.

Kennedy PGE. (1988b) Neurological complications of human immunodeficiency virus infection. Postgrad Med J 164:180–7.

Kennedy PGE. (1989) Human immunodeficiency virus in-fection of the nervous system. Curr Opin Neurol Neurosurg 2:191–4.

Lane MC, Masur H, Edgar LC et al. (1983) Abnormalities of B-cell activation and immunoregulation in patients with the acquired immunodeficiency syndrome. N Engl J Med 309:453–8.

Lantos PL, McLaughlin JE, Scholtz CL et al. (1989) Neuro-pathology of the brain in HIV infection. Lancet i:309–11.

Latest AIDS Figures. (1990) Lancet 335:119.

Letvin NL, Daniel MD, Sehgal PK et al. (1986) Induction of AIDS-like disease in macaque monkeys with T-cell tropic retrovirus STLV-III. Science 230:71–3.

Levy RM, Bredesen DE, Rosenblum ML. (1985) Neuro-logical manifestations of the acquired immunodeficiency syndrome (AIDS): experience at UCSF and a review of the literature. J Neurosurg 62:475–95.

Maier H, Budka H, Lassmann H et al. (1989) Vacuolar myelopathy with multinucleate giant cells in the acquired immunodeficiency syndrome (AIDS). Light and electron microscopic distribution of human immunodeficiency virus (HIV) antigens. Acta Neuropathol 78:497–503.

McArthur JC. (1987) Neurologic complications of AIDS.

Medicine (Baltimore) 66:407–37.

McDonald WI, Weber J. (1989) Meeting report HIV and the nervous system. MRC-INSERM 19–20 January 1989, Paris, France. AIDS 3:465–7.

Monte de la SM, Ho DD, Schooley RT et al. (1987) Subacute encephalomyelitis of AIDS and its relation to HTLV-111 infection. Neurology 37:562–9.

Navia BA, Price RW. (1987) The acquired immunodeficiency syndrome dementia complex as the presenting or sole mani-festation of human immunodeficiency virus infection. Arch Neurol 44:65–9.

Navia BA, Jordan BD, Price RW. (1986a) The AIDS dementia complex: 1. Clinical features. Ann Neurol 19:517–24.

Navia BA, Cho E-S, Petito CK et al. (1986b) The AIDS dementia complex: 2. Neuropathology. Ann Neurol 19:525–35.

Nelson JA, Reynolds-Kohler C, Oldstone MBA et al. (1988) HIV and HCMV coinfect brain cells in patients with AIDS. Virology 165:286–90.

Petito CK, Navia BA, Cho E-S et al. (1985) Vacuolar myelopathy pathologically resembling subacute combined degeneration in patients with the Acquired Immuno-deficiency Syndrome. N Engl J Med 312:874–9.

Petito CK, Cho E-S, Lemann W et al. (1986) Neuropathology of acquired immunodeficiency syndrome (AIDS): an autopsy review. J Neuropathol Exp Neurol 45:635–46.

Portegies P, de Gans J, Lange JMA et al. (1989) Declining incidence of AIDS dementia complex after introduction of zidovudine treatment. Br Med J 299:819–21.

Price RW, Navia BA. (1987) Infections in AIDS and in other immunosuppressed patients. In: Kennedy PGE, Johnson RT, ed. Infections of the nervous system. Butterworths: Sevenoaks 248–73.

Price RW, Brew B, Sidtis J et al. (1988a) The brain in AIDS: central nervous system HIV-1 infection and AIDS dementia complex. Science 239:586–91.

Price RW, Sidtis J, Rosenblum M. (1988b) The AIDS dementia complex: some current questions. Ann Neurol 23 (suppl):s27–33.

Resnick L, Berger JR, Shapshak P et al. (1988) Early pen-etration of the blood–brain-barrier by HIV. Neurology 38:9–14.

Shaw GM, Harper ME, Hahn BH et al. (1985) HTLV-III infection in brains of children and adults with AIDS encephalopathy. Science 227:177–82.

Snider WD, Simpson DM, Nielsen S et al. (1983) Neurological complications of acquired immune deficiency syndrome: analysis of 50 patients. Ann Neurol 14:403–18.

Sonnerborg AB, Ehrnst AC, Bergdahl SKM et al. (1988) HIV isolation from cerebrospinal fluid in relation to immuno-logical deficiency and neurological symptoms. AIDS 2:89–94.

Surtees R, Hyland K, Smith I. (1990) Central-nervous-system methyl-group metabolism in children with neurological complications of HIV infection. Lancet 335:619–21.

Tross S, Price RW, Navia B et al. (1988) Neuropsychological characterization of the AIDS dementia complex: a pre-liminary report. AIDS 2:81–8.

Wiley CA, Schrier RD, Nelson JA et al. (1986) Cellular localization of human immunodeficiency virus infection within the brains of acquired immune deficiency syndrome patients. Proc Natl Acad Sci USA 83:7089–93.

Yarchoan R, Thomas RV, Graffman J et al. (1988) Long-term administration of 3'-azido-2', 3'-dideoxythymidine to patients with AIDS-related neurological disease. Ann Neurol 23 (suppl):s82–7.

Subject Index